Symposium Transsonicum III

International Union of Theoretical and Applied Mechanics

J. Zierep, H. Oertel (Eds.)

Symposium Transsonicum III

IUTAM Symposium Göttingen, 24.-27. 5. 1988

Springer-Verlag
Berlin Heidelberg New York
London Paris Tokyo

Prof. Dr.-Ing. J. Zierep
Institut für Strömungslehre und Strömungsmaschinen
Universität Karlsruhe TH
Kaiserstraße 12
D-7500 Karlsruhe

Prof. Dr.-Ing. habil. H. Oertel
Institut für Theoretische Strömungsmechanik
Deutsche Forschungs- und Versuchsanstalt
für Luft- und Raumfahrt
Bunsenstraße 10
D-3400 Göttingen

ISBN 3-540-50202-5 Springer-Verlag Berlin Heidelberg NewYork
ISBN 0-387-50202-5 Springer-Verlag NewYork Berlin Heidelberg

Library of Congress Cataloging-in-Publication Data
Symposium Transsonicum (3rd : 1988 : Göttingen, Germany)
Symposium Transsonicum III : IUTAM symposium, Göttingen, 24.-27. 5. 1988 /
J. Zierep, H. Oertel (eds.).
At head of title: International Union of Theoretical and Applied Mechanics.
ISBN 0-387-50202-5
1. Aerodynamics, Transonic--Congresses.
I. Zierep, Jürgen.
II. Oertel, Herbert.
III. International Union of Theoretical and Applied Mechanics.
IV. Title.
TL571.S9 1988
629.132'304--dc19 89-4092

Offsetprinting: Mercedes-Druck, Berlin; Bookbinding: B. Helm, Berlin
2161/3020 5 4 3 2 1 0 Printed on acid-free paper

Scientific Commitee

J. Zierep (Chairman)
S. Berndt
J. Délery
I. Imai
R.C. Lock
O.S. Ryzhov
R. Seebass

Honorary Chairmen

K. Oswatitsch
K.G. Guderley

Editors of the Proceedings

J. Zierep
H. Oertel jr.

Local Organization

H. Oertel jr.
H.G. Hornung

Conference Secretary

I. Bertram: Tel.(0551) 709-2275
Telex 96839 dfvgoed
Telefax (0551) 7092101

Sponsored by:

British Aerospace
CONVEX
CRAY
DAAD
Daimler Benz
DFG
Dornier
IUTAM
Kreissparkasse Göttingen
Lufthansa
MBB
MTU
Rheinmetall GmbH
Stifterverband für die
Deutsche Wissenschaft

Preface

In a session held on August 30, 1986 the general council of IUTAM agreed to the organization of a third Symposium Transsonicum. The members of the scientific committee were: S. Berndt (Sweden), J. Délery (France), I. Imai (Japan), R.C. Lock (Great Britain), O.S. Ryzhov (USSR), R. Seebass (USA), and J. Zierep (FRG). The local organization was taken over by H. Oertel jr. and co-workers of the Institute for Theoretical Fluid Mechanics of the DFVLR in Göttingen, with the cooperation of H. Hornung (USA).

With the enthusiastic cooperation of these persons it became possible to put together an updated program including a wide range of topics within the subject of transonic flows - certainly a fascinating field of fluid mechanics.

The aim was to continue the tradition of the two preceeding Symposia Transsonica, 1962 in Aachen and 1975 in Göttingen. These had been mainly influenced by the personality of Professor Oswatitsch, whose name cannot be separated from transonic flow. Going through the Springer Proceedings of the past two meetings and comparing the contents with the present program, we recognize the following trends: in 1962 the analytical methods (hodograph-, parabolic-, characteristic-, integral equation method) dominated. Already in 1975 the center of interest shifted markedly to Computational Fluid Dynamics (CFD). These methods are now the dominating ones. The largest section of this volume is dedicated to CFD. To avoid large discussions about accuracy and reliability, a workshop was organized before this meeting (September 1987). A well-defined boundary value problem, the flow over a wing in a wind tunnel, was formulated. All groups and institutions we know, that had developed suitable Navier-Stokes computer programs, were invited to participate in solving this flow problem numerically. All theoretical results were collected by W. Kordulla and H. Sobieczky who compared them with each other and with experiments done in Göttingen. This was the important first step to test all available numerical methods - an effort which should be continued in the future.

Main topics besides the above are: Asymptotics and Theory, Stability and Transition, Vortical Flows, Turbulent Shock - Boundary Layer Interaction, and Unsteady Flows. Surely, this collection is not complete, but it contains the most important subjects in this field today.

We have to thank IUTAM for the decision to support this meeting, as well as all the different sponsors being mentioned by name in the program. Further, we have to thank the DFVLR for being the host of this meeting. We thank the MPI of Fluid Mechanics for permitting the use of their lecture hall, and the university and the town of Göttingen for their helpful cooperation. We thank the Springer Publishing Company for publishing the proceedings and T. Hefer, B. and J. Srulijes for the collaboration in the difficult task of revising the English texts. Last but not least, we thank all the persons involved in the organization of the Symposium, particularly the members of the Institute for Theoretical Fluid Mechanics of the DFVLR. Thanks to their engagement, this Symposium led not only to an important scientific exchange but took place in a friendly and cooperative atmosphere.

Göttingen, September 1988

J. Zierep

H. Oertel jr.

Contributors

Ahmad, J.	Department of Mechanical Engineering University of California Davis, California USA
Arnal, D.	ONERA/CERT Aerothermodynamics Department Toulouse, FRANCE
Basler, D.	DFVLR Institut für Experimentelle Strömungsmechanik Göttingen, FRG
Benoit, B.	ONERA/CERT Aerothermodynamics Department Toulouse, FRANCE
Bohning, R.	Institut für Strömungslehre und Strömungsmaschinen Universität Karlsruhe Karlsruhe, FRG
Breuer, M.	RWTH Aerodynamisches Institut Aachen, FRG
Bütefisch, K.A.	DFVLR Institut für Experimentelle Strömungsmechanik Göttingen, FRG
Bushnell, D.M.	NASA Langley Research Center Hampton, Virginia, USA
Chen, C.-L.	Sterling Software Palo Alto, California, USA
Cheng, H.K.	Department of Aerospace Engineering University of Southern California Los Angeles, California, USA
Chokani, N.	Department of Engineering University of Cambridge Cambridge, UK

Chow, C.-Y.	Department of Aerospace Engineering Sciences University of Colorado Boulder, Colorado, USA
Cole, J.D.	Rensselaer Polytechnic Institute Troy, New York, USA
Cook, L.P.	University of Delaware Newark, Delaware, USA
Dallmann, U.	DFVLR Institute for Theoretical Fluid Mechanics Göttingen, FRG
Das, A.	DFVLR Institute for Design Aerodynamics Braunschweig, FRG
Délery, J.	ONERA Aerodynamics Department Chatillon, FRANCE
Desai, S.S.	Aeronautical Development Agency Bangalore, INDIA
Diesperov, V.N.	USSR Academy of Sciences Computing Center Moscow, USSR
Doerffer, P.	Polish Academy of Science Institute of Fluid Flow Machinery Gdansk, Poland
Dohrmann, U.	Institut für Strömungslehre und Strömungsmaschinen Universität Karlsruhe Karlsruhe, FRG
Dor, J.B.	ONERA/CERT Aerothermodynamics Department Toulouse, FRANCE
Drougge, G.	FFA, The Aeronautical Research Institute of Sweden, Bromma, SWEDEN
Ehrenstein, U.	DFVLR Institute for Theoretical Fluid Mechanics Göttingen, FRG
Elsenaar, A.	NLR Nationaal Lucht- en Ruimtevaart-laboratorium Amsterdam, NETHERLANDS

Fischer, T.M.	DFVLR Institute for Theoretical Fluid Mechanics Göttingen, FRG
Girodroux-Lavigne, P.	ONERA Chatillon, FRANCE
Goodsell, A.M.	M.I.T. Computational Fluid Dynamics Laboratory Cambridge, Massachusetts, USA
Guiraud, J.-P.	Laboratoire de Modélisation en Mécanique Université P&M. Curie Paris, FRANCE
Hänel, D.	RWTH Aerodynamisches Institut Aachen, FRG
Hafez, M.	Department of Mechanical Engineering University of California Davis, California, USA
Hall, M.G.	Royal Aerospace Establishment Farnborough, Hants, UK
Harvey, W.D.	NASA Langley Research Center Hampton, Virginia, USA
Hirose, N.	National Aerospace Laboratory Jindaiji-Higashi Chofu, Tokyo, JAPAN
Hirschel, E.H.	Messerschmitt-Bölkow-Blohm GmbH, München, FRG
Hitzel, S.M.	DORNIER GmbH Numerische Strömungsmechanik Friedrichshafen, FRG
Holst, T.L.	NASA Ames Research Center Applied Computational Fluids Branch Moffet Field, California, USA
Hornung, H.G.	Graduate Aeronautical Laboratories California Institute of Technology Pasadena, California, USA

Ide, H.	Rockwell International Science Center Thousand Oaks, California, USA
Jungbluth, H.	Institut für Strömungslehre und Strömungsmaschinen Universität Karlsruhe Karlsruhe, FRG
Kluwick, A.	Institut für Strömungslehre und Wärmeübertragung Technische Universität Wien Wien, ÖSTERREICH
Koller, F.	Institut für Strömungslehre und Wärmeübertragung Technische Universität Wien Wien, ÖSTERREICH
Kordulla, W.	DFVLR Institute for Theoretical Fluid Mechanics Göttingen, FRG
Krogmann, P.	DFVLR Institut für Experimentelle Strömungsmechanik Göttingen, FRG
Kumar, A.	National Aeronautical Laboratory Bangalore, INDIA
LeBalleur, J.C.	ONERA Chatillon, FRANCE
Mack, L.M.	Jet Propulsion Laboratory California Institute of Technology Pasadena, California, USA
Malecki, R.E.	M.I.T. Computational Fluid Dynamics Laboratory Cambridge, Massachusetts, USA
Malik, M.R.	NASA Langley Research Center Hampton, Virginia, USA
Meier, G.E.A.	Max-Planck-Institut für Strömungsforschung Göttingen, FRG
Meyer, F.	DFVLR Institute for Theoretical Fluid Mechanics Göttingen, FRG

Mignosi, A.	ONERA/CERT Aerothermodynamics Department Toulouse, FRANCE
Miwa, H.	National Aerospace Laboratory Jindaiji-Higashi Chofu, Tokyo, JAPAN
Monnoyer, F.	Messerschmitt-Bölkow-Blohm GmbH, München, FRG
Morice, Ph.	ONERA Aerodynamics Department Chatillon, FRANCE
Müller, B.	DFVLR Göttingen, FRG
Murman, E.M.	M.I.T. Computational Fluid Dynamics Laboratory Cambridge, Massachusetts, USA
Narayan, K.Y.	National Aeronautical Laboratory Aerodynamics Division Bangalore, INDIA
Oertel, H.	DFVLR Institute for Theoretical Fluid Mechanics Göttingen, FRG
Paisley, M.F.	Royal Aerospace Establishment Farnborough, Hants, UK
Rangarajan, R.	National Aeronautical Laboratory Aerodynamics Division Bangalore, INDIA
Ravichandran, K.S.	National Aeronautical Laboratory Aerodynamics Division Bangalore, INDIA
Rizzi, A.	FFA, The Aeronautical Research Center of Sweden Bromma, SWEDEN
Rubbert, P.E.	BOEING USA
Ryzhov, O.S.	USSR Academy of Sciences Computing Center Moscow, USSR
Schmatz, M.A.	Messerschmitt-Bölkow-Blohm GmbH, München, FRG

Schnerr, G.	Institut für Strömungslehre und Strömungsmaschinen Universität Karlsruhe Karlsruhe, FRG
Schwamborn, D.	DFVLR Institute for Theoretical Fluid Mechanics Göttingen, FRG
Seraudie, A.	ONERA/CERT Aerothermodynamics Department Toulouse, FRANCE
Seshadri, S.N.	National Aeronautical Laboratory Aerodynamics Division Bangalore, INDIA
Shankar, V.	Rockwell International Science Center Thousand Oaks, California, USA
Singh, J.P.	National Aeronautical Laboratory Aerodynamics Division Bangalore, INDIA
Sinha, U.N.	National Aeronautical Laboratory Aerodynamics Division Bangalore, INDIA
Sobieczky, H.	DFVLR Institute for Theoretical Fluid Mechanics Göttingen, FRG
Squire, L.C.	Department of Engineering University of Cambridge Cambridge, UK
Stanewsky, E.	DFVLR Institut für Experimentelle Strömungsmechanik Göttingen, FRG
Strykowski, P.	DFVLR Institute for Theoretical Fluid Mechanics Göttingen, FRG
Takanashi, S.	National Aerospace Laboratory Jindaiji-Higashi Chofu, Tokyo, JAPAN
Thiede, P.	Messerschmitt-Bölkow-Blohm GmbH Unternehmensgruppe Transport- und Verkehrsflugzeuge Bremen, FRG

Van Dalsem — NASA
Ames Research Center
Applied Computational
Fluids Branch
Moffet Field, California, USA

Wagner, B. — DORNIER GmbH
Numerische Strömungsmechanik
Friedrichshafen, FRG

Wanie, K.M. — Lehrstuhl für Strömungsmechanik
Technische Universität München
München, FRG

Contents

Session 1 — CFD - Applications

W. Kordulla H. Sobieczky	Summary and Evaluation of the Workshop "Numerical Simulation of Compressible Viscous-Flow Aerodynamics"	3
P.E. Rubbert	Use of CFD and Experiment in Transonic Aircraft Design	19
N. Hirose S. Takanashi	Computational Transonic Aerodynamics Research in Japan	33
J.C. Le Balleur P. Girodroux-Lavigne	Viscous-Inviscid Strategy and Computation of Transonic Buffet	49
M.A. Schmatz F. Monnoyer K.M. Wanie E.H. Hirschel	Zonal Solutions of Three-Dimensional Viscous Flow Problems	65
D. Hänel M. Breuer	On the Computation of Transonic Viscous Flows	75
M.F. Paisley M.G. Hall	Improvements in the Formulation and Numerical Solution of the Euler Problem for Swept Wings	87
Ph. Morice	Transonic Computations by Multidomain Techniques with Potential and Euler Solvers	97
S.S. Desai R. Rangarajan U.N. Sinha K.S. Ravichandran J.P. Singh	On the Development and Application of Software based on the Framework of Cartesian Coordinates	109
D. Schwamborn P. Strykowski H. Oertel jr.	Numerical Simulation and Physical Modelling of Transonic Trailing Edge Flow	121
G. Schnerr U. Dohrmann	Theoretical and Experimental Investigation of 2-D Diabatic Transonic and Supersonic Flow Fields	131

Session 2	Asymptotics and Theory	
O.S. Ryzhov	Asymptotic Methods in Transonic Flow Theory	143
J.D. Cole L.P. Cook	Some Asymptotic Problems of Transonic Flow Theory	157
J.-P. Guiraud	Transonic Degeneracy in Systems of Conservation Laws	171
H.K. Cheng	Transonic Equivalence Rule: Unresolved Aspects of Asymptotic and CFD Studies	179
A. Kluwick F. Koller	Perturbation Analysis of Unsteady Transonic Flow through Cascades	189
V.N. Diesperov	Viscous Transonic Flow in the Vicinity of a Corner Point	199

Session 3	Stability and Transition	
L.M. Mack	Stability of Three-Dimensional Boundary Layers on Swept Wings at Transonic Speeds	209
D.M. Bushnell M.R. Malik W.D. Harvey	Transition Prediction in External Flows via Linear Stability Theory	225
T.M. Fischer U. Ehrenstein F. Meyer	Theoretical Investigation of Instability and Transition in the DFVLR-F5 Swept-Wing Flow	243
D. Arnal	Transition Prediction in Transonic Flow	253

Session 4	Vortical Flows	
E.M. Murman A.M. Goodsell R.E. Malecki	Transonic Vortical Flow	265
A. Elsenaar K.A. Bütefisch	Experimental Study on Vortex and Shock Wave Development on a 65° Delta Wing	281
S.M. Hitzel B. Wagner	Computations Relating to the Vortex Flow Experiment	291
A. Rizzi G. Drougge B. Müller	Navier-Stokes and Euler Solutions for Vortex Flow over a Delta Wing	305

A. Kumar A. Das	Computation of Vortex Flow on a Delta Wing at Transonic Speed	317
K.Y. Narayan S.N. Seshadri	Vortical Flows on the Lee Surface of Delta Wings	329
M. Hafez J. Ahmad	Numerical Simulation of Rotational Flows	339
Session 5	Turbulent Shock - Boundary Layer Interaction	
J. Délery	Recent Basic Studies on Transonic Shock-Wave/ Turbulent Boundary-Layer Interactions	357
P. Thiede P. Krogmann	Passive Control of Transonic Shock/Boundary Layer Interaction	379
R. Bohning H. Jungbluth	Turbulent Shock - Boundary-Layer Interaction with Control Theory and Experiment	389
N. Chokani L.C. Squire	Passive Control of Shock - Boundary-Layer Interactions: Numerical and Experimental Studies	399
C.-Y. Chow C.-L. Chen T.L. Holst W.R. Van Dalsem	Transonic Inviscid-Viscous Interactions over Porous Surfaces	407
J.B. Dor A. Mignosi A. Seraudie B. Benoit	Wind Tunnel Studies of Natural Shock Wave - Separation Instabilities for Transonic Airfoil Tests	417
U. Dallmann P.Doerffer	Three-Dimensional Flow Separation caused by Normal Shock-Wave / Turbulent Boundary-Layer Interaction	429
Session 6	Unsteady Flows	
G.E.A. Meier	Unsteady Phenomena	441
V. Shankar H. Ide	Unsteady Aeroelastic Computations for Flexible Configurations at Transonic and Supersonic Speeds	465
E. Stanewsky D. Basler	Mechanism and Reynolds Number Dependence of Shock-Induced Buffet on Transonic Airfoils	479
N. Hirose H. Miwa	Computational and Experimental Research on Buffet Phenomena of Transonic Airfoils	489
H.G. Hornung	Concluding Remarks	501

CFD – Applications

Summary and Evaluation of the Workshop "Numerical Simulation of Compressible Viscous-Flow Aerodynamics"

W. Kordulla and H. Sobieczky

DFVLR
Institute for Theoretical Fluid Mechanics
D-3400 Göttingen, FRG

Summary

The present paper describes the transonic DFVLR-F5 wing experiment which has been designed to provide the data needed by the computational analyst for code validation purposes. Thus, the experimental data allow to formulate analytically the corresponding boundary value problem for the integration of the Navier-Stokes equations. Based on the experiment a workshop took place. Results of the numerical simulations are compared with each other and with the experimentally observed data, to date only pressure measurements. The discussion of the comparisons and of the difficulties associated with the numerical simulations leads to suggestions for further experimental and theoretical work.

Introduction

Aerodynamic research and development is traditionally based on wind tunnel experiments, and is only more recently - with the advent of super computers with high speed and large memory - being complemented with numerical analysis. Wind tunnel flow simulations in the high speed transonic flow regime suffer from the interaction between the model and the wall, such that a substantial part of the wind tunnel technology is aimed at developing wall corrections which have, of course, to be tailored specifically to each wind tunnel. Present computer codes, on the other hand, are increasingly efficient for the simulation of external flows and for the representation of real flight past more and more complicated configurations. However, local phenomena based on viscous effects, recompression shocks, vortex generation and the interaction between all of these are still represented in an inaccurate fashion. This includes the modelling of transition and of turbulence which is lagging considerably the development of numerical methods.

Note that in research-type application the Navier-Stokes equations have been integrated numerically to simulate the transonic flow past planes with realistic shape, both transport and fighter airplanes at various angles of attack. Although such results represent major achievements, it is clearly understood that the accuracy of these computations in terms of boundary-layer properties, e.g. wall shear stress, is in doubt due to the lack of sufficient mesh resolution, and of the aforementioned deficiencies. However, such solutions illustrate the potential of numerical flow field simulations. To date there is nothing like a black-box solver. Instead, the numerical solutions have to be adapted to the particular flow field under consideration with respect to the grid choice and to the necessary physical modelling, and this will be true for some time to come. Hence, it is obvious that a validation process is needed with respect to the simulation of a flow field which is typical for the kind of flows of interest.

J. Zierep, H. Oertel (Eds.)
Symposium Transsonicum III
IUTAM Symposium Göttingen, 24.-27. 5. 1988

Such a validation process may include the comparison of results obtained by one method, with those produced by different other solvers. In any case the assistance of a well-defined experiment is required to judge the quality of the numerical solution in terms of integral and local quantities. By "well-defined experiment" an experiment is understood which allows the numerical analyst to properly formulate the boundary value problem for the wind tunnel flow. The underlying idea in doing so is, of course, inherently based on the assumption that the ability to predict the flow field in a wind tunnel correctly also comprises the ability to predict the flow field existing in free flight conditions.

Workshops have been carried out in the past to validate computer codes by comparing solutions with each other or with experimental data. None of the experiments used for comparison was, to the authors' knowledge, a dedicated experiment providing the computational analyst with unique boundary conditions. In the transonic flow regime wind tunnel wall ventilation simulates free flight conditions only up to a certain degree, with some of the remaining wall interference effects being corrected with empirical calibration data. Two recent workshops dealing with two-dimensional viscous flow simulations are most noteworthy. One workshop dealt with laminar low Reynolds number transonic and supersonic flows past airfoils, and through a nozzle with double throat [1]. Laminar flow was chosen to avoid the difficulties associated with turbulence modelling. The other workshop then concerned the simulation of turbulent transonic flows past airfoils [2]. Except for the nozzle test cases, a fair amount of scatter is observed in the comparisons of the results with each other.

The present paper reports on a workshop which is based on an experiment dedicated to the needs of the computational analyst. The experiment is unique in some sense because it provides all boundary condition data needed to integrate the Navier-Stokes equations. This experiment is described first, in the following. Next, the experimental, physical and numerical difficulties associated with the experiment and its simulation will be discussed, before selected results furnished by the workshop contributors are presented. The paper ends with concluding remarks and with a description of the anticipated continuation of the work.

The DFVLR-F5 Wing Experiment

For the steady compressible subsonic flow with isolated local supersonic flow pockets occurring e.g. in present transport aircraft aerodynamics, it is possible to define a boundary value problem for the corresponding flow simulation. Thus, for code validation purpose one has to determine the corresponding boundary values in the experiment. The simplest shape of the control volume is given by the wind tunnel walls, by plane entrance and exit surfaces, and, of course, by the surface of the considered wing. This was the basic idea when the DFVLR-F5 wing experiment was conceived. In order to provide unique boundary conditions on the tunnel walls, these were closed completely so that simple no-slip conditions prevail for the integration of the Navier-Stokes equations. Remember that walls with ventilation do not offer a practical way of determining the out- or inflow conditions on those walls. Adiabatic surface conditions complement the no-slip conditions. This assumption is a good approximation since the used wind tunnel is of the continuously running type. The boundary conditions in entrance and exit planes require detailed local measurements of the corresponding flow quantities.

The analytic approximation of the measured data in the control volume surfaces then allows the analytic definition of the numerical boundary value problem.

Figure 1 displays the chosen experimental setup to scale in a planform view. Half-model technology [3] with a splitter plate has been used to make sure that the wing flow is not drowned in the comparatively thick tunnel wall boundary layers (roughly 90 mm), see also [4,5,6]. The approximation of the measured data in the entrance and exit planes is described in [5,6], and will not be discussed here. Note, however, that the given entrance conditions include all flow profiles, while in the exit plane only the pressure is provided analytically.

The selection of the aerodynamic configuration for the test case was motivated by keeping some of the practical considerations but also allowing some simplifications. High aspect ratio wings exhibit large portions of quasi-2-D swept wing flow so that 2-D profile design methodology can be applied there. Special interest is in section pressure distributions which allow to maintain laminar flow. But the design of wing roots and tips requires the understanding of 3-D transonic flow phenomena produced by the interaction of the wing flow with viscous body flow and with tip vortices.

A wing-body fairing with fillets e.g. is the aerodynamic designer's answer to circumvent root flow problems associated with strong horse-shoe vortices. The wing chosen for the experiment, see figure 1, exhibits an extreme fairing of the wing - splitter plate junction where the splitter plate is a simple approximation for a fuselage. The wing itself has an aspect ratio of 9, and is in its main part defined by sections resulting from an airfoil designed to be shock free at a free stream

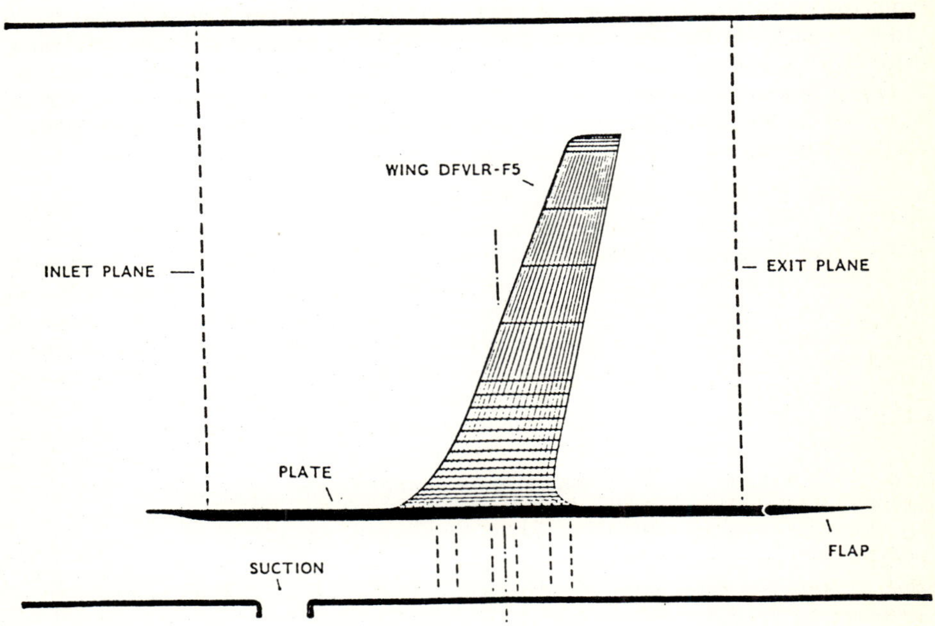

Fig. 1: Schematic plan form view of the experimental setup of the DFVLR-F5 wing experiment (to scale).

Mach number of 0.78. The leading edge sweep angle is 20°, such that a free flight Mach number of roughly 0.82 is expected. In order to avoid the difficulties associated with simulating the severe rear loading observed with modern supercritical profiles the sections are symmetrical. The resulting pressure distribution should, however, allow a substantial part of laminar flow on the wing surface. The surface of the wing is defined by analytical functions with only a few parameters in a small computer code [6].

The flow conditions were chosen such that unsteady flow phenomena were definitely avoided. Preliminary experiments then led to the choice of a center line Mach number of 0.82 with a range of the angle of attack α of $-2° < \alpha < 2°$ at a Reynolds number per meter of 10^7. The flow measurements were thus made at $\alpha = 0°$ and 2°. The negative angle of attack is necessary because for each cross section in question the majority of pressure taps is installed on the upper surface of the wing, and only three holes are positioned on the lower surface for control purpose. The transition line has been obtained by acenaphtene visualizations and by approximating the resulting pictures with a smooth curve fit [5,6]. Owing to the lack of time, the corresponding experiment was done quickly, and the quality of the acenaphtene observation could have been better. Figure 2 illustrates the approximations carried out by the second author. In future experiments it will be important to obtain a clean line for the separation of laminar and turbulent flow. Thus, the conditions needed for the numerical simulation of the DFVLR-F5 wing are complete.

Some Problems Associated With the Numerical Simulation

When carrying out the numerical integration of the Navier-Stokes equations in order to simulate the flow within the control volume, some difficulties became evident. These are caused in part by the lack of physical modelling, by the way of approximating the experimental data, by insufficient experimental resolution, and also by the global meshes, preferred by the participants of the DFVLR-F5 wing workshop.

Starting the discussion with the modelling of transition, it is surprising to realize that the manner how turbulence is switched on, once a line of transition is known, is important (remember that transition does not occur instantly but gradually). Three different answers can be obtained depending on how the eddy viscosity (all participants used the Baldwin-Lomax model) is switched on. One may assume to have turbulent flow from the leading edge, switch on the turbulent eddy viscosity across the distance given by several cells at the transition line, or switch on the model across one cell width only. Chaderjian reports that if transition is performed across five cells or more in computational space, the solution would not change, see his contribution in [7]. The position of the blending region with respect to the transition "line" may also be of importance. Although the effects are perhaps felt only locally in the flow in question, e.g. because of the extreme fairing, the proper modelling of turbulence is certainly important. The scalar Baldwin-Lomax approximation needs to be adapted to the various situations, e.g. near wing - fuselage junctions, for the interaction of wake flow and surface boundary layers, or for tip flows.

Another difficulty is encountered for the simulation of the flow in the wind tunnel if no-slip boundary conditions are considered on the tunnel walls and if an appropriate resolution of the corresponding boundary layers becomes necessary. Whenever a global mesh is being used, the desire to resolve the boundary layers on the wing and at the same time on the splitter plate will produce extreme skewness at the wing - splitter plate junction for a surface-fitted mesh. This problem can be overcome by the use of zonal grids. The test case calculations indi-

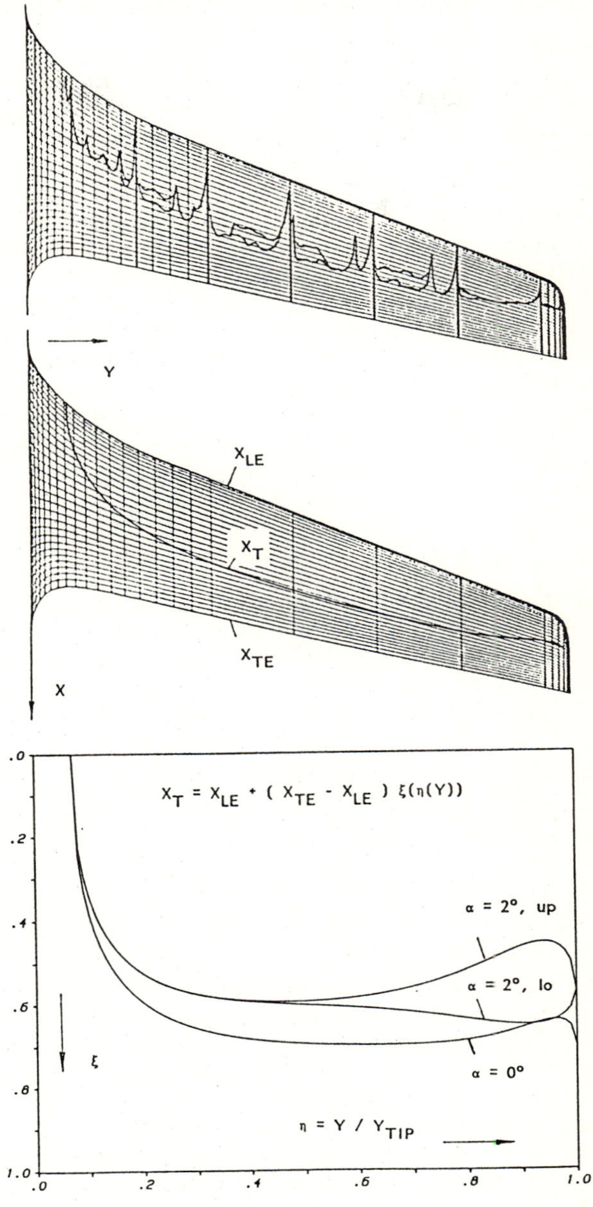

Fig. 2: Approximation of the experimentally observed line of transition and the analytical approximation.

cate that for the situation in question a coarser grid may be appropriate near the splitter plate if this plate is treated as inviscid. This is possible because of the thin boundary layer which does not seem to largely influence the flow on the wing owing to the smooth fairing. The desire to keep the efforts with respect to man power and computation low, made the majority of participants employ inviscid boundary conditions on the tunnel walls and on the splitter plate.

The inviscid treatment of the walls, however, puts the validity of the comparison simulations in jeopardy because the prescribed experimental exit pressure distribution is no longer compatible with the flow in the entrance plane, see figure 1. The corresponding assumption of uniform inflow falsifies the mass flow in the tunnel. Note that the modelling of the flow variables in inlet and exit planes by the second author neglects the dependence of the exit pressure distribution on both the angle of attack of the wing and also on the vertical direction [5,6]. This gave rise to complaints of the workshop participants. Chaderjian experienced a slightly unsteady flow solution in the lifting case if the given downstream pressure was prescribed while using inviscid tunnel wall conditions, see [7]. Obayashi et al., see [7], obtained streamwise c_p-distributions with kinks near the entrance and exit planes although they employed no-slip conditions on all walls. A close inspection of the experimental pressure distributions reveals, indeed, a two-dimensional pressure field in the exit plane instead of the suggested one-dimensional one. It is evident that the exit pressures need to be represented sufficiently well because of the proximity of the trailing edge, see figure 1.

Selected Workshop Results

Based on the described DFVLR-F5 wing experiment a workshop took place in Göttingen from September 30 to October 2, 1987. Six of the participants who performed computations provided a written version of their oral presentations, see [7], and furnished also their raw solutions as well as a filled-in questionnaire which enabled some evaluation of the workshop [8]. Table 1 summarizes these individuals and groups, indicating also the chosen test cases. The computational effort involved in the corresponding flow simulations is reflected in the size of the grids used by the participants, see table 2. Notice that the topology of the meshes, generated in general algebraically, is typically of C-type (or H-type) in the streamwise direction in order to facilitate the use of an algebraic, scalar turbulence model in the wake. Only Lindeberg et al. employ an O-type grid wrapped around the wing in chordwise direction which necessitates some modification of the calculation of the lengths needed in the Baldwin-Lomax model.

The used numerical methods consist of three implicit and three explicit ones, the latter based on various versions of the Runge-Kutta finite volume scheme. Five out of the six contributions rely on the concept of local time stepping to accelerate the convergence of the solutions to steady state. The employed vector computers are versions of CRAY, CYBER and FUJITSU. Only in the case of the CRAY computers the used word length is 64 bits, otherwise 32 bits. The computation time per point or cell and iteration is typically between 10 and $30 \cdot 10^{-6}$ sec, and the number of iterations or time steps, needed for convergence, ranges from 1000 for implicit to 10 000 for explicit methods.

The main purpose of performing Navier-Stokes computations is to determine the forces acting on aerodynamic configurations, in particular the drag. This cannot

Table 1: List of individuals and groups contributing to the comparison of results

Participant	Affiliation	Test Cases Done	
		Wind tunnel	Free air
Chaderjian	NASA, Ames Research Center, USA	$\alpha = 0°, 2°$ (A1, A2)	$\alpha = 0°, 2°, 5°, 8°$ (B1,B2,B3,B4)
Lindeberg, Rizzi, Müller	FFA, Sweden	-	B1, B2
Obayashi, Fujii, Gavali	NAL, Japan / Amdahl, USA	A1, A2	B1 to B4
Schmatz	MBB, München, FRG	A2	-
Schwamborn	DFVLR, Göttingen, FRG	A1, A2	B1, B2
Vatsa, Wedan	NASA and Vigyan Res. Inc., Langley Research Center, USA	A1, A2	B1, B2

Table 2: Some information on the grids used by the participants

Participant	Grid Topology	Grid Dimensions			
		Wind Tunnel		Free Air	
		total	one wing surface	total	one wing surface
Chaderjian	H-H (4 zones)	337.000 -	55x35	415.000 -	55x35
Lindeberg, Rizzi, Müller	O-O	-	-	387.737 193x49x41	97x41
Obayashi, Fujii, Gavali	C-H	768.825 201x75x51	81x55	522.801 201x51x51	76x41
Schmatz	C-H	568.000 144x82x52	56x47	-	-
Schwamborn	C-O (4 blocks)	272.000 200x40x34	81x40	384.000 200x40x48	81x40
Vatsa, Wedan	C-O	260.337 161x33x49	65x33	286.209 177x33x49	65x33

be achieved yet in a routine and accurate fashion, even for simple wing shapes. Table 3 reflects the scatter of the results in terms of the integral quantities lift and drag coefficients [8]. Unfortunately, the setup of the experiment does not allow force measurements. The computed results are not very conclusive. Notice that the scatter because of different grids, methods, etc. is larger than the difference due to different turbulence assumptions, see the values for the wind tunnel test cases by Obayashi et al..

It was mentioned earlier that to date only pressure measurements have been carried out, in particular on the wing surface and on the splitter plate. In the following these data are compared with the predicted ones. Note that the pressure

Table 3: C_L and C_D obtained for the different flow cases

Cases → Participant ↓	A 1	A 2	B 1	B 2	B 3	B 4
Chaderjian	0.000066 0.01783	0.2245 0.02964	0.000093 0.00760	0.2298 0.01444	0.3438[1]) 0.03469	0.4017[1]) 0.07017
Lindeberg, Rizzi, Müller	-	-	0.00016 0.0053	0.2430 0.0126	-	-
Obayashi, Fujii, Gavali	- 0.000042 0.0282	0.196 0.0414	-	-	-	-
Obayashi, Fujii, Gavali[1])	- 0.000015 0.032	0.169 0.0429	- 0.000021 0.0361	0.191 0.0396	0.312 0.0547	0.411 0.0884
Schmatz	-	0.214 0.053	-	-	-	-
Schwamborn	0 0.0115	0.278 0.0241	0 0.0120	0.273 0.0221	-	
Vatsa, Wedan	N/A[2])	N/A	N/A	N/A	-	-

1) Fully turbulent flow is assumed

2) N/A: not available

data derived from the solutions provided by the participants have to be interpolated in some (linear) fashion to obtain the values at the locations of the pressure taps [8]. There are 10 chordwise wing sections with pressure taps, the first one nearly on the splitter plate and the tenth one nearly at the tip. The figures 3 and 4 display c_p-distributions at two span stations (8.5 and 80 percent span) for non-lifting and lifting free air flow simulations, respectively. Note that the wind tunnel data are simulated well by the free air simulations over large chordwise portions of the wing sections, in particular also near the trailing edge. This is, in particular, true in the immediate neighborhood of the splitter plate. This indicates that the thin boundary layer on the splitter plate hardly interacts with that one on the wing surface, due to the extreme fairing of the wing - plate junction. For the sections shown, the scatter of the various solutions is very low except for the solution of Obayashi et al.. One discrepancy is obvious in the comparison with the experimentally observed pressure data (note that Schwamborn's results have been chosen as comparison solution for convenience), which holds for all following

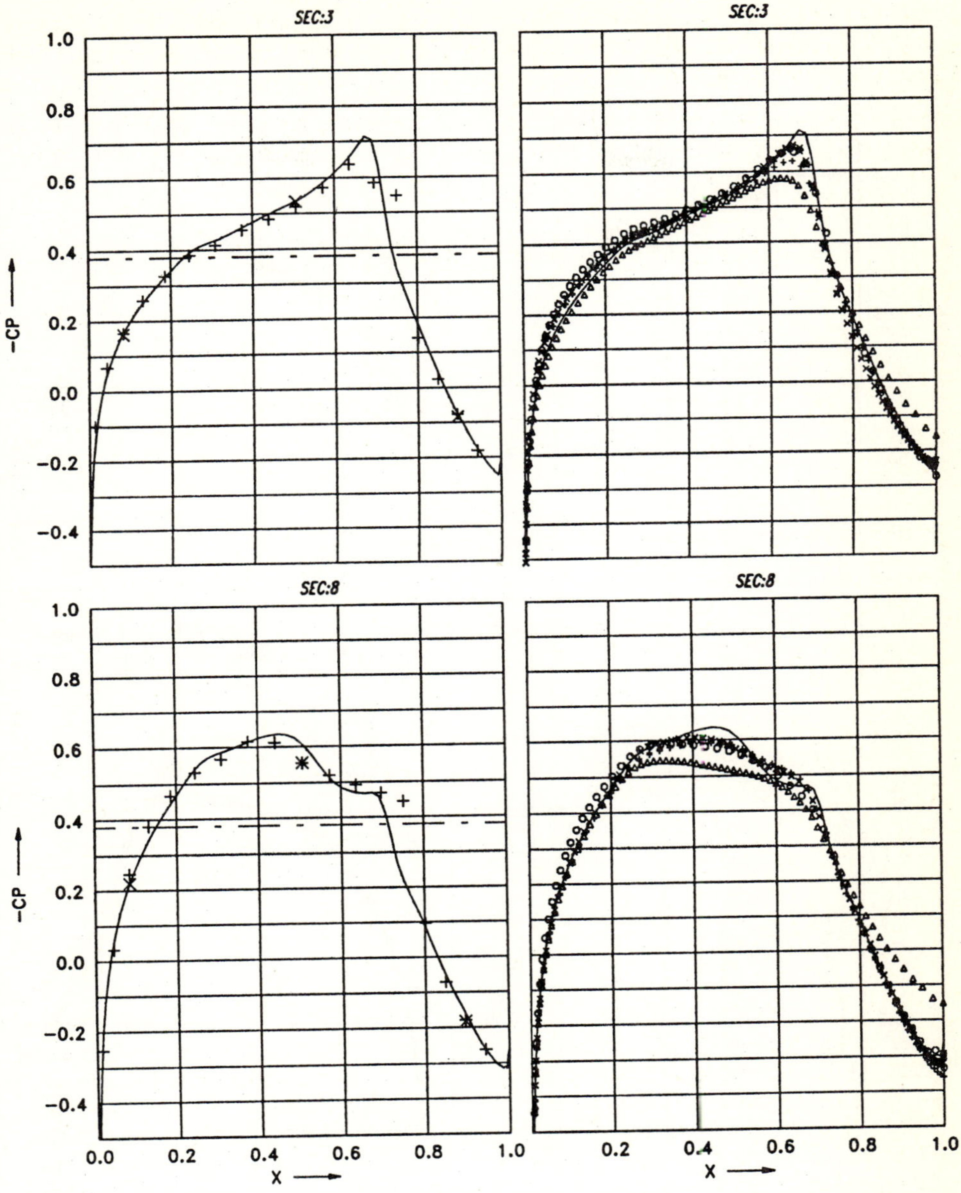

Fig. 3: The chordwise c_p-distributions on the wing surface for non-lifting free air simulations (8.5 and 80 % span). Left: experimental data are given by symbols. Right: symbols indicate participants: o Chaderjian, Δ Obayashi et al., x Lindeberg et al., + Vatsa/Wedan. The solid lines are Schwamborn's results.

comparisons as well. The experimentalists interpret the slight chordwise pressure drop and subsequent pressure plateau at about mid-chord position as laminar flow separation (E. Stanewsky, DFVLR, Göttingen, private communication). This is not predicted by the participants, instead an over expansion is obtained. This

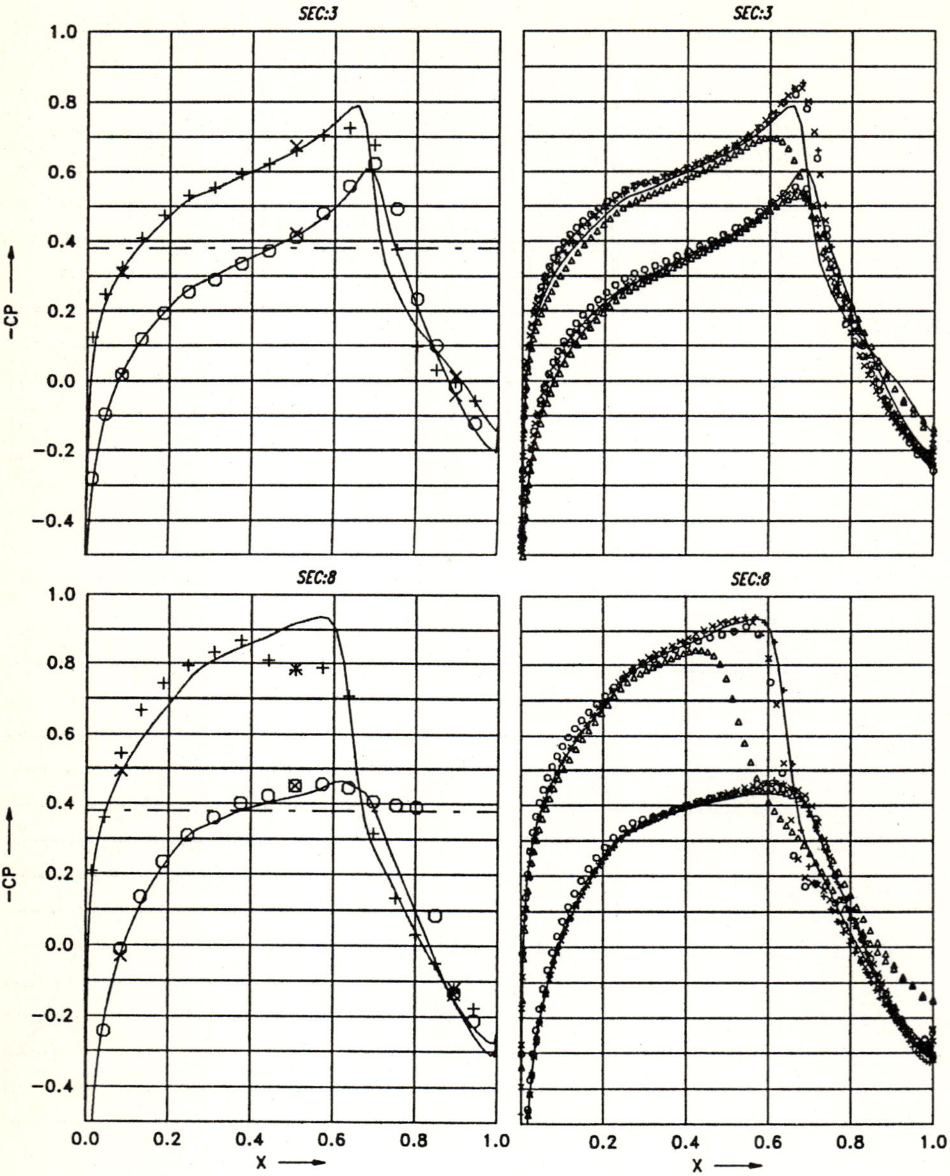

Fig. 4: The chordwise c_p-distributions on the wing surface for lifting free air simulations (α = 2°, 8.5 and 80 % span). Left: experimental data are given by symbols. Right: symbols indicate participants: o Chaderjian, Δ Obayashi et al., x Lindeberg et al., + Vatsa/Wedan. The solid lines are Schwamborn's results.

is visible, here, only at section 3, but is present over most of the wing span, see [8]. This behaviour is more pronounced in the lifting flow case, see figure 4.

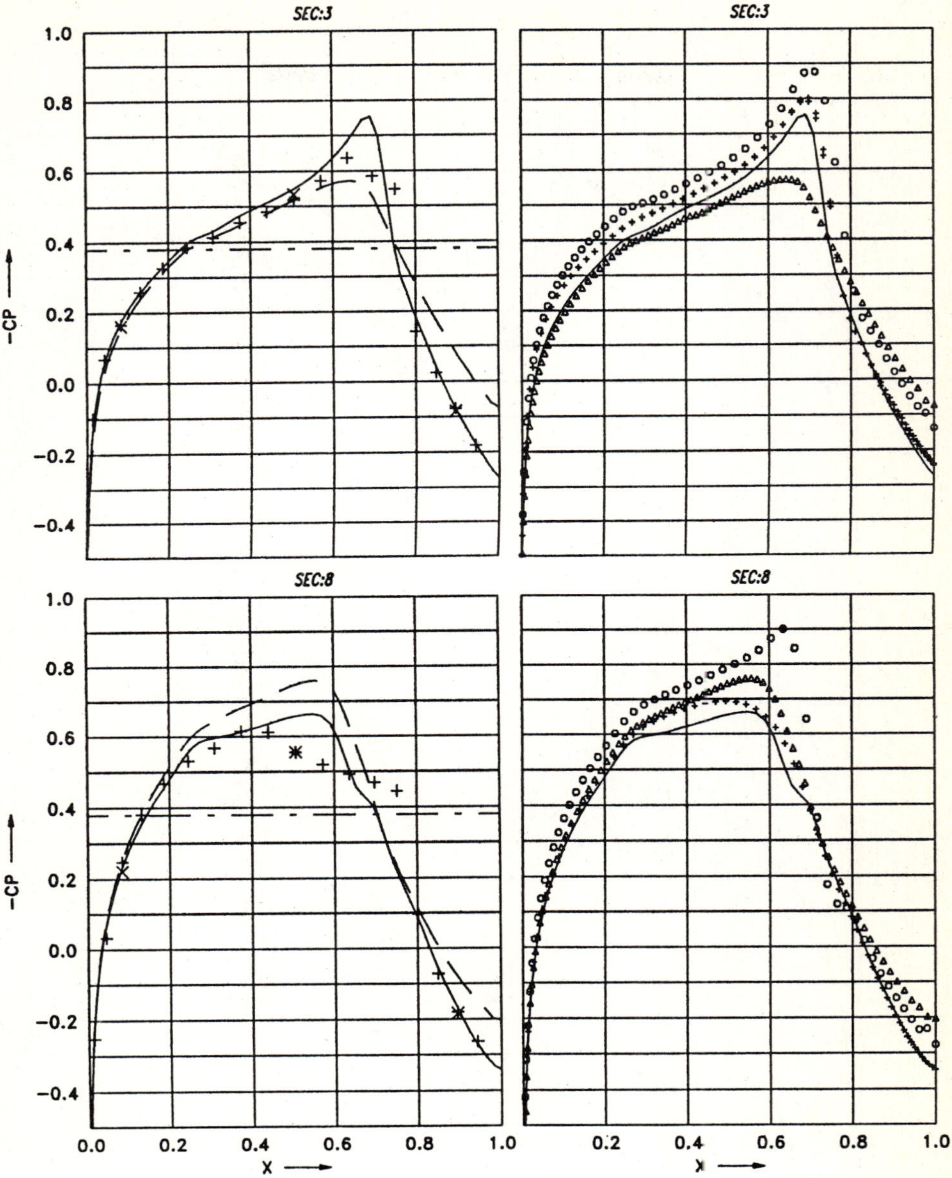

Fig. 5: The chordwise c_p-distributions on the wing surface for non-lifting wind tunnel simulations (8.5 and 80 % span). Left: experimental data are given by symbols, broken lines correspond to Obayashi's results (viscous walls). Right: symbols indicate participants: o Chaderjian, Δ Obayashi et al., x Lindeberg et al., + Vatsa/Wedan. The solid lines are Schwamborn's results (inviscid walls).

In figures 5 and 6 the results of the wind tunnel flow simulations are compared with the experimental data. These are now compared on the left not only with

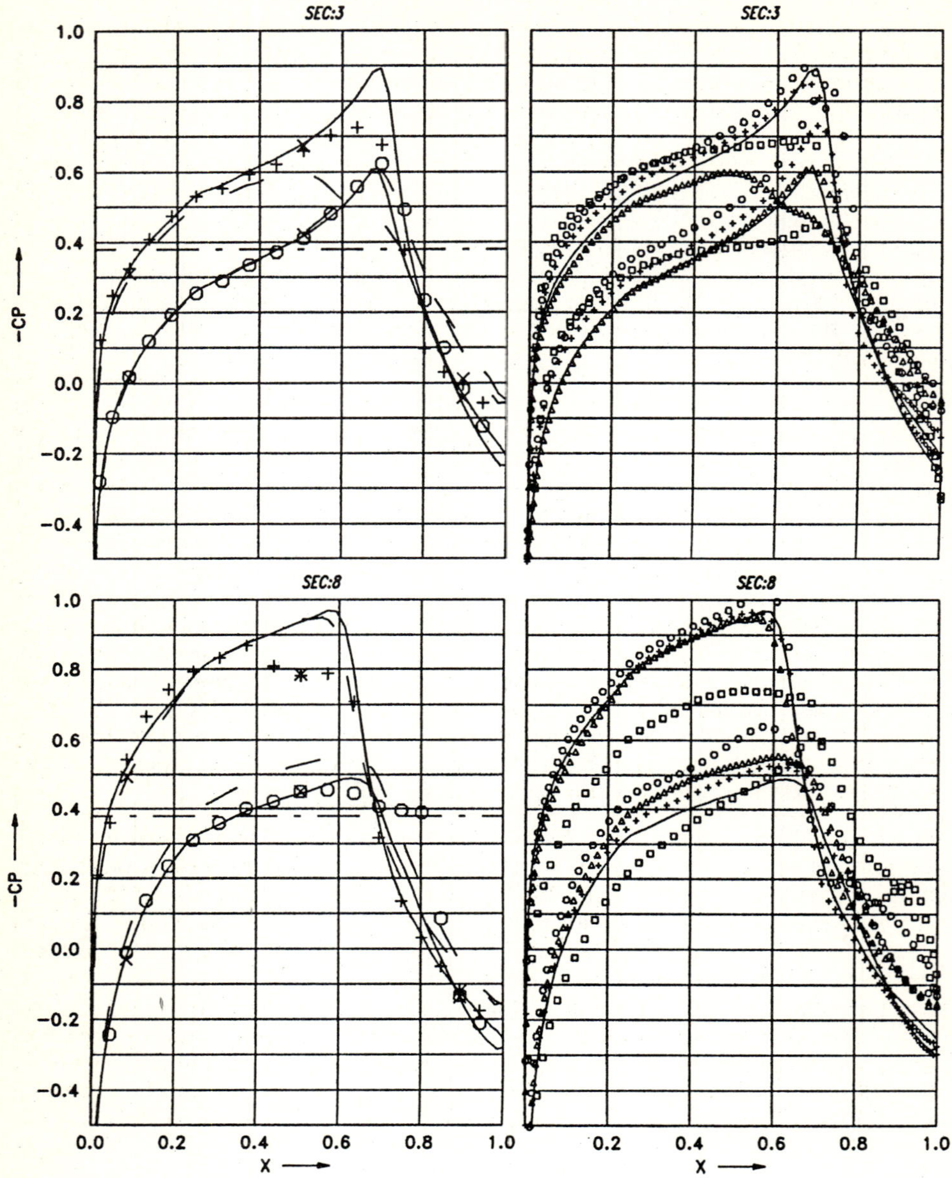

Fig. 6: The chordwise c_p-distributions on the wing surface for wind tunnel simulations for an angle of attack of 2°. Left: experimental data are given by symbols, broken lines correspond to Obayashi's results (viscous walls). Right: symbols indicate participants: o Chaderjian, Δ Obayashi et al., x Lindeberg et al., + Vatsa/Wedan. The solid lines are Schwamborn's results (inviscid walls).

those obtained by Schwamborn who assumed inviscid wall boundary conditions, but, in addition, with those of Obayashi et al. who assumed the appropriate no-slip boundary conditions on all surfaces. In spite of the better boundary conditions, the agreement is not improved in total. Note that the scatter is somewhat larger than for the free air simulations, in particular for lifting flow. This is certainly influenced by the different choices of the exit boundary conditions, since some participants allowed the exit pressure to evolve as a function of the solution inside the control volume, see also [8]. The major difference between wind tunnel and free air simulations is the increase of the over expansion due to the constraints of the walls, see also [9]. Note that Schmatz is the only other participant, besides Obayashi et al., who considered viscous wind tunnel wall boundary conditions. The discrepancy between the corresponding two solutions ($\alpha = 2°$) is fairly large, though.

Figure 7 illustrates the influence of the exit plane boundary conditions in terms of the streamwise pressure distribution on the splitter plate at selected sections (one, 13, at 0.2 m above the non-lifting wing center line and the the other two symmetric with respect to that line, at a distance of 0.05 m). The flow case is for $\alpha = 2°$. Shown are the experimental data in comparison with the results of Obayashi et al. and Schwamborn. While the former use the prescribed pressure distribution, Schwamborn allows the pressure to evolve and, in fact, obtains a two-dimensional distribution in the exit plane (note that a one-dimensional one is suggested by the second author [5,6]). Assuming that the influence of the different wall boundary conditions is not the only reason for discrepancies, one observes that the different pressures at the exit plane induce upstream differences as well. While the solution ahead of the wing is nearly identical, discrepancies develop above the wing root section (sections 13, 16). Note that the comparison of the solutions with each other shows moderate discrepancies, except for the regions near leading and trailing edges.

Although oil flow visualization techniques have not yet been employed in the experiment to determine the skin friction line pattern, the theoretical ones have been compared with each other in [8]. This is useful because skin friction line patterns are much more sensitive to the quality of the numerical solution than the pressure because the computation of skin friction vectors involves the differentiation of flow profiles. Note that the integration of the skin friction lines (in forward and backward direction) is performed for all data sets with the same computer code and the same set of integration parameters [10]. Figure 8 shows as an example the comparison of the different simulations of the lifting wind tunnel flow. Some scatter can be observed on lower and upper surface of the wing although the inviscid wall simulations compare well on the upper surface.

Concluding Remarks

The transonic DFVLR-F5 wing experiment is shown to provide, in principle, the basis for the correct analytical formulation of the corresponding boundary value problem associated with the numerical integration of the Navier-Stokes equations. The measurement and the analytical modelling of the conditions in the entrance and, in particular, in the exit plane of the specified control volume need, however, more attention in future follow-up experiments. Crucial is the close proximity of the exit plane to the wing trailing edge. Another critical issue is the

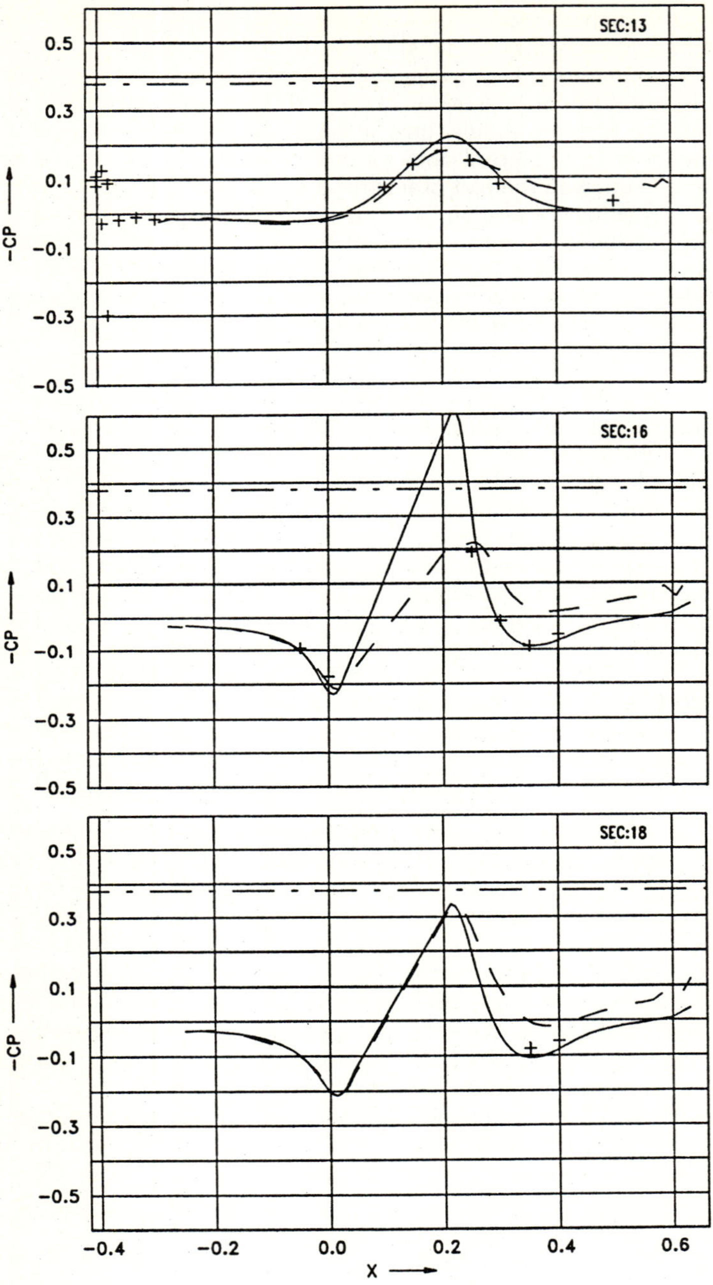

Fig. 7: Selected streamwise c_p-distributions on the splitter plate for the wind tunnel simulation for an angle of attack of 2° (see table 3). Experimental data are given by symbols. The broken lines correspond to the results of Obayashi et al., and the solid lines to those of Schwamborn.

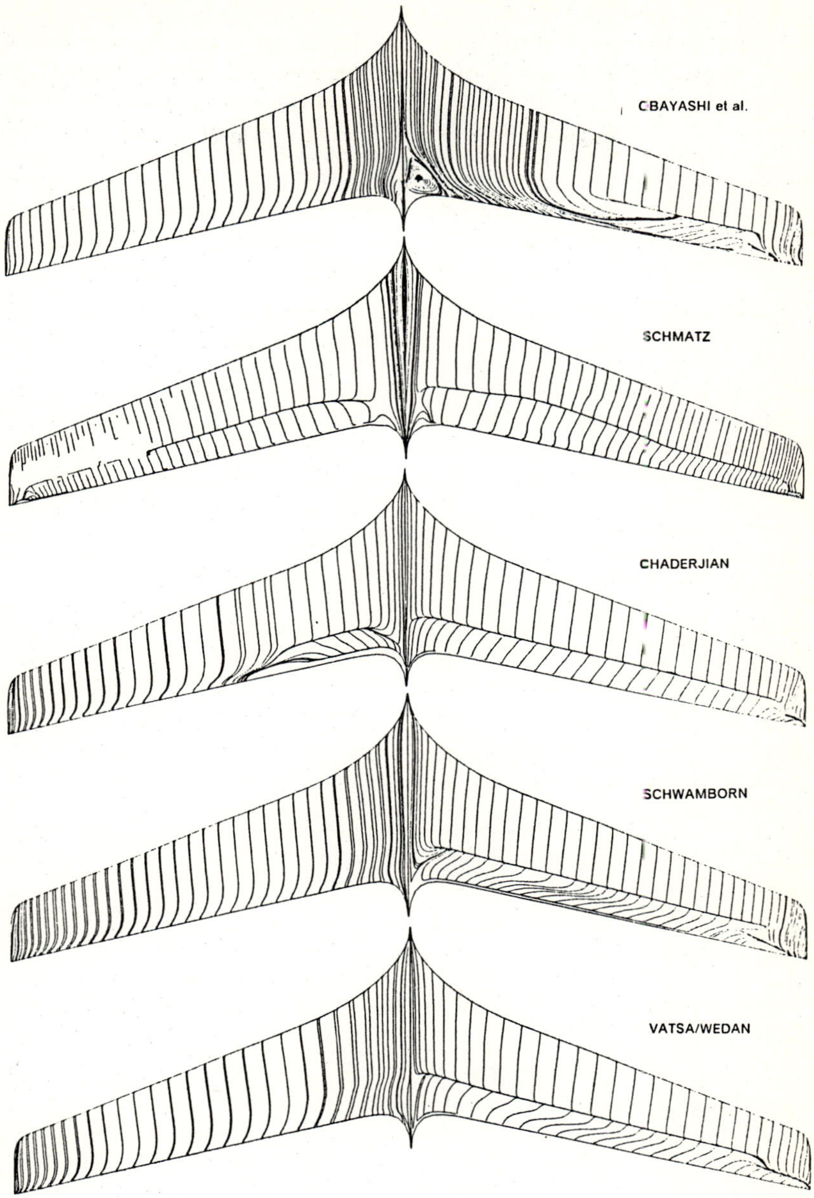

Fig. 8: Comparison of the predicted skin friction line patterns for the wind tunnel flow simulations for $\alpha = 2°$. Left: lower wing surface, right: upper wing surface.

evaluation of a unique line of transition. From the technical point of view all participants but one employ a global grid with associated skewness problems near the wing - plate junction because of the extreme fairing there. All solutions rely on the scalar turbulence model after Baldwin and Lomax which is certainly not appropriate everywhere in the wind tunnel flow field although the corresponding effects may remain local, e.g. near the wing - plate junction. It is pointed out that the manner how turbulence is "switched on" varies from participant to participant with large consequences in the results. Some scatter is observed in integral as well as local flow quantities, the sources of which are hard to trace back because the basic elements of the solutions are too different from each other.

A continuation of the experiment is planned with the current deficiencies in mind, in particular the need for more boundary layer information such as skin friction line pattern or velocity profiles. Another workshop would then be useful if all participants are able to consider viscous walls. Then one mandatory test case is needed where all parameters of the solution with respect to boundary conditions, turbulence model, transition process and also the grid are clearly defined. This would enable to really compare the efficiency and accuracy of the various methods. There would still be room for the participants to turn on their individual tuning kits or physical modellings to obtain improved solutions.

References

[1] Bristeau, M.O.; Glowinski, R.; Periaux, J.; Viviand, H. (eds): Numerical Simulation of Compressible Navier-Stokes Flows. A GAMM-Workshop. Notes on Numerical Fluid Mechanics, Vol. 18, Vieweg Verlag, Braunschweig, 1987.

[2] Holst, T.L.: Viscous Transonic Airfoil Workshop - Compendium of Results. AIAA Paper 87-1460, 1987.

[3] Hefer, G.; Sobieczky, H.; Tusche, S.: Verbesserung der Halbmodelltechnik im Transsonischen Windkanal Göttingen. Jahrbuch der DGLR, Bd. II, 87-083, 1987.

[4] Sobieczky, H.; Hefer, G.; Tusche, S.: DFVLR-F5 Test Wing Experiment for Computational Aerodynamics, in [7].

[5] Sobieczky, H.: DFVLR-F5 Test Wing Configuration - The Boundary Value Problem, in [7].

[6] Sobieczky, H.: DFVLR-F5 Test Wing Configuration for Computational and Experimental Aerodynamics. DFVLR IB 221-87 A 01, 1987.

[7] Kordulla, W.(ed.): Numerical Simulation of the Transonic DFVLR-F5 Wing Experiment - Towards the validation of viscous wing flow codes. Proceedings Notes on Numerical Fluid Mechanics, Vol. 22, Vieweg Verlag, Braunschweig, 1988.

[8] Kordulla, W.: Evaluation of the Workshop, in [7].

[9] Kordulla, W.; Schwamborn, D.; Sobieczky, H.: The DFVLR-F5 Wing Experiment - Towards the Validation of the Numerical Simulation of Transonic Viscous Flows. AGARD CP-437, Paper 14, 1988.

[10] Vollmers, H.: Rechnerprogramme zur Integration von Linien in einem Vektorfeld. DFVLR-IB 221-87 A 02, 1987.

Use of CFD and Experiment in Transonic Aircraft Design

P. Rubbert

Boeing Commercial Airplane Co.
Seattle, Wa 98124
USA

INTRODUCTION

The aerodynamic design of an aircraft involves the total flight envelope - flight ranging from takeoff and landing at low speeds to cruise at transonic Mach numbers, and to flight at the maximums speeds, Mach numbers, angles of attack and yaw that the aircraft will forseeably encounter. The design issues, the parameters of interest, and the dominant fluid physics can be vastly different for these different regimes of flight.

The type of design information and insight that is available from CFD or from experiment depends strongly on the dominant aerodynamic physics and the associated region of the flight envelope. And so, the relative roles of CFD and experiment for design in one part of the flight envelope can be very much different from their roles in another part of the envelope. The present paper will describe today's roles of CFD and experiment for the design of subsonic commercial transport aircraft in the various parts of the flight envelope.

OVERVIEW OF DESIGN OBJECTIVES

A typical flight envelope of a jet transport aircraft is displayed in Figure 1. At the lowest Mach numbers, the conditions for design occur at high lift conditions (the upper crosshatched region in the figure) with the high lift system (flaps) deployed. Simply stated, design desires are to maximize C_{Lmax} and its associated buffet boundary in the landing flap configuration, and to provide high L/D in the takeoff flap configuration to reduce airport noise (rapid climb) and to provide the required climb and maneuver margins in the event of an engine failure. In addition, airplane stability and control effectiveness about all axes is certainly a design issue, and includes handling characteristics at high angles of attack above the stall.

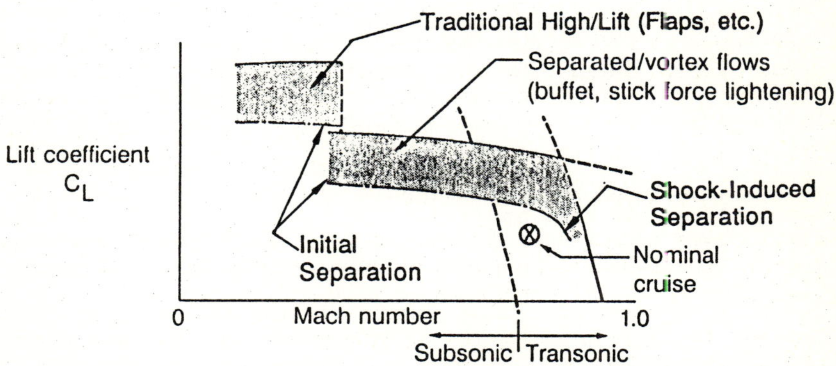

Figure 1 Typical Flight Envelope of a Transport Airplane

J. Zierep, H. Oertel (Eds.)
Symposium Transsonicum III
IUTAM Symposium Göttingen, 24.-27. 5. 1988

During climb, descent and loiter the aircraft is in the subsonic region of Figure 1, but in a clean configuration with flaps stowed. In this regime L/D is important, but not to the extent that it assumes at cruise Mach number. Of design importance are the buffet boundary and phenomena, such as stick force lightening, that may be encountered at the higher lift coefficients.

Cruise design is strongly focused on maximizing L/D in the transonic regime. But, cruise design is also strongly impacted by the shock wave movements and flow separations that will occur in a Mach overspeed condition and/or at high angles of attack and which require both acceptable stability and adequate control effectiveness to overcome phenomena such as Mach tuck or pitchup.

STRENGTHS AND LIMITATIONS OF CFD

Most of the success stories in CFD applications to date for commercial transports relate to areas of the flight envelope in which the fluid physics is relatively benign such that viscous effects fit within the assumptions of boundary layer theory. Computational methodology for simulating flows that are dominated by viscous effects does not yet have a major role, principally because of limitations in accuracy, reliability, problem setup and computing costs, and limited ability to capture the widely differing aerodynamic length scales that characterize such flows.

A major strength of CFD is its ability to produce a finely detailed, "microscopic" understanding of the flow, and to support the study of cause-and-effect relationships. These can range from drag decomposition where the contributions of vortex drag, wave drag, and profile drag are computed separately, to component buildups which enable a study of the flow changes as nacelles, struts, engine power setting, flap track fairings or whatever are systematically added. And, a very unique and powerful capability of CFD is its applicability for inverse design, wherein the designer specifies the aerodynamic pressures to be achieved and the CFD code computes the geometrical contouring of the airplane surface that will produce those pressures.

To be most effective, a computational method must faithfully represent the dominant fluid physics, must be able to deal with complex airplane geometries, be capable of providing solutions in a timely manner, and be affordable [1]. That is a tall order which few CFD codes satisfy completely. At Boeing we work with a collection of CFD codes that meet these desires to varying degrees. Some codes are "production" tools that are fully validated and in wide use throughout the company by expert and non-expert users alike. Engineering setup time for these codes is generally small. There are "expert user" codes, which may have more general and more advanced capabilities, but which may require special user skills and greater setup time. And there are "emerging" codes which are undergoing active development and yet may be used for certain engineering applications.

Effective CFD applications demand effective geometry tools and graphics tools. Geometry tools are essential for the preparation of the inputs to the various CFD codes. Three-dimensional graphics running on suitable workstations allows the inspection of surface and field grids prior to execution, and provides the graphical means to understanding the frequently massive output from a typical CFD solution. The "timeliness" of CFD is very closely related to the quality of the geometry and graphics tools available.

Primary tools fitting this need at Boeing include the interactive, three-dimensional geometry system known as the Aero Grid and Paneling System (AGPS), [2]. AGPS provides an efficient means of defining any three-dimensional surface or aircraft component. These surfaces can be combined to represent a complete aircraft configuration. AGPS has built-in user-programmable features for extracting geometric data in the proper form for use by CFD codes. Paneling, surface grid generation, and grid distribution at block boundaries for input to three-dimensional grid generation codes are all possible. AGPS can also display CFD results in the form of three-dimensional objects with color representing the value of some flow property. In addition, the graphics workstations in our CFD laboratory host graphics codes such as the NASA-developed PLOT 3D [3] and GAS codes, and our internally-developed RPD and PIX graphics codes.

STRENGTHS AND LIMITATIONS OF WIND TUNNEL TESTING

A well conducted wind tunnel test can be an extremely efficient and effective way to get data which is beyond the reach of computational methods alone. The cost per flight condition (one Mach, one angle of attack) is frequently cheaper than a CFD computation, and it is common to run a broad spectrum of Mach numbers and angles of attack, something that in many cases would be prohibitively expensive with CFD. A wind tunnel experiment contains all of the physics, albeit usually at a Reynolds number that is lower than flight and frequently with a very imperfect simulation of the airplane's propulsion system. And there are issues involving boundary layer tripping and transition, wall interference, model support interference, flow noninformity, etc., that must be dealt with.

One generally learns different things from an experiment than from CFD. The wind tunnel today is by far the major source of information involving separated flows such as prevail at the perimeter of the flight envelope. It is easy to measure overall forces and moments in an experiment, whereas the calculation of absolute values of drag and accurate moments is frequently difficult and sometimes impossible with today's CFD codes. It is also experimentally easy to do surface flow visualization and obtain a surface impression of the viscous flow behavior and shock wave positions. Measurement of surface pressures is regularly done, but the cost increases with the number of pressure taps. And, one can only measure surface pressures where one had the foresight to build them into the model. Today it is common to do an _a priori_ CFD analysis of an experimental test setup in order to determine, among other things, where to place the pressure measurement ports and other instrumentation, and also trip strips.

A major limitation of experiments, other than usually having to test at the wrong Reynolds number, is the difficulty of measuring the flow in the field away from the surface. There are capabilities such as wake imaging, LDVs, hot wires, etc., but the cost of surveying large portions of the flow field can be very high and the information obtained is usually incomplete. CFD can provide detailed flow directions, pressure, temperature, density, swirl, etc everywhere in the field, usually at no added computational cost. But this is of maximum value only if the CFD correctly models the dominant physics, and when the dominant physics involves separated flow, today's CFD codes have very severe limitations.

THE COMPLEMENTARY ROLES OF CFD AND EXPERIMENT

The differing strengths and limitations of CFD and experiment have led to strongly complementary roles for each in the design process. One enhances the information obtained from the other, and together they provide a synergism whereby the design information obtained from complementary use of CFD and experiment is generally greater than the sum of the individual contributions of each.

Selection of the relative roles of CFD and experiment in a given design application depend on many factors - prior experience, the physics that is expected to dominate, cost, timeliness, risk assessment, and return on investment to name a few. We live in a world where we know how to do more than available resources or time allow , and part of the art of aerodynamic design is in judging how much is enough. This applies both to the use of CFD and to experiment.

The following section illustrates by example the use of CFD and experiment in several different aspects of aircraft design.

Designing the Experiment

An experiment may involve a ground-based facility such as a wind tunnel, or measurements made in flight test. In any event, CFD usually plays an integral role in designing an experiment. CFD is typically used to assess flow interference due to model supports, intrusive instrumentation, etc. It is used to make decisions on trip strip location, and to provide information on the type of instrumentation desired and where on the model surface or in the field to position the instrumentation.

A recent example from Rubbert [4] illustrates the complementary roles of CFD and experiment in designing and carrying out a test of the GE-UDF engine simulator in the Boeing Transonic Wind Tunnel. The test setup, Figure 2, featured an acoustic liner covering the wind tunnel walls and slots, which rendered impossible the customary setting of the tunnel Mach number via a pressure measurement in the surrounding plenum. We learned via CFD that the strut and turbine housing of the model had a significant effect on the Mach number at the propeller location and hence on measured engine performance. We ended up modeling the entire test setup with CFD and using CFD to establish the effective freestream Mach number in the wind tunnel, and to account and correct for Mach number differences at the propeller plane induced by the entire test setup. CFD thus played a central role in enabling the measurement of model engine performance.

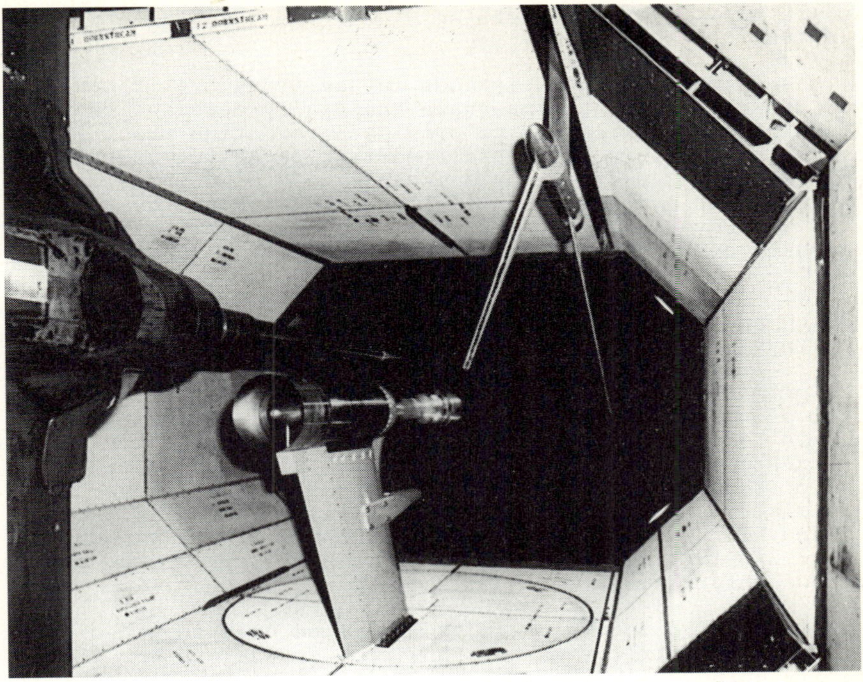

Figure 2 Engine Simulator In a Transonic Wind Tunnel

Low Speed Design

Design needs in this flight regime range from preliminary design to detailed design, and encompass both performance and stability and control. Bengelink [5] discusses the roles of CFD and experiment for this flight regime in some detail. The present discussion summarizes some of the highlights.

The use of CFD and/or experiment to support preliminary design requirements must be rapidly responsive to continually changing configuration variables and their effects on global aerodynamic characteristics. The tradeoffs between ease of use, accuracy, rapidity of turnaround and low cost must favor rapid turnaround and low cost. The design technology that has evolved to meet these critieria involves easy-to-use inviscid 3-D CFD methods combined with baseline experimental data or generic empirical data as described in Bengelink [5].

The detailed design of high lift systems involves both 2-D and 3-D computational methods and 2-D and 3-D wind tunnel experiments. 2-D codes that model subsonic and transonic inviscid physics and boundary layers, with free or fixed transition, are used for detailed design of airfoil and flap section shapes. They also serve as one means of learning how the flow will scale between typical wind tunnel Reynolds number and flight. Two-dimensional experiments are used to validate CFD-based designs, to acquire information on Reynolds number effects, and to systematically search through a matrix of leading and trailing edge flap defections, gaps, etc.

Three-dimensional CFD methods in use today are primarily inviscid and are used to integrate the highly detailed section designs into a 3-D wing and to provide information about induced drag, which is a very large and important parameter in this flight regime. 3-D experiments include both wind tunnel and flight test and are used principally to confirm the overall designs. However, it is not unusual to use flight test as the final means of "fine tuning" the flap settings.

Stability and Control Design Issues near the Perimeter of the Flight Envelope.

A primary issue in all flight regimes is stability and control of the airplane. The range of possible problems and issues is large. Many problems of interest near the perimeter of the flight envelope involve "off-design" (high angle-of-attack/sideslip) flow fields with complex viscous/vortex interactions, highly distorted wakes and separated flows. Present CFD codes are incapable of dealing with the full range of these problems for real airplane geometries, and experiment remains the primary source of information.

However, CFD is beginning to make some contributions to stability and control work for flow regimes wherein the flow is primarily attached. Figures 3 and 4, taken from Bengelink [5], show wind tunnel and inviscid CFD predictions of horizontal stabilizer and elevator effectiveness, and surface pressure distributions. Figure 5 shows that neither CFD nor wind tunnel measurement (at less than flight Reynolds number) accurately predict the experimental results measured in flight. The primary reason is that the boundary layers are too thick at typical wind tunnel Reynolds numbers, and too thin (zero) in the inviscid CFD analysis. However, the CFD and wind tunnel results together form a basis for interpolating between them to achieve a better a priori estimate of full scale flight results than could be gotten either from wind tunnel or CFD. Figure 6 from Bengelink [5] shows such an interpolation scheme. Application of this interpolation technique to the Boeing 767 resulted in the prediction shown in Figure 7. The predicted full scale values of both stabilizer and elevator effectiveness are in quite good agreement with flight test data over the entire speed range up to Mach 0.84. In all cases shown, the error in predicted values is less than 5%. This example shows the synergism that can be obtained by the complementary use of CFD and the wind tunnel.

Cruise Wing Design

CFD has totally transformed the process of wing design. Our standard computational tools [1] today include full potential and Euler transonic wing/body/strut/nacelle codes coupled in an integrated system of programs which include grid generation, wing 3-D laminar and turbulent boundary layer calculations, semi-empirical shock/boundary layer interaction, displacement thickness modeling, and various interface codes. The resulting system of codes is very user-friendly and requires minimal setup time. It is used both for preliminary design and for detailed design work. A

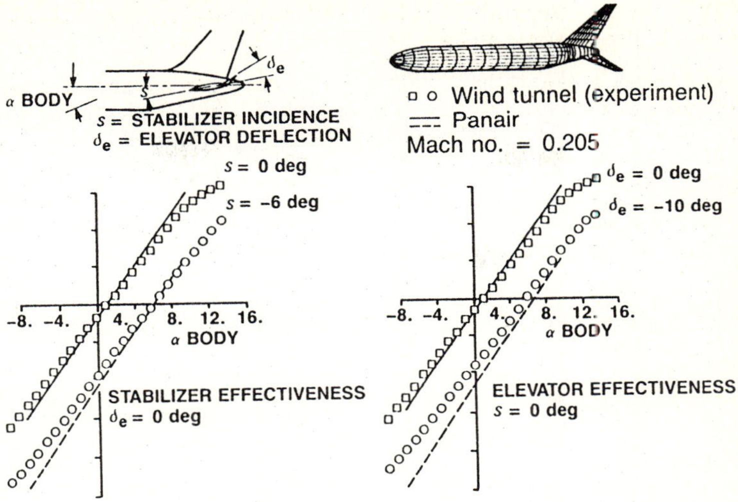

Figure 3 Panair Prediction of Boeing 767 Stabilizer and Elevator Effectiveness

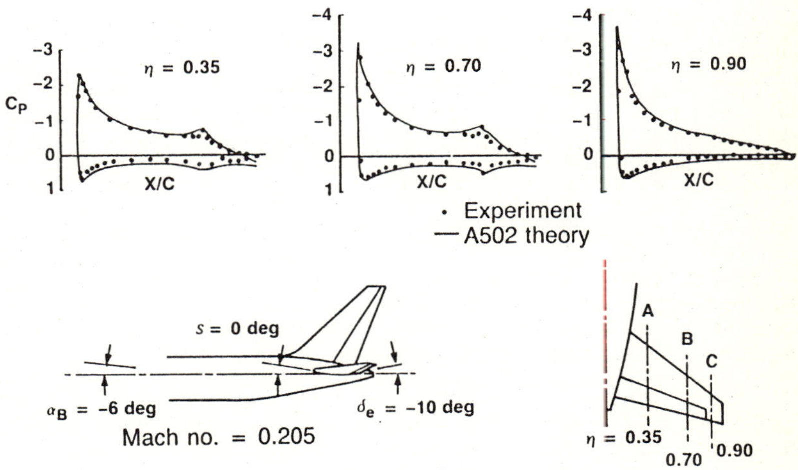

Figure 4 Panair Pressure Distribution Test-Theory Comparison for a 767 Horizontal Stabilizer

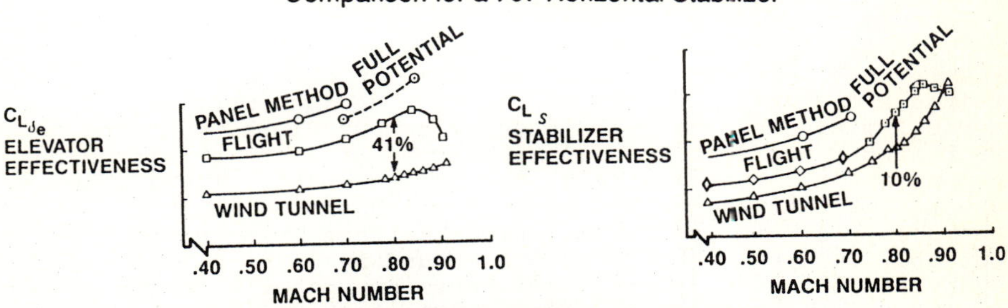

Figure 5 Inviscid CFD, Wind Tunnel, and Inflight Measurements of Effectiveness (767-200)

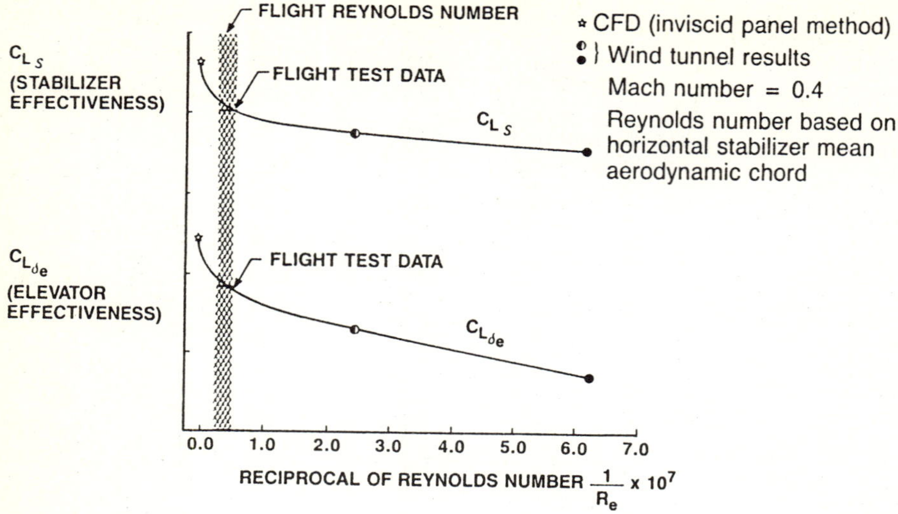

Figure 6 Procedure for Interpolating Between Wind Tunnel and CFD

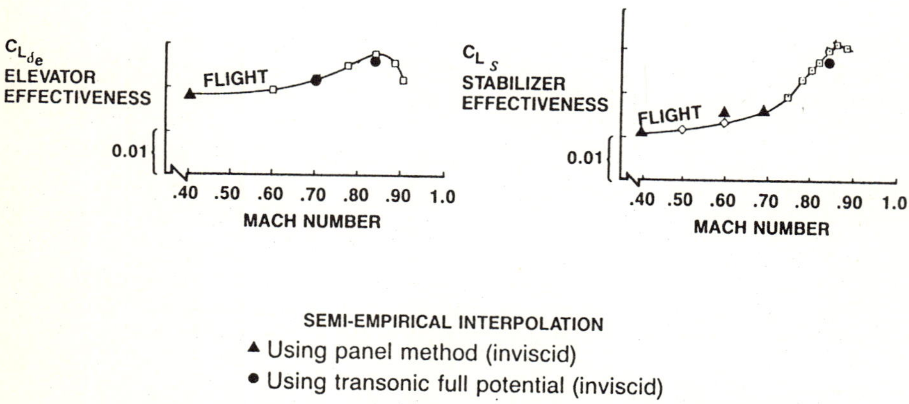

Figure 7 Stabilizer and Elevator Effectiveness Predictions Compared with Flight Measurements

version of the system for inverse wing design in the presence of the body is also available.

Figure 8 shows the level of agreement between computed results and test data that are typical of this system. Many things are examined in the course of a design exercise. A typical wing design will be checked computationally at the design Mach number and C_L. It will be checked at increments in Mach number and C_L above and below the design point. It will be checked at both flight and wind tunnel Reynolds numbers. Features that are monitored include surface pressure distributions, isobars, shock strength, boundary layer parameters, spanwise load distribution, thickness and other geometrical considerations, and drag (although the latter can be somewhat unreliable).

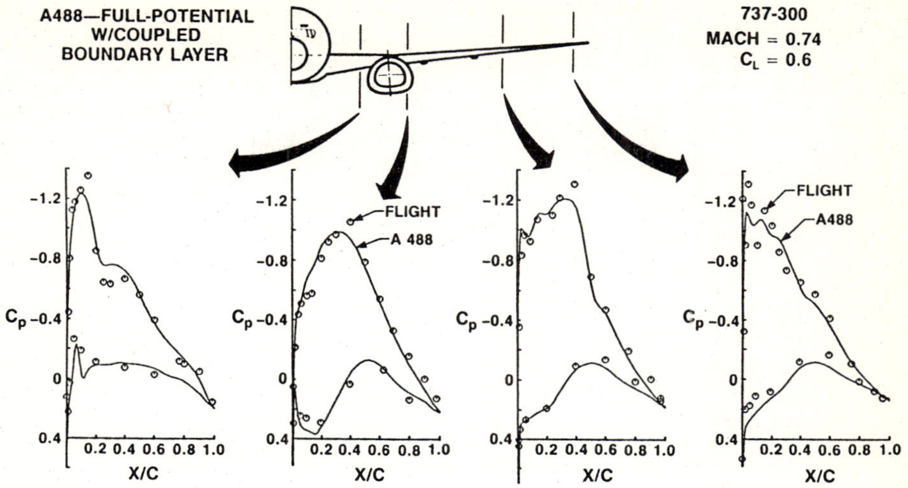

Figure 8 Comparison of CFD Predictions Pressure Distribution Measured in Flight -- 737-300

The automated nature of the transonic/viscous analysis and design system allows many iterations in developing a highly refined wing design in a timely manner. The ability to do the inverse problem (calculate surface shape, given a specified wing pressure distribution) in a manner totaly consistent with analysis (calculate pressure distribution, given shape) has given the designer a very powerful tool for achieving superior wing designs.

The role of the wind tunnel in wing design today is principally for validation and confirmation in the neighborhood of the cruise regime. At the perimeter of the flight envelope, the wind tunnel remains the principle source of information.

Engine/Airframe Integration

Successful aerodynamic integration of engine and airframe is primarily a detailed design task where the objectives are to keep the boundary layers on all neighboring surfaces attached and well-behaved,to keep local supervelocities low and to prevent or minimize shock wave formation, to control induced drag, and to create an inlet flow environment that is compatible with the operating characteristics of the engine. CFD is the design tool of choice for such work, although its usefulness for some transonic applications remains less than desirable because of the inability of many of today's transonic codes to handle truly arbitrary combinations of geometry and the multiple aerodynamic length scales associated with some engine/airframe installations. The role of wind tunnel testing has become mainly one of confirmation.

Both CFD and experimental measurements that sometimes involve complex and costly engine simulators are used to sort out the several issues that are involved in accurate determination of installed thrust and thrust-minus-drag.

Some of the earliest extensive use of CFD for the engine/airframe integration of pylon-mounted turbofans was carried out by Boeing [6], which led to the close-coupled, low drag engine installations on the 707/CFM-56, the 757, 767 and 737-300 and 737-400 series of airplanes. The most remarkable of these is the 737-300 installation, shown in Figure 9.

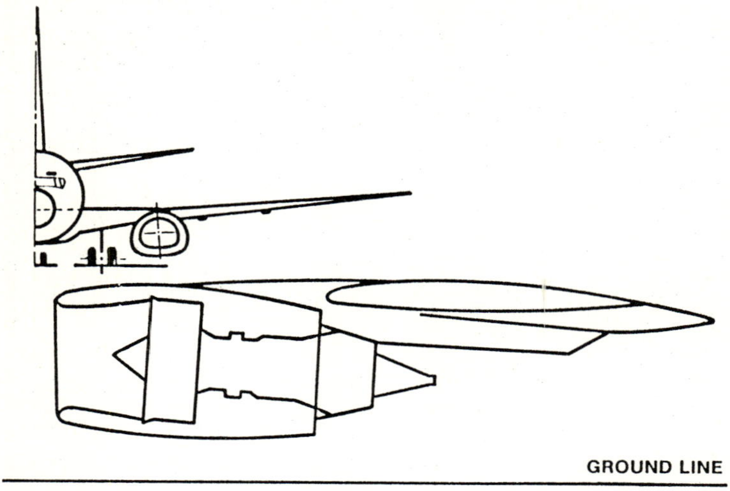

Figure 9 737-300 Nacelle Installation

More recently, CFD played a central role in the design of an aft-mounted propfan installation. Figure 10, from Yu et al. [7], shows detailed Euler calculations that feature the ability to systematically build up the propulsion installation by separately adding the strut, the nacelle, and the actuator disk, to arbitrarily specify power setting, and to compute the aerodynamics of the installation at yaw and with differential power settings on right and left sides of the airplane. One interesting and effective application of CFD on a similar configuration was to design aft fuselage contours that were crafted computationally to minimize the variation in flow angularity of the airstream entering the propfan disk.

CFD and Flight Test

CFD has been used extensively in support of flight test in several different ways. One example was a comprehensive computational analysis that was carried out concurrently with the precertification flight test development of the 737-300 aircraft. The original goal of the CFD study was to enhance understanding and provide guidance to the flight development program. One aspect of this was low speed flight (flaps down). For that purpose, 18 complete configuration CFD analyses [8] were carried out over a period of 3 months; they included different flap settings,

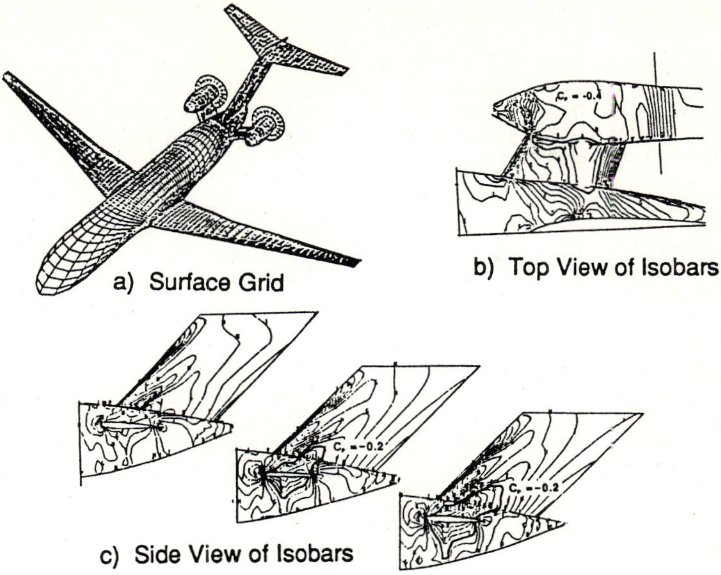

Figure 10 Euler Analysis of a Propfan Engine Installation

wind tunnel and flight wing twist, flowthrough and powered nacelle simulations, free air and wind tunnel walls, ground effect, sealed and slotted flaps, and other geometric variations. The computational model for the baseline takeoff flap configuration is shown in Figure 11.

That work really paid off. It provided an understanding of the cause of a "surprise" that showed up during flight test and the guidance to fix it. It explained and clarified the limitations of prior low-speed wind tunnel test techniques. And it ended up doing a good job in predicting the L/D of the airplane, Figure 11b.

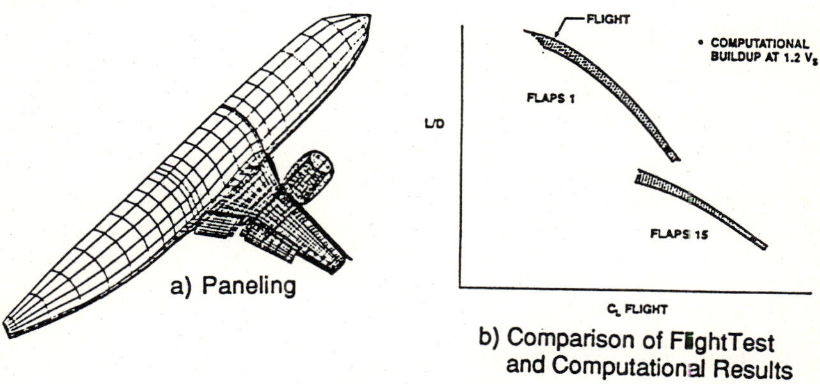

Figure 11 737-300 High Lift Configuration

CFD And Airplane Certification Flying

Certification is required whenever a change is made to an airplane that can affect its performance or safety. Most of the time flight testing is required, but sometimes analysis is sufficient. Over the years, CFD has come to be a growing source of information in this area.

One recent example [4] of such use of CFD had to do with venting and overflow of toilets. There is an FAA requirement that possible efflux from toilet drain masts on an airplane not impinge and possibly freeze on the fuselage. When Boeing recently stretched the 767 airplane to create the 767-300, CFD was used to establish that the flow about one of the two toilet drain masts was essentially unchanged from the 767-200 (which had been certified earlier). Boeing and the FAA both accepted the computational result and did not require flight certification of that drain mast. However, CFD did show that the flow about the other drain mast was different, and so that one required a certification flight test.

Role of CFD and Experiment in "Special Situations"

There are many special situations that arise in airplane design, some of which are addressed purely with experimental means, some with CFD alone, and some making use of both. One area of design that remains purely experiment-based is the determination of optimum patterns of vortex generator placement on an airplane. A typical design task might be to achieve a certain improvement in aileron effectiveness or airplane pitch characteristics at off-design conditions by means of vortex generators while minimizing the increment in cruise drag associated with vortex generators. That is usually carried out as a cut-and-try flight test operation.

An example of a design task carried out with CFD alone was the design of the CFD-crafted contours of the wing tip pod of the E6-A version of the Boeing 707 shown in Figure 12. There was enough confidence in the ability of CFD to design a cruise drag shape having low drag that a decision was made not to wind tunnel test this design. Later a flight test program was carried out which included an assessment of the off-design aerodynamic characteristics.

In another special situation, there arose a requirement to determine a suitable location on the fuselage of the 747 aircraft to position a small vane that would measure the angle of attack of the airplane. The significant issue was to find a location on

Figure 12 E6-A Tip Pod on Boeing 707

the fuselage during low speed, flaps-down flight where the local angle of attack of the flow, as measured by a small vane, would be invariant to sideslip. This was done by using CFD to create a map on the fuselage surface of the rate of change of flow angle with respect to sideslip, and selecting a location for the vane where this rate of change with sideslip was zero.

A wind tunnel test was subsequently carried out for verification. One source told me that during the test the initial thinking was that the instrumentation had broken; there was no discernible change in angle of attack as measured by the vane when the model was yawed.

A COMMENT ON WORKING RELATIONSHIPS

A discussion of the use of CFD and experiment in transonic aircraft design would not be complete without mention of the desirability of a close working relationship between the people who develop CFD methods and those who use both CFD and experiments in the design of airplanes, namely the engineering design community. The emergence of CFD as a major component of aircraft design technology has caused the Boeing engineering design community to pay close attention to the source of new CFD capabilities, to formulate and make known their needs and desires for improved capabilities, and to be supportive of CFD R & D. This in turn has caused Boeing CFD researchers and code developers to better understand the needs of the engineering design community and to be more responsive to those needs. The net result is a more focused and more effective CFD R & D activity, and more effective application of new CFD capabilities.

CONCLUDING REMARKS

Today the uses of CFD and experiment in transonic aircraft design are strongly complementary. CFD and experimental methods each have unique strengths and limitations, and each tends to be used for what it does best.

The applicability of one or the other depends strongly on the flight regime. The forte of CFD today for transport design applications is in that portion of the flight envelope in which steady, attached flow prevails, such as at the cruise condition. In this regime, the use of experiment is primarily for design confirmation.

At the perimeter of the flight envelope where extensive regions of separated and unsteady flows prevail and where stability and control as related to flight safety are primary issues, experimental methods continue to be the primary source of design information. But a role for CFD is beginning to emerge in this area.

The interrelationship between the use of experiment and CFD is dynamic and changing yearly. As new CFD capabilities emerge, they enter into and change the aerodynamic design process. And so does the emergence of new experimental measurement techniques. The current trend is for CFD to have an ever-increasing role in airplane design, and this trend will prevail for a long time into the future.

References

1. Tinoco, E. N., "Transonic CFD Applications at Boeing," Presented at the Transonic Symposium, NASA Langley Research Center, Hampton, VA., April 19-21, 1988

2. Snepp, D. K. and Pomeroy, R. C., "A Geometry System for Aerodynamic Design," AIAA Paper 87-2902, September, 1987.

3. Buning, P. G. and Steger, J. L., "Graphics and Flow Visualization in Computational Fluid Dynamics," AIAA Paper 85-1507-CP, July, 1985

4. Rubbert, P. E., "Design Applications of CFD at Boeing," Presented at the NAS Dedication, NASA Ames Research Center, November, 1986.

5. Bengelink, R. L., "The Use of Computational Fluid Dynamics in Predicting Full Scale Aircraft Aerodynamics," Presented at the International Symposium on Aeronautical Science and Technology (ISASTI 86), Jakarta, Indonesia, June 1986

6. Rubbert, P. E. and Tinoco, E. N., "Impact of Computational Methods on Aircraft Design" (Invited Paper), AIAA Paper 83-2060, August, 1983.

7. Yu, N. J., Kusunose, K., Chen, H. C., and Sommerfield, D. M., "Flow Simulations for an Aft-Mounted Propfan Using Euler Equations," SAE Technical Paper 861718, Long Beach, CA, October, 1986.

8. Tinoco, E. N., Ball, D. N., and Rice II, F. A., "PAN AIR Analysis of a Transport High-Lift Configuration," Journal of Aircraft, Vol. 24, No. 3, March, 1987.

Computational Transonic Aerodynamics Research in Japan

N. Hirose and S. Takanashi

National Aerospace Laboratory
7-44-1 Jindaiji-Higashi
Chofu, Tokyo 182, JAPAN

Abstract

Transonic aerodynamics research has been extensively conducted at NAL, JAPAN. The progress in CFD application is remarkable. N-S codes for airfoil, wing, and full-configuration have been developed. The codes revealed various viscous effects which were not predicted by past inviscid codes and were validated with wind tunnel testing results and proved to be excellent to apply as practical design tools for aircraft design such as the Boeing-Japan 7J7. A versatile wing design code was combined with various flow solvers and used for practical purpose. The paper covers most of the recent CFD work in this field.

1. Introduction

Japan has a long tradition of aerodynamics and fluid dynamics research. In the area of theoretical transonic aerodynamics, classical works of Tomochika-Tamada and I. Imai are well-known. When computers came into use, computer simulation of fluid flow was tried by several people. In its early days, computer power was very limited, and the number of large computers accessible to researchers was also limited. The National Aerospace Laboratory (NAL) was one of the institutions, besides a few universities, where computers were available.

NAL has been engaged in developing computational transonic aerodynamics for the design of aircraft during the past 20 years. The computer codes developed are now being applied to the aerodynamic design of the Japanese civil transport YXX (Boeing-Japan 7J7), etc. As the advanced flow simulation codes became available, code validation evaluations with the existing codes and wind tunnel experiments were extensively made. The computational aerodynamics is now an established new discipline besides the existing theoretical and experimental aerodynamics in the research of transonic flow. In the present paper, only the practical transonic CFD-development for aircraft pursued at NAL, Japan, is briefly described although various academic research works have been done at the other academic institutions.

2. Historical Aspects

2.1 1960s

When the experimental works of Pearcey and Whitcomb revealed the practical possibility of transonic airfoil design in the 1960s, both the experimental and the

J. Zierep, H. Oertel (Eds.)
Symposium Transsonicum III
IUTAM Symposium Göttingen, 24.-27. 5. 1988

theoretical research for transonic airfoils and wings was in progress at NAL. The computer was used to solve the transonic flow equation numerically by a hodograph method by the senior author (Takanashi). The details of those early day works can be found in the ICAS-74 paper by Shigemi [1]. Computational fluid dynamics (CFD) using finite difference methods was investigated mainly for the supersonic blunt body and jet flow problems in the 60s and early 70s at NAL. In those days, transonic flow application of CFD was not practical although we noticed Magnus-Yoshihara's early pioneering work as an important one.

2.2 1970s

Murman's success of the TSD type-dependent relaxation method and the practical application of the method to the C-141 wing analysis by Ballhaus et al. gave us practical interest to develop transonic aerodynamic analysis and design tools utilizing CFD in the 70s. Transonic airfoil and wing analysis codes based on TSD and full potential equations similar to Garabedian-Korn's and Bailey-Ballhaus' were developed. The latter required 10 hours of CPU time on then-available FACOM-230-75 (almost equivalent to CDC6600) to obtain one data-point of an NAL-designed transonic swept wing with favorable agreement with wind tunnel experiment. An Euler analysis code for airfoils was also developed and applied to both steady and unsteady flow by Ishiguro. Isogai developed a transonic flutter analysis code based on TSD equations. Takanashi developed an efficient wing analysis method using 3-D transonic integral equations. An efficient and effective airfoil design method was developed by Ishiguro. When the surface pressure distribution was specified, the code iteratively solved the airfoil geometry including viscous effects utilizing inverse and analysis steps for full potential equations. Kamiya developed his transonic airfoil design methodology, utilized the aforementioned CFD tools to design supercritical airfoils and clarified the applicability of the tools. In 1979, the High Reynolds Number 2-D Transonic Wind Tunnel was introduced to the transonic aerodynamics research. The wind tunnel provides high Reynolds number flow up to $4 \cdot 10^7$ and presents high-quality experimental data of airfoils to researchers and designers at industries. The details are described in the ICAS-80 paper by Kamiya and Hirose [2] and the ICAS-82 paper by Takashima [3].

2.3 1980s

The 7th ICNMFD, Stanford, revealed that CFD outgrew the engineering status just as FEM did. The potential codes are used easily, and revolutionary advancement will not be made even in Navier-Stokes (N-S) solvers. Only high-speed computers and massive complicated grid-generation were needed although turbulence modeling remains unsolved. At NAL, the trends of research followed similar to the ones in other countries. N-S codes for airfoil, wing, and full-configuration were developed and applied to the practical computations. Improvements o the N-S and Euler schemes were made. New design methods were developed and applied to various flow regimes. To enhance the code development and applications, supercomputers were installed at NAL. The new computer facility is named "Numerical Simulator (NS)" and consists of FACOM VP400, VP200 and Front-end processor M780 with various graphic terminals. Some of the early achievements are described in Kamiya's paper at ISCFD, Tokyo [4].

3. Navier-Stokes Analysis of Airfoil, Wing, and Complete Aircraft

3.1 Airfoil Analysis

Time-averaged Navier-Stokes analysis for high Reynolds number transonic flow has been one of the major research topics. A N-S code was pursued rather than an Euler code because Euler codes are not worth while for transport type aircraft aerodynamics in which the shock wave is not so strong compared with military aircraft. Also viscous interaction will be inevitable when the shock wave becomes strong. As the selection of the scheme type is different for Euler and viscous flow, the N-S approach was selected.

The two-dimensional airfoil analysis code NSFOIL based on the Beam - Warming - Steger IAF algorithm developed by Kawai and Hirose [5] has been applied to various supercritical airfoil analyses for practical purposes. High-Reynolds 2-D Wind Tunnel data was used for code validation. The computation of one data point requires less than 8 min. and the code is now used as a preliminary analysis of transport wing designs before wind tunnel testing by industries. Fig. 1 shows some results of such application. A flow field for an airfoil with aileron deflection is shown in (a), the aileron effect $dC_l/d\alpha$ vs Mach number is compared with experiment in (b). Blunt trailing edge flow of supercritical airfoils was analyzed using fine mesh distribution in this particular region, and Kármán vortex shedding was discovered in (c). Recent 2-D analysis covers Laminar Flow Control (LFC) and Natural Laminar Flow (NFL) application, and evaluation of more accurate turbulence modeling and transition is the pacing item.

3.2 Wing and Complete Aircraft Analysis

Obayashi [6] et al. proposed an LU-ADI factorization scheme for N-S equations. This gives an efficient improvement to the original IAF scheme, since the algorithm is reduced to the scalar bidiagonal inversions suitable to supercomputers. Fujii and Obayashi [7] developed the 3-D version code. New ideas of scheme improvement are incorporated every day, and the latest version includes a TVD upwind using Row's averaging. A MUSCL interpolation is applied to obtain smooth shock- and high-order accurate resolution. Meanwhile, Takanashi [8] developed an efficient and versatile grid-generation method for realistic 3-D configurations. Analytic functions to map the physical space covering a basic cylindrical body and an unswept wing into the cubic computational space are used. To accomodate to the realistic aircraft with swept wing, horizontal and vertical tails and canards, grid adjustment and clustering are performed using algebraic functions. The grid generation only requires 20 sec. Some of the latest works by Takanashi et al. [9] are shown in Fig. 2. (a) is the surface grid for a T-tailed transport configuration with rear-prop-fan. To accurately evaluate viscous effects, fine mesh clustering is made over the entire body surface although not shown in this figure. Preliminary N-S analysis for wing-body-T-tail is shown in (b) - (d) at several flow Mach numbers and angles of attack. Wing surface pressure distributions C_l vs α and $C_l - C_d$ curves show excellent agreement with wind tunnel results. CPU time for a 221x74x62 grid computation requires 6 to 20 hours on VP400 depending on the flow conditions. Further code validation study is presently being done. Another application of the present code is the flow analysis of space-planes. The N-S code covers low-speed to hypersonic flow, and grid generation is easy to handle. The leading edge vortex at large angle of attack will be the only one item to care about. Fig. 3 shows an example of supersonic flow. Parametric flow analysis at various Mach numbers is being developed.

3.3 Unsteady Wing Analysis

Another 3-D N-S research is Nakamichi's [10] unsteady flow analysis past a swept wing in pitching motion for an aeroelastic problem. A 3-point implicit time-accurate version of the Beam - Warming diagonalized scheme for a time-dependent moving 3-D grid was applied to the NORA swept wing with low aspect ratio about which unsteady aerodynamic conditions were measured in various European wind tunnels in 1978 by Lambourne. Computations were made for Mach numbers $0.9 \sim 1.1$, Reynolds number $6 \cdot 10^{6}$, angle of attack 4°, amplitude of pitching motion 1°, and reduced frequencies $0.238 \sim 0.428$. A 161x29x35 grid point computation required about 7 - 10 hours on VP400 for each case. Fig. 4 shows the mesh used and the density contours at one span station. Vortex shedding from the up-drooped leading edge interacts with the shock wave. Good agreement with experiment was obtained compared to Isogai's full potential code USTF3 [11], although more detailed work remains for quantitative agreement. Here Isogai's USTF3, itself, is excellent work to be mentioned.

3.4 3-D Fan Analysis

A 3-D N-S analysis application to aero-engin problems was made by Nozaki and Nakahashi et al. [12]. A 3-D flow field past a rotating turbine and fan cascades were computed applying a scheme similar to the previous ones. Fig. 5 shows a 3-D view of surface density contours around a strongly twisted fan blade. The hub wall rotates with the blades, while the outer casing remains fixed. The inlet Mach number is (0.61 at hub, 0.62 at tip) and the outlet Mach number is (0.72, 0.57). The Reynolds number is $5 \cdot 10^{5}$. (b) shows the density contour at the tip section. A strong trailing shock wave from the suction surface reaches the mid-chord of the pressure surface of the neighbouring blade and produces shock-induced separation. On the suction surface, the shock wave location advances to mid-chord at mid-span station, and the flow is separated. At the hub region 3-D flow separation is observed. Although not shown here, good agreement with experiment was obtained in the case of the FJR 710 R&D fan engine turbine cascade flows.

In all of the N-S analyses, the Baldwin-Lomax turbulence model is used.

4. Euler Analysis of Aircraft and Component

4.1 Fan-Jet Engine Simulation

Hirose, Asai and Ikawa [13] analyzed the transonic flow field around a fan-jet engine at angle of attack. Turbine Powered Simulator wind tunnel tests were also made for the same geometry. Fig. 6 shows grid and pressure contours for Mach number 0.9, angle of attack 6°, and mass flow ratio 0.8. A shock wave appears on the leeside of the intake nacelle, and fan-jet and core-jet interaction is visible. At some distance from the nozzle exit, a pair of horse-shoe like vortices appears owing to cross flow. The details are not clear because of crude grid spacing. The surface pressure distribution agreed with experiment. It was confirmed that the temperature effect of core-jet to outside flow is negligibly small as far as TPS testing is concerned.

4.2 Complete Quiet-STOL Research Aircraft ASKA Configuration

A complete configuration analysis of the NAL Quiet-STOL "ASKA " was made by Takanashi and Sawada [14] applying the second order Roe's Rieman solver. A multi-block transformation technique was used to transform this complicated geometry to a single structured computational space. To construct a good distribution, an interactive procedure was deviced rather than automatic grid generation. Fig. 7 (a) shows the surface grid of the aircraft. The effects of four-jet, twin-jet and clear wing geometries were analyzed. 162x103x76 grid points were used. (b) shows the isobar pattern on the surfaces, and (c) shows shock patterns on the upper surface of the wing at Mach number 0.8. It is clear that the flow between the nacelles is choked and complex shock waves are formed.

5. Viscous-Inviscid Interaction

The viscous-inviscid interaction method is a fast and easy analysis tool to predict the aerodynamic performances of the airfoil and wing in the engineering point of view although the N-S analysis has been exploited recently among researchers. Matsuno [15] developed a predictor-corrector Crank-Nicolson scheme to solve 3-D compressible turbulent boundary layer equations over a swept wing, which was called code BLAY. The scheme is vector-oriented. The code was incorporated with various inviscid wing analysis codes including Jameson's FLO27, FLO30 and similar homemade potential codes. The iterative interaction effects of transonic wings including forward-swept wings for transport have been evaluated extensively at NAL and industries [16]. Also, the code served for swept wing design which is mentioned in the following section.

To advance in more refined studies, Matsuno [17] developed an advanced interaction method for airfoils, which incorporates turbulent wake thickness and curvature effects as well as the improved algebraic turbulence model for the lifting wake region. The result for the RAE2822 is shown in Fig. 8. The velocity profiles on the airfoil and wake show excellent agreement with Cook's experiment. The drag-polar curve also presents good agreement.

6. Design Method and its Applications

One of the most practical and versatile design methods for transonic swept wings was developed by Takanashi [18]. The method is a residual-correction type procedure and is based on the solution of the transonic integral equation of a potential perturbed around a known solution of the flow past a wing. When the pressure distribution on a wing with fixed planform is specified as the target, an arbitrary flow solver used gives an initial pressure distribution for an initially guessed geometry. The difference between the target and the initial distributions is related with the geometry correction through the integral equation. Therefore, the geometry correction can be solved easily and the corrected geometry is used for the flow solver to obtain the new pressure distribution. The procedure continues iteratively until the target pressure distribution is realized. The significant feature of the method is that (1) the geometry correction step is completely independent of the flow solver in the analysis mode, i.e., any flow solver can serve in the analysis mode. (2) a mixed inverse/direct calculation is possible, i.e., only arbitrarily specified parts of the span station can be designed. (3) the design covers the entire wing section from leading edge to trailing edge. (4) it is robust, even a strong shock wave is specified. (5) the computing time of the design mode is negligibly short.

The present code, "WINDES", was linked with various flow solvers such as Jameson's FLO22, FLO27 (with BLAY), FLO30, Boppe's WIBCO, and used as research tool for designing practical wings at NAL and industries. Fig. 9 (a) shows an example of a forward-swept wing design. For high lift design at high Mach number, a strong shock wave is formed at the wing root, and the geometry to alleviate the shock strength becomes an unrealistic shape. In present, the authors are investigating how to solve this problem. The leading edge extension is one device for such a treatment. Fig. 9 (b) shows the effect of L.E. extension to a forward-swept wing. By moderate L.E. extension, the shock wave can be diffused spanwise and the strength can be reduced. A BLAY analysis revealed favorable viscous surface streamlines flowing inward. The applicability of the design method to other than potential codes was demonstrated by Hirose, Takanashi and Kawai [19]. The N-S analysis code for airfoils was linked with WINDES. Supercritical airfoils with strong shock wave and weakly-separated boundary layer were modified utilizing the present method. Fig. 9 (c) shows an example. A low-speed thick airfoil for LFC was also designed easily. When the work was done, it required more than 10 hours of CPU time. But now it can be done in less than 20 minutes. Another application is the cascade design version. Fig. 9 (d) shows a thin compressor blade design at a high Mach number of 0.9. A thick blade design which is more difficult to apply was also tried successfully.

7. Symposiums and CFD Workshop

There are no specialist meetings for transonic research because there are quite few researchers devoted purely to this field, even at NAL. Various fluid dynamics meetings, however, are held quite often. The first CFD Workshop, JSME, was held in 1987. Among several basic flow problems, transonic airfoil analysis was the theme of the workshop which was organized by Hirose. Applicants from industry and universities presented their results for NACA0012, RAE2822, and GK-75-06-12 obtained by their N-S and Euler codes. Comparison in detail clarified various aspects of the present status of CFD application.

NAL hosts the Symposium on Aircraft Computational Aerodynamics (SACAD) annually since 1983. NAL is promoting cooperative researches on CFD and transonic aerodynamics with industries and universities. Some of the works in this paper were the products of these cooperations. The relations with universities were not tight, unlike the ones in the USA or in Europe. The same was true for industries, except wind tunnel testing. One major reason for this is the small size of the Japanese aerospace industry. But recent progress in this area is stimulating more close relations between NAL, industries, and universities. The authors hope that there will be more cooperation in the near future.

8. Concluding Remarks

Transonic aerodynamics research in Japan, especially in the field of CFD, is reviewed. The historical background and recent activities, mainly in the field of practical applications, at NAL are presented. The authors are afraid that all of the excellent works especially done outside of NAL were not referred to in this paper owing to space limitations.

References

[1] Shigemi, T., ICAS-P. #74-12, (1974).

[2] Kamiya, N. and Hirose, N., ICAS-P. #80-11-1, (1980).

[3] Takashima, K., ICAS-P. #82-5-4-4, (1982).

[4] Kamiya, N., Proc. Int. Symp. on CFD Tokyo (ISCFD Tokyo), (1985).

[5] Kawai, N. and Hirose, N., NAL-TR-816, (1984).

[6] Obayashi, S. and Kuwahara, K., J. Comp. Phys., 63, No.1, (1986).

[7] Fujii, K. and Obayashi, S., AIAA P. 86-1831, (1986).

[8] Takanashi, S. and Isawa, T., Proc. 4th NAL SACAD, NAL SP-7, (1986).

[9] Takanashi, S., Obayashi, S., Matsushima, K. and Fujii, K., AIAA P.87-2410, (1987).

[10] Nakamichi, J., AIAA P.87-1158, (1987).

[11] Isogai, K., NAL-TR-706T, (1982).

[12] Nozaki, O., Nakahashi, K. and Tamura, A., 87-TOKYO-IGTC-43, (1987).

[13] Hirose, N., Asai, K. and Ikawa, K., NAL TR- to appear, (1988).

[14] Takanashi, S. and Sawada, K., Proc. Int. Powered Lift Conf., Santa Clara, (1987).

[15] Matsuno, K., AIAA P.81-1020-CP, (1981).

[16] Hirose, N., Matsuno, K., et al., Proc.16th JSASS Ann. Meet., (1985).

[17] Matsuno, K., NAL TR-873, (1985)

[18] Takanashi, S., J. Aircraft, 22, No.8, p.655-660, (1985).

[19] Hirose, N., Takanashi, S. and Kawai, N., AIAA J., 25, No.3, 353-359 (1987).

Figures

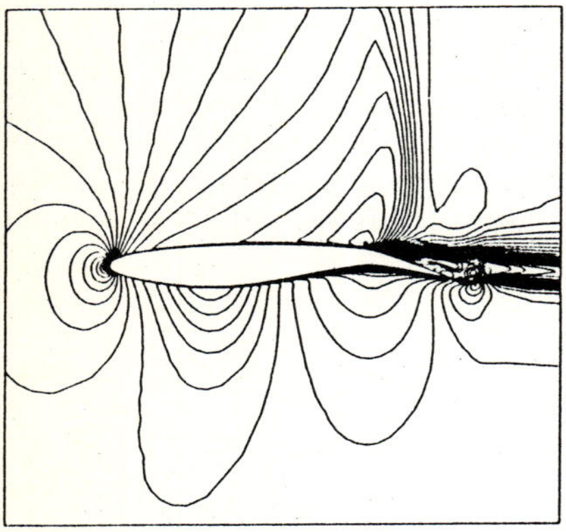

(a) Flowfield for aileron deflection

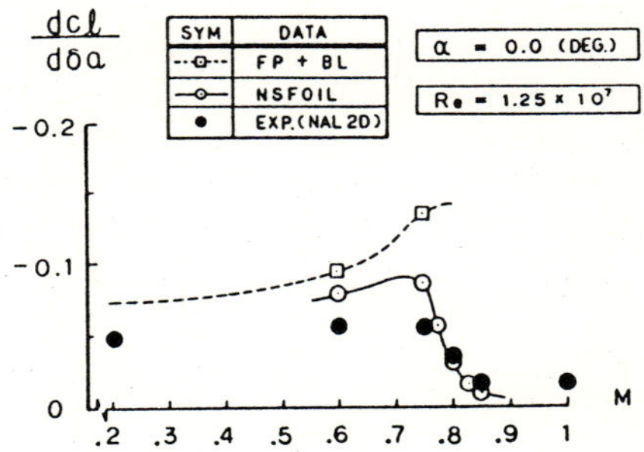

(b) Aileron effect $dC_l/d\delta a$ vs. $M\infty$
○ computation, ● experiment

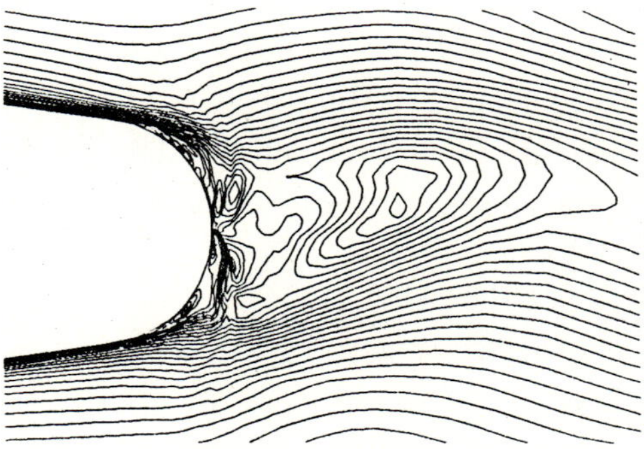

(c) Detailed blunt trailing edge flow of NACA0012

Fig. 1 N-S analysis of airfoils

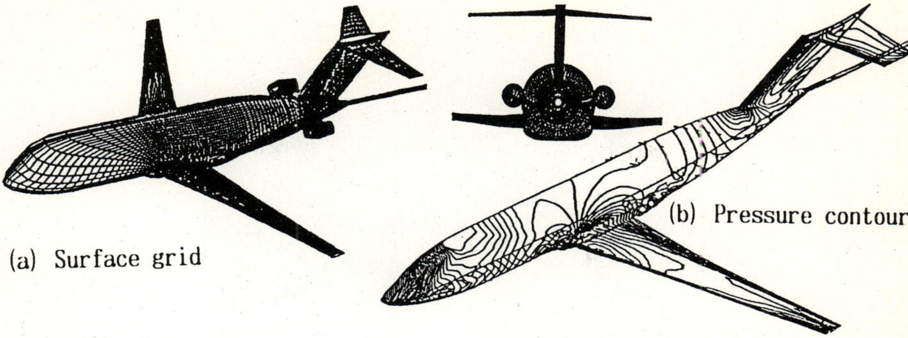

(a) Surface grid

(b) Pressure contours

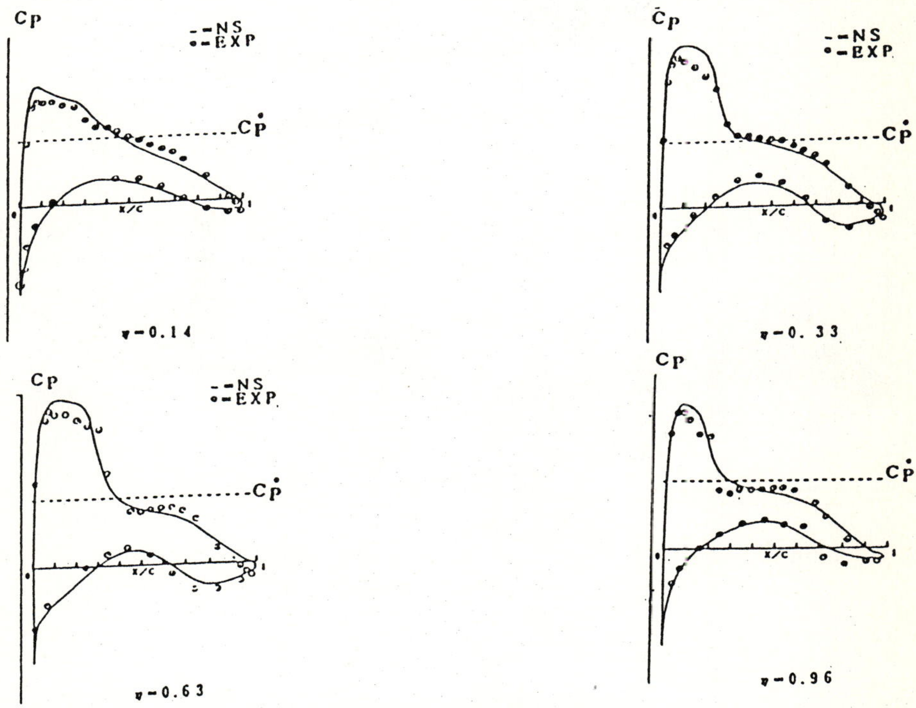

(c) C_p distributions on wing

(d) $C_l-\alpha$, C_l-C_d curves

Fig.2 N-S analysis of complete aircraft

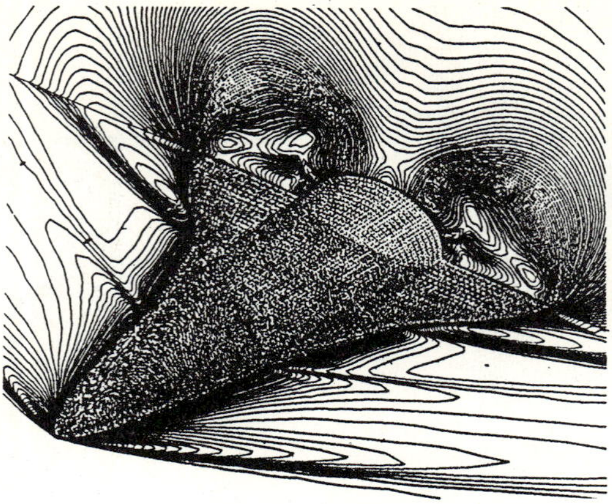

Fig.3 Flow past SPACEPLANE

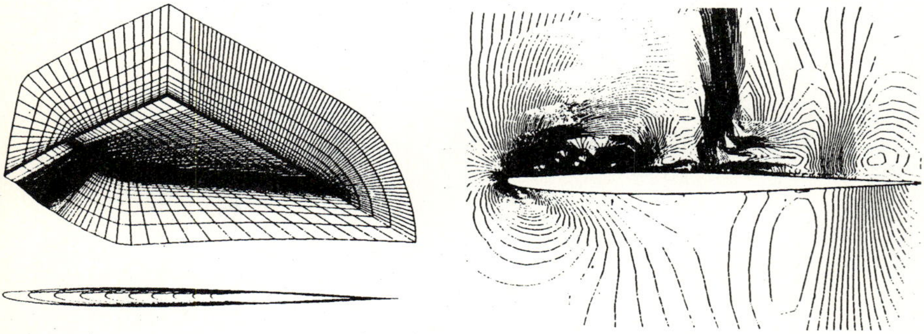

(a) Grid for analysis (b) Density contours

Fig.4 Unsteady analysis of NORA wing

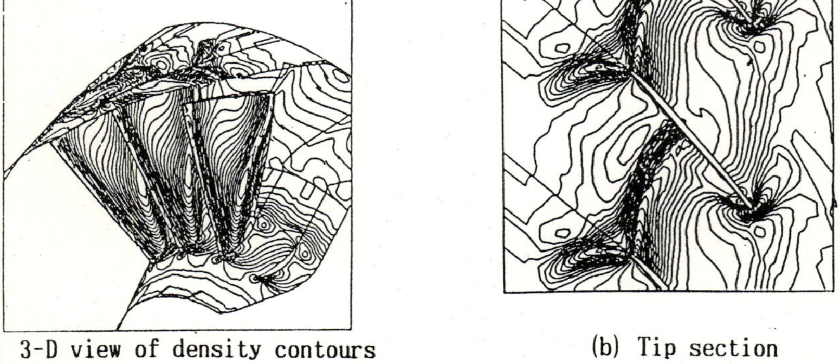

(a) 3-D view of density contours (b) Tip section

Fig.5 3-D fan-blade analysis

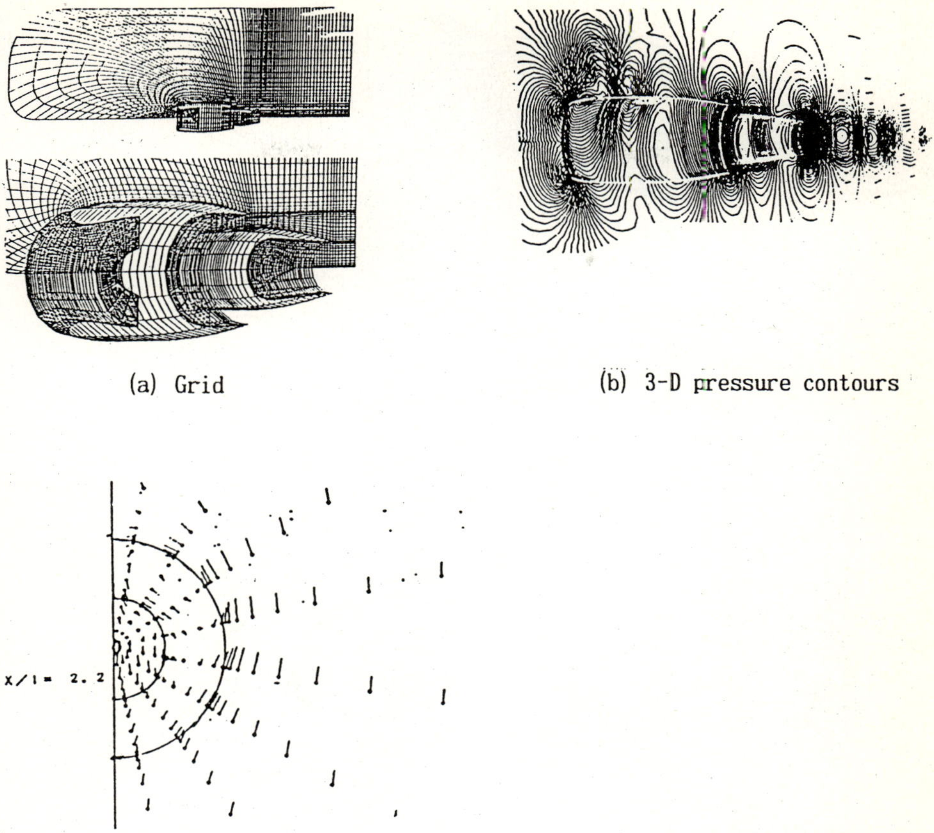

(a) Grid

(b) 3-D pressure contours

(c) Velocity vectors of exhaust jet

Fig. 6 Euler analysis of fan-jet engine

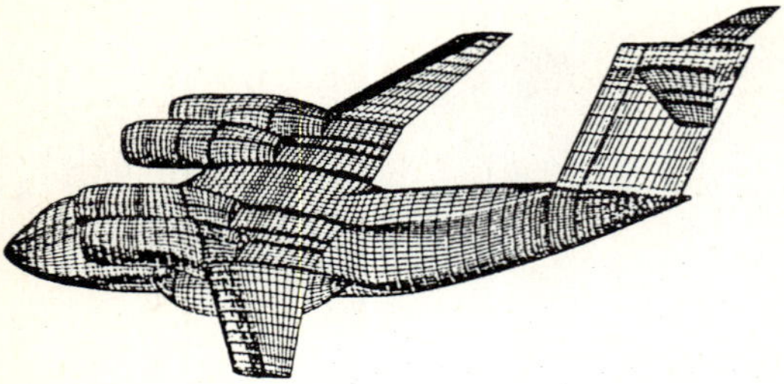

(a) Surface grid

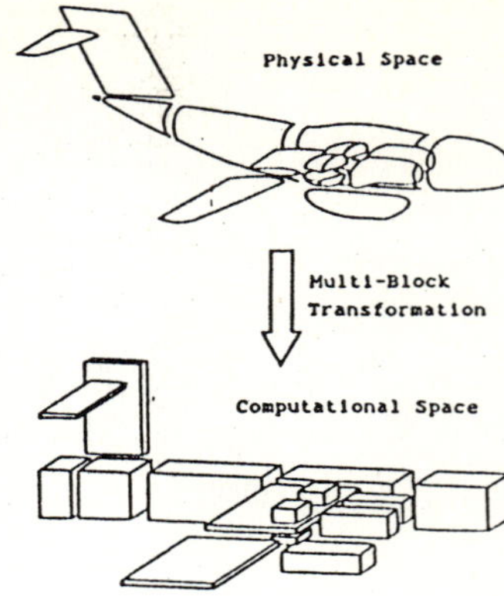

(b) Multi-block transformation

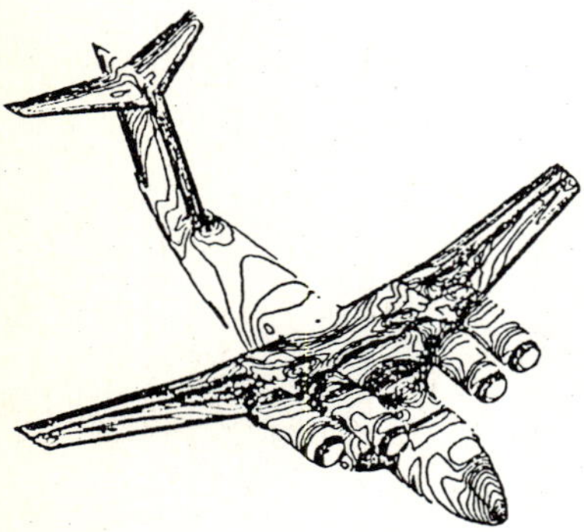

(c) 3-D pressure contours

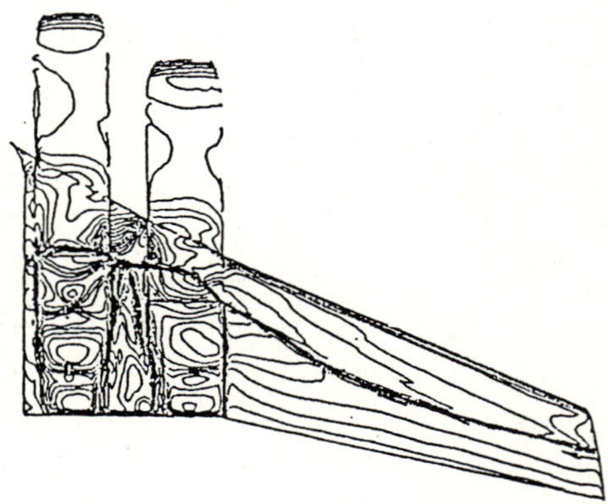

(d) Shock patterns on wing

Fig.7 Euler analysis of NAL Quiet-STOL" ASKA (飛鳥)"

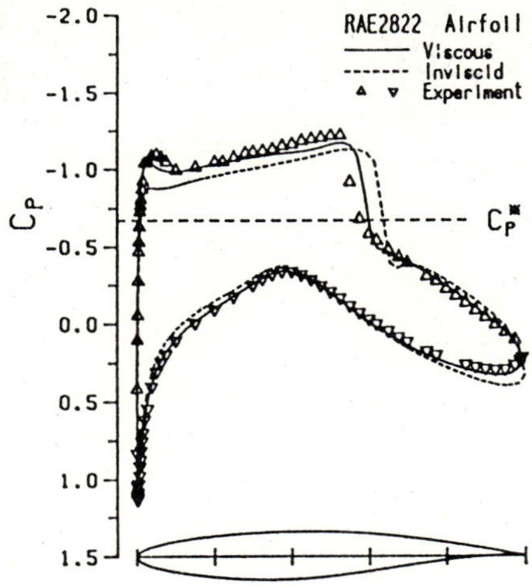

(a) Result for RAE2822

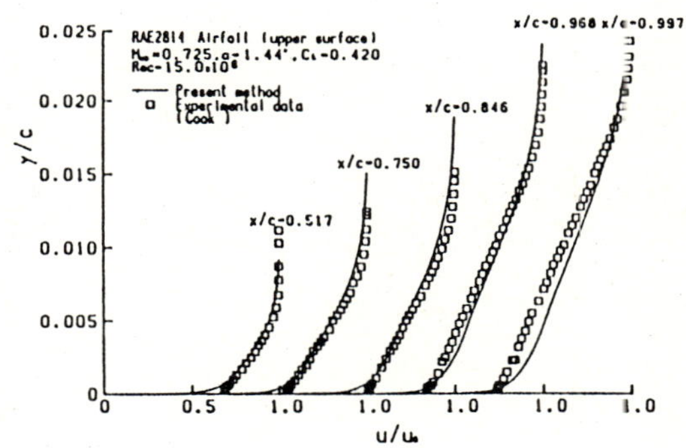

(b) Velocity profiles

Fig. 8 Advanced interaction method

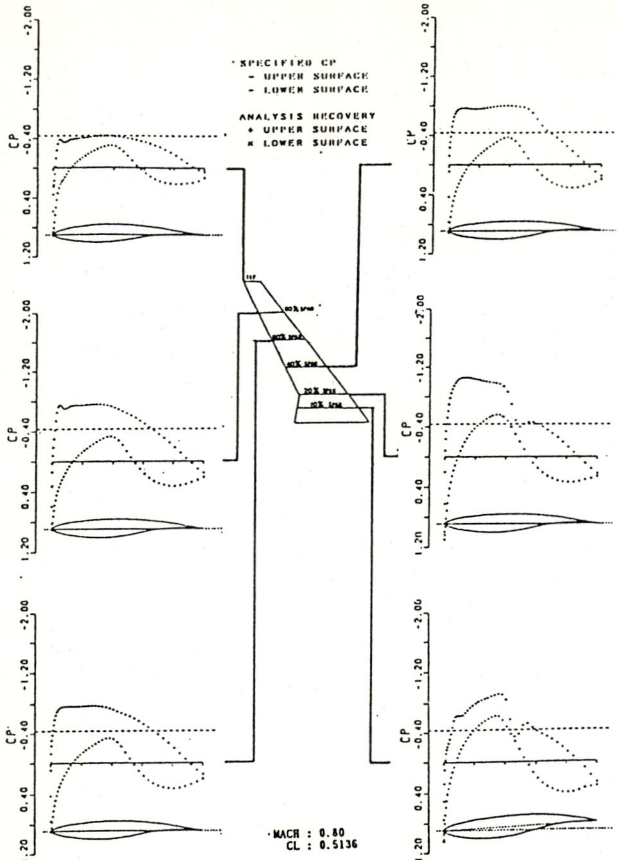

(a) Forward-swept wing design

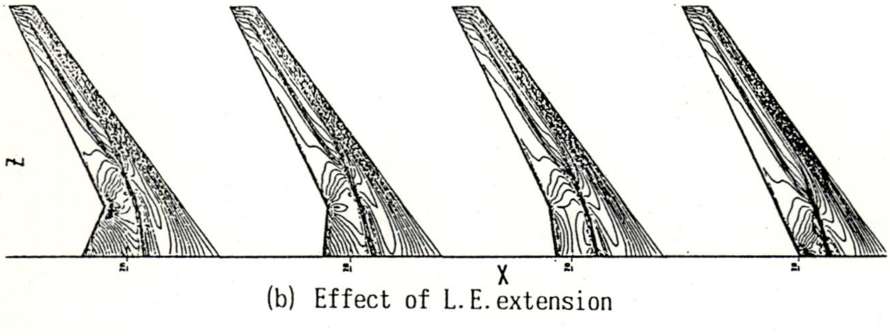

(b) Effect of L.E. extension

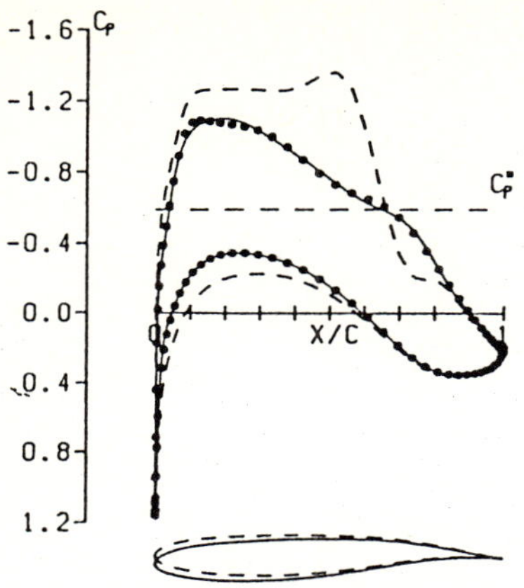

(c) N-S airfoil design

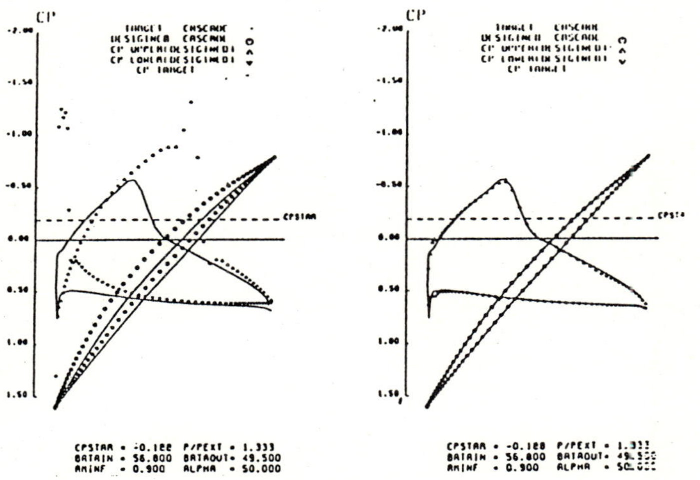

(d) compressor blade design

Fig. 9 Examples of new design method

Viscous-Inviscid Strategy and Computation of Transonic Buffet

J.C. Le Balleur and P. Girodroux-Lavigne

ONERA
BP 72 - 92322 Chatillon Cedex, FRANCE

Abstract

After summarizing the viscous-inviscid strategy, the progress in non-boundary-layer formulations and in numerical coupling techniques, an interacting time-dependent thin-layer method is presented, and shown capable of computing time-consistently the transonic buffet over airfoils. The method approximates the Defect Formulation theory with a modelling of the instantaneous viscous velocity profiles, and fully recovers the viscous upstream influence from a time-consistent viscous-inviscid coupling, converged at each time-step. Results are shown both for a supercritical and an NACA0012 airfoil. The computed buffet-onset, and buffet time-evolutions, are compared with available experiments versus incidence and Mach number.

Viscous-inviscid interaction methods were frequently believed to be restricted to the Prandtl equations and to calculate very mild and simple flow separations.

The progress, however, has changed these restricting views and the viscous-inviscid strategy appears more and more as a general computational technique applicable to any viscous flow equation in high Reynolds number aerodynamics.

The technique is now capable of computing the complex separation processes, such as the massive separations and stalls [1], the shock wave-boundary layer interactions at high speed [2], and the unsteady separations with transonic buffet [3].

1. Viscous-Inviscid Strategy

The progress relies first on the viscous-inviscid splitting. It is now possible to define a unique formulation, the "Defect Formulation", Le Balleur [4,7,1], which is valid both for solving Navier-Stokes equations as well as boundary-layer-like equations.

With this tool, the solution of Navier-Stokes equations with a viscous-inviscid strategy would be possible; the advantage would be to maintain the implicit-marching numerical techniques of the boundary-layer methods, which are optimal when the Reynolds number increases, whereas the disadvantage would be the requirement of a viscous-inviscid coupling solver. With this tool also, the usual question of selecting approximate thin-layer equations at high Reynolds number, which is never avoided even with full Navier-Stokes solvers, owing to the trun-

J. Zierep, H. Oertel (Eds.)
Symposium Transsonicum III
IUTAM Symposium Göttingen, 24.-27. 5. 1988

cation errors introduced by the very high aspect ratios of the cells in tractable grids, can be handled more rationally. A continuous change of the approximation level can be operated, from the simplest boundary-layer-like to the Navier-Stokes levels, and the approximation is made grid-independent. This grid-independency of the approximations is important, when we realize, for example, how illusory it is to discuss the hypothetical gains of Navier-Stokes solutions with respect to parabolized Navier-Stokes solutions, while maintaining grids which are coarse, or even insufficient, for simply resolving the Prandtl equations.

At the lower level of approximation, namely first order boundary-layer-like, the "Defect Formulation" also avoids a full elimination of the Euler equations in viscous regions, in analogy with parabolized Navier-Stokes equations, in such a way that a non-zero normal pressure gradient is maintained within the viscous layers. This point is important and maintains the capability of first-order methods to describe the flows where shock waves penetrate viscous layers, as in turbulent viscous interactions, see Le Balleur [4,5,6,7,1,8]. It was shown, moreover, that the matching condition used for coupling the overlaying viscous and pseudo-inviscid fields of the "Defect Formulation" fully removes the breakdown of the interacting boundary-layer theory of Crocco and the occurrence of fictitious "Supercritical Layers" or "Critical Points" at transonic as well as at hypersonic speeds, Le Balleur [5]. This insures the full capability of the "Defect" theory in shock wave-boundary layer interactions, even at first order, and the full recovery of the viscous upstream influence of the NS equations.

These properties of the first-order-thin-layer approximation of the "Defect Formulation" are also maintained when integrating the equations normal to the viscous layer. In this way, the simplest level of the integral thin-layer equations, in "Defect" form is found to be less approximate than the Prandtl equations, Le Balleur [4,7,9,1], and to provide a fully general tool which can be made valid also for massive separations, Le Balleur [1]. In this case, an advantage of integral equations is that the thin-layer approximation is no longer assumed tangent to the wall or to the grid, but tangent to the free shear layers. This makes the approximation fully grid-independent, allows to close the integral equations with a thin-layer modelling of viscous velocity profiles in the correct reference frame of the free shear layers [1], and offers, however, the simplicity of being solved along the walls or wake center-lines without adapting the grid to the distance between the viscous shear layers and the walls.

2. Viscous-Inviscid Coupling Solvers

The progress in the viscous-inviscid strategy relies secondly on the viscous-inviscid coupling which has to be strong and requires speficic numerical algorithms. The rational development of coupling algorithms is a difficult task which has been initiated rather late. It is still probably the less advanced component of such methods, which mainly restricts the domain of applications and convergence rates and which has to be carefully dissociated from the domain of validity of thin-layer equations.

For steady separated flows, if we summarize, after understanding the limitations of the "Inverse" coupling methods, the "Semi-Inverse" method of Le Balleur [6] was first suggested. The method is an "explicit" update formula for the normal inviscid velocity at the wall, which couples iteratively a direct inviscid problem with an inverse solution of the viscous problem. The method was based on a linearized original stability analysis, and the "Semi-Inverse" update, deduced from the exact coupling at maximal wave number, was found stable for all wave numbers; the method can be used successfully for steady and massively separated airfoil flows [1,7], and for steady shock-boundary layer interactions at transonic or supersonic speeds [6,2]. A simplified Semi-Inverse method which is linearly equivalent at subsonic speed, but without stability control, has been defined later by Carter [10] and used widely for small separations. An alternative way, at subsonic-transonic speed, is the "Quasi-Simultaneous" method suggested later by Veldman [11]. The method is based on a marching technique where the viscous problem is directly solved coupled, but coupled with a numerically approximated inviscid operator, for the ellipticity to be treated iteratively with a Gauss-Seidel discretization and update.

For unsteady separated flows, the viscous-inviscid coupling still has to be strong, and, moreover, strong at each time-step, which requires the convergence of a coupling algorithm within each time-step [9,12,13]. This penalty is necessary to solve time-consistently the strong viscous-inviscid interaction, which guarantees to recover a full viscous upstream influence within each time-step, and a full capability in unsteady shock-boundary layer interactions. The "Semi-Implicit" method of Le Balleur, Girodroux-Lavigne [12,13], which first gave access to a time-consistent strong-coupling, is used in the present computation of transonic buffet. At each time-step, each coupling iteration of the "Semi-Implicit" method first maintains an uncoupled viscous solution, switched in direct or inverse modes, that deduces the numerical viscous-operator of the coupling; the coupling iteration then uses a marching technique where the discrete ADI unsteady inviscid-operator is coupled with a relaxation update for this viscous-operator, see section 4.

3. Governing Equations

3.1 Full Defect Formulation: Navier-Stokes Equations

Denoting with a bar the real viscous flow, and without a bar an overlaying pseudo-inviscid flow, considering the contravariant velocity components $\overline{u}^j$ in curvilinear coordinates x^j (X^i), the Jacobian J = det $(\partial X^i/\partial x^j)$, the Christoffel symbols Γ^i_{jk}, the metric tensor g_{ij}, the stress tensor t^{ij}, the pressure $\overline{p}$ and density $\overline{\rho}$, the "Defect Formulation" solves the difference between the Euler and Navier-Stokes equations, and strongly couples the Euler an Navier-Stokes fields with a smooth matching outside viscous regions, thus providing viscous boundary conditions to determine the pseudo-inviscid flow:

$$\frac{\partial}{\partial t}\cdot J\begin{pmatrix}\rho-\overline{\rho}\\ \rho u^i-\overline{\rho u}^i\end{pmatrix}+\frac{\partial}{\partial x^j}\cdot J\begin{pmatrix}\rho u^j-\overline{\rho u}^j\\ \rho u^i u^j-\overline{\rho u}^i u^j+t^{ij}\end{pmatrix}+$$

$$+\Gamma^i_{jk}\cdot J\begin{pmatrix}0\\ \rho u^j u^k-\overline{\rho u}^j\,\overline{u}^k+t^{jk}\end{pmatrix}+J\,g^{ij}\,\frac{\partial}{\partial x^j}\begin{bmatrix}0\\ p-\overline{p}\end{bmatrix}=0\quad i,j,k=1,2,3 \tag{1}$$

$$0=\lim_{x^3\to\infty}\begin{pmatrix}u^i-\overline{u}^i\\ p-\overline{p}\\ \rho-\overline{\rho}\end{pmatrix}\quad,\quad i=1,2,3$$

Real Flow: $\overline{u}^i,\overline{p},\overline{\rho}$

Pseudo-inviscid Flow: u^i,p,ρ

3.2 First-order thin-layer approximation

After assuming thin-layer Navier-Stokes approximations with the steep gradients along x^3, and after showing than x^3 has to be normal to the interacting inviscid streamlines or to the interacting displacement surface, Le Balleur [1], the Defect equations are reduced at first order / δ:

$$\frac{\partial}{\partial t}\cdot J\begin{pmatrix}\rho-\overline{\rho}\\ \rho u^i-\overline{\rho u}^i\end{pmatrix}+\frac{\partial}{\partial x^j}\cdot J\begin{pmatrix}\rho u^j-\overline{\rho u}^j\\ \rho u^i u^j-\overline{\rho u}^i\,\overline{u}^j\end{pmatrix}+$$

$$+\Gamma^i_{lk}\cdot J\begin{pmatrix}0\\ \rho u^l u^k-\overline{\rho u}^l\,\overline{u}^k\end{pmatrix}=\frac{\partial}{\partial x^3}\begin{pmatrix}0\\ Jt^{i3}\end{pmatrix} \tag{2}$$

$$\frac{\partial}{\partial x^3}(p-\overline{p})=0+\cdots\qquad i=1,2\quad j=1,2,3\quad k,l=1,2$$

3.3 First-order Defect Integral Equations

A straight x^3-integration of (2) provides the fully equivalent integral equations. The deviation of the displacement surface $x^3=0$ from the wall or wake center-line, when the separated regions are thick, have been shown to be handled easily in these integral equations, Le Balleur [1]. For simplicity, we assume here that $x^3\simeq 0$ is roughly the wall. Denoting q the modulus of the velocity:

$$\left[\frac{\partial}{\partial t} \cdot J\rho \begin{pmatrix} \delta^0 \\ q\delta^i \\ \delta - \delta^0 \end{pmatrix} + \frac{\partial}{\partial x^j} \cdot J\rho \begin{pmatrix} q\delta^j \\ q^2\theta^{ij} + qu^i\delta^j \\ u^j\delta - q\delta^j \end{pmatrix} + \right.$$

$$\left. + \Gamma^i_{jk} \cdot J\rho \begin{pmatrix} 0 \\ q^2\theta^{ij} + qu^i\delta^k \\ 0 \end{pmatrix} = J\rho \begin{pmatrix} u^3 \\ q^2 \frac{Cf^i}{2} + u^i u^3 \\ qE \end{pmatrix} \right]_{(x^1, x^2, 0, t)}$$

$$\bar{p}\,(x^1, x^2, x^3, t) = p\,(x^1, x^2, x^3, t) + (p - \bar{p})_{2^{nd}order} + \cdots \qquad i, j, k = 1, 2 \qquad (3)$$

$$J\rho \begin{pmatrix} q\delta^j \\ q^2\theta^{ij} + qu^i\delta^j \\ \delta^0 \end{pmatrix}_{(x^1, x^2, 0, t)} = \int_0^\infty J \cdot \begin{pmatrix} \rho u^j - \overline{\rho u}^j \\ \rho u^i u^j - \overline{\rho u}^i \bar{u}^j \\ \rho - \bar{\rho} \end{pmatrix} \cdot dx^3 \,,$$

$$E = \left[\frac{1}{\rho q} \cdot \frac{\frac{\partial}{\partial x^3}\left(t^{i3}\right)}{\frac{\partial}{\partial x^3}\left(u^i - \bar{u}^i\right)} \right]_{(x^1, x^2, \delta, t)}$$

These equations are less restrictive than the Prandtl equations because $\bar{p}$, p, ρ, u^j are x^3-dependent. In addition to the integral mass and momentum equations, the system (3) includes the entrainment equation, which is a collocation with the local $x^1 - x^2$ momentum equations of (2) along $x^3 = \delta_{(x^1, x^2, t)}$.

3.4 Integral Closure

For turbulent flow, the integral equations are closed with the turbulent velocity profiles modelling previously suggested for 2D and 3D viscous layers, Le Balleur [7,9], in attached or massively separated flows, see also [1]:

$$\bar{u}^i = u^i - q\left[\underbrace{W^i_{(a^1, a^2)} \cdot F\left(\frac{x^3}{\delta}\right)}_{\text{law of the wake}} + \underbrace{T^i_{(a^1, a^2)} \cdot \frac{1}{k}\left(\frac{Cf}{2}\right)^{\frac{1}{2}} \log\left(\frac{x^3}{\delta}\right)}_{\text{law of the wall}} \right]$$

$$+ \text{(laminar layer)} \ldots \qquad (4)$$

$$F(x) = F^*\left(\frac{x - x^*}{1 - x^*}\right), \quad F^*(x) = \left(1 - x^{1.5}\right)^2, \quad x^* = x^*\left(\frac{\delta^i}{\delta}\right), \quad i = 1, 2$$

This wall-wake modelling generalizes Cole's idea with an extension to separated flows, with a different wake function F^* for attached flows, and with a new wake function F for separated flows. For F, the closure x^* (δ^i/δ) is deduced from the analysis of boundary layer singularities at separation, of characteristic cones and upstream influence recovery in separated flows, and also from the capability of describing the limiting case of the 2D free mixing layers in massively separated flows, see [1]. T^i is the unit vector parallel to $(u^i - qW^i)$, and Cf is deduced from the law of the wall in T^i direction. This modelling is assumed for the instantaneous profiles in unsteady flows, where the three computed unknowns δ, a^1, a^2 are t -dependent. Numerically, the velocity profiles are used with a discretization of 19 or 37 nodes, optimized in x^3-direction, within the defect layer. The instantaneous entrainment E defined in (3) is deduced from the profile modelling (4) with a mixing-length closure, and with additional transport equations for turbulence, Le Balleur [7,9,1], but only the mixing length was used here. The density is deduced from the velocity with the adiabatic approximation $\bar{h}_t = h_t$. For laminar regions, at last, the closure relations are deduced from self-similar solutions.

3.5 Pseudo-inviscid Field in Two Dimensions:

The unsteady potential small disturbance equation is presently used and discretized with an ADI method [15] that involves an X-sweep with upwind supersonic biasing, and an upwind Y-sweep. Denoting Φ the perturbation potential, ω the pulsation, U_∞, M_∞ the velocity and Mach number at infinity, γ the specific heat ratio, $H_{(x,t)}$ the airfoil contour, $k = \omega c/U_\infty$, Γ the circulation:

$$k^2 M_\infty^2 \frac{\partial^2 \Phi}{\partial t^2} + 2k M_\infty^2 \frac{\partial^2 \Phi}{\partial x \, \partial t} =$$

$$= \frac{\partial}{\partial x}\left[\left(1 - M_\infty^2\right)\frac{\partial \Phi}{\partial x} - \frac{1}{2} M_\infty^2 \left[(\gamma + 1) M_\infty^2 + 3\left(1 - M_\infty^2\right)\right]\left(\frac{\partial \Phi}{\partial x}\right)^2\right] + \frac{\partial^2 \Phi}{\partial y^2} \tag{5}$$

$$\text{wall:}\ \frac{\partial \Phi}{\partial y} = \frac{\partial H}{\partial x} + k\frac{\partial H}{\partial t} + \left(\frac{v}{q}\right)_{(x,0,t)},$$

$$\text{wake:}\ < \frac{\partial \Phi}{\partial y} > = \left(\frac{v}{q}\right)_{(x,0^+,t)} + \left(\frac{v}{q}\right)_{(x,0^-,t)}, \quad k\frac{\partial \Gamma}{\partial t} + \frac{\partial \Gamma}{\partial x} = 0 .$$

4. Method of Solution in Two Dimensions

4.1 Viscous Problem

After closure, the viscous Defect integral system is written:

$$\left\{ \frac{1}{q} C^i_{j(a,m,\delta)} \frac{\partial \hat{f}^j}{\partial t} + A_{j(a,m,\delta)} \frac{\partial \hat{f}^j}{\partial x} = b^i \right\}_{(x,0,t)} ,$$

$$\hat{f}^j = \begin{bmatrix} \hat{m} \\ \hat{h}_t \\ \delta \\ a \end{bmatrix} , \quad b^i = \begin{bmatrix} \hat{v}/q \\ Cf/2 \\ E \end{bmatrix} \quad i = 1,3 \ , \ j = 1,4 \tag{6}$$

where the unknowns are the reduced Mach number $\hat{m} = 0.5(\gamma - 1)\hat{M}^2$, transpiration velocity $\hat{v}/q$ and total enthalpy $\hat{h}_t$ in the inviscid flow at the wall, plus the boundary layer thickness δ and a shape parameter $a = (\delta'/\delta)_{incompressible}$. The three unknowns $(\hat{m}, \hat{v}/q, \hat{h}_t)$ have to be coupled to the corresponding quantities of the pseudo-inviscid problem m, v/q, h_t.

Before coupling at each time-step, uncoupled viscous calculations are performed with a prescribed enthalpy $(\hat{h}_t = h_t)$. For attached flows, the viscous system is solved marching in direct mode $(\hat{m} = m)$. As the marching solution is no longer possible in direct mode beyond separation $(a > 0.40)$ where upwind characteristics appear in (6), Le Balleur [5], an Inverse mode is switched $(\hat{v}/q = v/q)$ in separating or separated zones $(a > 0.28)$, and marched in free-stream direction. The viscous upstream influence is recovered at convergence of the coupling step, where we will have together $(\hat{h}_t = h_t,\ \hat{m} = m,\ \hat{v}/q = v/q)$. The numerical scheme is implicit in time and space, is 4-nodes box-like in $x - t$ plane, and a Newton method solves the non-linearities of C^i_j, A^i_j, see references [7,12,13].

4.2 Coupling Problem

The coupling is achieved with the "Semi-Implicit" method explained in [12]. From system (6), a numerical elimination of the pure viscous unknowns δ, a, where the discretization scheme has to be involved, provides the following relation, called "viscous influence" function, which connects the inviscid unknowns $(\hat{v}/q,\ \hat{m},\ \hat{h}_t)$ of the viscous problem, which summarizes the numerical viscous operator at time level n and node $(i, 1)$, and where the viscous stiff non-linearities or singularities are included in the four coefficients α_j, β_j:

$$R^n_{i,1}\left(\frac{\hat{v}}{q}, \hat{f}^j\right) = \left[\frac{\hat{v}}{q} + \alpha_j \frac{\partial \hat{f}^j}{\partial t} + \beta_j \frac{\partial \hat{f}^j}{\partial x} \right]^n_{i,1} = 0 , \tag{7}$$

$$\hat{f}^j = [\hat{m}, \hat{h}_t] \quad j = 1, 2$$

As the inviscid unknowns $(v/q,\ m,\ h_t)$ of the inviscid problem are not coincident with $(\hat{v}/q,\ \hat{m},\ \hat{h}_t)$, so long as the strong coupling is not converged, non-zero coupling residuals $R^n_{i,1}(v/q, m, h_t) \neq 0$ are deduced from [7]. The "Semi-Implicit" coupling algorithm, see [12,13] for details, is a relaxation technique for these residuals, which requires an iteration ν at time n and node i :

$$\left[\frac{\Delta v}{q} + \alpha_j^v \frac{\partial}{\partial t}\Delta f^j + \beta_j^v \frac{\partial}{\partial x}\Delta f^j\right]_{i,1}^n = -\overline{\omega}[R^v]_{i,1}^n \quad j = 1,2 \quad \overline{\omega} = 1$$

$$\Delta v = v^{v+1} - v^v \ , \ \Delta f_n^j = (f^j)^{v+1} - (f^j)^v \tag{8}$$

$$[R^v]_{i,1}^n = \left[\frac{\hat{v}}{q} - \frac{v^v}{q}\right]_{i,1}^n + \left[\alpha_j^v \frac{\partial}{\partial t}\left(\hat{f}^j - f^{j,v}\right) + \beta_j^v \frac{\partial}{\partial x}\left(\hat{f}^j - f^{j,v}\right)\right]_{i,1}^n$$

The conditioning terms Δv, Δf^j on the left hand side are written in the variable of the inviscid solver. They are discretized with an upwind scheme for $\partial^2\phi/\partial x\,\partial t$, a 3-nodes centered Gauss-Seidel scheme based on iterations v and $v + 1$ (at same time n) for $\partial^2\phi/\partial x^2$. The residuals R are discretized 3-nodes centered. The relaxation v is converged at each time-step, which insures consistency and upstream influence. The "Semi-Implicit" method, summarized in Fig. 1, initiates the time loop with the inviscid X-sweep which is not influenced by v/q at the new time level. Uncoupled viscous calculations, switched in direct or inverse modes, are then performed. The coefficients α_j^v, β_j^v and the residues R^v of the semi-implicit equation (8) are then deduced. This semi-implicit relaxation (8) is then solved from upstream to downstream with the inviscid Y-sweep. The procedure is iterated at a given time-step until convergence of the viscous step (6) and coupling step (8), which capture the full viscous upstream influence within a coupling of Veldman, Houwink [11,14]. The "Semi-Implicit" method maintains uncoupled direct-inverse viscous calculations and, in addition, provides the time-consistency of the coupling.

5. Computation of Transonic Buffet

The time-consistent viscous-inviscid interaction method has been found to compute satisfactorily the unsteady separations and unsteady shock-boundary layer interactions in forced oscillations [12], and also to be capable of distinguishing between steady separated flows and buffet in transient calculations around a fixed supercritical airfoil [12,3]. The analysis of such a computation of transonic buffet has been extended to an NACA0012 airfoil and to the influence of Mach number.

5.1 Numerical Conditions

After recognizing that a minimal grid resolution has to be used, see [3], a Cartesian H-grid of 160x100 nodes in the X-Y plane is used for the pseudo-inviscid field, with 100 coupling nodes on each side of the airfoil and 30 along the wake-cut, the viscous velocity profiles being discretized with 19 nodes in normal direction (37 for the wake). The far-field conditions of undisturbed flow are located at ± 20 chord lengths in Y, at -20 and +10 chords in X. The stretching in Y provides square cells near the wall. This discretization is capable of capturing the physical scale of the turbulent viscous-inviscid interaction that induces separation at the foot of the shock wave (Δx lower than δ) for the middle and rear positions of the oscillating shock; but the mesh is believed to be a minimal one, and still rather poor for viscous interaction at the upstream positions of the shock.

The initial conditions correspond to a uniform flow. A transient viscous calculation is performed time-consistently for several thousands of time-steps while and after increasing incidence and thickness of the airfoil from zero within 32 time-steps. With these conditions the maximal CFL numbers are rough y 10-20. Owing to the inviscid approximation, the viscous layer does not begin at the stagnation point but near the leading edge on the upper and lower surfaces. The calculations are mainly turbulent, and the 1-2 equation model for turbulence transport is not used. The results have then to be analyzed within the constraints of the grid, of the algebraic turbulence model, and without a careful modelling of the transition location.

5.2 Shock-induced Transonic Buffet

Fig. 2 displays the history of instantaneous iso-Mach contours over one cycle of buffet, for the NACA0012 and the supercritical airfoil. When the lift increases, the supersonic pocket increases, the boundary layer is thin, the shock-motion is inviscid-like and the almost normal shock moves slowly to the rear with increasing intensity. At the rear position, the intensity is strong enough to separate the boundary layer, the normal shock is transformed into an oblique shock which moves upstream more quickly and with a viscous-like behaviour. When the intensity of the shock as well as the extent of the supersonic region decrease, the boundary layer reattaches and the shock begins to increase again as a normal shock, growing and moving downstream. The behaviour of the phenomena, and the hysteresis observed for the separation and the shock in its up and back motions seem to reproduce well the experimental observations of Dor et al. [16] for the supercritical airfoil RA16SC1. The phenomena are similar for the supercritical and NACA0012 airfoils, except for the amplitude, which is higher on the NACA0012.

5.3 Supercritical RA16SC1 Airfoil

Buffet configurations for this airfoil were reported in the ONERA experiments of Benoit [17] and Dor et al. [16]. The results of [17] at Mach number 0.732 have been compared to the present calculations at Mach 0.723, as recommended for wind-tunnel corrections, and at the high Reynolds number $4.2 \cdot 10^6$. The time-interval of 0.25 seconds was computed with 3400 time-steps, the second half of which has been used to analyze the signals of pressure at the wall.

The envelope of the lift curve, fig. 3, shows that steady solutions are computed until 3° incidence, in good agreement with experiment, and that buffet is obtained at incidences 3.5° and higher. The order of the amplitude of the shock-motion is 20% chord, but the lift fluctuation is moderate. The incidence of buffet onset is a little higher than the experimental one, but the envelope of pressure distributions at buffet onset, fig. 4, is quite similar in computation and experiment. A similar agreement is found for the wall pressure signals at different locations, fig. 5 , and for the spectrum, summarized by the distribution versus chord and frequency of the power spectral density of the unsteady pressure on the upper side, fig. 6. The computed spectrum which has been shown insensitive to the numerical initial conditions [3], indicates, however, that coarser grids can influence the spectrum [3] and also that the fundamental frequency with the present finest grid is higher than experiment at buffet onset (but note the comparison is not at same incidence).

5.4 NACA0012 Airfoil

The NASA Ames buffet experiments of Mc Devitt, Okuno [18] on the NACA0012 airfoil habe been compared with the present new calculation, performed at the highest Reynolds number of 10^7, and at the four Mach numbers 0.725, 0.775, 0.800. In each case, increasing incidences have been computed until buffet onset. A time-interval of around 0.12 seconds, computed with 2400 time-steps, has been used more often, and was sometimes doubled to check the signals. Figure 7 displays the results for 3 of the Mach numbers, summarized by the time-histories of the lift.

The main difference as compared with the supercritical airfoil is the very large amplitude of the lift fluctuations at buffet onset, fig. 7, with also a larger amplitude of the shock motion, which is shown around 40% chord on the envelope of pressure fluctuations, fig. 8. A very good comparison with experiment is obtained for the buffet boundary versus Mach number and incidences, fig. 9, the buffet onset being sudden and well-defined within the 0.5° step used for increasing the incidence, fig. 7. Only the highest Mach number computed 0.800 deviates from experiment. The computed time-histories of the pressure fluctuations at different locations in chord of the upper surface are also found very similar to the experiment, fig. 10. The qualitatively very different behaviours of the signals at the different stations and their levels are very well reproduced. As for the supercritical airfoil, the computed fundamental reduced frequency of the lift fluctuations, which is $k = 2\pi fc/U_\infty = 0.35$ at Mach = 0.775, is, however, found to deviate somewhat from the experimental value of 0.44. The reason for the deviations in fundamental frequency is not known, and grid refinements as well as effecs of turbulence models have to be checked. However, the fundamental frequency, fig. 7, seems to depend strongly on incidence and on the fact that the incidence is well in the buffet domain or close to the buffet boundary, which would explain the difficulty of the comparison with experiment for a given and always imperfect turbulence model.

6. Conclusion

Even with simplified thin-layer equations, the viscous-inviscid method and Defect Formulation provide an efficient computational tool for the numerical simulation of the transonic buffet at high Reynolds numbers if the strong viscous-inviscid interaction is converged time-consistently at each time-step. An interacted modelling of the instantaneous viscous velocity profiles is found convenient and capable of describing both vorticity in steady flows and unsteady vortex shedding within the time-dependent viscous layers. With this closure, interacted Defect integral equations can describe the transonic buffet and the unsteady shock-boundary layer with separation, as do the direct Navier-Stokes solvers, with an advantage in higher grid-resolutions, in lower numerical viscosity, and in efficiency at high Reynolds numers.

Results are obtained both for a supercritical and an NACA0012 airfoil. They are in overall very good agreement with experiment for the description of the buffet phenomena, for the prediction of buffet onset, including the influence of Mach number, for the behaviours and amplitudes of the time-histories of pressure fluctuations, as well as for their spectral behaviours. Some discrepancy is noticed in comparing the fundamental frequencies with experiments. The influence of the sensitivity of such comparisons with the incidence, with the different model of turbulence and transition, and with the grid resolutions has to be investigated.

Study financially supported by French STPA.

References

[1] Le Balleur, J.C.: Viscous-inviscid interaction solvers and computation of highly separated flows. - "Studies of Vortex Dominated flows", chap. 3, p.159-192, Proceed. ICASE symp. NASA Langley Field, USA, (July 9-10, 1985), Hussaini and Salas ed., Springer-Verlag 1987, (or ONERA-TP 1986-4).

[2] Le Balleur, J.C., Blaise, D.: Computation of separated internal flows and shock wave-boundary layer interactions by viscous-inviscid interaction. - La Recherche Aerospatiale 1985-4, p. 211-227, English and French editions, (July 1985).

[3] Girodroux-Lavigne, P., Le Balleur, J.C.: Unsteady viscous-inviscid interaction method and computation of buffeting over airfoils. Proceed. IMA/SMAI Conf. "Computational Methods in Aeronautical Fluid Dynamics", Reading, England (6-8 April 1987), Springer-Verlag 1988, to appear, (or ONERA-TP 1987-58).

[4] Le Balleur, J.C.: Computation of flows including strong viscous interactions with coupling methods. - AGARD-CP-291, General Introduction, Lecture 1, Colorado Springs (1981), or ONERA-TP 1980-121.

[5] Le Balleur, J.C.: Viscous-inviscid flow matching: Analysis of the problem including separation and shock waves. - La Recherche Aerospatiale 1977-6, p. 349-358 (Nov.1977). French, or English transl. ESA-TT-476.

[6] Le Balleur, J.C.: Viscous-inviscid flow matching: Numerical method and applications to two-dimensional transonic and supersonic flows. - La Recherche Aerospatiale 1978-2, p.67-76 (March 1978). French, or English transl. ESA-TT-496.

[7] Le Balleur, J.C.: Strong matching method for computing transonic viscous flows including wakes and separations. Lifting airfoils. - La Recherche Aerospatiale 1981-3, p.21-45, English and French editions, (March 1981).

[8] Le Balleur, J.C., Peyret, R., Viviand, H.: Numerical studies in high Reynolds number aerodynamics - Computers and Fluids, Vol. 8, no 1, p. 1-30, (March 1980).

[9] Le Balleur, J.C.: Numerical viscid-inviscid interaction in steady and unsteady flows. - Proceed. 2nd Symp. Numerical and Physical Aspects of Aerodynamic Flows, Long-Beach, (1983), chapt.13, p. 259-284 T. Cebeci ed., Springer-Verlag, 1984, (or ONERA-TP 1983-8).

[10] Carter, J.E.: A new boundary layer interaction technique for separated flows. - AIAA Paper no 79-1450, 4th Comp. Fl. Dyn. Conference, (July 1979.

[11] Veldman, A.E.P.: New, quasi-simultaneous method to calculate interacting boundary layers. - AIAA J. Vol. 19, no 1 (1981), p. 79-85.

[12] Le Balleur, J.C., Girodroux-Lavigne, P.: A semi-implicit and unsteay numerical method of viscous-inviscid interaction for transonic separated flows. - La Recherche Aerospatiale 1984-1, p. 15-37, English and French editions, (Jan. 1984).

[13] Le Balleur, J.C., Girodroux-Lavigne, P.: A viscous-inviscid interaction method for computing unsteady transonic separation. - Proceed. 3rd Symp. Numerical and Physical Aspects of Aerodynamic Flows, Long-Beach (Jan. 1985), T. Cebeci ed., Springer-Verlag (1986).

[14] Houwink, R., Veldman, A.E.P.: Steady and unsteady separated flow computations for transonic airfoils. - NLR MP 84028U, or AIAA Paper 84-168, Snowmass, Colorado (June 1984).

[15] Couston, M. Angelini, J.J.: Solution of nonsteady two-dimensional transonic small disturbances potential flow equation. - ASME Conf., San Francisco (1978) or J. of Fluids Engin., vol. 101, no 3, (1979).

[16] Dor, J.B., Mignosi, A., Seraudie, A., Benoit, B.: Wind Tunnel Studies of Natural Shock Wave - Separation Instabilities, for Transonic Airfoil Tests. - Proceed. IUTAM Symposium Transsonicum III, Göttingen, Germany (24-27 May 1988), Springer-Verlag, to appear.

[17] Benoit, B., Legrain, I.: Buffeting Prediction for Transport Aircraft Applications Based on Unsteady Pressure Measurements - AIAA CP 875 - 5th Applied Aerodynamics Conference, Monterey, 17-19 August 1987.

[18] Mc Devitt, J.B., Okuno, A.F.: Static and Dynamic Pressure Measurements on a NACA0012 Airfoil in the High Reynolds Number Facility. NASA Technical Paper 2485, (June 1985).

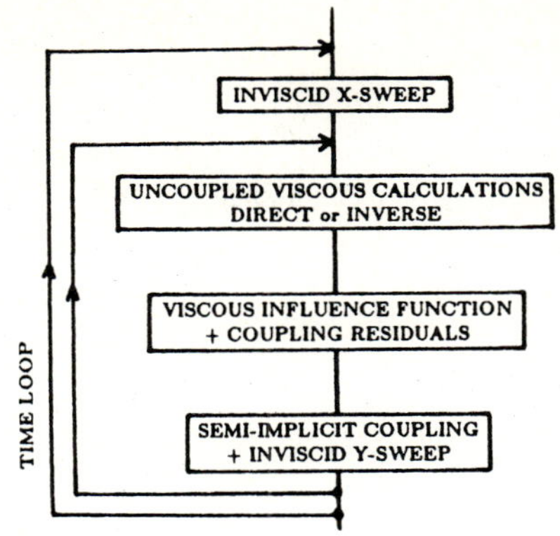

Fig. 1 Semi-Implicit coupling algorithm.

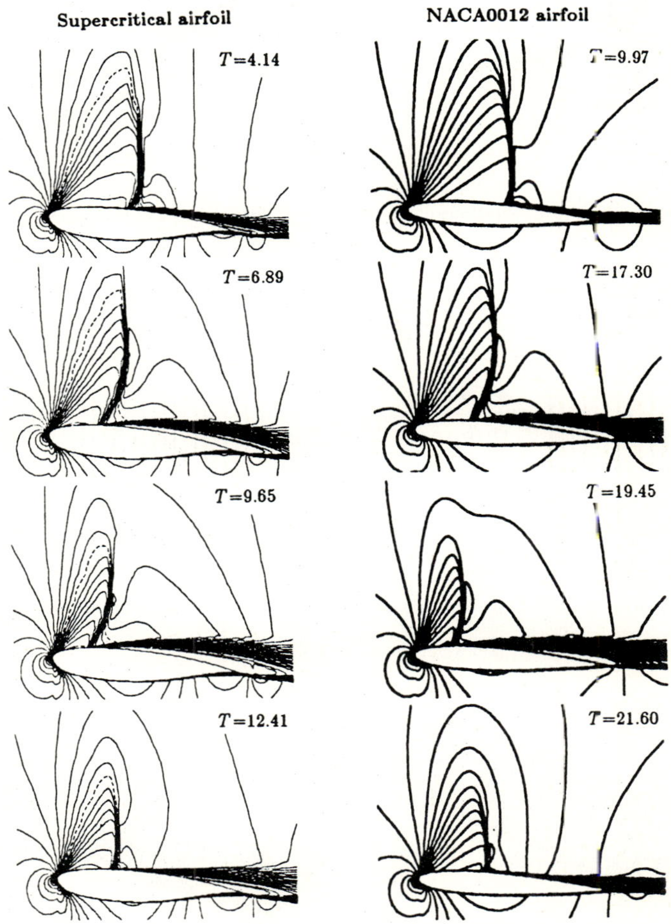

Fig. 2 Instantaneous iso-Mach contours over one cycle of buffet:

a) RA16SC1 airfoil ($M_\infty = .723$, $\alpha = 4^o$, Re = 4.2 10^6)

b) NACA0012 ($M_\infty = .750$, $\alpha = 4^o$, Re = 10^7).

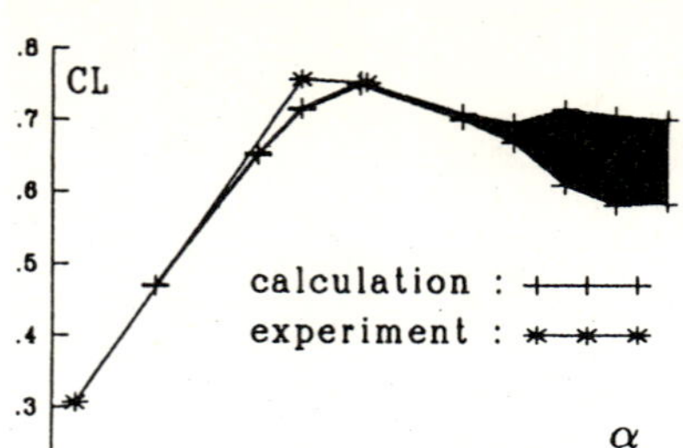

Fig. 3 RA16SC1 airfoil. Envelope of the lift curve ($M_\infty = .723$, Re = 4.2 10^6).

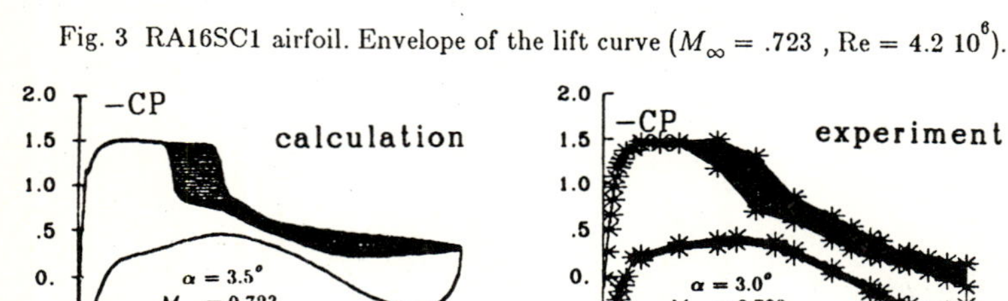

Fig. 4 RA16SC1 airfoil. Envelope of the pressure distributions at buffet onset.

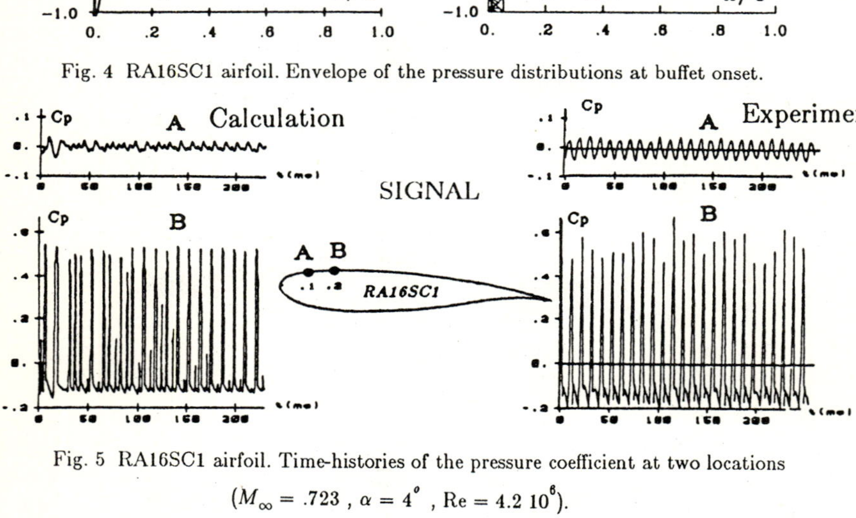

Fig. 5 RA16SC1 airfoil. Time-histories of the pressure coefficient at two locations ($M_\infty = .723$, $\alpha = 4^\circ$, Re = 4.2 10^6).

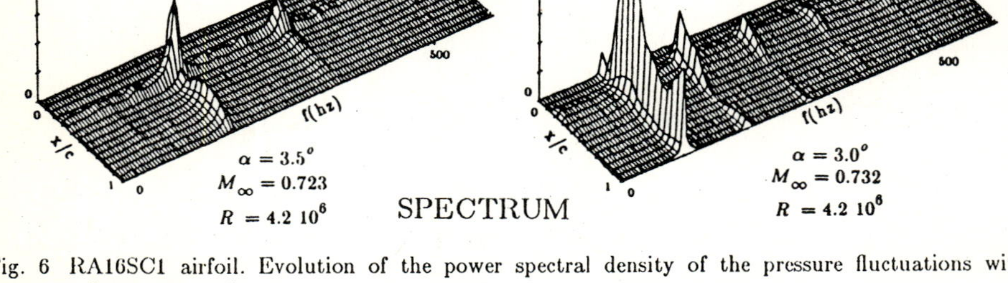

Fig. 6 RA16SC1 airfoil. Evolution of the power spectral density of the pressure fluctuations with frequency (upper surface).

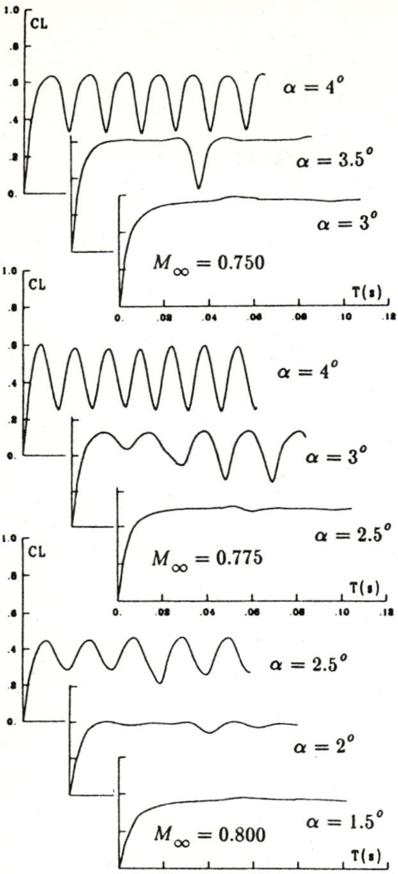

Fig. 7 NACA0012 airfoil. Time-histories of the lift. Buffet onset (Re = 10^7).

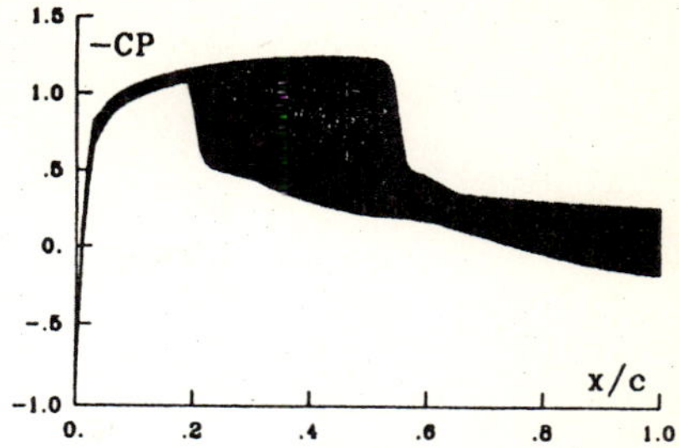

Fig. 8 NACA0012 airfoil. Envelope of the pressure fluctuations on the upper surface ($M_\infty = .775$, $\alpha = 4^\circ$, Re = 10^7).

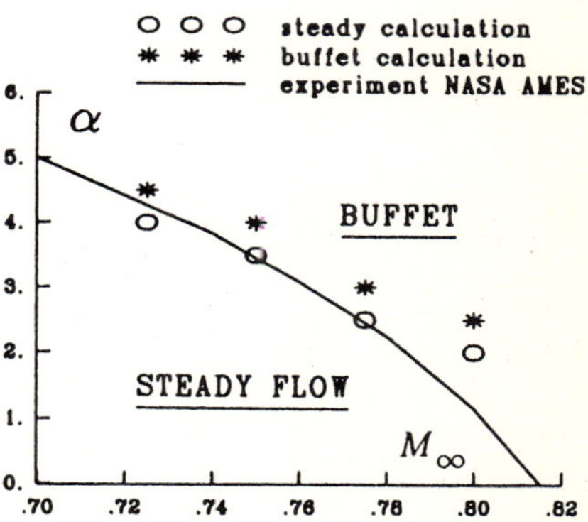

Fig. 9 NACA0012 airfoil. Computed and experimental buffet boundary (Re = 10^7).

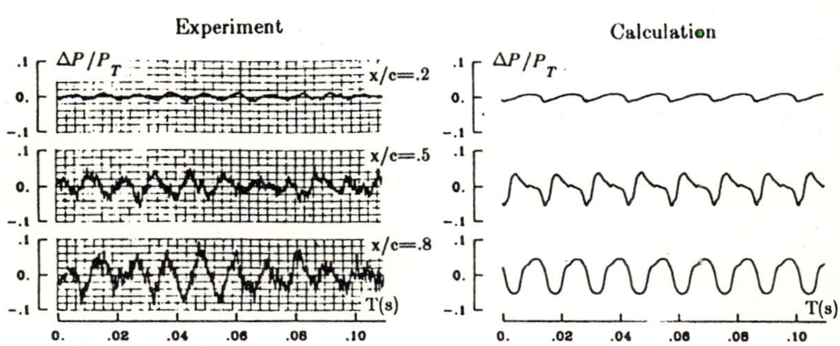

Fig. 10 NACA0012 airfoil. Time-histories of the pressure fluctuations at 3 locations on upper surface ($M_\infty = .775$, $\alpha = 4^\circ$, Re = 10^6).

Zonal Solutions of Three-Dimensional Viscous Flow Problems

M.A. Schmatz*. F. Monnoyer*, K.M. Wanie**, E.H. Hirschel*

* Messerschmitt-Bölkow-Blohm GmbH, FE122,
Postfach 801160, D-8000 München 80, FRG

** Lehrstuhl für Strömungsmechanik, Technische Universität
München, Arcisstr. 21, D-8000 München 2, FRG

Summary

A method is summarized for the calculation of three-dimensional viscous flows about general configurations, namely a close coupling procedure (CCPNS) in which the full Navier-Stokes equations are solved only in regions of strong viscous-inviscid interaction. In the remaining part of the flow, the coupled Euler/boundary-layer equations are employed. CCPNS is an option in the governing Navier-Stokes code NSFLEX to reduce computation cost. Several applications demonstrate the efficiency of the method. Good agreement with full NSFLEX results is achieved.

Introduction

The importance of numerical fluid mechanics has been growing significantly in the last few years. In particular, methods for the solution of the complete Navier-Stokes equations have been designed, and the development of large vector computers has made it possible to compute the viscous flow about realistic configurations. However, systematic application of the codes to the design of new aircrafts is still limited by the extremely high costs of computer time. In the design process the flow often is predicted only by use of the Euler equations or the potential equations. A boundary-layer solution is performed thereupon to get a glance at the skin-friction distribution and to assess separation properties of the flow. The displacement of the inviscid flow by the boundary layer is often neglected, or is accounted for by a distribution of equivalent sources at the wall. Since the resulting equivalent inviscid flow does not present very steep gradients at the wall, the Euler grid can be rather coarse. On the other hand, the boundary-layer equations describe the near-wall viscous flow. They can be solved on very fine meshes with space-marching procedures. Such simple viscous-inviscid coupling procedures can be applied to configurations of practical interest at moderate costs. However, it only applies to regions of weak viscous-inviscid interaction. The Navier-Stokes equations, or other

J. Zierep, H. Oertel (Eds.)
Symposium Transsonicum III
IUTAM Symposium Göttingen, 24.-27. 5. 1988

adequate models, must be solved in regions of strong viscous interaction, such as separated flows or shock-boundary layer interaction. In Fig. 1 the principle regions of weak and strong interaction are given for transonic airfoil flow. The so-called zonal solution concept, in which both procedures are combined in such a way that the Navier-Stokes equations are solved only in regions where the boundary-layer theory does not apply, was introduced in [1] and extensively applied to two-dimensional flow cases in [2,3]. Compared to other zonal approaches, the present method is the first one to couple Euler, boundary-layer and Navier-Stokes solutions. In [4], Navier-Stokes equations are solved in the regime near the wall and Euler equations elsewhere. Another approach is described in [5] for fortifying the convergence and accuracy of a Navier-Stokes solution on a coarse mesh with an underlying boundary-layer solution on a finer mesh.

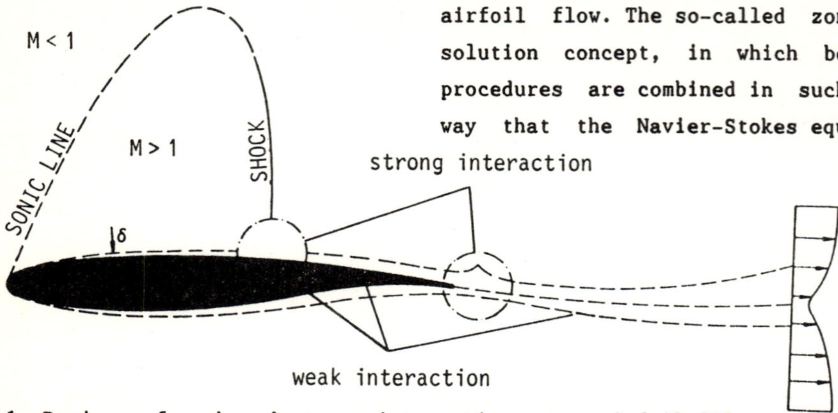

Fig. 1 Regions of weak and strong interaction on an airfoil [1]

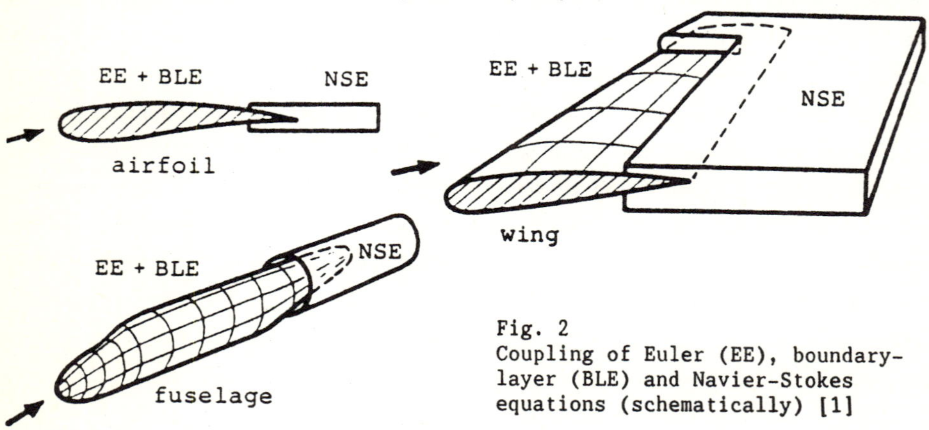

Fig. 2
Coupling of Euler (EE), boundary-layer (BLE) and Navier-Stokes equations (schematically) [1]

The present zonal method has been extended to three-dimensional flows in [6], results for a subsonic fuselage type flow are given in [7]. The code is an extension of the well-proven NSFLEX code [8,9] and is highly vectorized. The boundary-layer code SOBOL [10], which solves the second-order boundary-layer equations is incorporated in the method as a subroutine. In the following the main principles and some applications of this close coupling procedure for the Navier-Stokes equations (CCPNS) are summarized, see also Fig. 2. The main benefits of the CCPNS method are [1]: reduction of storage and computer time, improvement of the solution for the boundary-layer region, avoidance of hydrodynamic instability phenomena. Note that CCPNS is an option in the full Navier-Stokes solver NSFLEX. If, for instance, strong interaction regions are located too closely to each other, one should omit the zonal option, since then the gain is too small.

Zonal solution method

The governing equations are the time-dependent Reynolds-averaged compressible Navier-Stokes equations. Conservation laws are used with body-fitted arbitrary coordinates ξ, η, ζ with Cartesian velocity components u, v, w:

$$\mathbf{U}_t + \mathbf{E}_\xi + \mathbf{F}_\eta + \mathbf{G}_\zeta = 0 , \tag{1}$$

where **U** is the solution vector $\mathbf{U} = J\ (\rho, \rho u, \rho v, \rho w, e)^T$. **E**, **F**, **G** are the fluxes normal to ξ=const., η=const., ζ=const. faces. J the Jacobian of the inverse mapping, is the volume of the finite volumes. Effective transport coefficients are introduced with the Boussinesq approximation. The second-order boundary-layer equations are derived by an order-of-magnitude analysis of the steady Navier-Stokes equations. Note that the Navier-Stokes momentum equations have to be transformed to locally monoclinic coordinates ξ, η, ζ [12] to get the boundary-layer equations. Terms up to the order of $1/\sqrt{Re}$ are kept instead of order one terms as in first-order theory [10,11]. They differ from the first-order equations by the presence of several terms, all of them including the surface curvature. The momentum equation normal to the wall does not reduce to the zero pressure gradient assumption. Calculations yield slightly inaccurate predictions if first-order theory is used while second-order boundary-layer results compare very well with the full Navier-Stokes solution [3].

To solve Eq. (1) an implicit finite-volume procedure, which allows large time steps, is applied to the unfactored equations [2,8,9]. The method is based on a well-proven Euler solver [13]. The first order in time discretized implicit form of Eq. (1) reads

$$(\mathbf{U}^{n+1} - \mathbf{U}^n)/\Delta t + \mathbf{E}_\xi^{\,n+1} + \mathbf{F}_\eta^{\,n+1} + \mathbf{G}_\zeta^{\,n+1} = 0 . \tag{2}$$

A Newton method can easily be constructed for U^{n+1} by linearizing the fluxes in Eq. (2) about the known time level n leading to

$$\Delta U/\Delta t + (A^n\Delta U)_\xi + (B^n\Delta U)_\eta + (C^n\Delta U)_\zeta = -(E_\xi + F_\eta + G_\zeta)^n = \mathbf{RHS} \ , \quad (3)$$

where **A**, **B**, **C** are the Jacobians of the flux vectors **E**, **F**, **G**. **ΔU** is the time variation of the solution and therefore the update is

$$U^{n+1} = U^n + \Delta U \ . \quad (4)$$

Eq. (3) is solved with a relaxation technique with special care to vectorization. To construct a relaxation method it is necessary to have a diagonal dominant system of equations. For inviscid flux calculations, therefore, a linear locally one-dimensional Riemann solver (Godunov-type differencing) is employed at the finite-volume faces. Diffusive fluxes at the cell faces are built with central differences. Eq. (3) has to be solved approximately at every time step. A point Gauss-Seidel technique is used (upper index ν). At every i,j,k location the variation in time can be calculated from:

$$\mathbf{DIAG}^n_{i,j,k}\ \Delta U^{\nu+1}_{i,j,k} = \omega\ \mathbf{RHS}^n_{i,j,k} + \mathbf{ODIAG}_{i,j,k} . \quad (5)$$

$\mathbf{DIAG}^n$ is a 5*5 matrix of the sum of the eigenvalue splitted inviscid and the viscous thin-layer Jacobians with **ΔU** as factor together with I/Δt. The term I/Δt strengthens diagonal dominance and therefore stability. The term **ODIAG** is a function of the i,j,k neighbouring **ΔU** vectors. ω is an underrelaxation parameter. By application of the so-called checkerboard scheme a high rate of vectorization is achieved. The following coupling steps have to be performed at every time step [2,3,6], see also Fig. 2:

- In the weak interaction region (EE+BLE) an equivalent inviscid flow (EIF) has to be calculated. Wall-normal mass fluxes (equivalent inviscid source distribution [12]) have to be prescribed at the body cell faces.
- A boundary-layer calculation in the EIF region delivers the equivalent inviscid source distribution and the flow profiles for the coupling of the EIF region and the Navier-Stokes region (NSE) at the artificial boundaries.
- The downstream coupling of the Euler/boundary layer with the Navier-Stokes solution is straightforward when second-order theory is used, since the boundary-layer profiles match the external inviscid flow profiles. The fluxes which enter the Navier-Stokes region are simply calculated from the equivalent flow and the boundary-layer profiles.
- For the upstream coupling of the Navier-Stokes solution with the Euler flow equivalent inviscid fluxes have to be constructed out of the Navier-

Stokes fluxes. For this purpose, the pressure is assumed equal at the boundary cell faces and the density is calculated with constant entropy normal to the wall from outside the boundary layer. The mass fluxes are linearly extrapolated from the neighbouring cells in the inviscid region.

The boundary layer is recalculated after some time steps with the code SOBOL (second-order boundary layers) [10,11] as a subroutine. The equations are integrated with an implicit finite-difference space marching method. As the boundary layer is computed on the same surface mesh as the Euler and the Navier-Stokes flow and as the overall grid is locally monoclinic in the vicinity of the body, only a minimum of interpolation work has to be done for the CCPNS method. Note that patched grids are used and no overlapping is necessary. The procedure is started with only a few Euler steps in the whole domain. Special grid generation procedures are used to get meshes best adapted to the CCPNS method. At the wall exponentially stretched surface normal coordinates are used in algebraically generated subblocks. An elliptic grid generation procedure follows algebraic mesh generation to cover the outer part with mesh points [8,14].

Selected applications

Here, some representative results of the CCPNS method are given. For the calculation of the turbulent eddy viscosity the Baldwin/Lomax turbulence model is used for the Navier-Stokes fluxes as well as for the boundary layer solution.

Firstly, results are shown for transonic airfoil flows, see also [6],

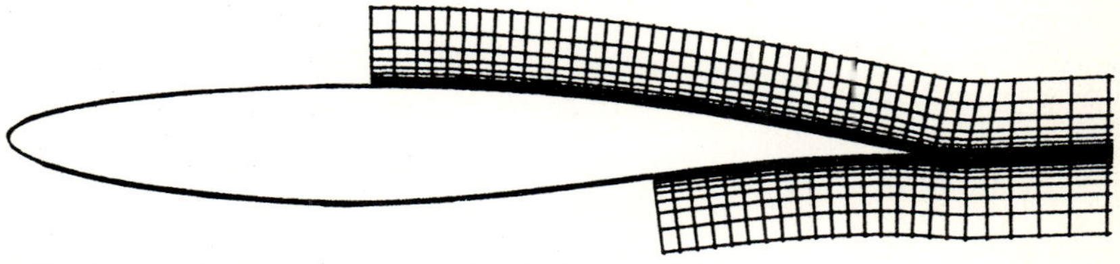

Fig. 3 Navier-Stokes part of zonal airfoil grid

namely the flow about the RAE2822 airfoil at freestream conditions $M_\infty=0.73$, Re=6,500,000, $\alpha=2.79°$. A C-type mesh is investigated, with zonal boundaries at 40% and 70% chordlength on the upper and the lower side, respectively (Fig. 3). The predicted c_p-distribution is compared with the experimental in Fig. 4. In comparison with a full Navier-Stokes solution, the CCPNS method requires half the computer time and yields nearly identical results.

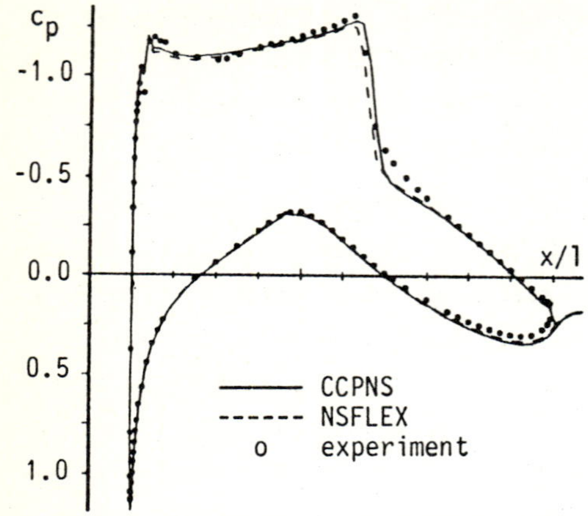

Fig. 4 Pressure distribution RAE2822

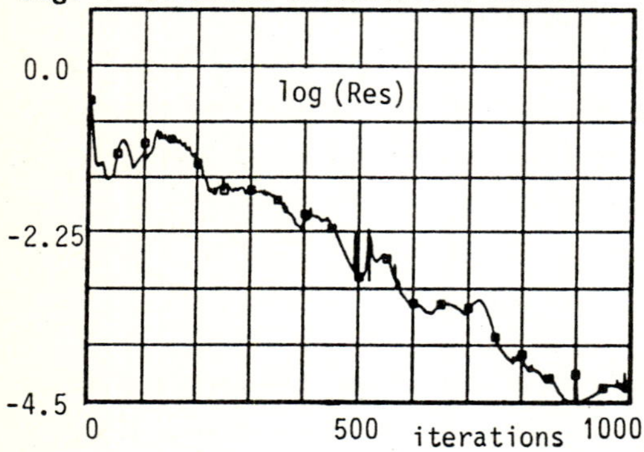

Fig. 5 Convergence history

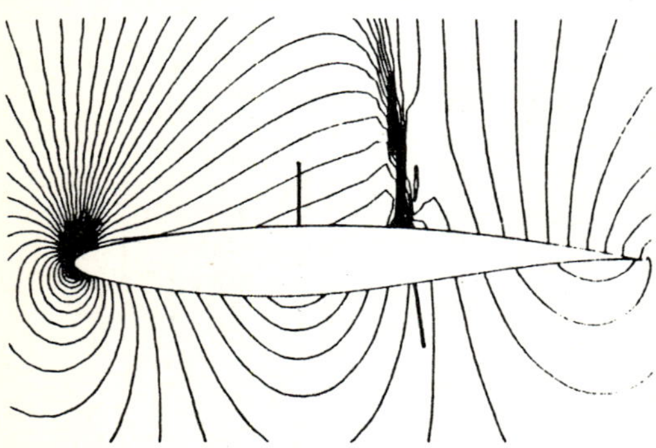

Fig. 6 Isobars

In Fig.5 the residual drop of the zonal method is plotted. Perfect smoothness of the isobars can be observed in Fig. 6. It is interesting to note that in some cases the Euler part of the solution appears to make the Navier-Stokes part considerably larger [6].

To verify the method for three-dimensional fuselage type flows a calculation of the subsonic flow about a 1:6 ellipsoid at 10° incidence [6,15] is shown with M_∞=0.17, Re=7,700,000. Fixed transition takes place at 10% of the body length. The surface mesh and the Navier-Stokes region are plotted in Fig. 7. Note that in this application the zonal boundary is not just an i=const. plane. At the upper part of the body, where separation is expected for higher angles of attack, the Navier-Stokes region is extended more upstream. In Fig. 8 the limiting streamline pattern is given. No separation is observed. The calculated and measured [16] pressure coefficients in the ellipsoid plane of symmetry are compared in Fig. 9. Two calculations with sting were performed. The first one is a CCPNS solution and the second one a full Navier-Stokes NSFLEX solution. Both show the same behaviour [6]. Good agreement with the experiment except at the rear part on the upper side is found. There, in the experiment, slight separation seems to occur. Another CCPNS solution without sting on an O-O-type mesh is shown in [7] for comparison. The surface pressure distribution is again smooth and no disturbances at the artificial boundaries enter the solution [6].

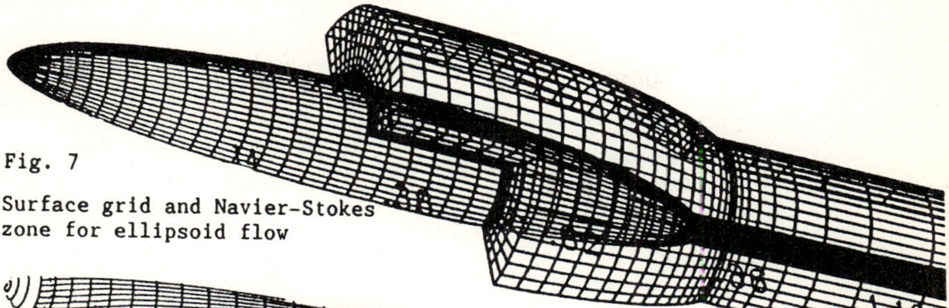

Fig. 7

Surface grid and Navier-Stokes zone for ellipsoid flow

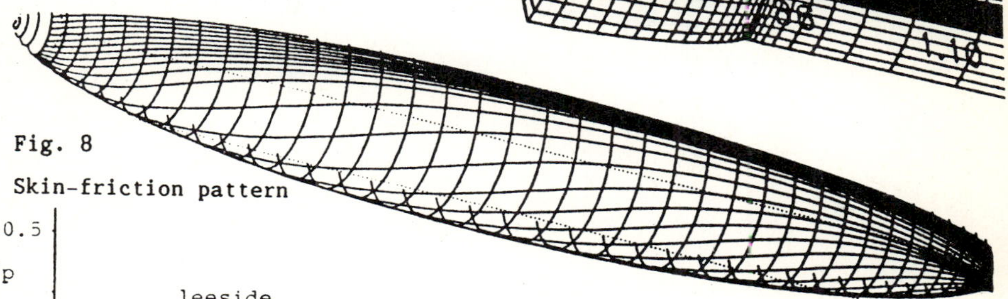

Fig. 8

Skin-friction pattern

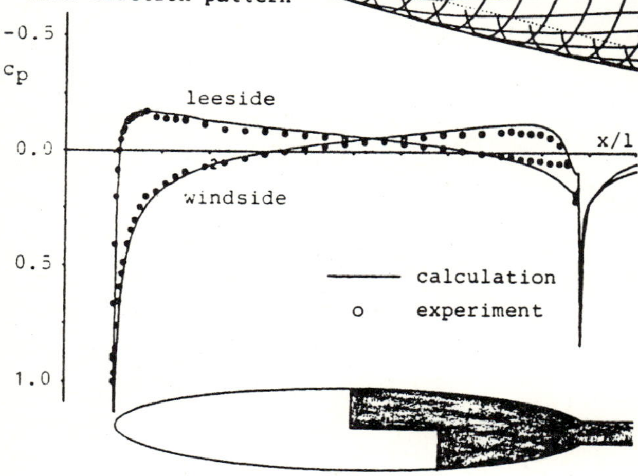

Fig. 9 Pressure distribution

The next application of the code is that to the transonic flow past lifting wings. The NSFLEX code, which solves the full Navier-Stokes equations and is the base for the CCPNS method, was applied initially to the DFVLR-F5 wing in a wind tunnel workshop test case with prescribed inflow and outflow conditions. The angle of

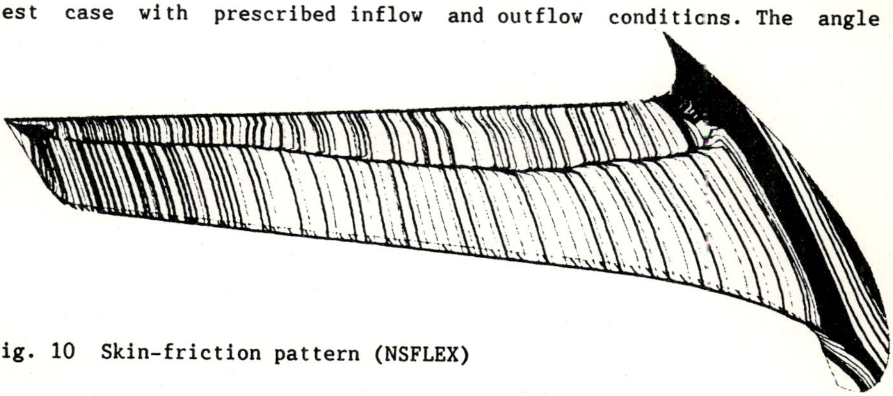

Fig. 10 Skin-friction pattern (NSFLEX)

attack is $\alpha=2°$, $M_\infty=0.82$, $Re=10^7$ [8]. A C-H-type grid with 568,000 cells was used. The complexity and the domains of weak and strong interaction can be

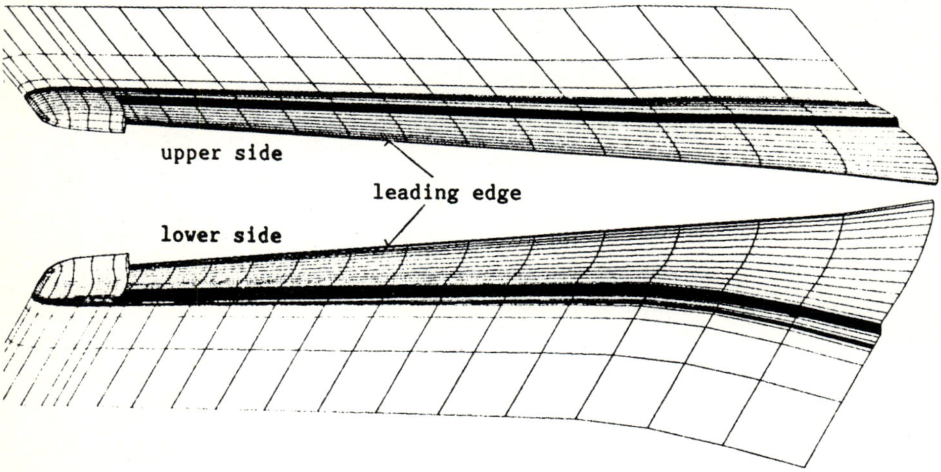

Fig. 11 Surface grid and Navier-Stokes zone for wing flow

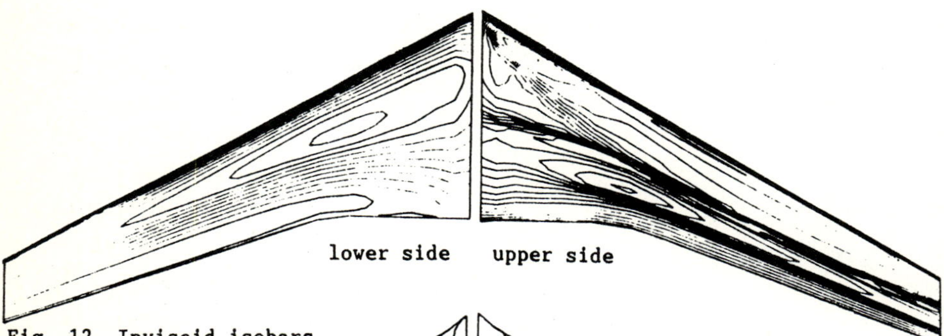

Fig. 12 Inviscid isobars

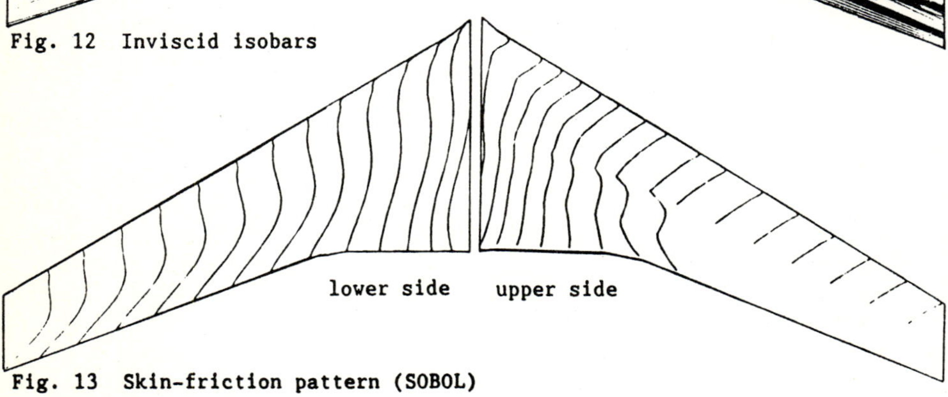

Fig. 13 Skin-friction pattern (SOBOL)

seen in Fig. 10. The computer time for this case was about 12 hours on a VP200 supercomputer. To reduce these costs the CCPNS method is developed now for wing flow problems. A special C-O-type grid generator was implemented for this purpose [17]. The Euler grid is a subset of the Navier-Stokes grid and therefore adaptive grid rezoning is easy to establish. This means that the different zones are found by the code itself during the convergence process [17]. Here, some preliminary results are shown for a generic transport aircraft wing. In Fig. 11 the different zones of flow modelling are obvious. With the inviscid Euler solution as input, Fig. 12, the boundary layer method SOBOL was applied. The resulting skin-friction lines on the wing surface are drawn in Fig. 13. On the upper side the strong shock causes the boundary layer to separate, whereas on the lower side the calculation is performed up to the trailing edge. In [17] the CCPNS method is extensively applied to a wing at different angles of attack.

Conclusions

A zonal method in which solutions of the Navier-Stokes equations are only solved in regions of strong viscous interaction is summarized. Representative results demonstrate the efficiency and accuracy of the method for transonic and subsonic flow cases. In well-suited cases the computation effort reduces drastically. In present, a method with adaptive grid rezoning for transonic wing design is being developed. A further development is the extension of the method to hypersonic flows.

Acknowledgement

The authors want to thank the Deutsche Forschungsgemeinschaft for supporting this work in part (Hi 342/1-6).

REFERENCES

[1] Hirschel, E.H., Schmatz, M.A.: Zonal solutions for viscous flow problems. Vol. 14, Notes Num. Fl. Mech., Vieweg, 1986, pp. 99-112.

[2] Schmatz, M.A.: Simulation of viscous flows by zonal solutions of the Euler, boundary-layer and Navier-Stokes equations. ZFW, Vol. 11, 1987, pp. 281-290.

[3] Wanie, K.M., Schmatz, M.A., Monnoyer, F.: A close coupling procedure for zonal solutions of the Navier-Stokes, Euler and boundary-layer equations. ZFW, Vol. 11, 1987, pp. 347-359.

[4] Flores, J., Holst, T.L., Kaynak, U., Gundy, K.L., Thomas, S.D.: Transonic Navier-Stokes wing solutions using a zonal approach. AGARD CP412, 1986, 30A-1 to 30A-12.

[5] Van Dalsem, W.R., Steger, J.L.: Using the boundary-layer equations in 3D viscous flow simulation. AGARD CP412, 1986, 24-1 to 24-12.

[6] Wanie, K.M.: Simulation dreidimensionaler Strömungen durch lokale Lösungen der Navier-Stokesschen Gleichungen. Doctoral Thesis, Technical University Munich, 1988.

[7] Monnoyer, F., Wanie, K.M., Schmatz, M.A.: Calculations of the three-dimensional viscous flow past ellipsoids at incidence by zonal solutions. Vol. 20, Notes Num. Fl. Mech., Vieweg, 1988.

[8] Schmatz, M.A.: Three-dimensional viscous flow simulations using an implicit relaxation scheme. Vol. 22, Notes Num. Fl. Mech., Vieweg, 1988.

[9] Schmatz, M.A., Brenneis, A., Eberle, A.: Verification of an implicit relaxation method for steady and unsteady viscous and inviscid flow problems. AGARD CP437, 1988, pp. 15-1 to 15-33.

[10] Monnoyer, F.: Calculation of 3D attached viscous flows on general configurations with second-order boundary layer theory. To be published in 1988.

[11] Monnoyer F.: Second-Order Three-Dimensional Boundary Layers. Vol. 13, Notes Num. Fl. Mec., Vieweg, 1986, pp.217-278.

[12] Hirschel, E.H., Kordulla, W.: Shear-flow in surface oriented coordinates. Vol. 4, Notes Num. Fl. Mec., Vieweg, 1981.

[13] Eberle, A.: Characteristic flux averaging approach to the solution of Euler's equations. VKI lecture series, Comp. fl. dyn., 1987-04, 1987.

[14] Schwarz, W.: Elliptic grid generation system for three-dimensional configurations using Poisson's equation. 1st Int. Conf. Num. Grid Generation in Comp. Fluid Dyn., Pineridge Press, 1986.

[15] Monnoyer, F.: Three-dimensional, laminar and turbulent, compressible boundary layers including the effect of surface curvature and application to zonal solution of the Navier-Stokes equations. To appear in: Proc. 12th DEA Meeting, May 3-4, Bethesda (Md), 1988.

[16] Meier H.U, Kreplin H.-P., Landhäußer A.: Wall Pressure Measurement on 1:6 Prolate Spheroid in the DFVLR 3mx3m Low Speed Wind Tunnel. DFVLR IB 222-86 A 04, 1986.

[17] Schmatz, M.A., Monnoyer, F., Wanie, K.M.: Numerical simulation of transonic wing flows using a zonal Euler/boundary-layer/Navier-Stokes approach. ICAS-paper 88-4.6.3, 1988.

On the Computation of Transonic Viscous Flows

D. Hänel, M. Breuer

Aerodynamisches Institut
RWTH Aachen
FRG

Summary

Transonic, viscous flows react very sensitive to small disturbances. In numerical solutions of the Navier-Stokes equations it was found that the numerical dissipation can significantly affect the temporal and spatial accuracy in viscous flows, as demonstrated in several examples. The validation of the numerical results by comparison with the experiment renders more difficult, since now uncertainties of both are superposed. In particular cases correlations between numerical and experimental results could be achieved only by "trial and error".

Introduction

Viscous,transonic flow is characterized by its strong sensivity to small disturbances,which is reflected in the large deviations within experimental and numerical results, as well. The main reason is found in the nonlinear wave behaviour of the transonic flow with subsonic and supersonic portions, and with smooth and discontinuous solutions. These effects are governed by the inviscid Euler equations.

In viscous flow, additional effects have to be taken into account like displacement effects of boundary layers and wakes or even interactions of separated and outer flows. In transonic flow these effects are amplified so that the flow becomes very sensitive to any small disturbances. In numerical simulations such small distrubances are not only caused by physical and geometrical factors but also by numerical errors arising from the discretization of the governing equations.

Therefore, in the first part of this paper, aspects of the discretization and their influence on the spatial and temporal accuracy of the numerical solution in viscous flow are considered. The presented examples show that an essential influence on the accuracy is caused by numerical dissipation of the discretization of the inviscid part of the Navier-Stokes equation. To achieve accurate and reliable results the influence of the discretization has to be carefully checked out.

J. Zierep, H. Oertel (Eds.)
Symposium Transsonicum III
IUTAM Symposium Göttingen, 24.-27. 5. 1988

The strong influence of discretization errors on the numerical solution requires the validation of the results by means of comparison with the experiment. However, since the experiments in transonic flows show a similar sensitive behaviour which results in remarkable deviations between the different experiments, the comparison and validation becomes a complicated matter. In the second part of this paper several problems are discussed by means of computational and experimental results for the supercritical CAST-7 airfoil.

Governing Equations and Discretization

The basic equations are the Navier-Stokes equations for a compressible gas. The algorithms for these equations are formulated conservatively in 2-D and 3-D curvilinear body fitted coordinates e. g. [1]. For the sake of brevity, the scheme will be described in the following for the 1-D case only. Then Navier-Stokes equations read

$$U_t + F_x = \frac{1}{Re} R_x \tag{1}$$

and their discretized version is

$$\frac{\Delta U}{\Delta t} + \frac{1}{\Delta x}\left(F_{i+1/2} - F_{i-1/2}\right) = \frac{1}{Re}\frac{1}{\Delta x}\left(R_{i+1/2} - R_{i-1/2}\right) \tag{2}$$

where $\Delta U/\Delta t$ is the discrete time derivative, defined by the method of solution and $F_{i \pm 1/2}$ and $R_{i \pm 1/2}$ are the Euler and viscous terms at the cell interfaces $i \pm 1/2$.

The viscous terms $R_{i \pm 1/2}$, which have the form $R = \hat{\mu} Uy$ are formulated centrally by e. g.

$$R_{i+1/2} = \hat{\mu}_{i+1/2} \frac{U_{i+1} - U_i}{\Delta x} \tag{3}$$

In the present computations different Euler solvers were applied. If central discretization is used then the flux $F_{i \pm 1/2}$ reads

$$F_{i+1/2} = F\left(\frac{U_i + U_{i+1}}{2}\right) + D^{(4)} + D^{(2)} \tag{4}$$

where D(4) and D(2) are the linear fourth-order, and the nonlinear second-order damping terms as formulated e. g. in [2]. To apply upwind differencing the Euler fluxes are split by van Leer's splitting concept [3] which results in

$$F = F^+ + F^- \tag{5}$$

Then the fluxes are given by

$$F^{\pm}_{i \pm 1/2} = F^{\pm}(U^{\pm}_{i \pm 1/2}) \tag{6}$$

A class of second/third-order accurate schemes ranging from central to fully upwind schemes can be constructed by assuming a piecewise linear or quadratic distributions for the extrapolated values $U^{\pm}_{i+1/2}$. To achieve an almost monotonic solution the extrapolation is limited by a TVD limiter. Details of the scheme are given in [1]. As solution methods, either an implicit relaxation scheme [1] for steady-state calculations or an explicit time-stepping scheme [4] for unsteady flows are used.

Influence of the Discretization Error on the Spatial and Temporal Accuracy

Computing viscous flows at high Reynolds numbers, the accuracy of the numerical solution is essentially influenced by the numerical damping, caused by the discretization of the inviscid, hyperbolic part (Euler terms) of the Navier-Stokes equations. The exact inviscid solution itself is non-dissipative, but numerical dissipation is always present and is used to suppress high-frequency errors and to capture weak solutions. In Navier-Stokes solutions, where viscous layers with strong gradients and curvatures occur, this dissipation becomes large and is added to the physical, viscous effects. The amount of the numerical dissipation depends on the type of Euler solver used. The classical central schemes like [2] with linear fourth-order and nonlinear second order damping terms seem to be less critical in that sense, since the linear damping can be held small per external parameters, and the nonlinear terms controlled by the curvature of the pressure remain small in shear layers where pressure changes are small.

More attention has to be paid to the so-called high-resolution Euler schemes within a Navier-Stokes solver. Such schemes are very well suited for capturing complex gasdynamical wave phenomena by using local approximate Riemann-solvers, higher order upwinding, and TVD flux limiters. Such schemes achieve a sharp shock resolution by a controlled numerical dissipation which, however, is also generated in shear layers due to the curvature of the profiles. These aspects are discussed in several papers published recently [5, 6]. The investigations have shown that the accuracy is influenced, on the one hand, by the different formulations of the Riemann solvers, and, on the other, by the way of constructing the higher order numerical fluxes, e. g. the flux limiter and the discretization. As an example of the latter, Fig. 1 shows the boundary layer profiles taken from a Navier-Stokes solution [6] for the flow over a flat plate with only the extrapolation of $U^{\pm}_{i+1/2}$ varied from central to upwind discretization. Compared with the central discretization ($\kappa = 1$), which agrees well with the Blasius solution, all upwind formulations of second-order accuracy ($\kappa = 0$) or even third order ($\kappa = 1/3$) tend to stronger dissipation.

A crucial demonstration of the influence of the discretization and of the limiters is given in Fig. 2, where the skin friction coefficient is plotted against the local Reynolds number for a flat plate boundary layer flow in the laminar-turbulent transition range. The turbulence and transition model used in this Navier-Stokes calculation is taken from [7]. The transition criterion uses the local eddy viscosity and becomes therefore very sensitive to errors in the flow gradient and to the location of the computed transition point, and thus is a strong indicator of the numerical dissipation. The three computational results in Fig. 2 are computed with second-order accurate schemes based on the flux splitting [1]. The hybrid scheme [8], which essentially has the properties of a central scheme, predicts the transition point sufficiently well, whereas the upwind-biased scheme without limiters predicts transition too far upstream. With limiter, however, no transition occurs as a result of the additional dissipation which reduces the velocity gradient. These few examples show how severe the influence of numerical dissipation on the spatial accuracy can be, even without consideration of the nonlinear transonic flow behaviour which generally amplifies these effects.

The temporal accuracy of a numerical solution for viscous flows is of the same importance as the spatial accuracy is, since many viscous flow problems become unsteady because of the strong interactions between viscous and inviscid flow. This effect is even stronger in transonic flows where separation is additionally provoked by shocks and amplified by the inviscid, transonic flow. An example of that is given in Fig. 3, where the isomach lines are shown for the laminar flow through a cascade of NACA-0012 profiles [9] with steady-state inflow. For a lower inflow Mach number, where the complete flow field is subsonic, the solution stays stationary (Fig. 3a). However, for a higher inflow Mach number supersonic regions occur and the laminar boundary layer separates due to the weak shocks and generates an oscillating flow (Fig. 3b).

The computation of such flows requires sufficient accuracy in space and time. But the numerical error can hardly be estimated since both the spatial and the temporal discretization errors influence the time resolution. Furthermore, due to different scale lengths in time and space, an analysis by means of scalar model equations is not sufficient to explain the numerical behaviour of a Navier-Stokes solution. Thus, numerical experiments with solutions of the Navier-Stokes equations are necessary to detect the influence of the numerical errors. Such experiments are done currently in the course of an investigation of trailing edge separation [10]. First results concerning the numerical effects in time-dependent solutions of the Navier-Stokes equation are published in [4]. In this investigation the Navier-Stokes equations are solved with a Runge-Kutta time-stepping scheme. Central discretization used in a cell-vertex grid system results in a second-order accurate algorithm in time and space. This method was used to study the time-dependent laminar flow over different airfoil geometries. As a typical result, Fig. 4 shows the instantaneous pattern of the velocity vectors for the separated flow around an NACA-0012 at high angle of attack. The periodical flow can be described quantitatively by the time history of the lift, which is a result of complex interactions between inertia forces and separated flow. To achieve a good time accuracy, about 2500 time steps per period were used However, the numerical experiments revealed that besides the time resolution the spatial discretization and its damping properties have a strong influence on the behaviour. To demonstrate this, the amount of the linear fourth-order damping term, which is of $0(\Delta x^3)$ compared with the physical terms of the equations, was varied in a range commonly used in steady-state calculations. In Fig. 5 the history of the lift is plotted for different values of the coefficient $\varepsilon^{(4)}$ of the damping term. The results show a very different behaviour concerning the frequency and the amplitude for every case which is also reflected in the corresponding flow pattern. Similarly large effects are also found by varying the grid spacing and grid arrangement.

The preceding results have demonstrated the influence of the numerical discretization on the accuracy, which becomes significant even in subsonic flow. A much stronger influence is expected in transonic flow and much more work has to be done to achieve sufficient accurate and reliable results.

Comparison with Experimental Results

The strong sensitivity of viscous transonic flow is also reflected in the results of wind tunnel measurements. Here, different tunnel conditions and other experimental factors may cause significant deviations between results of different facilities and of different experimental arrangements. Thus, the numerical simulation of experiments for data validation leads to a complicated task, since now all of the uncertainties of the computation and of the experiment are superposed. Some of these problems are discussed in the following by comparison of numerical and experimental results for the supercritical CAST-7 airfoil taken from a current research work [11] at the Aerodynamical Institute of the RWTH Aachen. The experiments were carried out in the small 7.5x40cm^2 wind tunnel by T. Franke [12]. For the computations, done by M. Breuer, the upwind relaxation method [1] is used to solve the thin-layer approximation of the Navier-Stokes equations. Herein, the turbulent flow is predicted by the Baldwin-Lomax model [7].

The considered CAST-7 airfoil represents a class of supercritical airfoils of high sensitivity. Near the design point ($\alpha = 0°$, Ma = .76) a small change of the angle of attack α (here from $\alpha = 0°$ to $\alpha = 0° 10'$) alters the pattern from nearly shock-free flow to flow with a remarkable shock, as shown in the interferograms in Fig. 6a and b. The numerical calculations exhibit a similar sensitive response to changes in α . Fig. 7a and 7b show the corresponding computed lines of constant density. However,the close agreement could be achieved only by trial and comparison since the correct boundary conditions of the experiment are unknown.

An important problem arises for the simulation of the fully turbulent flow at free flight conditions and very high Reynolds numbers. In general,the wind tunnels operate with lower Reynolds numbers and therefore transition has to be forced e. g. by use of a trip wire. However, such wires do not only force the transition but also induce finite pressure disturbances. In transonic flow both the transition and the pressure disturbances lead to significant deviations in the pressure distribution for free and forced transition,as shown in the experimental results (symbols Δ and 0) in Fig. 8. Numerically, the forced transition is simulated by switching on the turbulence model at the predicted transition point. Since the numerical way of forcing the transition is smoother, the resulting pressure distribution will be located somewhere in between the two experimental cases. In the same Fig. 8 two numerical pressure distributions for $\alpha = -.5°$ and $\alpha = 0°$ with forced transition are plotted against the experimental results. Both distributions cover the range between the forced and free transition, and no unique decision can be made which of the curves, if any at all, corresponds to the desired, undisturbed "free flight" condition.

Another critical flow condition occurs near the maximum lift of an airfoil where the flow tends to separate, and a strong interaction with the inviscid transonic flow takes place. Such a flow with $\alpha = 2°$, $Ma = .76$ and $Re_\infty = 1.5 \cdot 10^6$ was computed, and the resulting skin friction coefficient in Fig. 9 indicates flow separation downstream of the shock. The corresponding pressure distribution is plotted in Fig. 10 (full line) with the resulting lift coefficient of $c_L = .84$. But now the comparison with the corresponding experiment of the present study was not meaningful since the measured maximum lift was much lower, which is probably caused by the different strength of the flow separation. This fact again confirms the strong sensitivity of such flows to tunnel conditions, but for the same reason it is also possible to find results from other tunnels which agree with the computational lift value in the corresponding Mach and Reynolds number range. In Fig. 10 the computed pressure distribution is compared with the results of three different wind tunnel experiments [13]. The agreement with the DFVLR measurement is extremely good, but taking the other experimental results and the preceding discussion into account the close agreement can be considered only as a casual event. The discussed examples cover only a small part of the problems arising from the prediction of viscous transonic flow by experiments and computation. Similar influences, e. g. coming from wind tunnel adaption or 3-D effects have to be considered in further studies.

Conclusions

Viscous, transonic flow reacts very sensitive to small disturbances. This fact is reflected in the remarkable deviations of computational and experimental results. For numerical calculations, special attention has to be given to the discretization of inviscid terms of the Navier-Stokes equations. The numerical dissipation resulting from that can significantly influence the spatial and temporal accuracy of the solution, as demonstrated by several examples. Thus, comparison with experiments is required to validate the results. Since experiments suffer under the same sensitivity of the flow, the uncertainties of both are superposed in the comparison. This is demonstrated by a comparative study of results for supercritical airfoils, which shows that in particular cases correlations between numerical and experimental results can be achieved only by the "trial and error" method.

References

1. Schröder, W.; Hänel, D.: An unfactored implicit scheme with multigrid acceleration for the solution of the Navier-Stokes equations. Computers and Fluids, 15 (1987) pp. 313-336.
2. Jameson, A.; Schmidt, W.; Turkel, E.: Numerical solution of the Euler equations by finite-volume methods using Runge-Kutta time-stepping schemes. AIAA-paper 81-1959, 1981.
3. Van Leer, B.: Flux-Vector splitting for the Euler equations. Lecture Notes in Physics 170 (1982) pp. 507-512.
4. Dortmann, K.:Computation of viscous unsteady compressible flow about airfoils. Accepted paper for IC 11 NMFD, Williamsburg, VA, June 1988.
5. Van Leer, B.; Thomas, J. L.; Roe, P. L.; Newsome, R. W.: A comparison of numerical flux formulas for the Euler and Navier-Stokes equations. AIAA-paper 87-1104 CP, 1987.
6. Hänel. D.: On the accuracy of upwind schemes in solutions of the Navier-Stokes equations. AIAA-paper, 87-1105 CP, 1987.
7. Baldwin, B.; Lomax, H.: Thin layer approximation and algebraic model for separated turbulent flows. AIAA-paper, pp. 78-257, 1978.
8. Seider, G.: Berechnung kompressibler, reibungsbehafteter Strömungen mit einem hybriden Differenzenverfahren. Abh. aus dem Aerodynamischen Institut der RWTH Aachen, Heft 29, 1988.
9. Hänel, D.; Dortmann, K.; Schröder, W.: Numerische Berechnung reibungsbehafteter Innenströmungen. In: Luftfahrtforschung und Luftfahrttechnologie, 4. BMFT Statusseminar, 1986.
10. Krause, E. et al.: Hinterkantenumströmung in instationärer Strömung. Sonderforschungsbereich 25, Projekt B5, RWTH Aachen, 1988.
11. Krause, E. et al.: Wirbelstruktur im Nachlauf schallnah angeströmter Profile. Sonderforschungsbereich 25, Projekt B4, RWTH Aachen, 1988.
12. Franke, T.: Experimentelle Untersuchung schallnah angeströmter Tragflügelprofile in Kanälen geringer Meßkammerbreite. Dissertation an der RWTH Aachen, 1987.
13. Stanewsky, E.; Puffert, W.; Müller, R.; Batemann, T. E. B.: Supercritical airfoil CAST-7 surface pressure, wake and boundary layer measurements. AGARD Advisory Report No. 138.

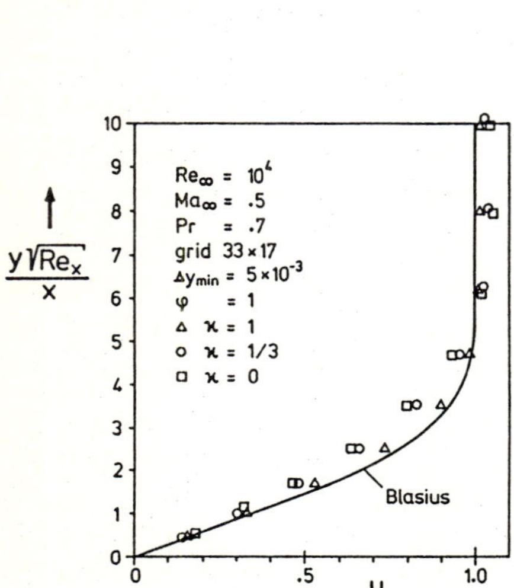

Fig. 1 Velocity profile in the boundary layer over a flat plate (Navier-Stokes solution from [6]). Comparison for different discretizations of the Euler terms without flux limiters.
$\kappa = 1$ central $0(h^2)$
$\kappa = 1/3$ upwind-biased $0(h^3)$
$\kappa = 0$ upwind-biased $0(h^2)$

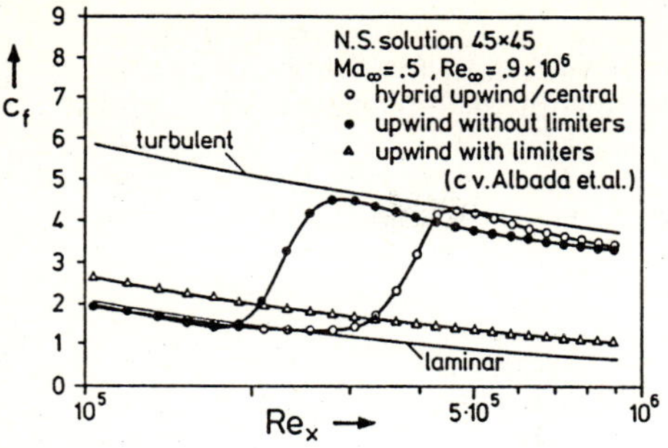

Fig. 2 Skin friction coefficient for the laminar/turbulent flow over a flat plate Navier-Stokes solution with different discretizations of the Euler terms.

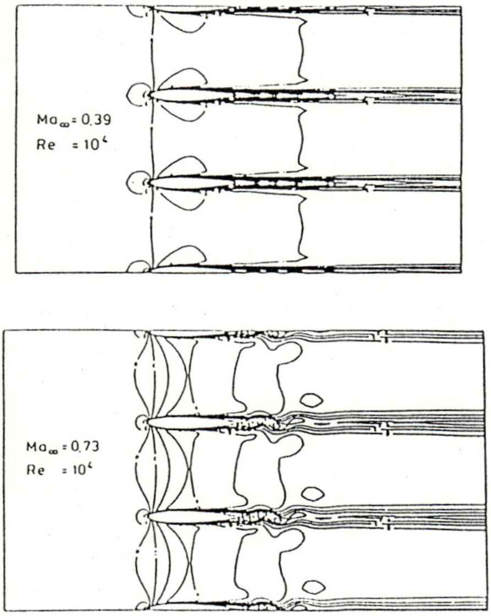

Fig. 3a and 3b Lines of constant Mach number for the viscous flow through a cascade of NACA-0012 airfoils.
3a) subsonic flow with $Ma_\infty = 0{,}39$
3b) transonic flow with $Ma_\infty = 0{,}73$

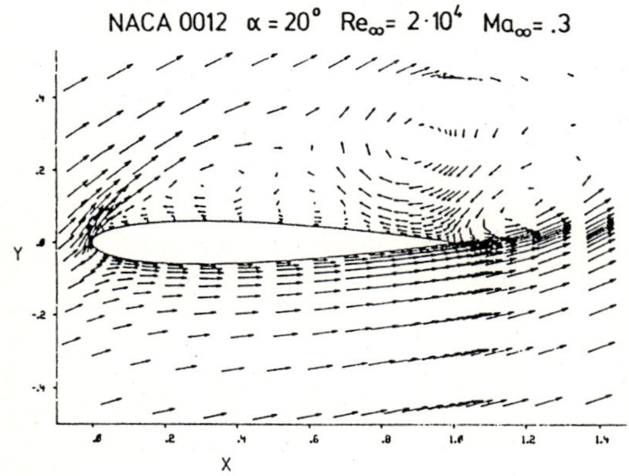

Fig. 4 Instantaneous pattern of the velocity vectors for the separated flow over an NACA-0012 at high angle of attack ($\alpha = 20°$). Navier-Stokes solution from [4].

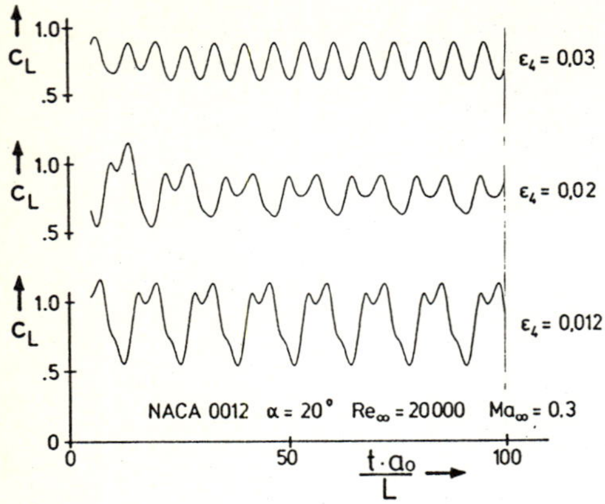

Fig. 5 Time histories of the lift coefficient for three values of the 4th order damping term (Legend see Fig. 4).

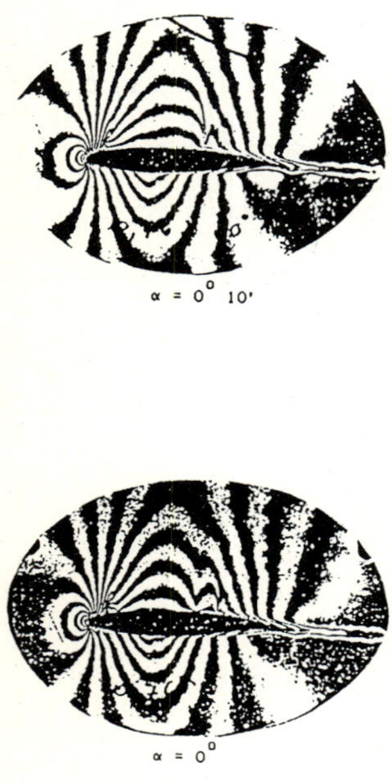

Fig. 6a and 6b Interferograms of the flow around a CAST-7 airfoil with $Ma_\infty = .76$ and $Re_\infty = 1{,}5 \cdot 10^6$, and with an angle of attack of $\alpha = 0°10'$ (6a) and $\alpha = 0°$ (Experiment [12]).

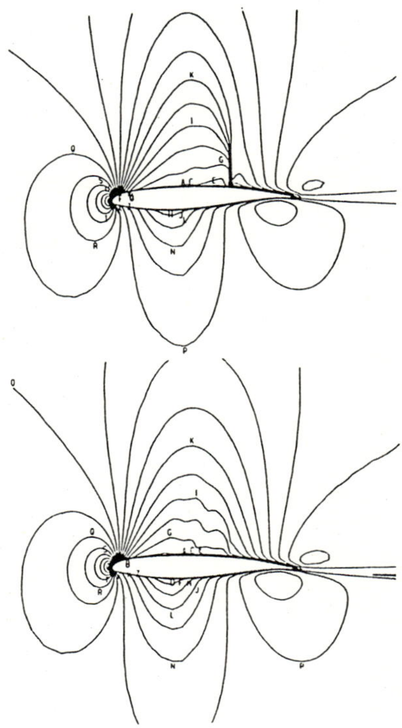

Fig. 7a and 7b Lines of constant density for the flow around a CAST-7 airfoil with $Ma_\infty = 0{,}76$ and $Re_\infty = 1{,}5 \cdot 10^6$, and with an angle of attack of $\alpha = 0°$ (7a) and $\alpha = -.5°$ (present Navier-Stokes solution).

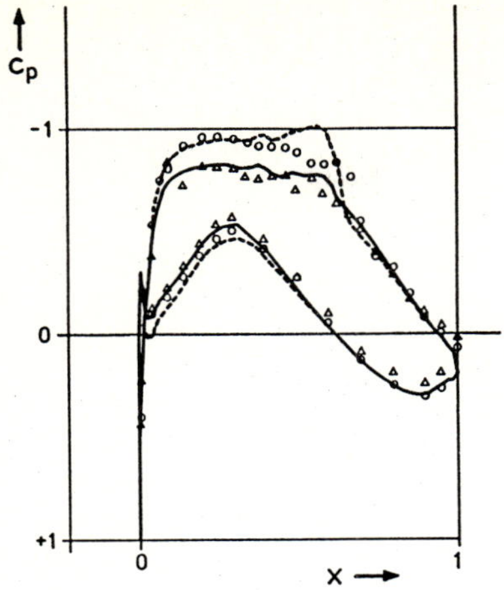

Fig. 8 Pressure distribution for the flow around a CAST-7 airfoil with $Ma_\infty = .76$ and $Re_\infty = 1{,}5 \cdot 10^6$. Comparison between experimental results [12] with free and forced transition, and present computational result with $\alpha = 0°$ and $\alpha = -.5°$ (prescribed transition point at 7%).

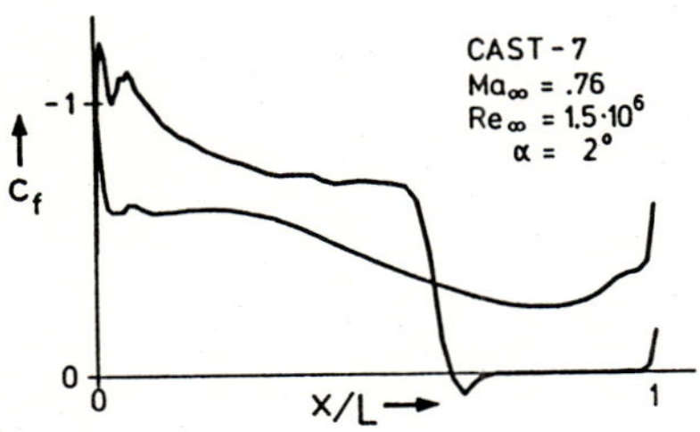

Fig. 9 Distribution of the skin friction for the lower and upper side of an CAST-7 airfoil near maximum lift (present Navier-Stokes solution).

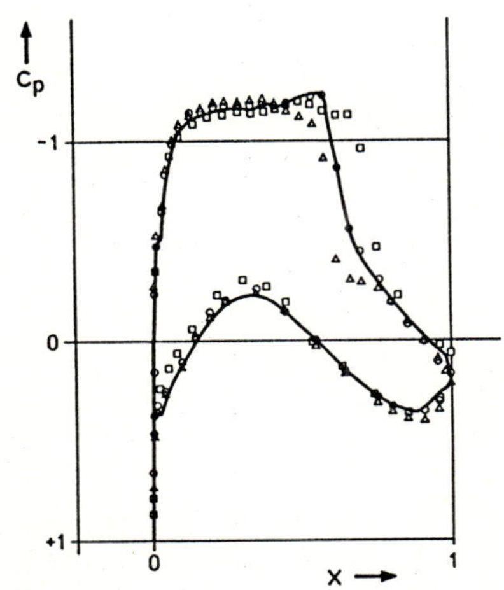

Fig. 10 Pressure distribution for the flow around a CAST-7 airfoil near maximum lift. Comparison of the Navier-Stokes solution (see also Fig.9) with experimental results from 3 different wind tunnels [13].

Improvements in the Formulation and Numerical Solution of the Euler Problem for Swept Wings

M.F. Paisley and M.G. Hall

Royal Aerospace Establishment
Farnborough
Hants, GU14 6TD
UK

Summary

A multigrid cell-vertex finite volume Euler method has been used to calculate steady inviscid transonic flow past the ONERA M6 wing. The treatment of the far-field boundary conditions includes the effect of velocity perturbations generated from Klunker's analytic asymptotic solution to the transonic small-disturbance equation. The geometry of the wing tip is modelled with three successively finer C-O grids. The results obtained show marked differences in comparison with those obtained on C-H grids and indicate that a shock-wave mechanism can contribute to the generation of tip vortices in compressible inviscid flow.

Introduction

The work of Ni [1] has been extended in two and three dimensions by Hall [2] and Salmond [3]. Numerical solutions to the Euler equations for transonic lifting flows past aerofoils and wings are obtained by approximating the steady flux balance for each computational cell using values of the flow variables stored at the vertices of each cell. A Lax-Wendroff time-stepping algorithm is used to update each variable, with boundary conditions where appropriate, and these changes in the solution are then used on coarser meshes to accelerate the convergence of the solution to the steady state. This paper describes some recent developments, mainly in the imposition of far-field boundary conditions and in the modelling of the flow around the wing tip.

The usual practise in calculating the three-dimensional flow past a wing has been to set conditions on the far-field boundary equal to the free-stream conditions. This assumption is loosely

J. Zierep, H. Oertel (Eds.)
Symposium Transsonicum III
IUTAM Symposium Göttingen, 24.-27. 5. 1988

justified by the knowledge that, except in a wake, a disturbance would decay more rapidly with distance in three dimensions than in two. Here, instead, we replace the free-stream conditions by a three-dimensional far-field solution. For "inflow" the conditions are derived from the transonic small-disturbance approximation of Klunker [4], while for "outflow" a perturbation pressure related to the cross-flow in the numerical solution is prescribed. The results are expected to be more accurate than those given by assuming free-stream conditions and to provide an indication of the range of validity of the latter assumption. Comparisons are shown for a range of distances from the wing to the far-field boundary.

It is becoming recognised (for example, see Ref. 5) that the resolution of flow details at the wing tip requires both the representation of the tip geometry and the use of an appropriate grid around the tip. We outline here the steps taken in generating C-O grids from C-H grids formed by spanwise stacking of two-dimensional planar C-grids. Results are shown for calculations performed on the ONERA M6 wing using both C-H grids and three successively finer C-O grids. We begin, however, with an outline of the basic numerical method.

Cell Vertex Finite Volume Euler Method

Since only the steady state is of interest, the unsteady Euler equations in three dimensions are expressed in conservation form as the isenthalpic system

$$\underline{u}_t + f(\underline{u})_x + g(\underline{u})_y + h(\underline{u})_z = 0 \qquad (1)$$

where the state vector $\underline{u}$ and the flux vectors $f(\underline{u})$, $g(\underline{u})$, $h(\underline{u})$ are given as follows:

$$\underline{u} = \begin{pmatrix} \rho \\ \rho u \\ \rho v \\ \rho w \end{pmatrix},\ f(\underline{u}) = \begin{pmatrix} \rho u \\ p + \rho u^2 \\ \rho uv \\ \rho uw \end{pmatrix},\ g(\underline{u}) = \begin{pmatrix} \rho v \\ \rho vu \\ p + \rho v^2 \\ \rho vw \end{pmatrix},\ h(\underline{u}) = \begin{pmatrix} \rho w \\ \rho wu \\ \rho wv \\ p + \rho w^2 \end{pmatrix}.$$

Here ρ is the density, u, v, w are Cartesian components of velocity and p is the pressure given by Bernoulli's equation

$$p = \frac{\rho}{\gamma}\left(1 - \frac{1}{2}(\gamma - 1)(u^2 + v^2 + w^2)\right)$$

where γ is the ratio of specific heats, taken to be 1.4. The equations have been non-dimensionalised with respect to a typical length, stagnation sound speed and stagnation pressure.

Integrating equation (1) over a control element of volume V and using the divergence theorem yields for the steady state the boundary integral (or residual)

$$\frac{1}{V}\iint (f,\ g,\ h)\ .\ \underline{ds} = 0\ . \tag{2}$$

Dividing our region of physical space surrounding the wing into hexahedral cells we seek to approximate equation (2) for each cell. The fluxes through each of the faces of a particular cell are calculated from the values of the flow variables stored at the vertices of that cell (see Fig.1) and are summed to give the net flux into the cell.

Given an initial flow the steady solution $\underline{u}$ satisfying equation (2) for all cells is obtained by a Lax-Wendroff time-marching algorithm. From equation (1) the change in the solution, $\delta\underline{u}^n = \underline{u}^{n+1} - \underline{u}^n$, in time-step Δt is given by

$$\delta\underline{u} = -\Delta t(f_x + g_y + h_z) - \frac{\Delta t^2}{2}\left\{\frac{\partial}{\partial x}\left(\frac{\partial f}{\partial \underline{u}}\frac{\partial \underline{u}}{\partial t}\right) + \frac{\partial}{\partial y}\left(\frac{\partial g}{\partial \underline{u}}\frac{\partial \underline{u}}{\partial t}\right) + \frac{\partial}{\partial z}\left(\frac{\partial h}{\partial \underline{u}}\frac{\partial \underline{u}}{\partial t}\right)\right\} . \tag{3}$$

The two parts of the change $\delta\underline{u}$ at the point P are calculated as follows (see Fig.2). The first term in expression (3) is given by averaging the residuals (2) for the eight cells surrounding P. The second term is recast by integrating over the inner hexahedron surrounding P formed by the centres of the eight cells. The divergence theorem again yields an integral over the surface of this hexahedron, evaluated as before, the values of the flow variables used being obtained as averages of the values at the vertices of the eight cells.

To prevent odd-even point decoupling and to capture shock-waves, an artificial viscosity is added to $\delta\underline{u}$. Convergence acceleration is achieved using local time-stepping and a multigrid procedure. Changes in the solution on the coarse grids are interpolated linearly to the finer grids as previously but are then smoothed by an averaging operator.

Far-field Boundary Conditions

In the present method the inflow condition is that the two

components of velocity in the plane locally tangential to the boundary are specified together with an isentropy condition for the density. To determine corrections to the free-stream condition in the far-field we firstly consider Klunker's expression for the transonic small-disturbance potential ϕ at a point $P(x, y, z)$. In the far-field Klunker shows that the dominant term is that representing the lifting effects of the wing, namely the following integral over the wing surface (see Fig.3)

$$\phi_p = \frac{z}{4\pi} \int_{\substack{\text{wing}\\\text{surface}}} \frac{\Delta u}{(y-\eta)^2 + z^2} \left(1 + \frac{x-\xi}{R}\right) dS \qquad (4)$$

where ξ, η are local coordinates on the wing and $R^2 = (x-\xi)^2 + \beta^2(y-\eta)^2 + \beta^2 z^2$ with $\beta^2 = 1 - M_\infty^2$. Δu is interpreted here as the difference in pressure between the upper and lower surfaces. Due to deficiencies in the potential model ϕ_p is only a good approximation away from the wake, and in fact is only used to generate perturbations to the velocities given at "inflow".

ϕ_p is differentiated to give expressions for the perturbations $(\delta u, \delta v, \delta w)$ from the free-stream velocities $(u_\infty, v_\infty, w_\infty)$. The resulting integrals are evaluated numerically for each point P on the "inflow" part of the far-field boundary. The modified velocities $(u_\infty + \delta u, v_\infty + \delta v, w_\infty + \delta w)$ are then used to calculate the two imposed components of the tangential velocity. If the perturbations are set to zero we regain the free-stream boundary conditions.

At outflow a boundary condition on pressure p is assumed which includes the effect of the component of velocity q_n normal to the free-stream direction. It can be shown that if variations in the function p/ρ^γ are small, and if at outflow

$$u^2 + v^2 + w^2 = q_\infty^2 + q_n^2 ,$$

where q_∞ is the free-stream velocity, then the Bernoulli equation can be approximated by

$$p = \left(p_\infty^{(\gamma-1)/\gamma} - \frac{\gamma-1}{2} q_n^2\right)^{\gamma/(\gamma-1)} .$$

This is the assumed condition.

Wing Tip Modelling

The stages in producing a C-O grid from the C-H grid comprising two-dimensional planar C-grids stacked spanwise are as follows:

(i) Discard the C-grids lying beyond the wing tip.

(ii) Complete the surface closure for the M6 wing using circular arcs as continuations of the wing generators. The (finest) surface grid has each arc subdivided into 32 equal segments.

(iii) As sketched in Fig.4, rotate the C-grid corresponding to the tip station 180° about an axis running chordwise through the leading and trailing edges. Equal subdivision of the arcs swept out provide the field grid. Lines of semi-polar singularity are left along the axis of rotation forward of the leading edge and aft of the trailing edge.

After these steps adjustments were made to produce the final grid. These included bending and packing of the C-grids to ensure smooth spanwise distributions, the addition of a further 16 C-grids on the main wing, and contracting the grid around the trailing edge of the tip where the tip radius is small. Thus, from an original C-H grid of 256 × 32 × 48 cells was produced a new C-O grid of 256 × 32 × 64 cells. Details of the medium grid (128 × 16 × 32) on the wing surface near the tip are shown in Figs.5,6.

Results

The numerical method was used in two forms, appropriate to the C-H and C-O grids, to calculate the flow past the ONERA M6 wing, with $M_{\infty} = 0.84$ and $\alpha = 3.06^{\circ}$. To evaluate the effect of the new boundary conditions, the far-field boundary was set at varying distances from the wing - 0.84, 1.69 and 3.37 spans (3, 5 and 10 root chords respectively) - and the calculations performed with and without the perturbations modifying the free-stream. The effect on C_L is shown in Fig.7 for the calculations on the medium C-H and C-O grids. With the free-stream conditions the variation in C_L is around 3% while for the perturbed conditions it is around 0.7%. All subsequent calculations were performed with the perturbed conditions with the boundaries set at 1.69 spans.

Next, we detail the comparison between the tip flows on the medium C-H and C-O grids. The spanwise lift distributions in Fig.8 show a marked spike at the tip for the calculation on the C-H grid, indicating a vortex. This is absent for the calculation on the C-O grid and examination of the cross-flow shows little sign of a vortex, although the cross-flows are much larger than for the C-H grid. Shown in Figs.9,10 are the upper surface Mach number contours in the tip region for the two cases. As might be expected, the flows are very similar inboard but quite different at the tip. For the calculation on the C-O grid the contours near the tip trailing edge indicate the presence of high velocities there.

Finally, we compare the calculations performed on the successively finer C-O grids. Fig.11 shows the value of C_L plotted against $(\text{number of cells})^{-2/3}$, the straight line indicating second-order accuracy. The spanwise lift distributions are plotted in Fig.12. Note the small kink at the tip for the finest grid; indeed (Fig.13), a plot of the cross-flow velocities obtained on this grid, very close to the trailing edge (98.5% chord), shows a region of recirculating flow.

The production of the recirculating flow, or tip vortex, can be explained by examining the cross-flow upstream. Having become sonic at about 89% chord the cross-flow Mach number reaches 3.0 at 96.5% chord. A shock in the cross-flow plane can be seen in the velocity vectors and in the contours of cross-flow Mach number in Figs.14 and 15 obtained on the finest grid. The corresponding plot of total vorticity contours in Fig.16 shows that while some vorticity is produced at the junction of the circular arc with the wing generator, more is produced through the shock-wave. This vorticity is convected downstream and may therefore be the origin of the vortex shown. Further work needs to be done to assess whether separation is occurring in addition.

Conclusions

Improved far-field boundary conditions and wing tip modelling have been proposed and tested. The boundary conditions allow the distance to the far-field boundary to be reduced considerably with little change in lift. Calculations on C-H and C-O grids show very marked differences in the flow at the wing tip, there being a distinct tip vortex on the C-H grids. Although cross-flows are much larger on the C-O grids, there is little sign of a tip vortex except for the finest grid. In this case a vortex appears to originate at least in part in vorticity generated by a shock in the cross-flow.

Acknowledgements

The authors thank Cray Research (UK) Ltd and AERE Harwell for providing assistance and computing facilities essential for the completion of the work described.

References

1. Ni, R.-H.: A multiple grid scheme for solving the Euler equations. AIAA J. vol.20 No.11 (1982) 1565-1571.

2. Hall, M.G.: Cell-vertex schemes for solution of the Euler equations. Proc. Conf. on Num Meth for Fl Dyn, Univ. of Reading, 1985, (Eds. K.W. Morton and M.J. Baines) Oxford University Press 1986, 303-345.

3. Salmond, D.J.: A cell-vertex multigrid scheme for solution of the Euler equations for transonic flow past a wing. Proc. ICNMFD 10, Beijing, China, 1986, (Eds. F.G. Zhuang and Y.L. Zhu) Springer-Verlag Lecture Notes in Physics No.264 1986, 549-553.

4. Klunker, E.B.: Contribution to methods for calculating the flow about thin lifting wings at transonic speeds - analytic expressions for the far-field. NASA TN D-6530 (1971).

5. Sacher, P.: Numerical Solutions for three-dimensional cases - swept wings. Chapter 7, "Test cases for inviscid flow field methods". AGARD Advisory Report No.211 (1985).

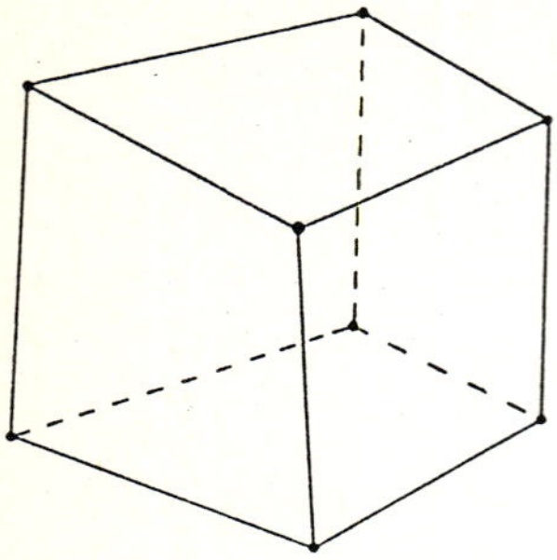

Fig 1 A single hexahedral cell

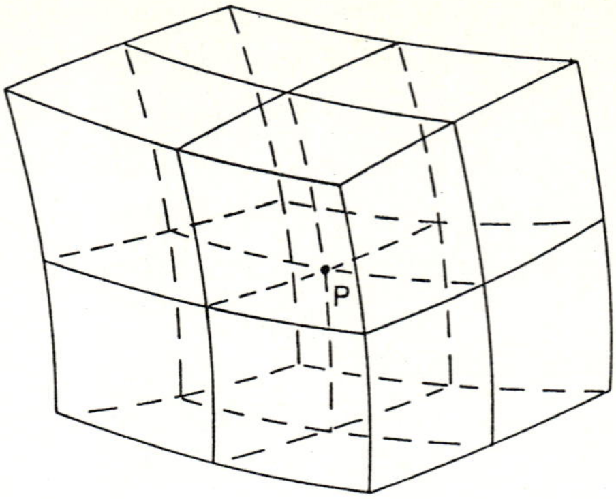

Fig 2 The eight cells surrounding the point P

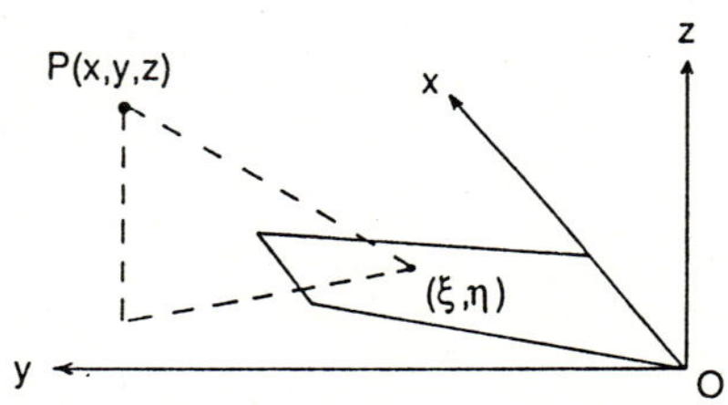

Fig 3 Wing and a boundary point P for Klunker's integral over the wing surface

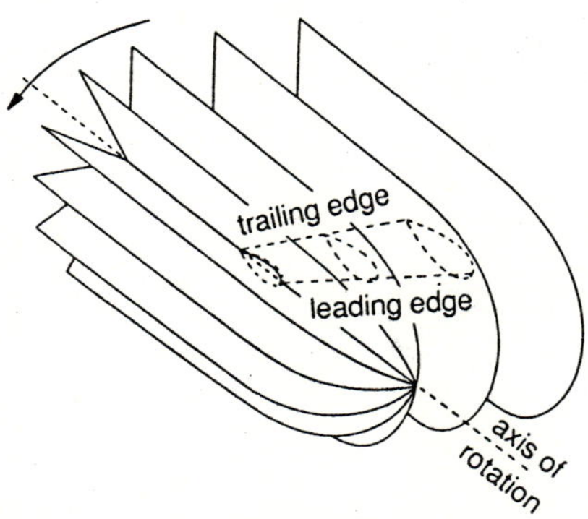

Fig 4 Sketch of construction of C-O grid from C-H grid by rotation

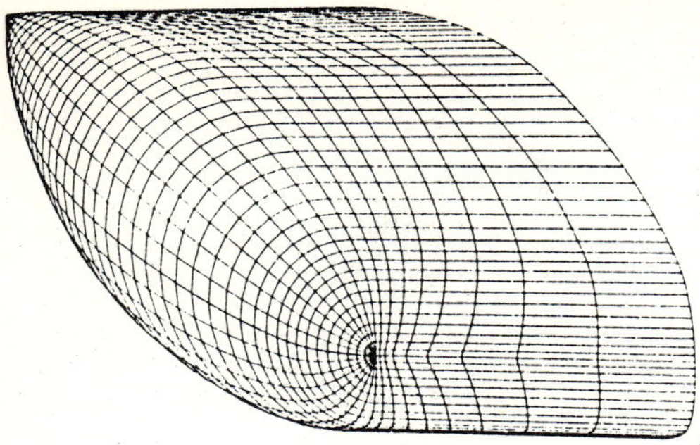

Fig 5 Surface grid at the wing tip (leading edge)

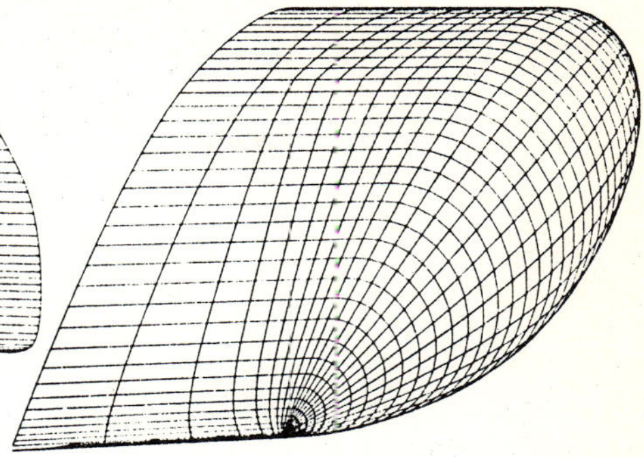

Fig 6 Surface grid at the wing tip (trailing edge)

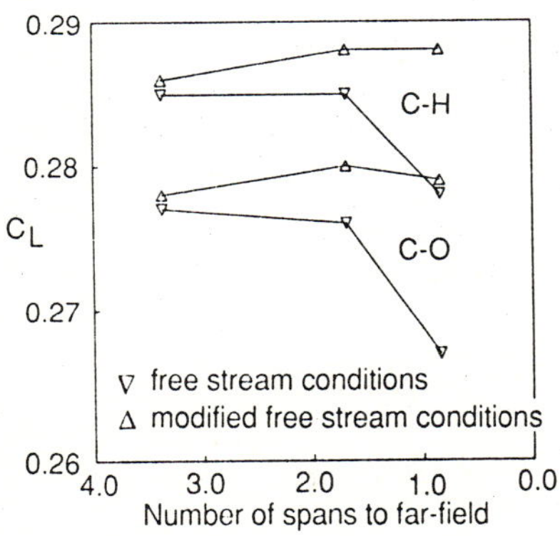

Fig 7 Variation of lift with position of far-field boundary

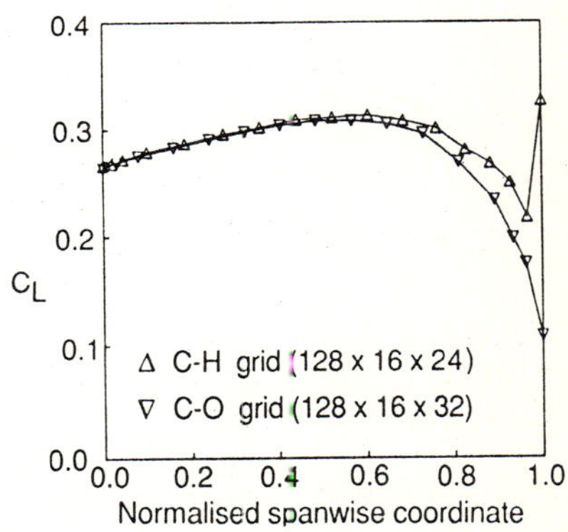

Fig 8 Spanwise lift distributions for calculations on the medium C-H and C-O grids

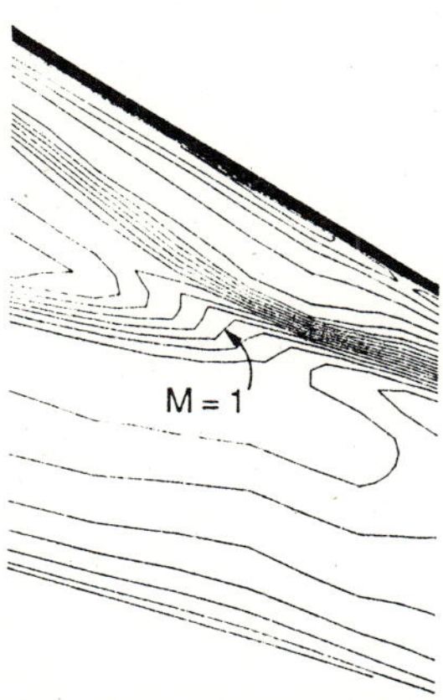

Fig 9 Mach contours on upper surface for the calculation on the medium C-H grid (Δ M = 0.04)

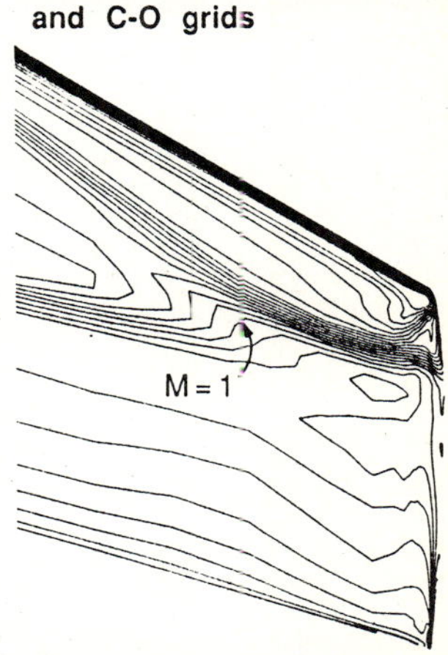

Fig 10 Mach contours on upper surface for the calculation on the medium C-O grid (Δ M = 0.04)

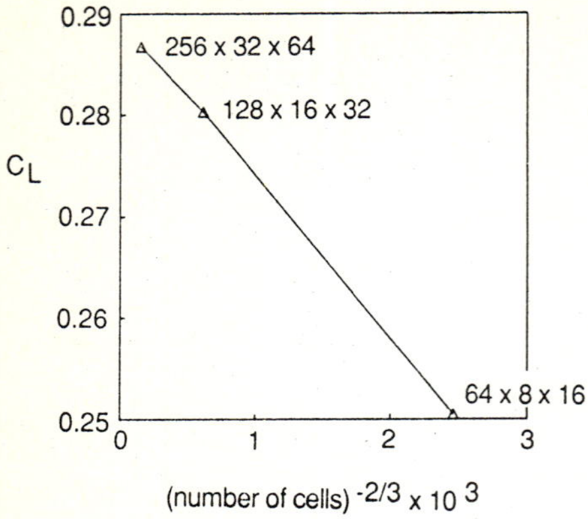

Fig 11 Variation of C_L with refinement of the C-O grid

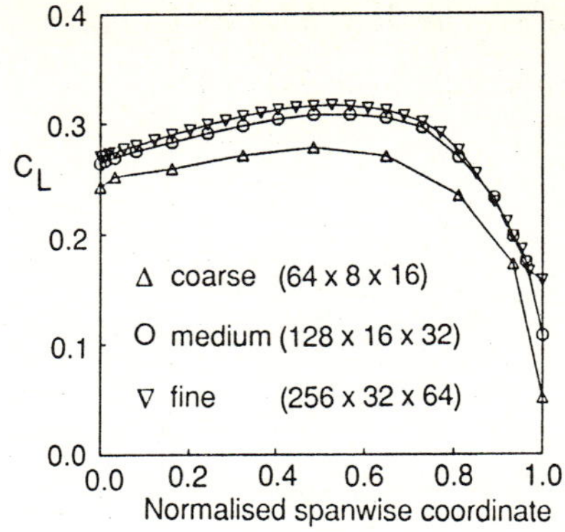

Fig 12 Variation of the spanwise lift distribution with refinement of the C-O grid

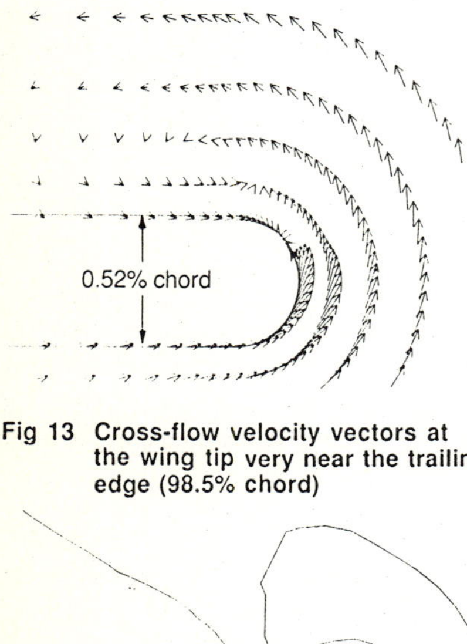

Fig 13 Cross-flow velocity vectors at the wing tip very near the trailing edge (98.5% chord)

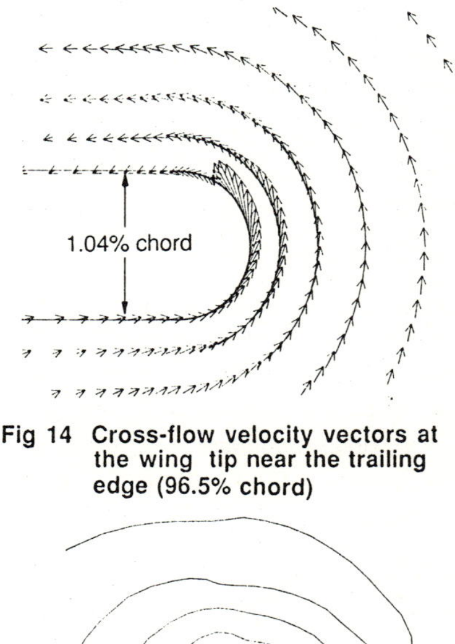

Fig 14 Cross-flow velocity vectors at the wing tip near the trailing edge (96.5% chord)

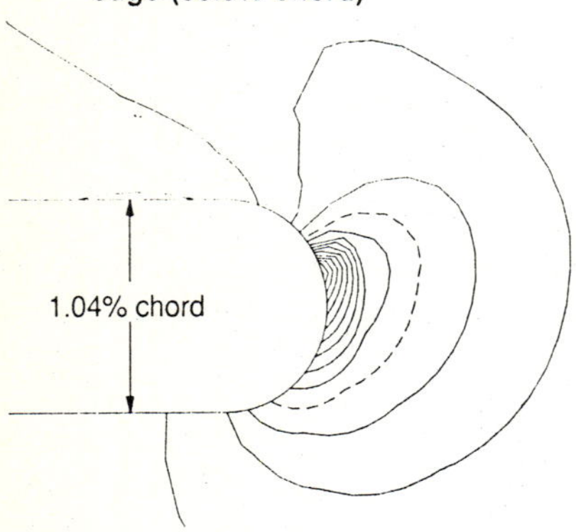

Fig 15 Cross-flow Mach contours at the wing tip near the trailing edge (96.5% chord) Dotted line is sonic and $\Delta M = 0.2$

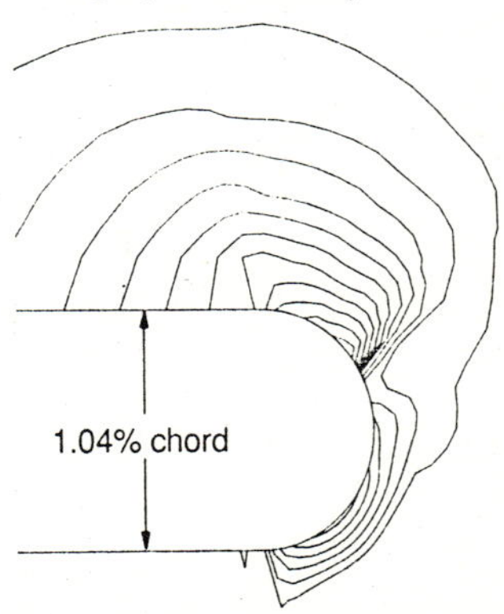

Fig 16 Vorticity ($|\Omega|$) contours at the wing tip near the trailing edge (96.5% chord)

Transonic Computations by Multidomain Techniques with Potential and Euler Solvers

Ph. Morice

ONERA
Aerodynamics Department
BP 72 - 92322 Chatillon, FRANCE

Summary

Some multidomain techniques are presented for the solution of potential and Euler transonic flows on multiblocked grids. The method for the transonic full potential calculations is founded on a substructured preconditioned conjugate gradient solution of linear subproblems in a finite element approximation. The Euler multiblock solution follows from an adaptation of characteristic relations to the treatment of interfaces for an upwind implicit linearized ADI non-iterative scheme. A coupling of a potential and of an Euler solver permits the three-dimensional calculation of a supersonic jet in a transonic outer flow with fitting of the jet surface.

1. Introduction

The recent increase in computing power (both for speed and for storage capacity) and the progress of solution algorithms have led to undertake numerical simulations for more and more complex flows past complicated aerodynamic shapes. Therefore, the mesh problem became more critical in the solution process. The multidomain or zonal approach where the computational domain is divided into several subdomains or zones is most often advocated as a remedy to this situation.

However, if the mesh problem is made less stringent, the effort is transferred to matching boundary techniques. Depending on the mathematical model (potential equation or Euler equations), the elliptic or hyperbolic type of the problem leads to distinct strategies. Without trying to give an exhaustive list of the earliest works on the subject, we can mention for elliptic problems the alternating Schwarz method [1] applied to the transonic full potential equation by Atta et al. [2,3], Yu [4] and Le [5], and for the "substructuring approach" in finite element methods, Morice [6], Bjorstad and Widlund [7]. For Euler equations, we refer to the early studies by Cambier et al. [8,9,10], Hessenius and Pulliam [11], Rai et al. [12,13,14], Lombard et al. [15].

In the present paper, some proposals and examples of techniques for zonal calculations both with potential and Euler solvers are given. After a discussion of various possible choices for the multidomain techniques in section 2, we present a multiblock solution for the transonic full potential equation currently studied at ONERA. Section 4 is devoted to a technique recently developed by Borrel [16]

J. Zierep, H. Oertel (Eds.)
Symposium Transsonicum III
IUTAM Symposium Göttingen, 24.-27. 5. 1988

for an implicit Euler solution on patched grids. Lastly, a tridimensional calculation coupling a potential and an Euler solver for a jet flow with slip surface fitting is presented in section 5.

2. The domain decomposition approach

As already said, the principal merit of the zonal approach is perhaps to make easier the mesh generation. However, many other motivations can obviously be considered. Namely, local grid refinements or grid adaptation, fitting of discontinuities, coupling of different sets of equations or of solution algorithms, better data management by blocks in the computer memory and parallel processing can justify the multidomain approach. According to the retained objectives, various choices are possible and the first distinction in zonal techniques must be made between patched (or non-overlapping) grids and overlapping (or overlaid) grids.

The latter category leads to more complex matching algorithms because of the multidimensional interpolation and book-keeping problems. Moreover, it is very difficult to find a mathematical analysis of the accuracy and convergence properties of these algorithms and, generally, they rest upon a relaxation process of explicit iterative type. However, the merit of overlaid grids is evident when applied to very intricate problems such as moving bodies with body fitted grids. If patched grid techniques are less general and necessitate more careful mesh generation, they can be devised on a more sound mathematical basis and they lead to simpler coding as the interpolation and the implicitness are concerned.

In this paper only patched grids and more precisely continuous grids at a patch boundary are considered, so that continuity of nodal values can trivially be obtained. By this choice, we hope that some general requirements for a zonal approach can be more easily fulfilled. In particular, we mean that for a given solution method in a one-block grid, it is desirable that, if we partition this grid into several zones and we apply the zonal algorithm, we could obtain, at each time step for an unsteady calculation or at convergence towards a steady solution, the same results as with the classical one-zone computation. To reach this ambitious objective it appears that it is more natural, firstly to define the patch-boundary scheme not directly at the continuous problem level but starting from the discrete set of equations, and secondly to keep a strong link between the numerical scheme defined for interior nodes and at boundary nodes. This will be illustrated in the two following sections.

3. Transonic full potential calculations in multiblock grids

3.1. Equations

The governing equations for the case of the full potential approximation are:

$$div\ (\rho\ \nabla\phi) = 0 \quad , \vec{V} = \nabla\phi \tag{3. 1}$$

$$\rho = [\ 1 + \frac{\gamma-1}{2} M_\infty^2\ (\ 1 - |\nabla\phi|^2\ )\ ]^{1/\gamma-1} \tag{3. 2}$$

where all the quantities are non-dimensionalized by freestream values of the density and the velocity. Two types of boundary conditions can be considered, a Dirichlet boundary condition:

$$\phi = g_D \quad on \quad \partial\Omega \qquad (\vec{V} \,.\, \vec{s} = dg_D \,/\, ds) \tag{3. 3}$$

or a Neumann boundary condition:

$$\rho \frac{\partial \phi}{\partial n} = g_N \qquad (\rho \, \vec{V} \,.\, \vec{n} = g_N) \tag{3. 4}$$

Both types of boundary conditions allow to put (3.1) under a weak formulation. Actually, we can assume that (3.1) is transformed intc a series of linear subproblems:

$$div \, (\, \rho^k \, \vec{\nabla}\phi^{k+1} \,) = 0 \tag{3. 5}$$

$$\rho^k = [\, 1 + \frac{\gamma-1}{2} M_\infty^2 \, (\, 1 - |\vec{\nabla}\phi^k \, |^2 \,)]^{1/\gamma-1}$$

$$\phi^{k+1} = g_D \quad or \quad \rho^k \frac{\partial \phi^{k+1}}{\partial n} = g_N \qquad on \, \delta\Omega$$

The weak formulation associated with (3.5) is approximated with hexahedral finite elements in I,J,K grid blocks, with constant density in each cell and a trilinear continuous approximation of the potential from its nodal values. At each iteration, a linear system must be solved:

$$A \; x = f \tag{3. 6}$$

where x represents the nodal values of theunknown ϕ^{k+1}; the right-hand side f comes from the non-homogeneous boundary conditions. The symmetric matrix A depends on cell values of ρ^k. It is positive–definite as long as a Dirichlet condition holds on a part of the boundary (ϕ^{k+1} must be given at least for one node). The iterative solution of (3.6) is necessary in 3D and we have developed a conjugate gradient method preconditioned by the incomplete Cholesky factorization (ICCG method) see [17].

The Preconditioned Conjugate Gradient (PCG) method is recalled here:

Assuming that $x^0,..,x^m$ are known from previous iterations, compute x^{m+1} by:

$$r^m = f - A \; x^m \tag{3. 7}$$

$$z^m = M^{-1} r^m \tag{3. 8}$$

$$q^{m+1} = z^m + \lambda^{m+1} \, q^m \quad , \quad \lambda^{m+1} = (\, r^m, z^m \,)/(\, r^{m-1}, z^{m-1} \,) \tag{3. 9}$$

$$x^{m+1} = x^m + \mu^{m+1} \, q^{m+1} \quad , \quad \mu^{m+1} = (\, r^m, q^{m+1} \,)/(\, A \; q^{m+1}, q^{m+1} \,) \tag{3.10}$$

where r^m is the residual, M the symmetric positive definite preconditioning matrix ($M = L \, L^t$ with L a lower triangular matrix resulting from the incomplete Choleski factorization of A).

Instead of the fix point iteration (3.5), it is possible to use the Newton method (at least in subsonic regime) or other linearization techniques leading, after discretization, to a linear system (3.6) which the multiblock solution strategy rests upon. In supersonic regions the density is classically upwinded [17] in order to get the proper numerical dissipation and a good shock resolution. Again, this has no influence on the linear system solution and does not change the multiblock solution.

3.2. Multiblock solution

We assume that the computational grid is made up by the union of I,J,K blocks with perfect "topological" continuity on the interface of two blocks. Alleviation of the "metric discontinuities" at the interfaces is a mesh generation problem for which various techniques are available. The arrangement of the blocks is assumed to be quite arbitrary without any prevailing natural numbering. The computational domain Ω is made of n zones ($\Omega = \bigcup_i \Omega_i$) and the set of interfaces is $\Gamma = \bigcup_{i \neq j} (\bar{\Omega}_i \cap \bar{\Omega}_j)$.

The vector x of the nodal unknowns in Ω can be split into two parts $x = (\dot{x}, \bar{x})^t$ where $\bar{x}$ represents the unknowns attached to the boundary Γ, and $\dot{x}$ represents those located at the interior nodes or at the nodes on $\partial\Omega$ with a Neumann boundary condition and not belonging to any block interface. The vector $\dot{x}$ can furthermore be decomposed into the n subsets of unknowns corresponding to nodes in Ω_i not belonging to any block interface $\dot{x} = (x_1, ..., x_i, ..., x_n)^t$. These two levels of decomposition induce corresponding decompositions for other vectors like f, r and for the matrix A with the obvious notations:

$$A = \begin{pmatrix} \dot{A} & B^t \\ B & \bar{A} \end{pmatrix} \quad ; \quad f = (\dot{f}, \bar{f})^t \quad ; \quad r = (\dot{r}, \bar{r})^t$$

$$A = \begin{pmatrix} A_1 & & & & B_1^t \\ & \cdot & 0 & & \cdot \\ & & & & \cdot \\ & & & \cdot & \cdot \\ 0 & & & A_n & B_n^t \\ B_1 & \cdot & \cdot & \cdot & B_n \; \bar{A} \end{pmatrix} \quad ; \quad \hat{A}_i = \begin{pmatrix} 0 & & & 0 \\ & & & \\ & A_i & & B_i^t \\ & & & 0 \\ & & & \\ 0 & B_i & 0 & \bar{A}_i \end{pmatrix}$$

$$\hat{f}_i = (0,...,0,f_i,0,...,0,\bar{f}_i) \quad ; \quad \hat{r}_i = (0,...,0,r_i,0,...,0,\bar{r}_i) \; ;$$

$$r_i = f_i - A_i x_i - B_i^t \bar{x} \quad ; \quad \bar{r}_i = \bar{f}_i - B_i x_i - \bar{A}_i \bar{x}$$

$$\sum_i \bar{A}_i = \bar{A} \quad ; \quad \sum_i \bar{f}_i = \bar{f} \quad ; \quad \sum_i \bar{r}_i = \bar{r} \tag{3.11}$$

$$\sum_i \hat{A}_i = A \quad ; \quad \sum_i \hat{f}_i = f \quad ; \quad \hat{r}_i = \hat{f}_i - \hat{A}_i x \quad ; \quad \sum_i \hat{r}_i = r \tag{3.12}$$

The matrix $\hat{A}_i$ corresponds to the restriction to Ω_i of the matrix A with Neumann boundary conditions on $\partial\Omega_i \cap \Gamma$, and $\hat{r}_i$ is the corresponding residual. The last property in (3.12) results directly from the additivity of the integrals over Ω_i in the weak formulation. This means that r (and therefore also $\bar{r}$) can be calculated by summing the contributions from each zone.

We see that in the algorithm (3.7)-(3.10) the only remaining difficulty to overcome for achieving a zonal PCG method is to choose the preconditioning matrix M. When there is a natural node numbering order for the whole grid it induces a natural block numbering (for example in a sliced partition of Ω), the extension of the ICCG (3.7)-

(3.10) is obvious. But our experience with the adaptation of this algorithm to a problem with three blocks not trivially connected was not so satisfactory.

The key idea to go further is to consider a problem defined for the interface node unknowns by eliminating the interior node unknowns. Indeed, from a given $\bar{x}$ we can compute $\dot{x}$ by n independent problems, one on each Ω_i with Dirichlet boundary conditions on $\Gamma \cap \partial\Omega_i$:

$$A_i\ x_i = f_i - B_i^t\ \bar{x} \qquad \text{or} \qquad \dot{A}\ \dot{x} = \dot{f} - B^t \bar{x} \tag{3.13}$$

By substitution we get the system :

$$S\ \bar{x} = \bar{g} \tag{3.14}$$

$$S = \bar{A} - B\ \dot{A}^{-1} B^t \quad ; \quad \bar{g} = \bar{f} - B\ \dot{A}^{-1} \dot{f} \tag{3.15}$$

This is the substructuring approach well-known in solid mechanics. We use the PCG method to solve (3.14) with, in (3.7)-(3.10), A replaced by S , M by $\bar{M}$, and x, r, f, z, q, by $\bar{x}$, $\bar{r}$, $\bar{g}$, $\bar{z}$, $\bar{q}$. In fact, the solution of (3.13) is not truly necessary during the iterations if the residual $\bar{r}^m$ is calculated in an incremental way:

$$\bar{r}^m = \bar{r}^{m-1} - \mu^m\ S\ \bar{q}^m \tag{3.16}$$

Then, only one Dirichlet problem for each zone is needed at each iteration:

$$q_i^{m+1} = -\ A_i^{-1}(\ B_i^t\ \bar{q}^{m+1}) \quad ; \quad S\ \bar{q}^{m+1} = \bar{A}\ \bar{q}^{m+1} + B\ \dot{q}^{m+1} \tag{3.17}$$

The choice of $\bar{M}$ is crucial for a good convergence rate. The first possibility $\bar{M}=\bar{A}$ proposed in [6] is not too bad when the structure of Γ is simple, but it is far from being optimal in general situations.

Following the idea presented in [7] of using one Neumann problem (for two zones) in the definition of $\bar{M}$, we suggested a quite general algorithm [18] which necessitates the solution of a Neumann problem in each zone at each iteration. This method is briefly described below, and more details could be found in [19].

1/ Define S_i the Schur complement of $\bar{A}_i$ in $\hat{A}_i$:

$$S_i = \bar{A}_i - B_i A_i^{-1} B_i^t \tag{3.18}$$

2/ Define D_i as a weighted restriction operator from Γ to $\Gamma \cap \partial\Omega_i$ with the property that $\sum_i D_i\ \bar{x} = \bar{x}$. A simple choice is:

$$D_i = diag\bar{A}_i\ (diag\bar{A})^{-1} \tag{3.19}$$

3/ Define $\bar{M}$ by:

$$\bar{M}^{-1} = \sum_i D_i^t\ S_i^{-1} D_i$$

Then the general PCG iteration giving $\bar{x}^{m+1}$ is :

$$\bar{r}^m = \bar{r}^{m-1} - \mu^m\ S\ \bar{q}^m \tag{3.20}$$

$$\bar{w}_i^m = diag\bar{A}_i\ .\ (diag\bar{A})^{-1}\ \bar{r}^m \tag{3.21}$$

$$\bar{z}_i^m = (S_i)^{-1}\ \bar{w}_i^m \tag{3.22}$$

$$\overline{z}^m = (diag\overline{A})^{-1} \sum_i diag\overline{A}_i \ . \ \overline{z}_i^m \tag{3.23}$$

$$\overline{q}^{m+1} = \overline{z}^m + \lambda^{m+1}\overline{q}^m \tag{3.24}$$

$$\overline{x}^{m+1} = \overline{x}^m + \mu^{m+1}\overline{q}^{m+1} \tag{3.25}$$

The step (3.22) is actually the solution in Ω_i of the Neumann problem :

$$\begin{pmatrix} A_i & B_i^t \\ B_i & \overline{A}_i \end{pmatrix} \begin{pmatrix} z_i^m \\ \overline{z}_i^m \end{pmatrix} = \begin{pmatrix} 0 \\ \overline{w}_i^m \end{pmatrix} \tag{3.26}$$

Therefore, for each zone at each iteration there are one Neumann problem (for the contribution to $\overline{M}$) and one Dirichlet problem (for the contribution to $S\ \overline{q}$) per zone to be solved. These problems are easily solved by the existing ICCG method.

Remarks.

-Other choices for D_i are possible such as $D_i = \overline{A}_i\ \overline{A}^{-1}$. The naive choice $D_i = S_i\ S^{-1}$ gives the solution at the first iteration. Anyway, with (3.19), the convergence rate appears very fast and almost independent of the fineness of the grid. A convergence study for a two-dimensional flow calculation past an airfoil has been made [19].

-Simple modifications must be made to the definition of S_i in case of a pure Neumann problem on Ω_i.

-The idea of using Neumann subproblems in combination with Dirichlet subproblems in a multiblock iterative scheme can be found also in Ecer et al. [20] . However, their very simple method seems based on a rather heuristic approach.

-A complete variational formulation corresponding to our discrete algorithm has been recently settled at the level of the continuous problem by Bourgat et al. [21].

4. Euler Multiblock Solution

4.1. One zone solution algorithm

To simplify the presentation we consider the system of Euler equations in two dimensions:

$$w_t + F_x + G_y = 0 \tag{4. 1}$$

$$w = (\rho, \rho u, \rho v, e)^t \tag{4. 2}$$

$$F = (\rho u, p + \rho u^2, \rho uv, (p + e)u)^t \tag{4. 3}$$

$$G = (\rho v, \rho uv, p + \rho v^2, (p + e)v)^t \tag{4. 4}$$

where ρ is the density, u and v the velocity components, p the pressure and e the total energy per unit volume. The state equation is given for a perfect gas by:

$$p = (\gamma - 1)(e - 0.5\rho(u^2 + v^2)) \tag{4. 5}$$

Considering a finite-volume discretization on a i,j structured grid with conservative variables taken at the nodes of the grid, the time evaluation of the mean value of w on a control volume surrounding the i,j node is :

$$w_{ij}^{n+1} = w_{ij}^{n} - \frac{\Delta t}{area_{i,j}} \int_{\partial\Omega ij} (Fn_x + Gn_y) d\sigma \tag{4.6}$$

The numerical fluxes across the control volume faces can be evaluated explicitly from surrounding values at time t^n giving first or second order of accuracy in (4.6). We consider here only the case of upwind schemes in which the evaluation of the flux takes into account the eigenvalues of A, the Jacobian matrix of the normal flux H:

$$A = \frac{\partial H}{\partial w} = R\Lambda R^{-1} \text{ and } H = (Fn_x + Gn_y)$$

where the columns of R are the eigenvectors of A and Λ is the diagonal matrix of the eigenvalues of A, the matrices $A^{\pm}$ are defined by $R\Lambda^{\pm}R^{-1}$ with $\Lambda^{\pm} = diag[1/2(\lambda_i \pm \Psi(\lambda_i))]$ where $\Psi(\lambda)$ results from an entropic correction of $|\lambda|$ [22]. An implicit first order scheme can be obtained by linearization of the normal flux at time t^{n+1} in (4.6). To maintain upwind evaluation one can take:

$$H_{i+1/2,j}^{n+1} = H_{i+1/2,j}^{n} + A_{i+1/2,j}^{+\;n} \Delta w_{i,j} + A_{i+1/2,j}^{-\;n} \Delta w_{i+1,j} \tag{4.7}$$

$$\Delta w_{i,j} = w_{i,j}^{n+1} - w_{i,j}^{n} \tag{4.8}$$

This leads to a linear system which is too expensive to solve and therefore is replaced by an approximate factorization of ADI type. The resulting implicit non-iterative scheme takes the following general form:

$$C^{\xi}\Delta\hat{w}_{i-1,j} + D^{\xi}\Delta\hat{w}_{i,j} + E^{\xi}\Delta\hat{w}_{i+1,j} = \Delta\overline{\overline{w}}_{i,j} \tag{4.9}$$

$$C^{\eta}\Delta w_{i-1,j} + D^{\eta}\Delta w_{i,j} + E^{\eta}\Delta w_{i+1,j} = \Delta\hat{w}_{i,j} \tag{4.10}$$

$$w_{ij}^{n+1} = w_{ij}^{n} + \Delta w_{ij} \tag{4.11}$$

where the matrices C^{ξ}, D^{ξ}, E^{ξ} and C^{η}, D^{η}, E^{η} are 4x4 block matrices attached to the node i,j and $\Delta\overline{\overline{w}}_{i,j}$ is the residual of a 2nd-order accurate explicit scheme. Before describing a patch-boundary algorithm, we consider first the treatment of the boundary conditions [16] at a left boundary (i=1). In (4.9), $\Delta\hat{w}_{0,j}$ is not known, we use, however, characteristic relations to set:

$$(R^{-1}\Delta\hat{w}_{0,j})_k = \begin{cases} (R^{-1}\Delta\hat{w}_{1,j})_k & \text{for } \lambda_k \leq 0 \\ 0 & \text{for } \lambda_k > 0 \end{cases} \tag{4.12}$$

this can be written in a compact form :

$$\Delta\hat{w}_{0,j} = RS^{-}R^{-1}\Delta\hat{w}_{1\,j} \tag{4.13}$$

where the kth entry of the diagonal matrix S^{-} is 1 if $\lambda_k \leq 0$ and 0 if not. The equation for $\Delta\hat{w}_{1,j}$ becomes :

$$(D^{\xi} + C^{\xi}RS^{-}R^{-1})\Delta\hat{w}_{1,j} + E^{\xi}\Delta\hat{w}_{2,j} = \Delta\overline{\overline{w}}_{1,j} \tag{4.14}$$

The same kind of modification is made for the other boundary points. After the ADI sweeps, a final correction is made in an explicit way according to the characteristic boundary conditions as proposed by Viviand and Veuillot [23]. Typically :

$$\overline{w}_{1,j}^{n+1} = RS^{-}R^{-1}w_{1,j}^{n+1} + RS^{+}R^{-1}w_{out}^{n+1} \tag{4.15}$$

where w_{out}^{n+1} is the prescribed outer value for w at time n+1.

4.2 Patch Boundary Algorithm

A patching technique has been derived directly from an extension of the boundary treatment with the explicit and implicit steps described above.

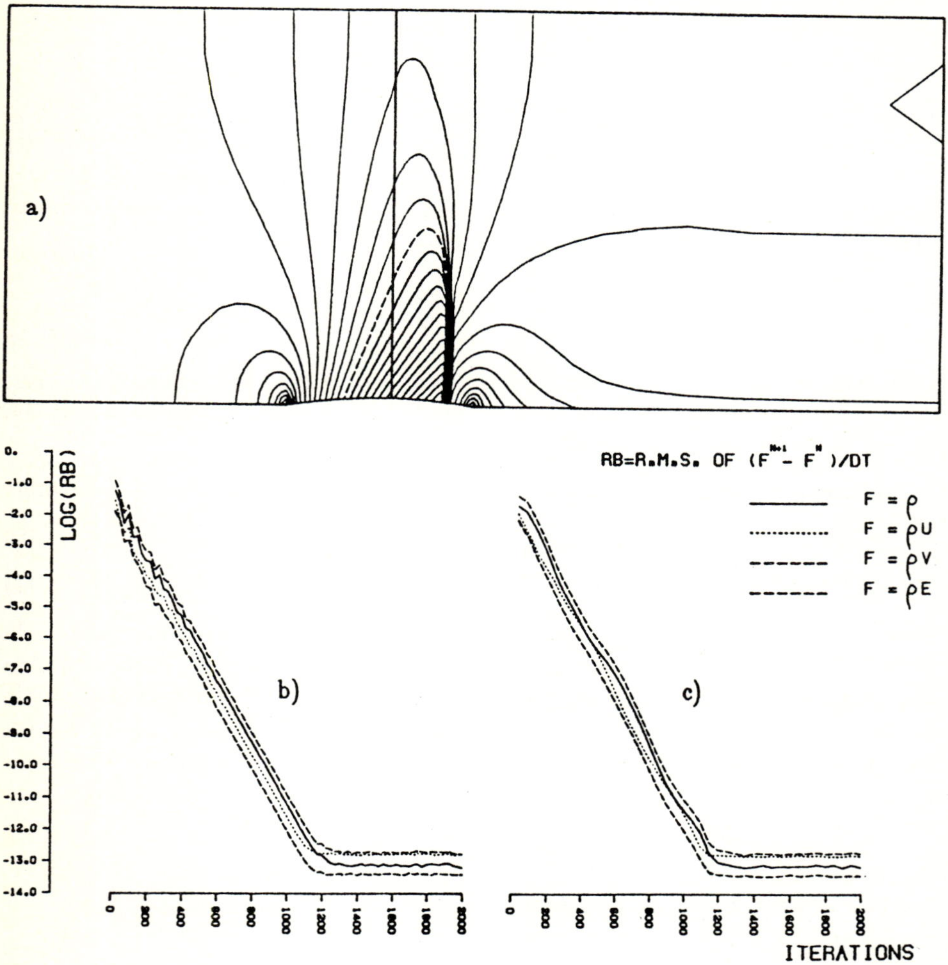

Fig.1 Transonic flow in a channel, $M_\infty = .85$, with a 72x21 grid. a) Mach number contours at convergence for a two-zone calculation. b) Convergence history for the two-zone case. c) Convergence history for the one-zone case.

Assuming, for a simplified presentation, that a one block grid is partitioned into two zones along the interface Γ: $i = I$. The proposed zonal method is summarized below for one time step [16]:

-1. Calculate $\Delta\overline{\overline{w}}^{1}$ and $\Delta\overline{\overline{w}}^{2}$ by an explicit second-order scheme for zone 1 and zone 2 with a zero-flux condition along Γ so that $\Delta\overline{\overline{w}}_{I,j}=\Delta\overline{\overline{w}}^{1}_{I,j}+\Delta\overline{\overline{w}}^{2}_{I,j}$ ensures conservativity for the explicit part of the scheme.

-2. Apply the first ADI sweep independently in zone 1 and zone 2 according to (4.9) with the following interface treatment:

$$C^{\xi}\Delta\hat{w}^{1}_{I-1,j}+(D^{\xi}+E^{\xi}RS^{+}R^{-1})\Delta\hat{w}^{1}_{I,j}=\Delta\overline{\overline{w}}_{I,j} \quad \textit{in zone } 1 \tag{4.16}$$

$$(D^{\xi}+C^{\xi}RS^{-}R^{-1})\Delta\hat{w}^{2}_{I,j}+E^{\xi}\Delta\hat{w}^{2}_{I+1,j}=\Delta\overline{\overline{w}}_{I,j} \quad \textit{in zone } 2 \tag{4.17}$$

After the second ADI sweep in the η direction where Δw^{1} and Δw^{2} are calculated independently according to (4.10), two values $w^{1}_{I,j}$ and $w^{2}_{I,j}$ are available at each interface node .

-3. Apply on the interface the characteristic relations in the form :

$$\overline{w}^{n+1}_{I,j}=RS^{+}R^{-1}w^{1,n+1}_{I,j}+RS^{-}R^{-1}w^{2,n+1}_{I,j} \tag{4.18}$$

Numerical tests have been carried out for the GAMM-Workshop channel flow test case[24] with one or two zone calculations with CFL 20. In case of two zones, the interface is located at the middle of the channel joining the upper and lower walls and crossing supersonic and subsonic parts of the flow. The converged solution is not perturbed by the interface and one can see in figure 1 that the convergence rates are quite similar giving same computation times. The requirement of " transparency " for the patch boundary algorithm is clearly fulfilled. More details and other results can be found in [16]. Although not fully implicit, the proposed method has some links with the one proposed by Lombard et al. [15] and the one given by Rai [14] which is perhaps less symmetric.

5. Coupling of potential and Euler solvers

One of the advantages of the zonal approach is the possibility of mixing different models. We present here the case of a supersonic propulsive jet in a transonic outer flow. For the inviscid case chosen here, Euler equations are mandatory in the jet whereas the full potential approximation is sufficient for the external flow which is very efficiently calculated by the ICCG finite element method mentioned above.

The stream surface bounding the jet is a flow discontinuity which must be fitted. This problem has been solved in the axisymmetric case with the outer flow computed in a stream function finite element method [25] and this technique has been extended in 3D along the same ideas.

The coupling technique is based on the following simple fix point iteration: by solving the potential equation for a given shape S^{n} of the jet, a pressure p^{n} is obtained. The jet flow is calculated by a pseudo-unsteady Euler method in inverse mode with the pressure p^{n} prescribed on the jet boundary which is moved in order to verify the slip condition. The new shape S^{n+1} is compared to S^{n} and, if needed, a new iteration is done.

In this iterative process, the potential flow is calculated in the direct mode around the nacelle and the jet. The only difficulty is to find "correct far-field" boundary conditions. This was done according to a somewhat heuristic superposition principle of elementary potential solutions: a uniform flow with an angle of attack α, a source and a sink singularity on the nacelle axis and transverse flows around the nacelle and jet inflow and outflow sections.

The inverse mode method for solving the 3D Euler equations has been worked out by Couaillier and Veuillot [26], and all the details on the numerical technique can be found there with an application to an isobaric jet. Only the main features of the method are given below. This Euler solver is based on the pseudo-unsteady system of equations for conservation of mass and momentum while the energy equation is replaced by the Bernoulli relation. This so-called H-system is integrated in time with an explicit Lax-Wendroff type scheme proposed by Ni. All the boundary conditions are treated by the characteristic relations [23]. The two distinctive features of the method are: firstly, that the pressure is prescribed and the normal fluid velocity is calculated during the time integration and secondly, that at each time step a new location of the surface Σ is computed and a new mesh is generated. The three characteristic relations associated with a transport of information from inside the jet or along the boundary permit to calculate the fluid velocity for the given pressure. The surface is then moved according to the normal fluid velocity. Numerical results are presented in figure 2 for a preliminary test case with the exhaust flow of a supersonic jet into an outer flow at a freestream mach number value of .8. The total pressure ratio is 5, giving a strong expansion at the nozzle exit.

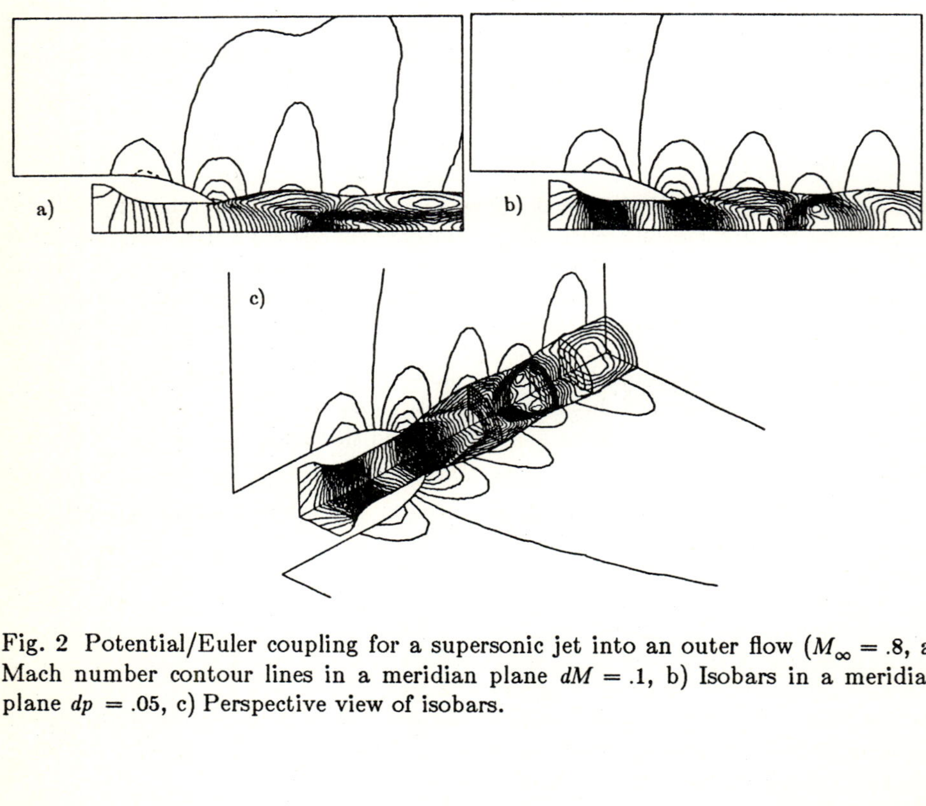

Fig. 2 Potential/Euler coupling for a supersonic jet into an outer flow ($M_\infty = .8$, a) Mach number contour lines in a meridian plane $dM = .1$, b) Isobars in a meridian plane $dp = .05$, c) Perspective view of isobars.

6. Conclusion

Zonal techniques have been presented for transonic inviscid steady flow calculations both with the full potential and the Euler model. In both cases, the methods apply to continuously patched (multiblocked) grids and have been devised in order to maintain the accuracy and the efficiency of the parent one-grid methods. In addition, a coupling technique between a potential and an Euler solver was illustrated on a problem of jet surface fitting and this application opens the way to the study of complex wing nacelle interaction.

Acknowledgments

The author wishes to thank his colleagues at the theoretical aerodynamics division 1, and particularly M. Borrel, V. Couaillier, Th. Dinand, Ph. Guillen and J. Maréchal for stimulating discussions and for their cooperation in the preparation of this paper.

References

[1] Schwarz H.A.: Gesammelte Mathematische Abhandlungen, Vol 2, 1860.
[2] Atta E.H.: Component adaptive Grid Interfacing. AIAA Paper 81-0382, January 1981.
[3] Atta E.H. and Vadyak J.: A Grid Interfacing Algorithm for Three-Dimensional Transonic flow Simulation for Complex Configurations. AIAA Paper 82-1017, June 1982.
[4] Yu N.J.: Transonic Flow Simulation for Complex Configurations with Surface Fitted Grids. AIAA Paper 81-1258, 1981.
[5] Le T.H.: Calcul d'écoulements transsoniques autour d'une configuration aile-fuselage dans une approche par domaines. La Recherche Aérospatiale, vol 1983-3, June 1983.
[6] Morice Ph.: Calcul parallelle et décomposition dans la résolution d'équations aux dérivées partielles de type elliptique. IRIA Report no 7201, June 1972.
[7] Bjorstad P.E. and Widlund O.: Iterative Methods for Elliptic Problems on Regions Partitioned into Substructures and the Biharmonic Dirichlet Problem. Conf. on Computing Methods in Applied Science and Engineering, Versailles 1983. Glowinski R. and Lions J.L. Eds, North-Holland Publ. Comp. 1984.
[8] Cambier L., Ghazzi W., Veuillot J.P. and Viviand H.: Une approche par domaines pour le calcul d'écoulements compressibles , in Computing Methods in Applied Science and Engineering, Versailles 1981. Glowinski R. and Lions J.L. Eds., North-Holland Publ. Company 1982. Also ONERA TP no 1981-143.
[9] Veuillot J.P. and Cambier L.: A Subdomain Approach for the Computation of Compressible Inviscid Flows. INRIA Workshop on Numerical Methods for the Euler Equations for Compressible Inviscid Flows. Rocquencourt, December 1983, Angrand F. et al. Eds, SIAM 1984. Also ONERA TP 1984-61.
[10] Cambier L., Dusson F. and Veuillot J.P.: Multidomain Method for the Euler Equations. Application to Overlapping Subdomains. La Recherche Aérospatiale no 1985-3. English edition 1985.
[11] Hessenius K.A. and Pulliam T.H.: A Zonal Approach to Solution of the Euler Equations. AIAA Paper 82-0969, January 1982.
[12] Rai M.M., Chakravarthy S.R. and Hessenius K.A.: Metric-Discontinuous Zonal Grid Calculations Using the Osher Scheme. International Journal of Computers and Fluids, vol 12, no 3, pp 161-175, 1984.

[13] Rai M.M.: A Conservative Treatment of Zonal Boundaries for the Euler Equation Calculations. AIAA Paper 84-0164, January 1984.

[14] Rai M.M.: An Implicit Conservative Zonal Boundary Scheme for Euler Equation Calculations. AIAA Paper 85-0488 January 1985.

[15] Lombard C.K. and Venkatapathy E.: Implicit boundary Treatment for Joined and Disjoint Patched Mesh Systems. AIAA Paper 85-1503 CP, July 1985.

[16] Borrel M.: Characteristic Boundary Conditions for Implicit Solution to the Euler Equations. La Recherche Aérospatiale, (in preparation).

[17] Bredif M.: Finite Element Calculation of Potential Flow around Wings. 9th ICNMFD, Saclay, June 1984, Springer Verlag 1985.

[18] Morice Ph. and Dinand Th.: Some Conjugate Gradient Methods for Elliptic Problems with Domain Decomposition. Presentation at the 1rst Internat. Sympos. on Domain Decomposition Methods for Partial Differential Equations. Paris January 1987.

[19] Dinand Th.: Domain Decomposition methods for solving elliptic problems. Application to the numerical simulation of transonic potential flows. Thesis in preparation.

[20] Ecer A. and Spyropoulos J.T.: Block-Structured Solution Scheme for Analysing Three-Dimensional Transonic Potential Flows. AIAA Paper 86-0510, Reno January 1986.

[21] Bourgat J.F., Glowinski R., Le Tallec P. and Vidrascu M.: Variational Formulation and Algorithm for Trace Operator in Domain Decomposition Calculations, 2nd Internat. Sympos. on Domain Decomposition Methods, UCLA Los Angeles, January 1988.

[22] Harten A.: High Resolution Schemes for Hyperbolic Conservation Laws. J. of Comp. Physics, vol 49, pp 357-393, 1983.

[23] Viviand H. and Veuillot J.P.: Méthodes pseudo-instationnaires pour le calcul d'écoulements transsoniques. ONERA Publication no 1978-4. English Version ESA TT 549, 1978.

[24] Rizzi A.and Viviand H.,Eds: Numerical Methods for the Computation of Inviscid Transonic Flows with Shock Waves. Notes on Numerical Fluid Mechanics, vol 3, Friedr. Vieweg & Sohn, 1981.

[25] Morice Ph., Cambier L. and Veuillot J.P.: Numerical Computation of Transonic Flow Past an Axisymmetric Nacelle. 14th ICAS Congress, Toulouse, September 1984. Also ONERA TP no 1984-91.

[26] Couaillier V. and Veuillot J.P.: Inverse Method for the Euler Equations. ICIDES II Pensylvania, October 1987. Also ONERA TP 1987-167.

On the Development and Application of Software based on the Framework of Cartesian Coordinates

S.S. Desai*, R. Rangarajan, U.N. Sinha,
K.S. Ravichandran, J.P. Singh

Aerodynamics Devision
National Aeronautical Laboratory
Bangalore, INDIA
*(on deputation: Aeronautical Development Agency,
Bangalore, INDIA)

Summary

The use of Cartesian grids in the development of aerodynamic application software underlies the work presented in this paper. Motivated by the challenge of having to manage with affordable computers of limited capacity of storage and speed, a concerted effort over the years with Cartesian grids has made it possible to develop useful application software to deal with transonic flow over aerofoils in design and analysis mode, including separated flow at low speeds, and to analyse axisymmetric flows as well as flow over wings and wing-body combinations. Several results are presented here for each of the above flow situations which clearly demonstrate the efficacy of the Cartesian grids for dealing with complex geometry shapes while simulating transonic flow in a full-potential formulation.

Introduction

It is well known that the conservative laws of mathematical physics take on their simplest form when cast in a Cartesian framework, except when symmetries exist in the problem, such as cylindrical or spherical. But the main difficulty to be faced in numerical work with Cartesian grids is in the implementation of the surface boundary conditions: flow variables are computed at grid points that are not necessarily the surface points; approximations in the neighbourhood of the surface, where invariably gradients in the flow variables are large, lead to failure of numerical computations. It has therefore become a very common practice to work with body fitted coordinates. Indeed, spurred by the advent of high speed and super computers with large memory, the past one decade has witnessed impressive growth in grid generation around complex aerodynamic configurations and has also witnessed many fine features of flow when this grid generation was used in conjunction with sophisticated flow solvers. However, in the area of applications, industries continue to experience many difficulties while using these developments. Firstly, the codes using sophisticated grid generation require more computer memory, because of the need to store the Jacobians and the metric coefficients of the transformations. Secondly, grid generation is not quite automatic, except for the 2-D case, and requires time and human interaction and expertise while handling complex shapes. Finally, for practical shapes encountered in engineering design and analysis, it may perhaps be impossible to find a transformation which

J. Zierep, H. Oertel (Eds.)
Symposium Transsonicum III
IUTAM Symposium Göttingen, 24.-27. 5. 1988

is good everywhere; it is impossible to escape several zones where the Jacobians can misbehave, which calls for a careful handling to limit the growth of the inaccuracies.

On the contrary, formulation with Cartesian grids requires a careful dealing with the inaccuracies at the boundaries, which the grid-generation methodology totally avoids. However, it has been our experience [1-8] that at a little price of extra analysis and attention and coding effort at the boundary points, a numerical simulation of equations of flow in a framework of Cartesian coordinates holds a promise for the development of cost-effective industry codes, with accuracies in no way inferior to those employing complex grid generation procedures. The pioneering work in this direction was done by Carlson [9] who used stretched Cartesian coordinates for the numerical solution, for the analysis and the design of 2-D transonic flow. Presently, one comes across interest reappearing in the use of Cartesian grids. Indeed looking to the number of papers on the use of Cartesian grids, which is slowly increasing, one perhaps cannot but agree with the suggestion of South [10] that "rectangular grids may provide the best alternative for dealing with flow over complex configurations".

The following presentation covers a brief description of the formulation, mainly following a potential approach, for dealing with numerical flow simulation in transonic regime. Many results are presented along with comparisons with available experimental results to bring out the usefulness of the approach.

Formulation of the problem:

The thrust on industry application of CFD in design and analysis of aerospace vehicles calls for the use of appropriate algorithms and numerical implementations that can deal numerically with a variety of approximations of the basic laws governing fluid flow, rather than flow simulation in its entirety through one basic flow model such as the Navier-Stokes equations; industry applications can only afford software codes that are fast (in repetitive applications over several design cycles), accurate, cost-effective and are, importantly, amenable to smaller computers. In this regard, a potential formulation is currently more popular, next to the panel methods. Though it looks inevitable that this scene should change in the very near future with Euler and Navier-Stokes solvers graduating to industry-standards, still there are situations where full-potential formulation should be more than adequate; it is not often the apparent completeness of the model that is important as it is the proper numerical implementation, tuned by extensive validations. Fig. 1 is reproduced here from Holst [11] to illustrate this point of view. It appears that excepting the situations where vortex dominance exists or where strong shocks, rotationality and shock-boundary layer interactions exist, a potential formulation is amply adequate.

Governing Equations in Stretched Cartesian Grids

The 3D full potential equation governing inviscid transonic flows with weak shock is given in terms of a perturbation potential ϕ , in a stretched Cartesian grid $\xi = \xi(x)$, $\eta = \eta(y)$, $\zeta = \zeta(z)$, in the form:

$$(\tilde{a}^2 - U^2) f (f\phi_\xi)_\xi + (\tilde{a}^2 - V^2) g (g\phi_\eta)_\eta + (\tilde{a}^2 - W^2) h (h\phi_\zeta)_\zeta$$
$$- 2\, UVfg\,\phi_{\xi\eta} - 2\, VWgh\,\phi_{\eta\xi} \quad 2\, WUfh\phi_{\xi\zeta} + \frac{\delta h \phi_\zeta}{z} = 0 \qquad (1)$$

where $\Phi = M_\infty(x \cos\alpha + z \sin\alpha + \phi)$, $\tilde{a} = \frac{a}{a} = \left(1 - \frac{\gamma-1}{2}(U^2+V^2+W^2-M_\infty^2)\right)^{\frac{1}{2}}$

$$f = \frac{d\xi}{dx}, \quad g = \frac{d\eta}{dy}, \quad h = \frac{d\zeta}{dz} \qquad U = M_\infty(\cos\alpha + f\phi_\xi), \text{ etc}$$

Equation (1) applies to 2D and axisymmetric flows when all the terms involving η -derivatives are dropped. Further, $\delta = 0$ for 2D and 3D flows and $\delta = 1$ and $\alpha = 0$ for axisymmetric case. Though it is generally believed that it is preferable to work with a conservative form of the governing equations from the point of view of ensuring accurate jump conditions across captured shocks, the present work is done with a non-conservative form as seen in Eq.(1). The past years have seen a considerable debate as to the proper form of the equations to be used for obtaining weak solutions containing discontinuity surfaces. The results of GAMM workshop [12] provide an interesting background in this regard; the non-conservative schemes produced results that had much less scatter from code to code than with the conservative equations and the results were closer to the Euler equations, though having slightly larger scatter than that seen for Euler Codes. Fig. 2 highlights a comparison of inviscid full-potential results from one of the codes developed in this study with two cases of those from published Euler results for flow over aerofoils. The closeness between the two results in the matter of shock position and its strength is particularly noteworthy. We have therefore continued to use the non-conservative form of the differential equations.

Stretched Cartesian Grid

The infinite physical space (x,y,z) is transformed into a finite computational domain (ξ , η, ζ) by a set of three mutually independent stretchings given below (see Fig 3). This mapping has the advantage of simplifying the application of far-field boundary conditions. It can also be arranged to cluster the grids near the body surface and around leading and trailing edges. The mutual independence of stretching functions results in storage economy in a computer.

x- ξ stretching: The region $-\infty < x < \infty$ is divided into three regions(Carlson[9]):

$$\infty < x \leq x_1 \quad : \quad x = x_1 + A_2 \tan(\pi/2 \,.\, \Delta\xi) + A_3 \tan(\pi/2.. \,\Delta\xi^3) \; ; \Delta\xi = \xi - \xi_1$$

$$x_1 \leq x \leq x_2 \quad : \quad x = a\,\xi + b\,\xi^3$$

$$x_2 \leq x < \infty \quad : \quad x = x_2 + A_2 \tan(\pi/2 \,.\, \Delta\xi) + A_3 \tan(\pi/2 \,.\, \Delta\xi^3) \; ; \Delta\xi = \xi - \xi_2$$

x_1 is located close to the leading edge, typically at 1% of the chordline for aerofoils and wings. $x_2=-x_1$ for 2D and axisymmetric cases. The parameters ξ_1 and ξ_2 are subject to the requirement that (a) the number of x-grid lines over the aerofoil or over the extent of the wing is equal to a specified fraction (typically 2/3) of the total x-grid lines, and (b) that the first grid point just behind the leading edge for a given wing planform be at a specified constant percent of the local chord, typically 1%. The latter requirement is easily met for straight wings but requires some attention for swept wings. In this case there will be an additional region which covers the leading edge portion of the wing, over which x=const. lines are placed at uniform intervals based on the sweep of the leading edge, Ravichandran et al. [13]. Planforms with kinks in the leading edge, typical of combat configurations, do pose problems, but it has been possible to get over these difficulties with slight modifications to the geometry in the leading edge portion. For 2D and axisymmetric cases $\xi_1 = \xi_2$. The parameters a and b are determined in all the cases by the requirement of continuity of the mapping function and its first derivative at the grid transition points x_1 and x_2. A_1 and A_3 are the free parameters set to A_1=0.15 A_3=3.87, Carlson [9]. For 3D cases b is set to zero consistent with (b) above.

z- ζ stretching $z = A_1 \tan(\pi/2\,\zeta) + A_4 \tan(\pi/2\,\zeta^3)$, where A_1 and A_4 determine the spacing of the z-grid lines near z=0 and infinity, respectively. For cases with shocks extending into the far field, we need to use A_4 between 1 and 2; otherwise, A_4 is generally set to zero. A_1 is chosen such that there is exactly one grid point between the contour and the z=0 line on the first x=constant at the nose portion. Generally, A_1=0.246 for 2D and 3D cases; for axisymmetric cases A_1 has been varied between 0.08 and 0.351, Singh [8].

y- η stretching (for 3D only): Here, $y = c\,\eta$, $0 \leq y \leq y_{tip}$ and $y = y_{tip} + C \tanh^{-1}(\eta - \eta_{tip})$, $y_{tip} \leq y < \infty$. This transformation maps $0 \leq y < \infty$ to $0 \leq \eta \leq 1 + \eta_{tip}$. η_{tip}, corresponding to y_{tip}, is determined by the fraction of total number of grid lines required to fall on the wing. C is determined by the requirement of continuity of the stretching across the $y = y_{tip}$ line.

Boundary Conditions:

Far-field boundary conditions are applied at true physical infinity and are as shown in Fig.4. The surface boundary conditions for the cases of analysis and design are summarized in Fig.5. In a Cartesian system the surface boundary conditions can only be satisfied by using a Taylor series expansion, since the body contour does not necessarily pass through the grid points (Fig.6). This is often the source of the difficulties in the Cartesian formulation; a simple reflection of the field values across the boundary into the "dummy" values inside the body, available to a body fitted orthogonal system, is no longer valid. In order to deal with the case when the first field point outside the body happens to be a supersonic point, we need two dummy values to be defined for a rotated difference scheme. While the statement of vanishing normal component of velocity in conjunction with Taylor series expansion can determine a value for the point 1, it is not obvious as to how to prescribe a value of the potential at the dummy point 2. A linear extrapolation makes the solution blow up in a very few iterations, at free stream Mach number and incidences for which KA happens to be a supersonic point. The reason for this instability is that the above linear extrapolation incorrectly restricts $\phi_{\zeta\zeta}$ to vanish at the first field point, when this derivative is discretized by a backward difference scheme as required by the rotated difference scheme. In the 2D and 3D codes we avoid the instability by a (compensative) explicit addition of the second derivative $\phi_{\zeta\zeta}$, now evaluated at KA through a centred differencing $\Delta\zeta^2 \phi_{\zeta\zeta KA} = \phi_{KA+1} - 2\phi_{KA} + \phi_1$. In the axisymmetric code the extrapolation $\phi_2 = 2.5\,\phi_1 - 2\,\phi_{KA} + 0.5\,\phi_{KA+1}$ has given better results; this extrapolation does not result in the vanishing of the second derivative.

Rotated Type-Differencing and Successive Over-relaxation

The discretization of the governing equation and the solution of the set of simultaneous equations through successive line over-relaxation follow the standard prescription used generally in transonic flow computations. These details have now become commonplace and are omitted here for reasons of space (see [1-5]).

Viscous Corrections:

A transpiration model is found convenient in that it modifies only the body boundary condition by the requirement of a normal velocity proportional to the growth of the boundary layer displacement surface; the body-grid intersections are not changed during viscous corrections. After allowing a reasonable evolution of the flow field from the initial uniform free stream condition, boundary layer computations are initiated and are interlaced with the potential computations after every ten inviscid interactions. The boundary layer and wake parameters for a given potential pressure distribution are obtained following the method of Green. A fourth-order Runge-Kutta scheme is used for integrating the boundary layer and entrainment equations (Desai and Kiske [14])

Results and Discussions

Figs.7-10 show the typical results obtained for the 2-D case. The viscous analysis results shown in Figs.7 & 8 compare well with the experiments for the two cases of standard aerofoils, viz. the NACA0012 and the RAE 2822 for subcritical and supercritical Mach numbers. Comparison for drag and for the boundary layer parameters is seen to be good. Fig.9 shows a good comparison with experiments for the low speed separated case, where the region of the flow field downstream of the separation point, detected in the boundary layer code, is computed in a design mode for a prescribed constant C_p downstream of the separation point. One case of the use of the code in a purely inverse mode for design of aerofoils is shown in Fig.10, where for the sake of illustration the contour of the NACA0012 is changed to support a given pressure distribution; this modified aerofoil is changed once more to recover the NACA0012 profile, for a consistency check of the design process. Aerofoils with or without shocks can be designed with this code for transonic operations.

Figs.11-12 give comparisons with experiments for the axisymmetric case. The axisymmetric code has been extensively used for an ongoing aircraft project in the evaluation of wave drag. As can be seen, wave drag variation through the transonic regime can be well predicted when viscous effects are taken into consideration.

Figs.13-17 summarize the capabilities of the Cartesian formulation in the transonic analysis of wings and wing-body combinations. A reasonable comparison with experiments is seen in all the cases computed.

It is to be noted that all these computations were carried out on a mainframe (UNIVAC 1100/70) with an estimated sustained rate of 0.3MFLOPS and a core of 256 kwords. For aerofoils and axisymmetric analysis typical grids were 49 x 25 reached through two grid refinements. For the 3-D cases typical fine grids were 71x14x25. Typical run times were 2-4 min of CPU for 2-D and axisymmetric case and 90 min of CPU for the 3-D case. These codes are now in general use through a software library, SOFFTS, with the National Aeronautical Laboratory.

Conclusions

Our experience so far has shown that efficient and accurate codes can be developed in the framework of Cartesian grids to deal with a variety of flows such as over aerofoils, axisymmetric bodies, wings and over wing-body combinations or application in developmental work in industries.

Acknowledgements:

We acknowledge with gratitude the constant support for this work from Prof. R. Narasimha, Director, and Mr M. Shivakumara Swamy, Head, Aerodynamics Division, National Aeronautical Laboratory, and the many opportunities for application of the software on an ongoing project by Dr T. S. Prahlad, Project Director, Technology Development, Aeronautical Development Agency, Bangalore. The credit for bringing the manuscript to the final form goes to Mr C. Sathiyaprakash, Aerodynamics Division and to the Drawing Office.

References

1. Desai, S.S.; Rangarajan, R.; Singh, J.P.; Ravichandran, K.S.: Full-potential flow computations using Cartesian grids. AIAA 8th Computational Fluid Dynamics Conference. Paper 87-1164-CP, Honolulu, 9-11 June 1987.

2. Rangarajan, R.; Desai, S.S.; Ramaswamy, M.A.: 2-D Transonic analysis and design program in Cartesian coordinates (TRADE-1). NAL AE-TM-6-83, Aug 1983.

3. Rangarajan, R.: Analysis and design of supercritical aerofoils. M.Sc.(Engg.) Thesis. Indian Institute of Science, Bangalore 1987.

4. Desai, S.S.; Rangarajan, R.: Prediction of viscous effects on an aerofoil using transpiration model for the effects of boundary layer and wake. NAL TM AE 8609, 1986.

5. Desai, S.S.; Rangarajan, R.: Viscous transonic flow over aerofoils using transonic full potential equation. AIAA 25th Aerosciences Meeting. Paper 87-0411. Nevada. 12-15 Jan 1987.

6. Singh, J.P.; Rangarajan, R.; Ramaswamy, M.A.: Computation of flow over aerofoil under high lift separated flow conditions. NAL PD AE 8531. Dec.1985.

7. Sinha, U.N.; Desai, S.S.: Calculation of wave drag at transonic speeds for bodies of revolution at zero incidence. NAL FM TM 1-83. Dec.1983.

8. Singh, J.P.: Inviscid axisymmetric flow computation in Cartesian coordinates. NAL AE TM 1084. Aug.

9. Carlson, L. A.: Transonic analysis and design using Cartesian coordinates, J.Aircraft, Vol.13, No.5, pp 342-356, 1976.

10. South, J.C.: Advances in computation transonics, Vol.4, 1985.

11. Holst, T.L.: Viscous transonic aerofoil workshop. Compendium of results. AIAA-87-1460, AIAA 19th Fluid Dynamics Conference, Honolulu, June 8-10, 1987.

12. Viviand, H.; Rizzi, A.(Eds.): Numerical methods for computation of inviscid transonic flow with shock waves; GAMM Workshop, 1981.

13. Ravichandran, K.S; Desai, S.S; Rangarajan, R.: Modification to the Cartesian grid in 3-D potential code for flow past swept wings, PD AE 8615, NAL, June 1986.

14. Desai, S.S.; Kiske, S.: A computer program to calculate turbulent boundary layers and wakes in compressible flows with arbitrary pressure gradient based on Green's Lag entrainment method. Bericht Nr.89/1982. Inst. Thermo-u Fluiddynamik, Ruhr University, Bochum, 1982.

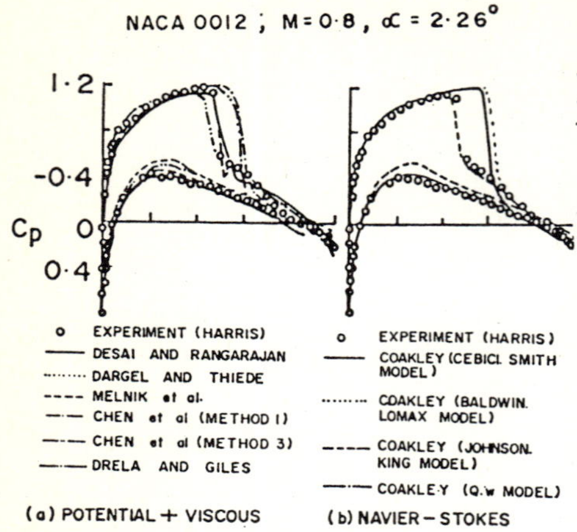

Fig. 1 Spread in the results from different analysis methods (Ref. 11)

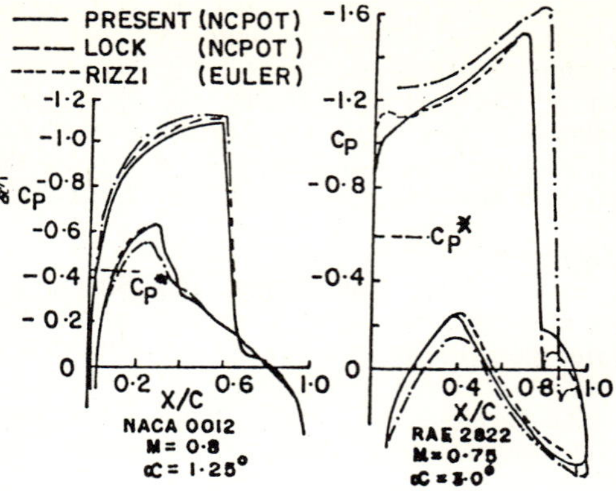

Fig. 2 Inviscid Aerofoil Results

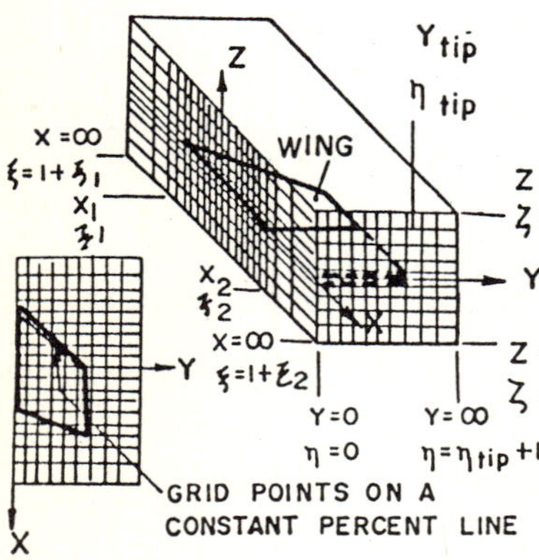

Fig. 3 Stretched Cartesian Grid

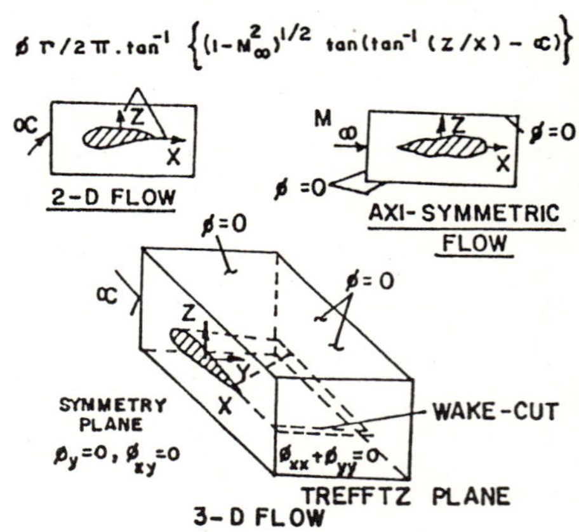

Fig. 4 Far-field Boundary conditions

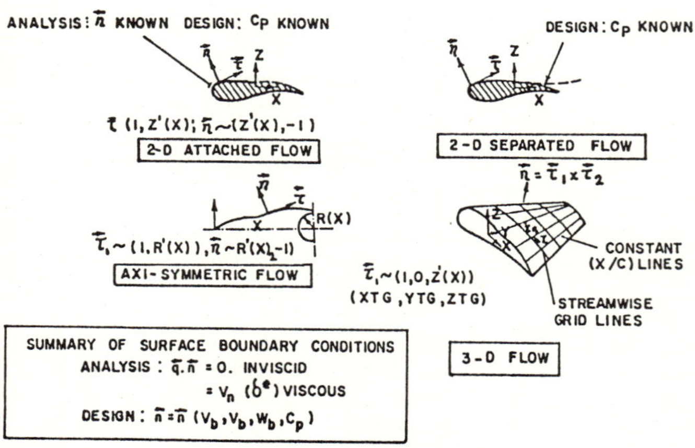

Fig. 5 Surface Boundary Conditions

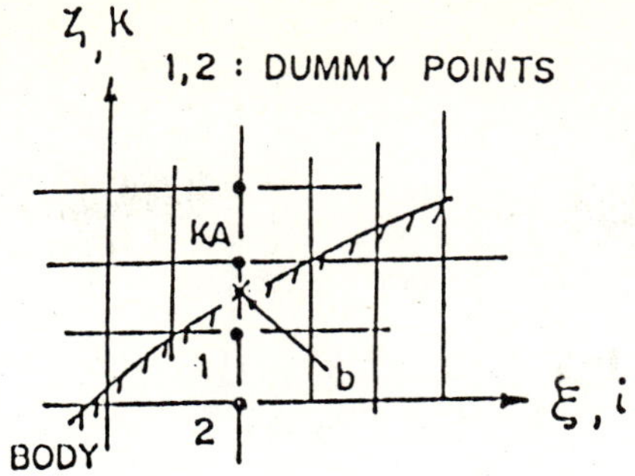

Fig. 6 The Dummy Points

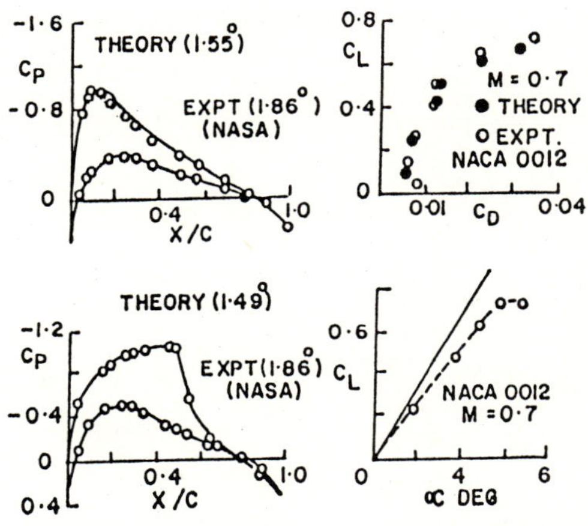

Fig. 7 Viscous Aerofoil Results: NACA 0012

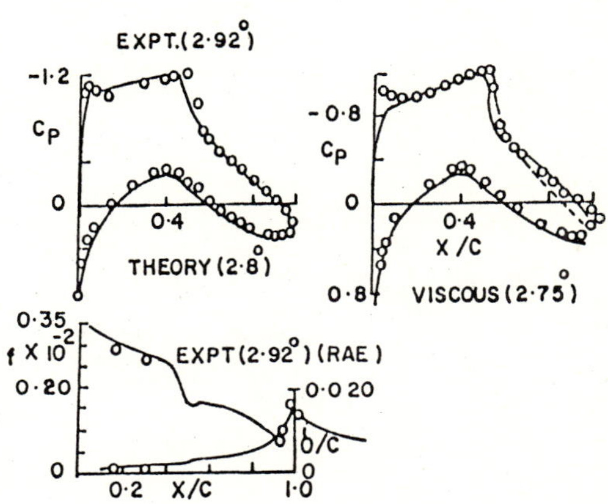

Fig. 8 Viscous Aerofoil Results: RAE 2822

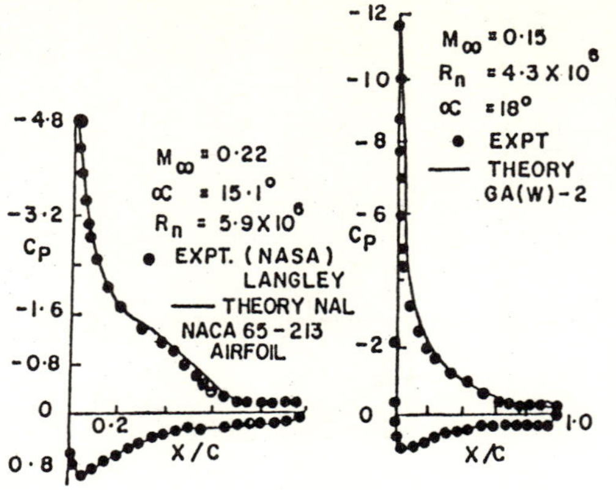

Fig. 9 Low-speed Separated Flow on Aerofoils

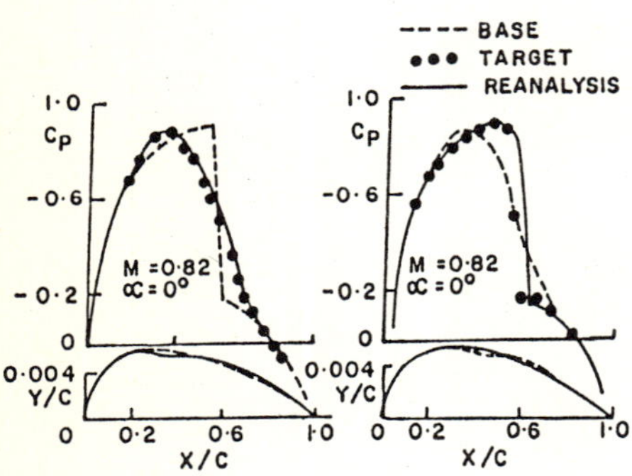

Fig. 10 Aerofoil Design

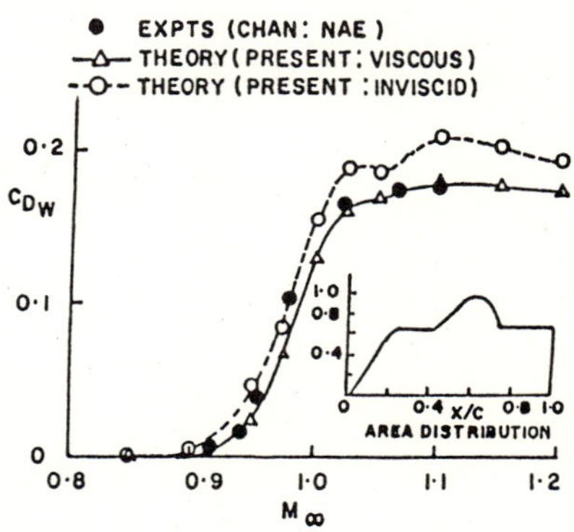

Fig. 11 Wave Drag for an Axisymmetric Body From Transonic Analysis

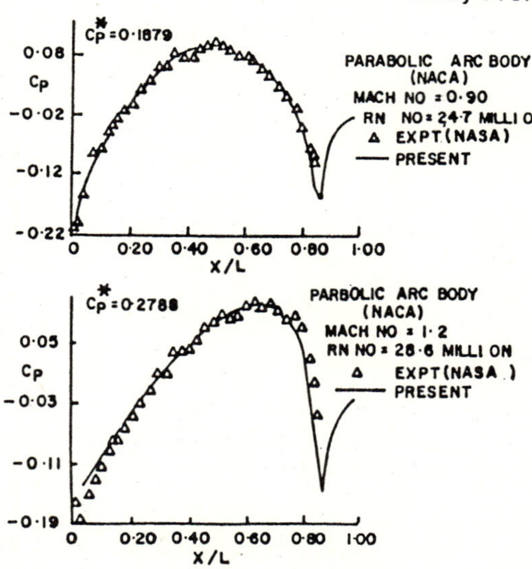

Fig. 12 Analysis of Parabolic Arc Body of Revolution

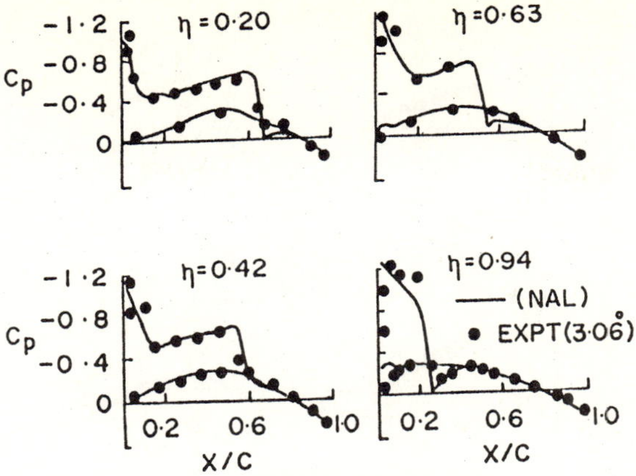

Fig. 13 ONERA M6 Wing :
$M = 0.84$, $°C = 3.06$, $AR = 3.8$, $\Lambda_{LE} = 30°$, $TR = 0.56$

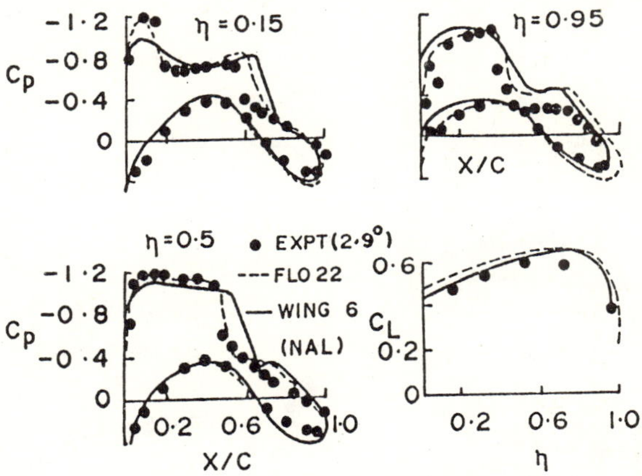

Fig. 14 Lockheed Wing - A:
$M = 0.82$, $°C = 1.5$, $AR = 8.0$, $\Lambda_{LE} = 37.25°$, $TR = 0.4$, $\theta_R = 2.76°$, $\theta_T = 2.04°$

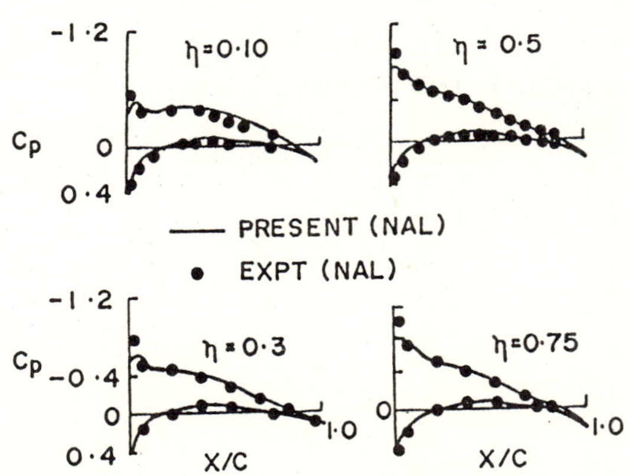

Fig. 15 Inviscid Results for Wing - G:
$M = 0.7$, $°C = 5.0$, $AR = 3.6$, $\Lambda_{LE} = 43.33°$, $TR = 0.46$

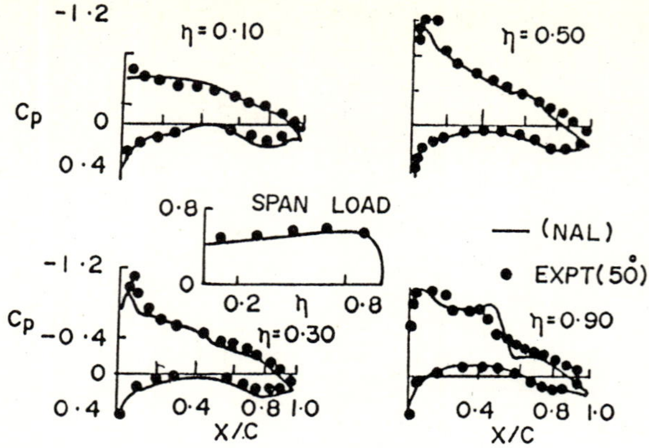

Fig. 16 Lockheed Wing - C:
$M = 0.82$, °$C = 6.0$, $AR = 2.6$, $\Lambda_{LE} = 45°$, $TR = 0.3$, $\theta_R = 2.38°$, $\theta_T = -5.79°$

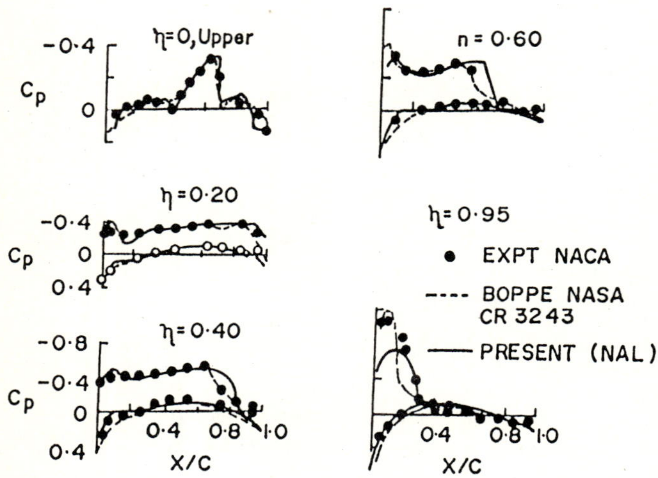

Fig. 17 Inviscid Results for L51F07 Wing-Body: $M = 0.93$, °$C = 4.0$

Numerical Simulation and Physical Modelling of Transonic Trailing Edge Flow

D. Schwamborn, P. Strykowski, H. Oertel jr.

DFVLR
Institute for Theoretical Fluid Mechanics
D-3400 Göttingen, FRG

Summary

Variations of the Baldwin-Lomax turbulence model are applied in the trailing edge region of the transonic DFVLR-F5 airfoil. We investigate to which extent the implementation of the turbulence modelling influences the local and global flow behavior. Further, computations are carried out to determine the importance of the viscous terms of the Navier-Stokes equations in the neighborhood of the trailing edge.

Introduction

The aim of this work is to answer specific questions about the flow details near the trailing edge of the transonic airfoil. Although the flow behavior near the trailing edge is of significant physical importance in its own right, we must in addition address the practically relevant questions regarding the influence of the trailing edge flow on global flow quantities such as lift and drag. Before we begin this discussion, however, we should consider what is meant by trailing edge 'modelling'. One approach of approximating the time-averaged flow in the trailing edge region would be accomplished by solving the full Reynolds-averaged Navier-Stokes equations. Assuming that a suitable grid can be found which captures the relevant flow scales of the problem we must first turn to the task of properly describing the turbulent flow behavior in this region. Secondly, assuming that the influence of the turbulence on the mean flow quantities is properly treated, we can then determine the importance of other terms in the Navier-Stokes equations, and, for instance, apply the thin-layer Navier-Stokes equations near the trailing edge. Proper trailing edge 'modelling', therefore, does not depend on the precise system of equations solved so long as these equations predict the physics of the flow correctly. Furthermore, if one is interested in the global flow behavior it is not at all obvious that the flow in the trailing edge region even needs to be correctly modelled! This of course depends on the extent to which the trailing edge influences lift and drag in comparison to, for instance, shock/boundary-layer interaction.

It appears that very little is known about the details of the flow in the trailing edge region, even if turbulence models exist which correctly predict the *local* as well as the *global* flow behavior. It is also unclear how important turbulence modelling is in comparison to other simplifying assumptions regarding the viscous terms of the Navier-Stokes equations.

We would like to address these and other issues for transonic flow past the DFVLR-F5 wing; here we consider an infinite swept wing at a chord Reynolds number of 1.5×10^6 and a Mach number of 0.82. The results to follow were performed for the non-lifting airfoil as well as for one with two degrees angle of attack. No attempt will be made to develop a new turbulence model for the trailing edge region. Instead we consider variations of the Baldwin-Lomax turbulence modelling [1] in the near-wake and investigate what influence these variations have on the local and global flow behavior. This approach was taken principally because of the

J. Zierep, H. Oertel (Eds.)
Symposium Transsonicum III
IUTAM Symposium Göttingen, 24.-27. 5. 1988

popularity of the Baldwin-Lomax turbulence model and also as a preliminary step toward the development of new turbulence modelling in the near-wake.

Numerical approach

For the discretization of the full Navier-Stokes equations we use the finite volume Runge-Kutta time stepping scheme [2,3]. The formulation is cell centered, i.e. the variables are constant within each cell of the mesh and placed in the middle of the cell and the fluxes across the cell faces are averaged from neighboring cells. Since the latter is equivalent to central differencing the use of artificial diffusion is necessary. The damping terms incorporated into the discretized equations are the usual blend of second and fourth order differences except that a weighting function is employed to avoid excessive artifical diffusion in the boundary-layer and in the wake.

After the spatial discretization of the Navier-Stokes equations, where a local coordinate transformation is used [4] for the derivatives in the physical diffusion terms, the resulting system of ordinary differential equations is integrated in time with a linearized four-stage Runge-Kutta scheme. Since we are, in general, only interested in steady state solutions, we accelerate the convergence with the help of local time stepping, and update the artificial and physical diffusion terms only once per time step resulting in a 50 per cent reduction in the execution time.

A set of boundary conditions is needed to have a well-posed problem. At the wing surface we use the no-slip and the adiabatic wall condition. The pressure is derived from the assumption of zero pressure gradient normal to the wall which is justified for the very small step sizes normal to the wall which are used in Navier-Stokes calculations for turbulent flows. At the far-field boundary we use one-dimensional Riemann invariants normal to the boundary in order to obtain boundary conditions. For the infinite swept-wing calculations carried out in this paper the variation normal to the spanwise boundaries is set equal to zero.

The method is cast in a blockstructured approach, i.e. the computational domain is divided into a number of subdomains called blocks, where only the data for one block must be in the central memory at a time. The advantages of this approach are that solutions for large-size grids can be obtained using only a relatively small amount of central memory (this is mainly important for real three-dimensional calculations), and that it is very flexible regarding the handling of complex geometries. The flexibility is partly due to the segmentation of the block faces, allowing the implement of different types of boundary conditions and different neighboring blocks at each block face. The information transport between blocks is enforced via a layer of dummy cells around each block, which correspond to the innermost cells of the respective neighbor. Since the boundary conditions are also fulfilled using the dummy cells, the algorithm for all cells within the block is the same and the number of segments on the block faces has no influence on the calculation.

The meshes used in this paper, although essentially two-dimensional, are obtained with an algebraic grid generator that handles global C-O or C-H meshes about wings and wing-body combinations [5]. To permit a quasi three-dimensional infinite swept-wing calculation there is no mesh variation in the spanwise direction -- where the mesh is only three cells wide -- and the sweep is achieved by yawing the oncoming flow. The mesh for a cross-section of the wing consists of 220x58 cells in chordwise and wall-normal directions, respectively; a view of the grid is given in figure 1. There are 158 cells on the wing surface and the first stepsize normal to the wing surface is about 10^{-5} chords. Where the boundary-layer is turbulent this is equivalent to an average value of $z^{+} \approx 2$ in the first cell center off the wall. The far-field is about 10 chords away from the wing.

Since the wing has a blunt trailing edge of a thickness of 0.5 percent of the chord, the grid generator yields a mesh with a slit in the wake of the wing. The slit is then filled with an appropriate rectangular mesh (see figure 2).

We begin this investigation by computing the global flow solution for the non-lifting airfoil. The converged global pressure distribution is shown in figure 3 along with the appropriate block boundaries discussed earlier. The computational domain is divided into three blocks: an inner block near the airfoil extending approximately 1/2 chord length in the wall-normal direction, an outer block incorporating the far-field boundary where only the Euler equations are solved and a block in the wake slit behind the blunt trailing edge (the wake block is not visible on the scale used in the figure). This pressure distribution agrees reasonably well with the three-dimensional calculations of Schwamborn [3] and the experimental results of Sobieczky et al. [6] assuming a spanwise location of about 60 per cent span where spanwise gradients are almost negligible.

The converged global solution is used to obtain the initial and boundary conditions for the more detailed trailing edge computations. The flow variables from the global solution are interpolated onto the trailing edge mesh over the domain indicated in figure 3; the trailing edge mesh incorporates approximately the same number of points as the global solution and extends from 90 per cent chord of the airfoil up to one chord length downstream of the trailing edge and 56 per cent chord in the flow-normal direction. Near the trailing edge the smallest stepsizes in the flow and the flow-normal direction are of the order of 10^{-5} and 10^{-6}, respectively.

Baldwin-Lomax turbulence modelling

The Baldwin-Lomax turbulence model has found reasonable success in describing the Reynolds stress distribution occurring in the turbulent boundary-layers of the airfoil as well as the turbulent wake far downstream. The question of interest here is whether the eddy viscosity distribution predicted in these two regions can be suitably matched near the trailing edge and thereby correctly describe the influence of the turbulent flow behavior occurring there. To answer this question we must first consider the eddy viscosity distributions predicted by the Baldwin-Lomax turbulence model in the boundary-layer and the far-wake regions, and then, incorporating the relevant flow scales of the problem, discuss how these two flow domains can be matched.

The boundary-layer and far-wake eddy viscosity distributions as well as one possible matching scheme are shown in figure 4. The eddy viscosity in the boundary-layer is reduced to zero near the wall, consistent with the expectation that disturbances there must vanish; since no such restriction is placed on disturbances in the wake, the eddy viscosity on the wake centerline is non-zero. It must be mentioned that although the Baldwin-Lomax model has been applied successfully in the far-wake there is essentially no difficulty, with small modification, in computing an eddy viscosity distribution in the near-wake. The problem, of course, is that we do not know what physical significance such computations might have (since, for instance, the near-wake pressure field is not considered in determining the eddy viscosity). Our approach, therefore, will be to assume that the Baldwin-Lomax turbulence model can be applied at any streamwise station and to proceed further by considering how the two different eddy viscosity distributions can be matched near the trailing edge.

To perform this matching we take into account the relevant scales in the problem. The first matching scheme is depicted in figure 4 and assumes that the length scale over which the eddy viscosity undergoes transition from boundary-layer to wake type is of one trailing edge thickness. After this distance the boundary-layer characteristics are forgotten and the wake distribution is applied thereafter downstream. The second matching scheme shown in figure 5 assumes that the eddy viscosity distribution from the boundary-layer will have an influence on the turbulence behavior downstream at least over a distance of the order of a boundary-layer thickness. Therefore the boundary-layer eddy viscosity distribution is propagated downstream about one boundary-layer thickness before it is relaxed to the wake distribution which applies further downstream.

The precise distance over which the relaxation occurs in the second matching scheme is a function of how the boundary-layer thickness is defined. Following Stock & Haase [7]we have chosen to base the boundary-layer thickness on the same length scale used to generate the eddy viscosity in the boundary-layer region. In the present context this results in a length scale of the order of approximately 5 trailing edge thicknesses. Due to the difficulty of precisely defining this boundary-layer thickness, we have also relaxed the eddy viscosity over distances approximately 10 times greater and have found no essential difference in the resulting flow field. This is not true, however, if the relaxation is performed over smaller distances as we will see shortly.

The spatial eddy viscosity distribution created by these two matching schemes is depicted in figure 6. The hatched regions are those locations where the eddy viscosity is 100 or more times the magnitude of the molecular viscosity. It is clear that the 'wake' matching results in a rapid 'filling-up' immediately downstream of the trailing edge whereas in the 'boundary-layer' matching, the upstream turbulence behavior is remembered and the wake is formed much further downstream. It should be mentioned that some care must be exercised when applying this algebraic turbulence model in the near-wake. Due to the complicated flow field caused by separation, several length and velocity scales can be obtained resulting in widely different eddy viscosity distributions; however, many of these produce converged solutions. We have tried to capture in our two matching schemes the extreme limiting cases from among the many we have investigated.

Results for the non-lifting airfoil

The pressure distributions obtained using the 'wake' and 'boundary-layer' matching schemes are presented together in figure 7 for zero angle of attack. The values for the iso-lines are the same so a direct comparison can be made between the two approaches. One can observe that the largest differences occur in the immediate neighborhood of the trailing edge and are in general greater downstream than upstream. The downstream influence can be detected in the velocity field up to the outflow boundary, amounting to a difference of 10 per cent in the centerline velocity one chord length downstream (200 trailing edge thicknesses). The upstream influence is much less noticeable and essentially negligible at the inflow boundary at 10 per cent chord length upstream of the trailing edge.

A more detailed view of the pressure field is provided in figure 8. Values of local minimum and maximum pressure locate the approximate position of the reverse flow region. The rapid transition used in the 'wake' scheme results in a recirculating region which remains very close to the trailing edge, whereas the recirculating flow is pushed downstream when the 'boundary-layer' matching is implemented. The downstream extent of the recirculating region is approximately 0.85 and 3.7 trailing edge thicknesses for the 'wake' and 'boundary-layer' schemes respectively.

From this precursory look it is already clear that the near-wake is strongly influenced by the precise application of the turbulence modelling. However, a more appropriate question might be: To which extent do these details influence the global flow, considering that the trailing edge thickness is only 0.5 per cent of a chord length? We have therefore, recomputed the global flow solutions with the implementation of the two trailing edge matching schemes. The block structure discussed earlier was modified to accommodate the two schemes, but the original mesh was applied. In this way a comparison could be made of the trailing edge modelling when a realistic global mesh was used, where the stepsizes in the vicinity of the trailing edge are about ten times larger than those in the local mesh. In addition, any differences in the two approaches would be minimized when trailing edge details are lost, thereby giving a conservative estimate to the actual differences that could be expected.

The new block structure together with the resulting pressure distributions are shown in figure 9. Differences in the two approaches are small and observable, in the figure, only downstream of the trailing edge. Drag calculations, however, reveal that the 'boundary-layer' matching results in a 3 per cent lower drag than the 'wake' matching. It is debatable whether a 3 per cent difference can be considered sizeable, although a relative comparison can be made between the influence of the turbulence modelling and a common assumption used in this region, namely the use of the thin-layer Navier-Stokes equations.

We begin this investigation by comparing the magnitude of the viscous terms of the full Navier-Stokes equations with the viscous terms used in the thin-layer approximation; such a comparison is shown in figure 10. The hatched regions are those in which the thin-layer viscous terms contribute more than approximately 90 per cent of the total magnitude of all viscous terms. It is again quite obvious that the two matching schemes produce different near-wake solutions, with the 'boundary-layer' scheme more closely satisfying the thin-layer approximation. Curves have also been included in figure 10 to indicate where the full and thin-layer terms are of equal magnitude. It appears that the thin-layer assumption is poorly satisfied outside the wake, however one must remember that the inertial terms in general are much larger than the viscous ones in this region.

To determine to which order the thin-layer assumption influences the total drag, we have computed the global flow solutions for both the full and thin-layer Navier-Stokes equations for the 'wake' matching scheme. These computations reveal that the drag is essentially uninfluenced by the thin-layer assumption in comparison to the full viscous equations (difference 0.2 per cent). Therefore, the 3 per cent difference found earlier would infer that variations in the turbulence model near the trailing edge have a much more significant influence on the global flow than the thin-layer assumption which is applied everywhere on the airfoil!

Results for the lifting airfoil

The investigation for the lifting airfoil -- angle of attack of 2 degrees -- follows the above treatment very closely. The converged trailing edge pressure solution is shown in figure 11 for the 'wake' and 'boundary-layer' matching schemes. The pressure contours reveal that the upstream influence on the upper surface of the airfoil is stronger than for the non-lifting case. Here we will not show a more detailed view of the near-wake region and simply mention that the reverse flow was influenced by the two schemes in a fashion similar to the non-lifting airfoil shown in figure 8. We turn instead to the global solutions which are shown in figure 12. It is apparent that the trailing edge modelling influences the pressure in a non-trivial way even upstream of the airfoil. The drag and lift computed using the 'boundary-layer' matching are 3 and 2 per cent lower, respectively, than the corresponding values obtained with the 'wake' scheme.

A comparison of the full and thin-layer Navier-Stokes terms is presented in figure 13. Consistent with the computations for the non-lifting airfoil, the 'boundary-layer' scheme adheres more closely to the thin-layer approximation. The computed global solutions confirm that the lift and drag are essentially the same when either the full or the thin-layer equations are solved. The difference is again of the order of 0.2 per cent.

Summary and conclusions

Variations of the Baldwin-Lomax turbulence model have been applied to the trailing edge region of the DFVLR-F5 swept-wing flow. The goal was not to develop a new turbulence model for the trailing edge, but rather to determine the influence that the application of a well-established model might have on the local and global flow behavior. The eddy viscosity distributions in the airfoil boundary-layer and in the far-wake were matched near the trailing edge to approximate the influence of the turbulence in that region. The matching was based on the interaction length scales of the trailing edge thickness and a boundary-layer thickness. Further, as a means of comparing integral quantities, the full and thin-layer Navier-Stokes equations were solved on the global domain.

Limited to the single trailing edge configuration of these computations, it is apparent that the precise method in which the trailing edge turbulence modelling is applied, will have a non-trivial influence on the flow field both locally and globally. The local influence is most dramatically seen in the velocity field, where the reverse flow behavior is strongly altered. The global solutions for both the lifting and non-lifting airfoils revealed that the trailing edge modelling could result in measurable differences in lift and drag. Although these integral quantities were changed only by at most 3 per cent, it was shown that the trailing edge modelling should not be ignored in comparison to the seemingly more global approximation of applying the thin-layer Navier-Stokes equations, which taken independently, had almost no measurable influence on the lift and drag.

Unfortunately no claim can be presently made as to which of the matching schemes is more correct, as it is not even clear if either produces physically reasonable results. With a lack of available experimental data on transonic trailing edge flows, and the complication that trailing edge geometries vary slightly, it appears that a general experimentally proven trailing edge model may be far off. The attitude that has been generally taken is that although trailing edge modelling is needed, the correct modelling of, for instance, shock/boundary-layer interaction, is more critical. Preliminary calculations made in the present study confirm this, where small changes in the turbulence modelling near the shock had a measurable influence on lift and drag of the order of 10 per cent or more.

References

[1] Baldwin, B.S., Lomax, M.: Thin Layer Approximation and Algebraic Model for Separated Turbulent Flows. AIAA Paper 78-257, (1978).

[2] Schwamborn,D.: Numerical Simulation of Compressible Navier-Stokes Flow in a Double Throat Nozzle. In: Notes on Numerical Fluid Mechanics, Vol.18,Vieweg Verlag, Wiesbaden,1987.

[3] Schwamborn,D.: Simulation of the DFVLR-F5 Wing Experiment using a Block Structured Explicit Navier-Stokes Method. To appear in: Notes on Numerical Fluid Mechanics Vol.22, Vieweg Verlag, Wiesbaden,1988.

[4] Deiwert, G.S.: Numerical Simulation of the High Reynolds Number Transonic Flows. AIAA J. 13 (1980), pp. 1354-1359.

[5] Sobieczky, H.: Geometry Generation for Transonic Design. In: Recent Advances in Numerical Methods in Fluids, Vol. 4, Ed. W.G. Habashi, Swansea: Pineridge Press, 1985.

[6] Sobieczky, H., Hefer, G., Tusche, S.: DFVLR - F5 Test Wing Experiment For Computational Aerodynamics. AIAA-87-2485, 1987.

[7] Stock, H.W., Haase, W.: The Determination of Turbulent Length Scales in Algebraic Turbulence Models for Attached and Slightly Separated Flows Using Navier-Stokes Methods,AIAA-87-1302, 1987.

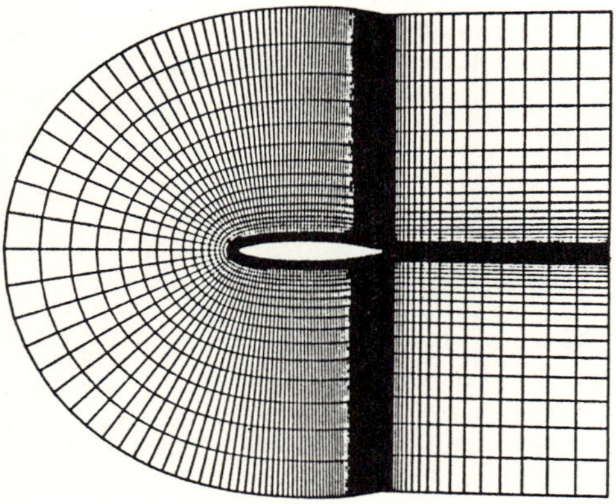

Figure 1: Global grid used to approximate the flow past the DFVLR-F5 wing.

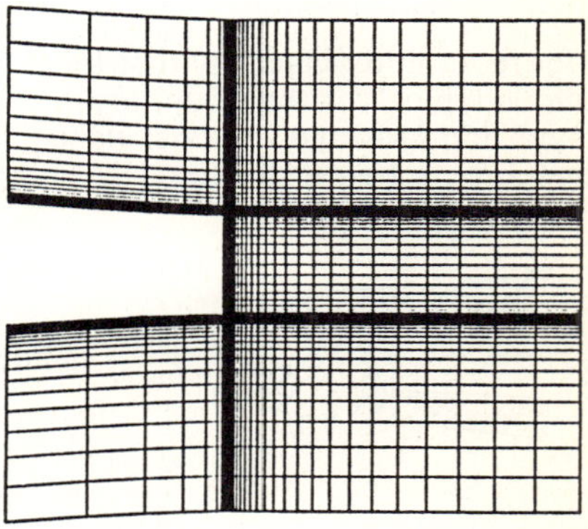

Figure 2: Detailed view of the DFVLR-F5 trailing edge grid.

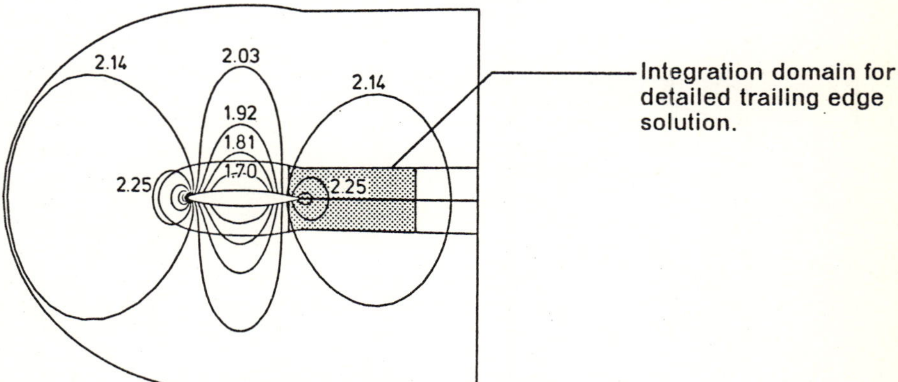

Figure 3: Pressure coefficient distribution for the infinitely swept-wing with no lift.

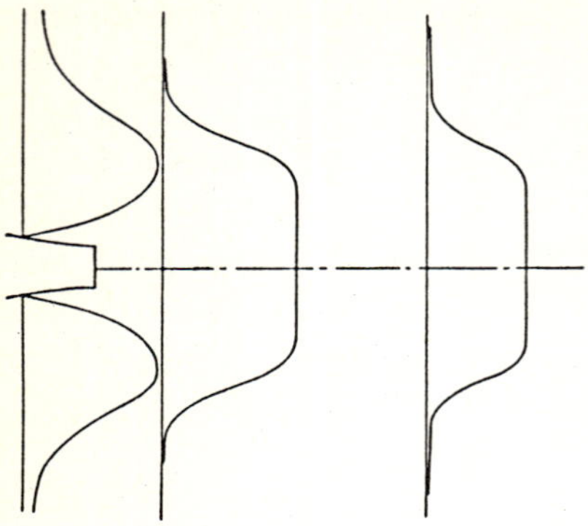

Figure 4: Eddy viscosity distributions typical for the boundary-layer and far-wake regions; wake matching scheme shown.

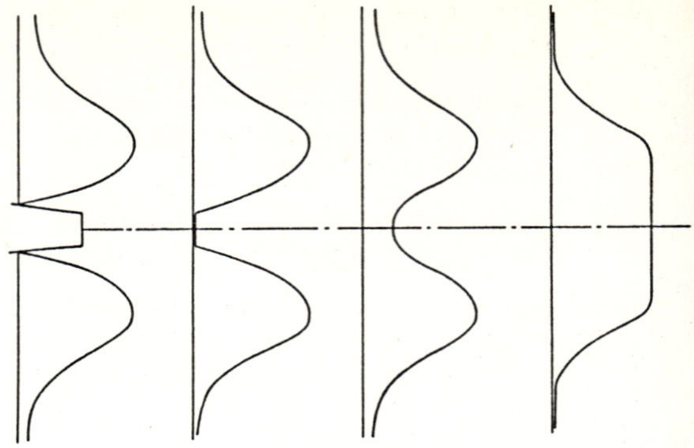

Figure 5: Eddy viscosity relaxation performed using the boundary-layer matching scheme.

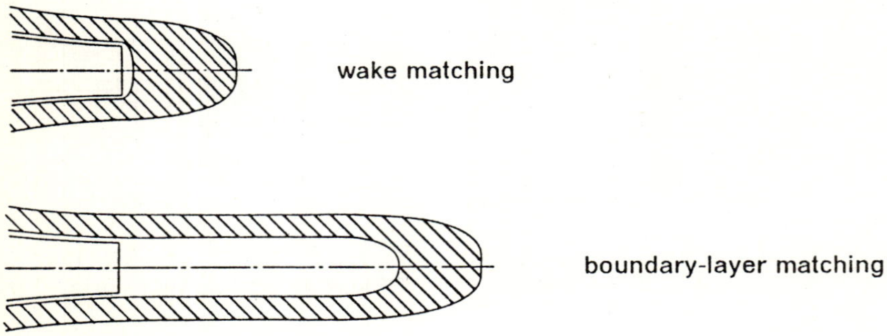

Figure 6: Spatial eddy viscosity distributions near the trailing edge.

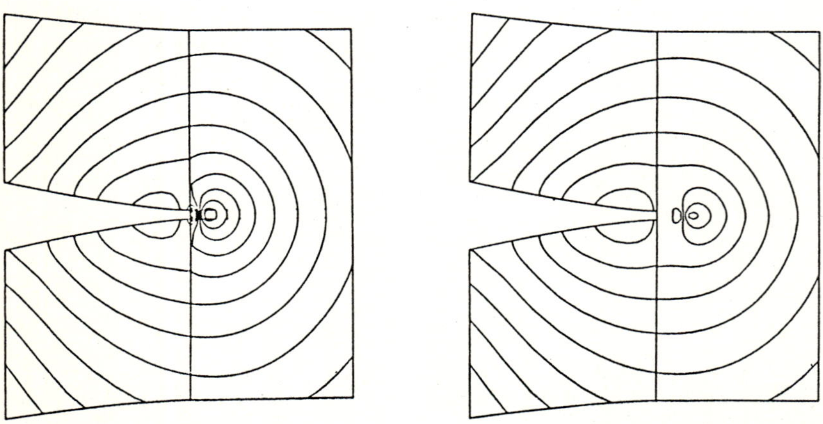

Figure 7: Pressure coefficient distribution obtained by applying both matching schemes; $\Delta c_p = 0.02$.

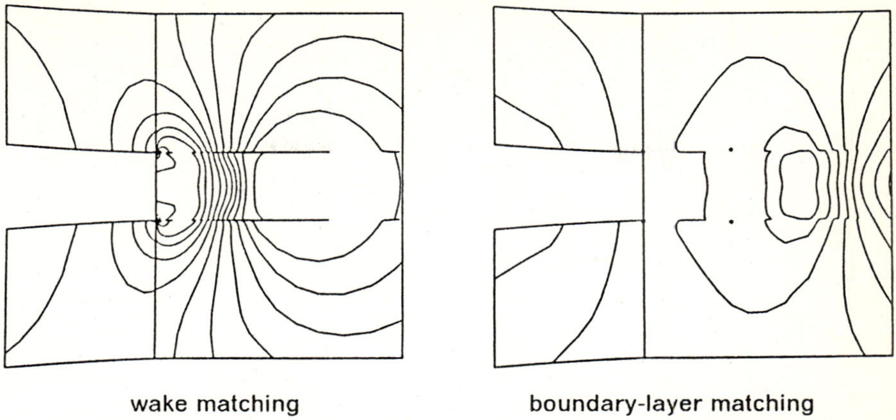

Figure 8: Detailed pressure distributions near the trailing edge; identical c_p values are used in both distributions, $\Delta c_p = 0.01$.

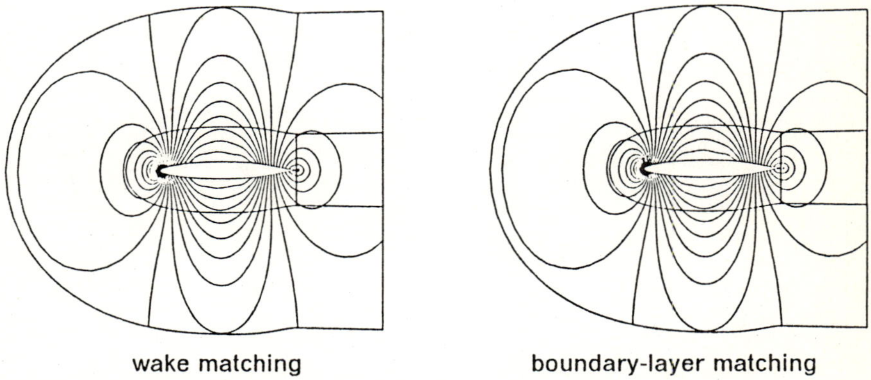

Figure 9: Global pressure distributions for the non-lifting airfoil.

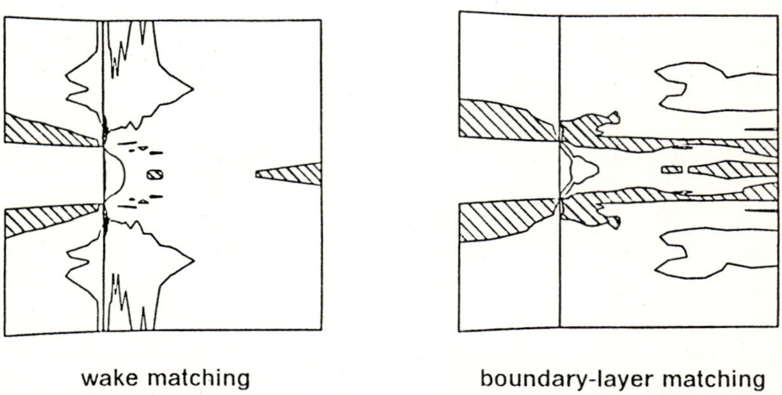

Figure 10: Comparison of the thin-layer viscous terms with all other viscous terms for the non-lifting airfoil.

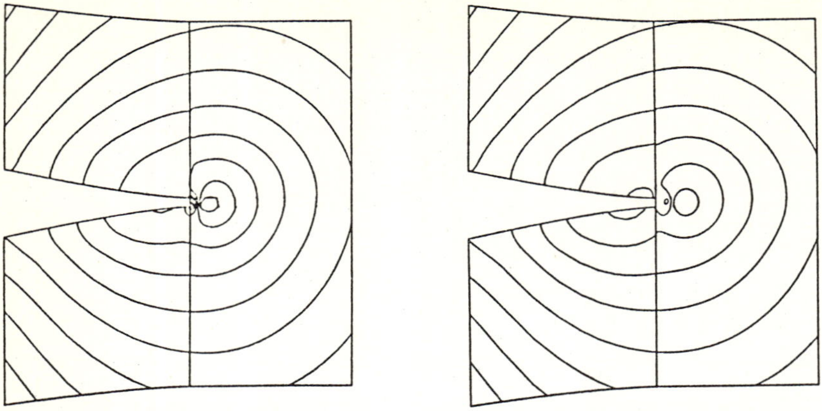

wake matching boundary-layer matching

Figure 11: Trailing edge pressure distribution for two degrees angle of attack; $\Delta c_p = 0.02$.

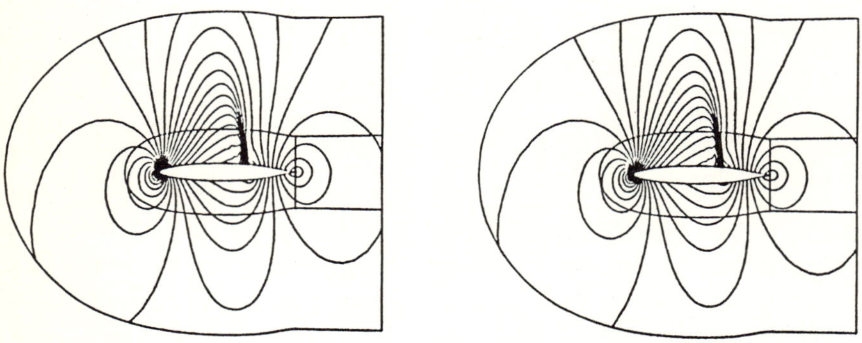

wake matching boundary-layer matching

Figure 12: Global pressure distributions for the airfoil at two degrees angle of attack.

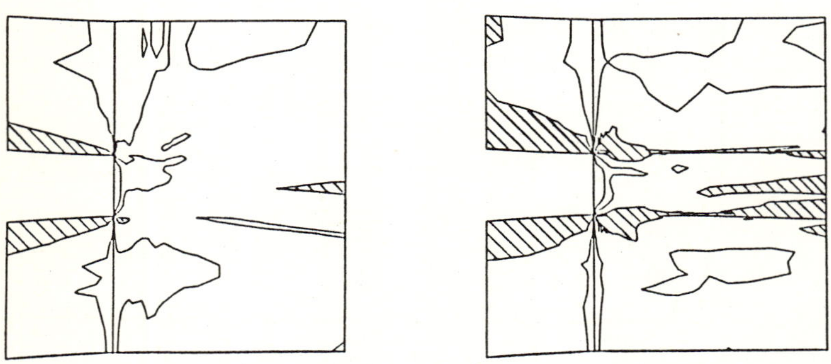

wake matching boundary-layer matching

Figure 13: Comparison of the thin-layer viscous terms with all other viscous terms for the airfoil at two degrees angle of attack.

Theoretical and Experimental Investigation of 2-D Diabatic Transonic and Supersonic Flow Fields

G. Schnerr and U. Dohrmann

Institut für Strömungslehre und Strömungsmaschinen
Universität (TH) Karlsruhe, FRG

Summary

Inviscid, stationary 2-D flows of moist air with nonequilibrium phase transition are calculated numerically using the Euler equation coupled with the classical nucleation theory. In Laval nozzles with supercritical heat addition the apparent normal shocks with local subsonic regions in the surrounding supersonic flow field are computed. Homogeneous condensation in local supersonic regions around profiles is calculated for the first time with detailed analysis of the 2-D structure for continuous change of state as well as for stationary double shock systems. For nozzles and profiles, all numerical results are well verified by experiments.

1. Introduction

A non-heat-conducting gas (moist air) which obeys the perfect gas equation of state is considered. During fast expansions with high cooling rates the vapour component relaxes and normally condensation first appears at Mach numbers greater than one (Oswatitsch [1]). Essential questions of this investigation are strong disturbances produced by the released latent heat, i.e. normal shocks with extended local subsonic regions in supersonic nozzles and stationary systems of two normal shocks on profiles ($M_\infty \leq 1$). Moreover the variation and sensitivity of pressure distribution, shock location on profiles and the structure of the wake with condensate are of interest. Earlier computational results (2-D) without subsonic regions or limited to weak compression are given by Bartlmä [2], Bratos/Meier [3] and Matsuo/Kawagoe/Sonoda/Sakuo [4]. First experimental results for profile flows with homogeneous condensation are due to Head [5], Schmidt [6] and Hiller/Meier[7].

2. Theory

We assume inviscid flow and use the Euler equation. First the basic adiabatic solution is calculated by the explicit time-dependent F-V-M (Finite-Volume-Method) of Eberle [8]. The extended diabatic scheme may be written in conservation form as:

$$\frac{\partial U}{\partial t} + \frac{\partial F}{\partial x} + \frac{\partial G}{\partial y} = Q$$

J. Zierep, H. Oertel (Eds.)
Symposium Transsonicum III
IUTAM Symposium Göttingen, 24.-27. 5. 1988

where

$$U = \begin{bmatrix} \rho \\ \rho u \\ \rho v \\ \rho e \end{bmatrix} ; \; F = \begin{bmatrix} \rho u \\ \rho u^2 + p \\ \rho uv \\ (\rho e + p)u \end{bmatrix} ; \; G = \begin{bmatrix} \rho v \\ \rho uv \\ \rho v^2 + p \\ (\rho e + p)v \end{bmatrix} ; \; Q = \begin{bmatrix} 0 \\ 0 \\ 0 \\ \rho L_c \frac{dg}{dt} \end{bmatrix}$$

in terms of gas density ρ, x- and y- velocity components u and v, total energy per unit mass e, latent heat of condensation or sublimation L_c and mass fraction g. **Q** denotes the energy supply by condensation. For description of the condensation process we use the classical nucleation theory of Volmer [9] and the Hertz-Knudsen-model for the droplet growth rate with the assumption of equal temperature of the droplets and the surrounding vapour. From Hill [10], the mass fraction of the new phase dg/dt in **Q** is given by the integro-differential equation

$$\frac{dg}{dt} = \frac{4}{3} \Pi \rho_c r^{*3} \frac{J}{\rho} + \frac{d\bar{r}}{dt} \int_{-\infty}^{t} 4 \Pi \rho_c r^2(t_1) \frac{J(t_1)}{\rho(t_1)} dt_1$$

with density of condensate (sublimate) ρ_c, critical droplet radius r*, droplet radius r, nucleation rate J, and droplet growth rate $d\bar{r}/dt$, which is related to a "surface-averaged" droplet radius $\bar{r}$. The first term on the right-hand side is the condensation rate due to the formation of new droplets, the following represents the mass fraction of all droplets created at a time t_1. According to this expression the droplet growth rate does not depend on the local droplet size. With the extended diabatic F-V-M solutions for stationary flows are calculated. Second-order accuracy is achieved without addition of artificial viscosity, with which the first apparent disturbances at condensation onset are well reproduced.

3. Results

It is common that the classical theory includes free parameters, in our case the surface tension $\sigma(T)$ - essential in the nucleation phase, i. e. the condensation onset - and the condensation coefficient α in the droplet growth model - essential for the heat release. $\sigma(T)$ and α are estimated empirically by comparison with well-defined nozzle experiments and then both are retained for all further calculations.

First numerical results are shown for a hyperbolic nozzle with a low critical temperature gradient of $(-dT/dx)^* = 5.1$ K/cm on the axis. Fig.1a,b give the nucleation rate J, the condensate mass fraction g/g_{max}, the pressure and frozen Mach number distribution for a relative stagnation humidity $\Phi_0 = 49.3\%$. Both, pressure (Mach number) and frozen Iso-Mach lines for $M_f \geq 1$, show a normal shock with a local subsonic region. This shock terminates the nucleation phase, but within the used theory no evaporation takes place because $\bar{r}$ remains larger than r*. At higher

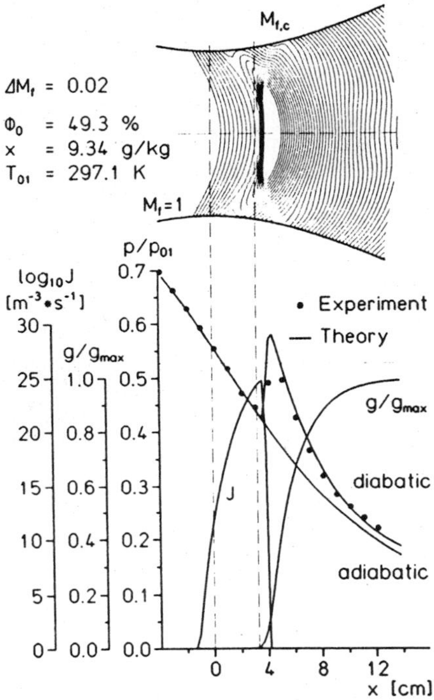

◄ a

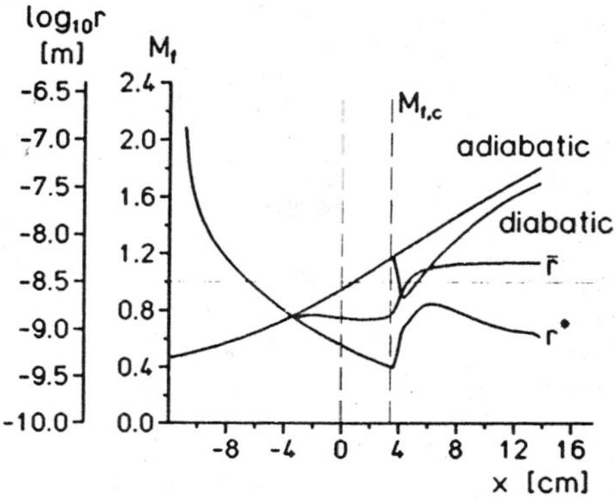

▲ b

c: Φ_O = 60.3%, x = 9.8 g/kg, T_{01} = 294.8 K
▼

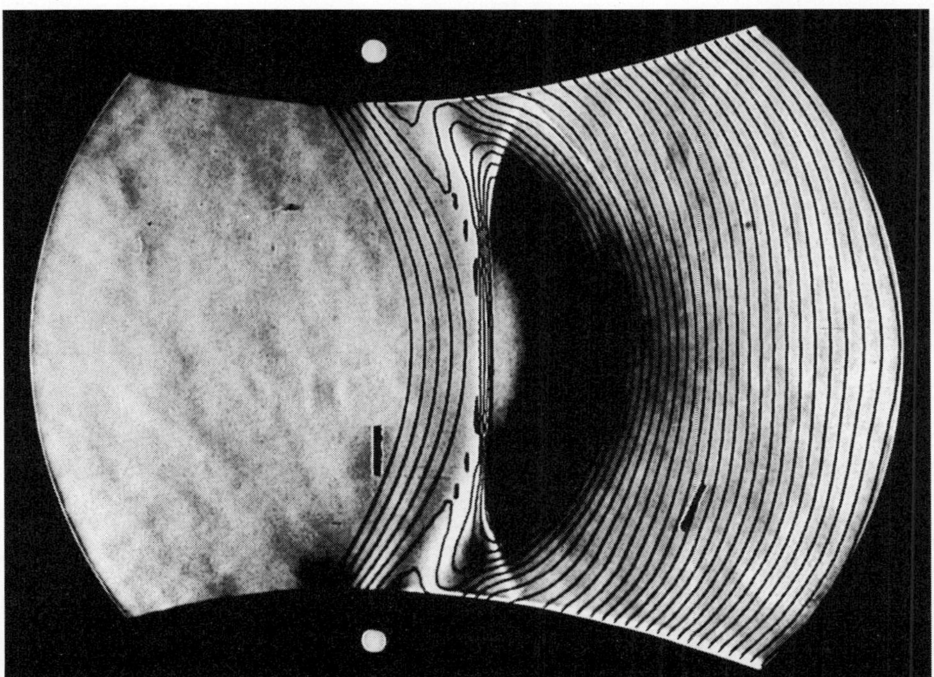

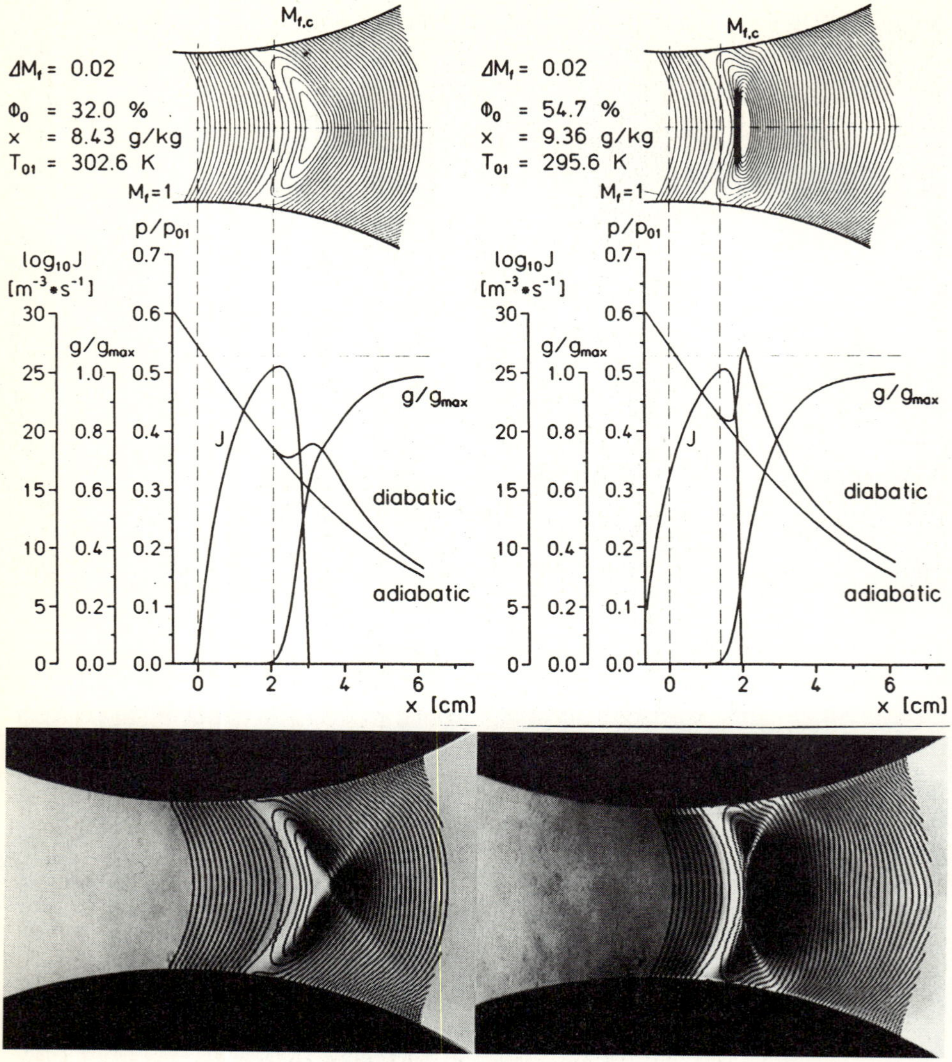

Fig. 2 Circular arc nozzle (-dT/dx)* = 12.1 K/cm

relative humidity (Φ_O= 60.3 %, Fig. 1c) the shock and the subsonic region grow and both are shifted towards the nozzle throat. The free parameters $\sigma(T)$ and α are evaluated using Schlieren pictures and pressure measurements in the nozzle. As a first check of these values, calculations were done in a second nozzle (circular arc) having a different time scale for the expansion, i. e. higher temperature gradient (-dT/dx)* = 12.1 K/cm within a large range of the relative humidities 32.0 % $\leq \Phi_O \leq$ 64.8 %. These results are shown in Fig. 2,3: a weak supersonic disturbance (''X -shock''- type) in Fig. 2(left) and a straight shock stretching nearly from wall to wall in Fig. 3. The Mach number $M_{f,c}$ at condensation onset decreases systematically,

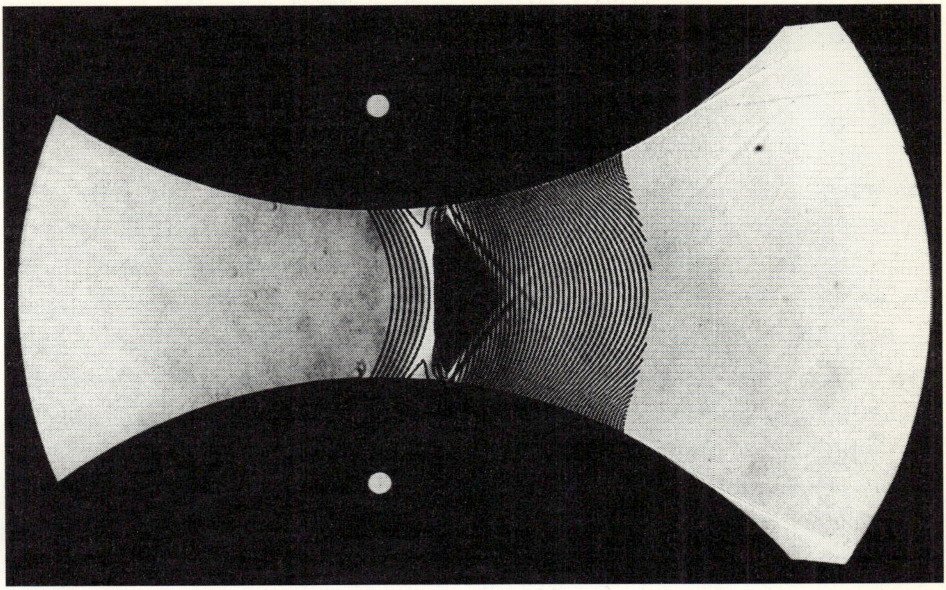

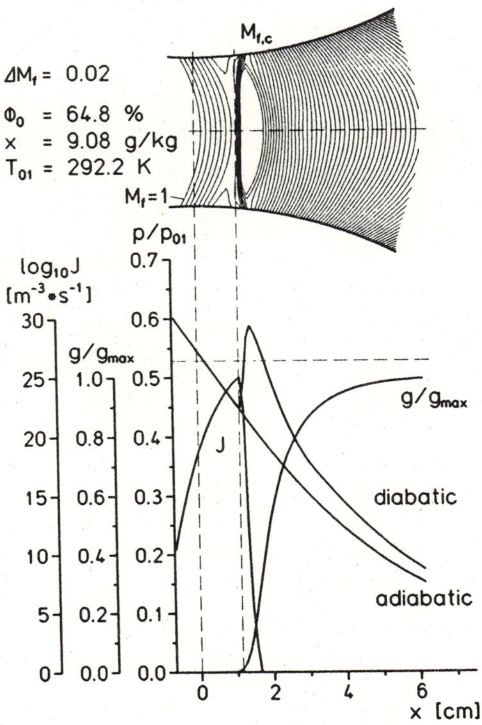

Fig. 3 Circular arc nozzle $(-dT/dx)^* = 12.1$ K/cm

following the empirical laws of Wegener [11], Zierep/Lin [12] and Schnerr [13]. At higher rates of heat addition, e.g. higher values of Φ_0, the flow becomes unsteady. Nozzles, showing this periodic oscillating flow, are investigated by Schmidt[14], Barschdorff[15], Zierep/Lin[16] and Frank[17].

Within the same range of parameters ($\sigma(T)$ and α), the flow around a circular arc profile (10 % thickness ratio) at a free stream Mach number $M_{f\infty} = 0.87$ was calculated. Fig.4 shows the comparison with the adiabatic flow at different relative humidities. Upstream of the profile the flow is supersaturated to a certain degree. With increasing supersaturation around the profile essential nucleation and condensation appear and produce a more or less strong compression in front of the rear shock. At low relative humidities or high condensation onset Mach numbers, only one shock with a smooth change in its curvature exists. For higher values of Φ_0 a second shock arises. Typical pressure distributions in the flow field with heat addition are presented in Fig. 5. As in nozzle experiments at higher values of Φ_0, the condensation onset is shifted to lower Mach numbers and the strength of the disturbance increases. Upstream of the condensation onset the local supersonic flow shows only a weak variation in comparison to the adiabatic case. The rear shock may be shifted in both directions, upstream at high onset Mach numbers (decrease of drag) and downstream to the trailing edge for lower ones.

Fig. 6 shows lines of constant condensate mass fraction for low and high values of Φ_0. On the profile surface a maximum of about $g/g_{max} = 0.85$ is reached. The droplets evaporate wholly or partly, passing the shock and the rear stagnation region. The wake shows a distinct distribution of condensate, no complete evaporation can appear because of supersaturation at free stream conditions. This result should be kept in mind for wake measurements. As a consequence of Hill's surface-averaged droplet radius the calculated values in the wake may be somewhat overestimated. By this numerical analysis, the structure of the visualized condensation sheets, shown in the experimental investigation of Hiller/Meier[7], can be explained immediately.

4. Comparison with experiments

The agreement of the calculated Iso-Mach lines with the Schlieren pictures of [13] can be realized immediately in Fig. 1c, 2 and 3. This is valid for all details especially for the condensation onset as well as for position and extention of shocks and local subsonic regions. Theoretical and experimentally determined onset Mach numbers agree well with those of Wegener/Pouring [18]. A sequence of Schlieren pictures with a flow over a circular arc (half) profile (10 % thickness ratio) shows Fig. 7. The typical weak change in curvature of one shock is reproduced as well as the stationary double shock system with subsonic flow between both. The last system represents a limiting case for steady flow at high values Φ_0, i. e. at low supersonic values of $M_{f,c}$. Within the measuring accuracy, the low deviation of adiabatic and diabatic pressure distribution

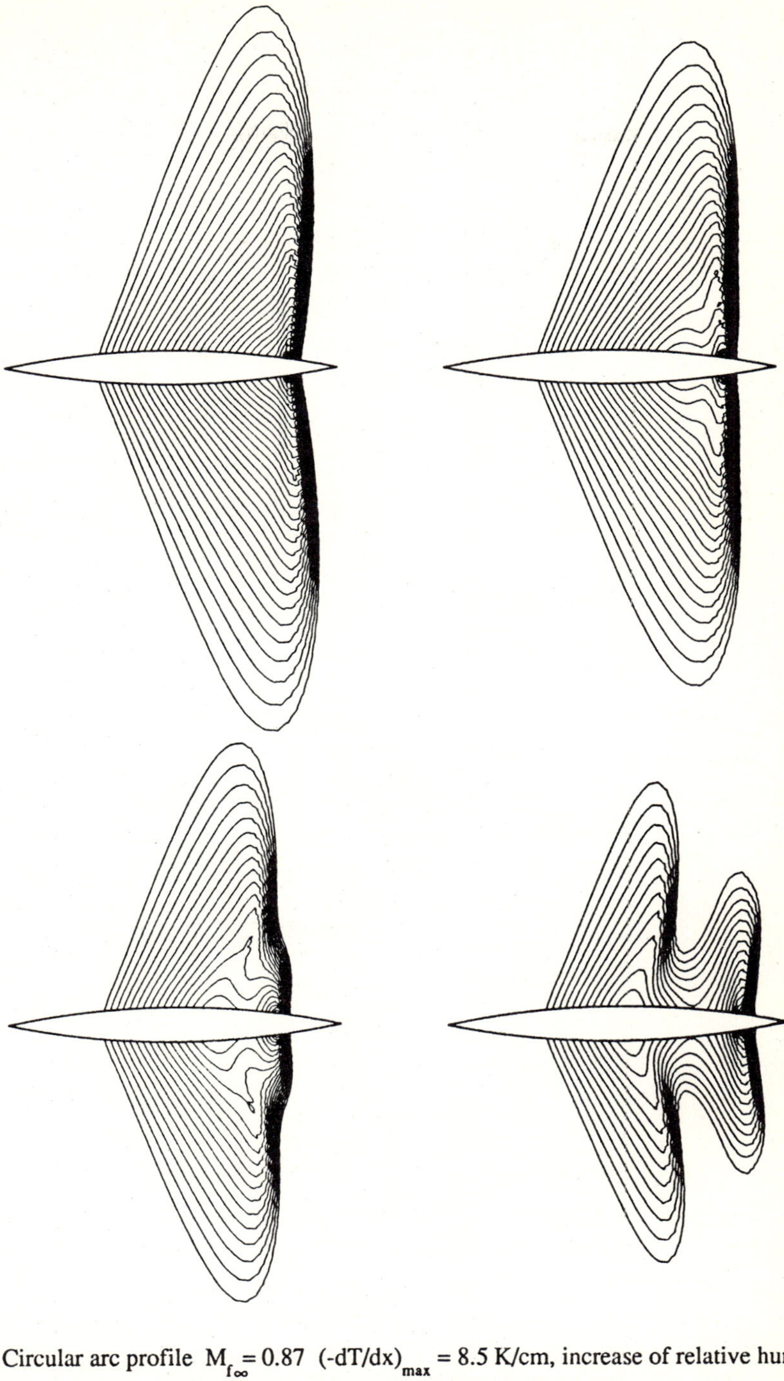

Fig.4 Circular arc profile $M_{f\infty} = 0.87$ $(-dT/dx)_{max} = 8.5$ K/cm, increase of relative humidity: $\Phi_o = 0\%, 40\%, 50\%, 65\%$ – from top left, Iso-Mach lines $M_f \geq 1$

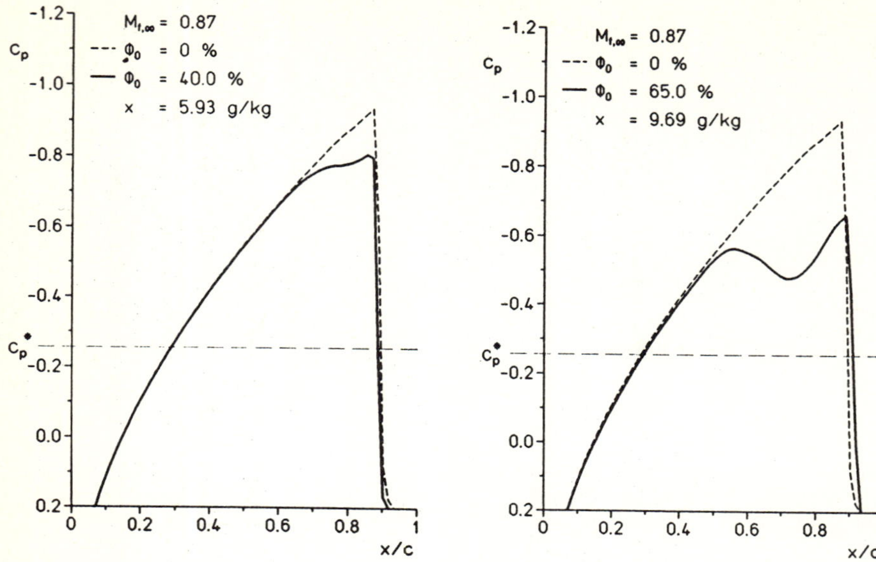

Fig.5 Pressure distribution around a circular arc profile $M_{f\infty} = 0.87$, adiabatic and diabatic – theory

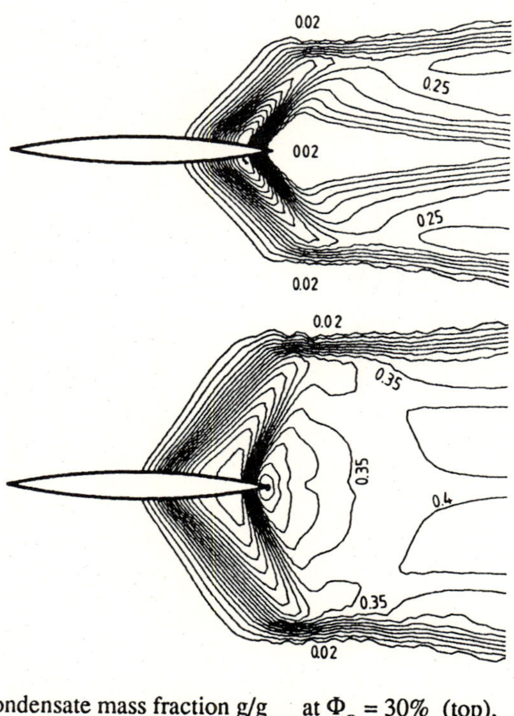

Fig.6 Condensate mass fraction g/g_{max} at $\Phi_o = 30\%$ (top), $\Phi_o = 60\%$ (bottom) – circular arc profile $M_{f\infty} = 0.87$

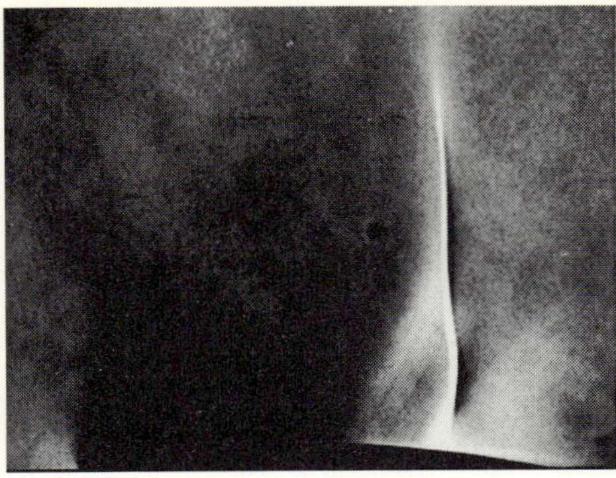

$M_\infty = 0.791$, $\Phi_o = 53.4\%$, x = 7.2 g/kg

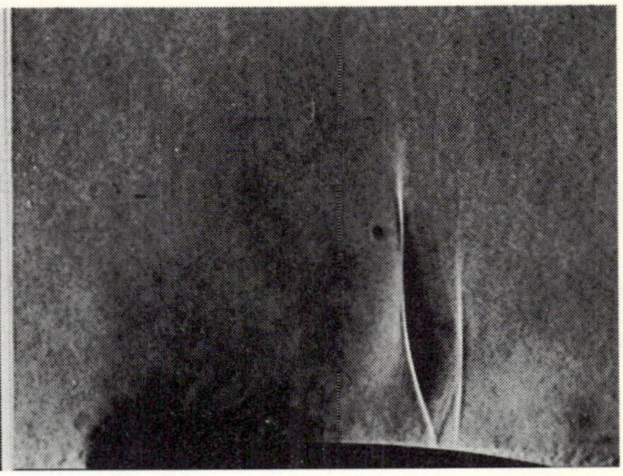

$M_\infty = 0.787$, $\Phi_o = 57.1\%$, x = 9.9 g/kg

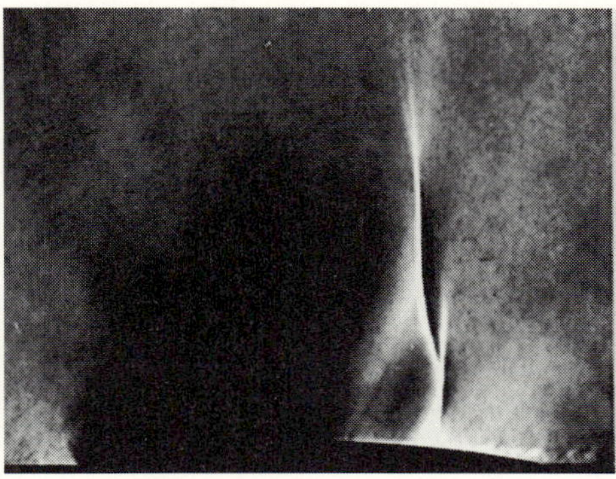

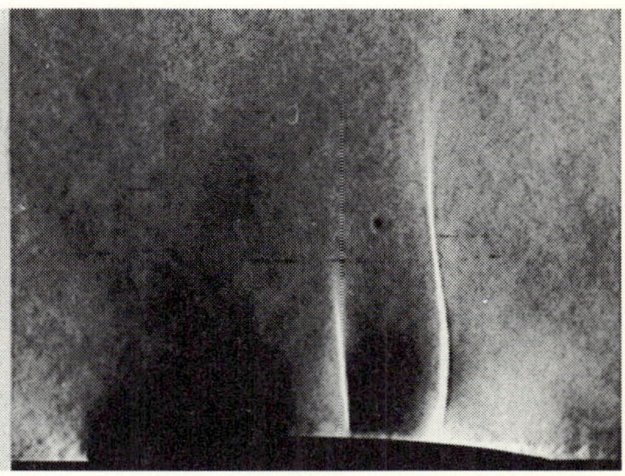

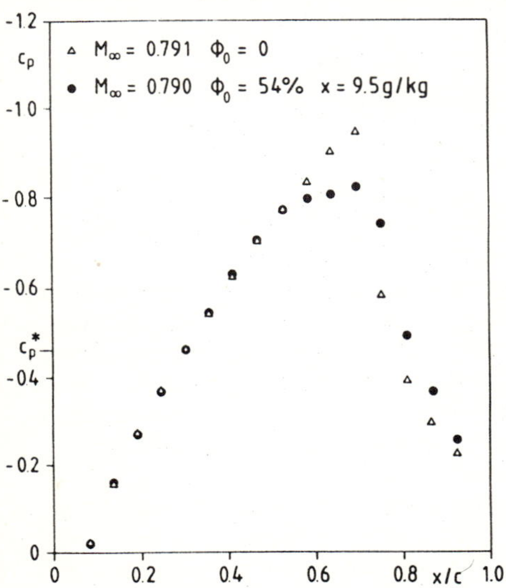

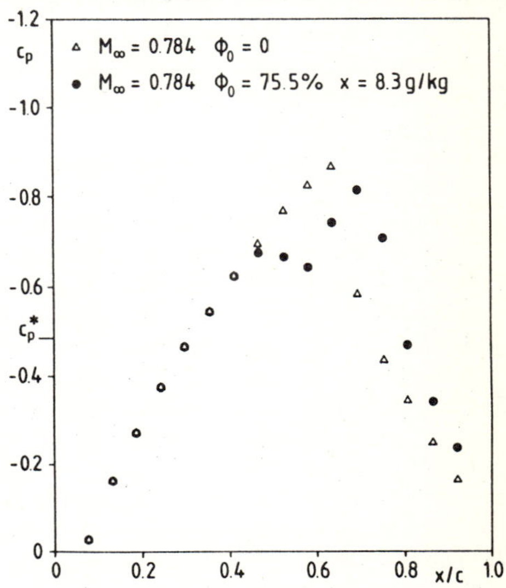

Fig.7 Pressure distribution around a circular arc (half) profile, adiabatic and diabatic – experiment

on the profile up to the condensation onset is confirmed. Further distributions qualitatively correspond to the theoretical results at slightly different parameters, due to test section interferences and boundary layer effects.

5. Conclusions

A numerical procedure is established for calculation of transonic flows with relaxation and heat supply. The observed experimental results are well reproduced for nozzles as well as for profiles. In the same manner the method can be used to study possible 2-D effects in cryogenic wind tunnels,for instance on the profile surface and in the wake downstream of profiles.

References

[1] Oswatitsch, K.: Kondensationserscheinungen in Überschalldüsen. ZAMM 22, 1-14 (1942)

[2] Bartlmä, F.: Ebene Überschallströmungen mit Relaxation. In: Applied Mechanics,ed.: H. Görtler, 1056-1060, Springer-Verlag (1966)

[3] Bratos, M., Meier, G. E. A.: Two-dimensional, two-phase flows in a Laval nozzle with nonequilibrium phase transition. Archives of Mechanics 28, 5-6, 1025-1037, Warszawa (1976)

[4] Matsuo, K., Kawagoe, Sh., Sonoda, K., Sakao, K.: Studies of Condensation Shock Waves. Bulletin of JSME, Vol. 28, No. 241, (1985)

[5] Head, R. M.: Investigations of Spontaneous Condensation Phenomena.Dissertation, Calif. Inst. of Techn., Pasadena, Calif. (1949)

[6] Schmidt, B.: Schallnahe Profilumströmung mit Kondensation. Acta Mech. II/2, 194-208(1966)

[7] Hiller, W.J., Meier, G.E.A.: Experimentelle Untersuchung einer transsonischen Potentialströmung um ein symmetrisches Tragflügelprofil. Max-Plank-Inst. f. Strömungsforschung, Bericht 10 (1971)

[8] Eberle, A.: A New Flux Extrapolation Scheme Solving the Euler Equations for Arbitrary 3-D Geometry and Speed. MBB/LKE 122/S/PUB/140 (1984)

[9] Volmer, M.: Kinetik der Phasenbildung. Steinkopff, Leipzig (1939)

[10] Hill, Ph. G.: Condensation of water vapour during supersonic expansion in nozzles. J. Fluid Mech. 25, 593-620 (1966)

[11] Wegener, P. P.: Water vapour condensation process in supersonic nozzles. J. of Appl. Phys. 25, 1485-1491 (1954)

[12] Zierep, J., Lin, S.: Bestimmung des Kondensationsbeginns bei der Entspannung feuchter Luft in Überschalldüsen. Forschung im Ingenieurwesen 33, 169-172 (1967)

[13] Schnerr, G.: Homogene Kondensation in stationären transsonischen Strömungen durch Lavaldüsen und um Profile. Habilitationsschrift Universität (TH) Karlsruhe, Fakultät für Maschinenbau (1986)

[14] Schmidt, B.: Beobachtungen über das Verhalten der durch Wasserdampfkondensation ausgelösten Störungen in einer Überschall-Windkanaldüse. Dissertation Universität (TH) Karlsruhe (1962)

[15] Barschdorff, D.: Kurzzeitfeuchtemessung und ihre Anwendung bei Kondensationserscheinungen in Laval-Düsen. Mitt. d. Inst. f. Strömungslehre und Strömungsmaschinen Universität (TH) Karlsruhe 6, 18-39 (1967)

[16] Zierep, J, Lin, S.: Ein Ähnlichkeitsgesetz für instationäre Kondensationsvorgänge in der Laval-Düse. Forschung im Ingenieurwesen 34, 97-99 (1968)

[17] Frank, W.: Condensation Phenomena in Supersonic Nozzles. Acta Mechanica 54, 135-156 (1985)

[18] Wegener, P. P., Pouring A. A.: Experiments on Condensation of Water Vapor by Homogeneous Nucleation in Nozzles. Physics of Fluids, Vol. 7, No. 3, (1964)

Asymptotics and Theory

Asymptotic Methods in Transonic Flow Theory

O.S. Ryzhov

USSR Academy of Sciences
Computing Center
Moscow 117333, USSR

Summary

The asymptotic approach to the investigation of flows in the transonic range of velocities is considered in the light of historical development in this branch of aerodynamics. Most studies in the past had ignored the viscosity and the heat conductivity of real air that made it possible to develop the nonlinear perturbation theory for the external potential region of the velocity field. In our days the emphasis has swung to processes taking place in the boundary layer, i.e. the internal region where the dissipative factors are of prime importance. Their role becomes decisive in the understanding and explaining of two fundamental phenomena: separation and stability, both showing specific features when the external velocity is close to the speed of sound.

External Potential Field

The basic assumption in the following analysis is that the Mach number M at each point of the external potential region does not deviate essentially from 1. Only two-dimensional flows are treated here, the problem of a finite aspect ratio wing being thus omitted. Let us designate the time by t^*, the Cartesian coordinates by x^* and y^*, the corresponding velocity components by u^* and v^*, the pressure by p^*, the density by ρ^* and the temperature by T^*. A small parameter ϵ is introduced formally; values of all variables on the surface, supposed to be thermally insulated and in the oncoming stream, are labelled by the subscripts w and ∞, respectively. Bearing in mind that for such a surface $T^*_w/T^*_\infty = (\kappa + 1)/2$, where κ is the ratio of specific heats, we define the independent nondimensional variables as

$$\frac{t^*}{(x^*_0/u^*_\infty)} = t_0 + \epsilon\, 2^{-\frac{1}{10}} C^{\frac{1}{10}} \lambda^{-\frac{9}{5}} \left(\frac{T^*_w}{T^*_\infty} \right)^{\frac{2}{5}} t \tag{1.1a}$$

$$\frac{x^*}{x^*_0} = 1 + \epsilon^3\, 2^{-\frac{3}{10}} C^{\frac{3}{10}} \lambda^{-\frac{7}{5}} \left(\frac{T^*_w}{T^*_\infty} \right)^{\frac{6}{5}} x \tag{1.1b}$$

$$\frac{y^*}{x^*_0} = \epsilon^2\, 2^{-\frac{7}{10}} C^{\frac{1}{5}} \lambda^{-\frac{8}{5}} \left(\frac{T^*_w}{T^*_\infty} \right)^{\frac{4}{5}} y \tag{1.1c}$$

and scale the desired functions in the following way:

J. Zierep, H. Oertel (Eds.)
Symposium Transsonicum III
IUTAM Symposium Göttingen, 24.-27. 5. 1988

$$\frac{u^*}{u^*_\infty} = 1 + \epsilon^2 2^{-\frac{1}{5}} C^{\frac{1}{5}} \lambda^{\frac{2}{5}} \left(\frac{T^*_w}{T^*_\infty} \right)^{-\frac{1}{5}} u \tag{1.2a}$$

$$\frac{v^*}{u^*_\infty} = \epsilon^3 2^{\frac{1}{5}} C^{\frac{3}{10}} \lambda^{\frac{3}{5}} \left(\frac{T^*_w}{T^*_\infty} \right)^{\frac{1}{5}} v \tag{1.2b}$$

$$\frac{(p^* - p^*_\infty)}{(\rho^*_\infty \; u^{*2}_\infty)} = \epsilon^2 2^{-\frac{1}{5}} C^{\frac{1}{5}} \lambda^{\frac{2}{5}} \left(\frac{T^*_w}{T^*_\infty} \right)^{-\frac{1}{5}} p \tag{1.2c}$$

If the dissipative processes in a boundary layer are fully neglected, the constant C from the Chapman viscosity law and the skin friction λ should be excluded from (1.1) and (1.2), whereas the use of κ becomes more preferable than T^*_w/T^*_∞.

Substitution of (1.1) and (1.2) into the Navier-Stokes equations leads to the conclusion that in the external region they reduce to the simpler nonlinear system [1 - 3]

$$- \frac{\partial u}{\partial t} - (K_\infty + u) \frac{\partial u}{\partial x} + \frac{\partial v}{\partial y} = 0 \tag{1.3a}$$

$$\frac{\partial v}{\partial x} = \frac{\partial u}{\partial y}, \qquad p = -u \tag{1.3b, c}$$

which admits a two-parameter group of similarity transformations if $K_\infty = 0$. Obviously, the latter parameter

$$K_\infty = \frac{M^2_\infty - 1}{\epsilon^2 2^{\frac{4}{5}} C^{\frac{1}{5}} \lambda^{\frac{2}{5}} (T^*_w/T^*_\infty)^{\frac{4}{5}}} \tag{1.4}$$

is a critical one for determining the regime of transonic flow.

Let the shape of an object be independent of time and given by the equation

$$\frac{y^*}{x^*_0} = \delta f(x) \tag{1.5}$$

Assuming that the relative thickness of the obstacle is

$$\delta = \epsilon^6 2^{-\frac{1}{10}} C^{\frac{3}{5}} \lambda^{-\frac{4}{5}} \left(\frac{T^*_w}{T^*_\infty} \right)^{\frac{7}{5}} \tag{1.6}$$

we write the slip condition on its surface as

$$v = \frac{df}{dx} \qquad \text{at} \qquad y = 0 \tag{1.7}$$

The boundary conditions at infinity read $u \rightarrow 0$, $v \rightarrow 0$ as $(x^2 + y^2)^{\frac{1}{2}} \rightarrow \infty$. Besides, discontinuous solutions must satisfy certain conditions across the shock fronts, which are omitted for brevity.

The estimation $\delta \sim \epsilon^6$ set by (1.6) is due to the fact that the small parameter ϵ enters the transformation for the longitudinal coordinate. This dependence can be removed by putting $x^*_0 = \epsilon^{-3} l^*$ into the expression for difference $x^* - x^*_0$ which follows from (1.1b). Corresponding changes will simultaneously take place in the right-hand sides of (1.1a) and (1.1c), namely, the scaling of time will involve ϵ^{-2} rather than ϵ, whereas ϵ^{-1} will appear in the stretching of the lateral coordinate instead of ϵ^2. As a result, the relative thickness of the obstacle, which can be now identified with a whole body, is estimated to be of the order of ϵ^3; nevertheless the boundary condition (1.7) still holds true. Substituting the latter estimation into (1.4) we may enunciate the similarity law of von Kàrmàn: the fields of ideal gas parameters around affinely-related bodies in transonic flows will be identical under the condition that the ratio $(M^2_\infty - 1) / \delta^{\frac{2}{3}}$ remains constant.

In the hodograph plane a solution of the simple wave type reads

$$v = \pm \frac{2}{3} \left[(K_\infty + u)^{\frac{3}{2}} - K_\infty^{\frac{3}{2}} \right] \tag{1.8}$$

It exists only if $u \geq - K_\infty$, whence follows that the local Mach number $M \geq 1$.

Similarity Solutions

In some cases it is preferable to deal with the single equation

$$- \frac{\partial^2 \varphi}{\partial t \, \partial x} - \left(K_\infty + \frac{\partial \varphi}{\partial x} \right) \frac{\partial^2 \varphi}{\partial x^2} + \frac{\partial^2 \varphi}{\partial y^2} = 0 \tag{2.1}$$

for the potential φ instead of the system (1.3a, b). The aforementioned invariance of the approximate equations with respect to a two-parameter group of similarity transformations as $K_\infty = 0$ ensures the existence of a wide class of similarity solutions describing steady flows with $M_\infty = 1$. These solutions have the form

$$\varphi = \varphi_0 (x, \; y) = y^m f_m (\xi), \qquad \xi = \frac{x}{y^n} , \qquad m = 3n - 2 \tag{2.2}$$

The properties of the function f_n for different n have been investigated by many authors whose results are reviewed in [4 - 6]. Let us present two examples that are typical of aerodynamics.

We begin with the construction of the symmetric velocity field near the center of the Laval nozzle. In this case n varies over a range of $2 \leq n < \infty$. It is natural to require that all the gas parameters may not have singularities of any type along the limiting characteristics coming into the point under consideration because discontinuities in spatial derivatives of velocity components and thermodynamic functions would have to arise at the walls which, however, can be considered arbitrarily smooth. At the same time one should keep in view possible occurrence of singularities on characteristics emanating from the nozzle center: their generation bears no relation to the special contouring of the walls. A shock originating here and spreading downstream cannot be excluded from the analysis for the same reasons.

According to [5], the requirements imposed above are insufficient to derive a unique value of n and the corresponding function f_n. If $n = 2$, the velocity field in the vicinity of the nozzle center is analytic and represents the "natural" asymptotics that have been pointed out for the first time by Meyer. In the flow with $n = 3$ the characteristics emanating from this point carry discontinuities in the third order derivatives of the gas parameters with respect to the coordinates. For $n = 11$ we have a solution including the shock originating on the sonic line and spreading downstream. When proceeding to the limit $n \to \infty$, Ovsjannikov was led to the nozzle with the straight sonic line. The flows constructed may also be treated as "natural" ones because their intrinsic singularities originate at the internal point of the velocity field where eq. (2.1) suffers parabolic degeneration for a steady potential. The question as to whether these flows can be realized in practice remains open up to now. Substantial difference in u-distributions (and, according to (1.3c), in p-distributions as well) for $n = 2$ and $n = 3$ is illustrated in Fig. 1. In the latter case there is a region with the adverse pressure gradient on the nozzle walls.

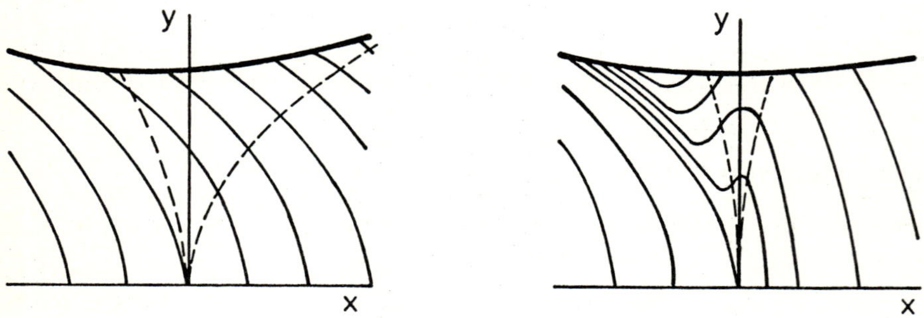

Fig. 1 The flow pattern in Laval nozzles: $a - n = 2$, $b - n = 3$. ——— u = const., - - - - - characteristics

The next example will be an airfoil in an ideal gas flow with $M_\infty = 1$. The problem is to derive the asymptotic behaviour of the velocity field at infinity, whence the range for n is confined by $0 < n < 1$. The subsonic part of the velocity field lies upstream of the body, the upper and lower branches of the sonic line and limiting characteristic spread to infinity. It is natural again to require the limiting characteristic to be free from any singularities in gas parameter derivatives with respect to the coordinates. According to Franckl' and Guderley the solution to this problem defines $n = 4/5$ uniquely. As it was shown later by Landau and Barish & Guderley, the supersonic region of the flow under consideration is crossed by a shock with both branches also spreading to infinity. The simplest parametric form

$$\xi = \frac{1}{6}\left(\frac{3}{4}\right)^{\frac{2}{5}} b^{\frac{1}{5}} \; \frac{8\zeta + 3}{\zeta^{\frac{2}{5}}} ,$$

$$f_{\frac{2}{5}} = \frac{1}{27}\left(\frac{3}{4}\right)^{\frac{1}{5}} b^{\frac{3}{5}} \; \frac{16\zeta^2 + 18\zeta + 27}{\zeta^{\frac{1}{5}}} .$$

$$b = 2^9\, 3^3\, 5^{-5}\, 11^{-2}\,(122 + 9\sqrt{3})$$

of the solution in the region downstream of the shock $\zeta = \zeta_s = (5\sqrt{3} - 8)/8$ is due to Landau and Euvrard.

Stabilization Law and Drag

To demonstrate how the asymptotic approach to flows around bodies builds a bridge between theory and experiment, let us turn to the expansion

$$\varphi = \varphi_0(x, y) + \Delta\,\varphi'(x, y)\ , \qquad \varphi' = y^{\mu}\,\chi_{\mu}(\xi) \tag{3.1}$$

where the basic solution is given by (2.2) and the correction term φ' is caused by the deviation of the oncoming stream velocity from the speed of sound. The problem is to determine the exponent μ that relates the coefficient Δ to the difference $M_\infty^2 - 1$.

A rather simple analysis based on the aforementioned invariance of the equation (2.1) with respect to a two-parameter group of similarity transformations yields $\Delta = |M_\infty^2 - 1|^{\frac{(5\mu-2)}{2}}$. To find the value of μ, we consider first the region upstream of the shock. Here the conventional requirement for the limiting characteristic to carry no singularities leads to $\mu = 8/5$. Taking into account this result, Diesperov, Lifshitz & Ryzhov [7] enunciated the stabilization law (or freezing effect): the gas parameters over an airfoil segment ahead of the shock deviate from the corresponding parameters in strictly sonic flow by a value proportional to $|M_\infty^2 - 1|^3$, where M_∞ can be greater as well as less than 1. A boundary-value problem for the correction potential has been recently explored by Cook & Ziegler [8] using the matched asymptotic expansion technique.

As for the region downstream of the shock, the appearance of eigenfunctions is possible there. As a result, the basic solution for $M_\infty = 1$ reads

$$\varphi_0 = y^{\frac{2}{5}}\, f_{\frac{2}{5}}(\xi) + y^{\frac{1}{5}}\, f_{\frac{1}{5}}(\xi) + \ \dots\ , \qquad f_{\frac{1}{5}} = \frac{C_{\frac{1}{5}}}{\zeta^{\frac{1}{10}}\,(1+\zeta)^{\frac{5}{6}}}$$

The exponent in the representation (3.1) for the correction potential is changed from $\mu = 8/5$ to $\mu = 3/5$ and the correction potential itself takes the form

$$\chi_{\frac{3}{5}} = C^{s}_{\frac{3}{5}}\, \chi^{s}_{\frac{3}{5}}(\xi) + C^{a}_{\frac{3}{5}}\, \chi^{a}_{\frac{3}{5}}(\xi)$$

$$\chi^{s}_{\frac{3}{5}} = \frac{\zeta + \frac{3}{4}}{\zeta^{\frac{3}{10}}\,(1+\zeta)^{\frac{5}{6}}}\ , \qquad \chi^{a}_{\frac{3}{5}} = \zeta^{\frac{1}{5}}\, F\left(2, \frac{1}{3}, \frac{3}{2}; -\zeta\right)$$

F being the hypergeometric function. As a consequence we have $\Delta = |M_\infty^2 - 1|^{\frac{1}{2}}$. Thus, in spite of the gas parameter stabilization over the leading part of an airfoil, the drag increment that occurs primarily owing to the shock movement towards the trailing edge obeys the law [7]

$$C_D(M_\infty) - C_D(1) \sim |M_\infty^2 - 1|^{\frac{1}{2}}$$

The properties of the transonic flows under discussion were revealed by Gal'perin, Gorsky, Kovalev & Christianovitch [9] when analyzing experimental data obtained as far back as 1944 - 1948. The most comprehensive account of the results of numerous relevant measurements was presented by Holder [10] in 1964. The asymptotic analysis allows to give the form of the quantitative estimations to the qualitative conclusions that could be drawn from experimental data. Recent calculations corroborate the above formulation of the stabilization law (see Fig. 2 taken from [7]).

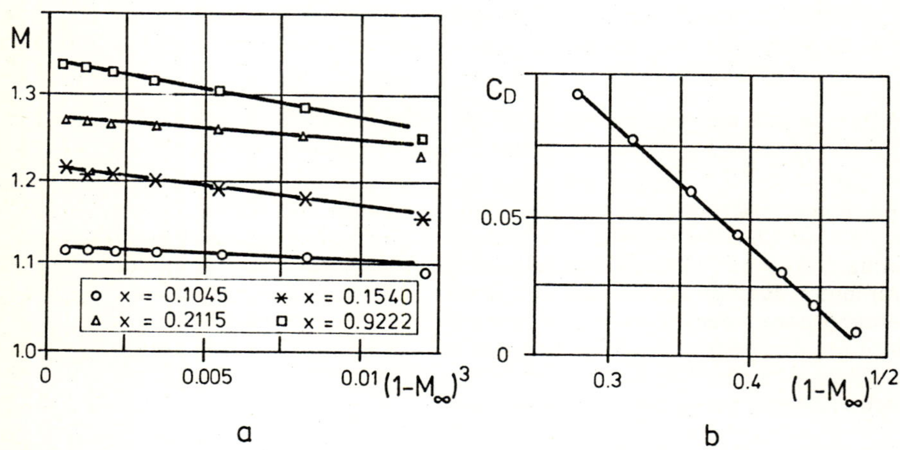

Fig. 2 The stabilization law: *a* - Mach numbers at fixed points ahead of a shock, *b* - drag variation

Interacting Viscous Sublayer

In this region the transformations of the time and longitudinal coordinates obey eq. (1.1a, b), whereas the lateral coordinate is introduced by

$$\frac{y^*}{x^*_0} = \epsilon^6\, 2^{-\frac{1}{10}} C^{\frac{3}{5}} \lambda^{-\frac{4}{5}} \left(\frac{T^*_w}{T^*_\infty}\right)^{\frac{7}{5}} y' \tag{4.1}$$

The desired functions are scaled by means of

$$\frac{u^*}{u^*_\infty} = \epsilon\, 2^{-\frac{1}{10}} C^{\frac{1}{10}} \lambda^{\frac{1}{5}} \left(\frac{T^*_w}{T^*_\infty}\right)^{\frac{2}{5}} u' \tag{4.2a}$$

$$\frac{v^*}{u^*_\infty} = \epsilon^4 2^{\frac{1}{10}} C^{\frac{2}{5}} \lambda^{\frac{4}{5}} \left(\frac{T_w^*}{T_\infty^*} \right)^{\frac{3}{5}} v' \tag{4.2b}$$

$$\frac{(p^* - p^*_\infty)}{(\rho^*_\infty u^{*2}_\infty)} = \epsilon^2 2^{-\frac{1}{5}} C^{\frac{1}{5}} \lambda^{\frac{2}{5}} \left(\frac{T_w^*}{T_\infty^*} \right)^{-\frac{1}{5}} p' \tag{4.2c}$$

Here the small parameter ϵ is related to the Reynolds number $R \to \infty$ as $\epsilon = R^{-\frac{1}{10}}$.

Substitution of (1.1a, b), (4.1) and (4.2) into the Navier-Stokes equations reduces them to the simpler Prandtl equations

$$u' \frac{\partial u'}{\partial x} + v' \frac{\partial u'}{\partial y'} = - \frac{\partial p'}{\partial x} + \frac{\partial^2 u'}{\partial y'^2} \tag{4.3a}$$

$$\frac{\partial u'}{\partial x} + \frac{\partial v'}{\partial y'} = 0, \qquad \frac{\partial p'}{\partial y'} = 0 \tag{4.3b, c}$$

for an incompressible laminar boundary layer. The absence of time implies that the viscous flow in the wall region is capable of adjusting instantaneously to low frequency perturbations propagating through the external potential region. To elucidate how this adjustment occurs, one needs to study processes in the greatest part of the boundary layer. Standard manipulations yield

$$p'(t, x) = -u(t, x, 0) \tag{4.4}$$

Allowing for (1.3c) and (4.3c). The boundary conditions

$$v = -\frac{\partial A}{\partial x} \quad \text{at} \quad y = 0 \tag{4.5a}$$

$$u' - y' \to A(t, x) \quad \text{as} \quad y' \to -\infty \tag{4.5b}$$

serve to close the system of the equations (1.3a, b) and (4.3a, b) by use of a new desired function A which describes the instantaneous displacement thickness. The condition of matching with the oncoming rotational flow is

$$u' \to y' \quad \text{as} \quad x \to -\infty \tag{4.6}$$

The boundary conditions on the body surface depend on the nature of the problem.

The extension of the free interaction theory covering steady transonic flows was done by Messiter, Feo & Melnik [11] and Bodonyi & Kluwick [12], the time dependence was included by Ryzhov [13]. Provided the conditions (1.7) and (4.5a) are similar, the function A defines at any time-step the contour of an effective body immersed into an ideal gas flow. The relative thickness δ in the contour equation (1.5) is of the same order of magnitude as the characteristic thickness of the viscous sublayer, K_∞ being the only similarity parameter.

Free Interaction in Supercritical Regime

Under the condition that the velocity of a steady uniform stream at infinity is supersonic, Neiland and Stewardson & Williams derived the non-trivial solution for a free interacting boundary layer, which satisfies the homogeneous boundary conditions

$$u' = v' = 0 \qquad \text{at} \qquad y' = 0 \tag{5.1}$$

This solution describes compressive flows with separation as well as expansive flows with singularity in the velocity field. The pressure disturbances in both types of flows decay exponentially upstream according to the asymptotic representation

$$p' = a\, e^{k_0 x}, \qquad k_0 = \left(\frac{3^{\frac{7}{6}}\, \Gamma\, \frac{2}{3}}{2\pi} \right)^{\frac{3}{4}} \tag{5.2}$$

brought out by Lighthill as long ago as 1953. As was mentioned in [12], this representation remains valid for transonic flows in the supercritical regime with $M_\infty > 1$ if the independent variables and desired functions are appropriately transformed. The disturbance attenuation becomes of an algebraic type if $M_\infty = 1$. In the last case we derive from (1.8) and (4.5a)

$$A = -\frac{2}{3} \int_{-\infty}^{x} [-p'(X)]^{\frac{3}{2}}\, dX \tag{5.3}$$

whence follows that [12]

$$p' = a(-x)^{-\frac{8}{3}}, \qquad a < 0 \tag{5.4}$$

for the velocity in the external potential fiel must be supersonic. Note that in view of (5.3) the asymptotic law (5.4) is in fact nonlinear.

Taking (5.2) and (5.4) as $x \to -\infty$ one gets the solutions of the Prandtl equations which can be treated as nonlinear eigenfunctions. Each of these functions is defined by the value of the similarity parameter K_∞. The existence of solutions for compressive flows is limited by the requirement that the velocity in the external potential field remain supersonic everywhere.

Trailing Edge Flows

Historically, it was the trailing edge flow for which the free interaction theory was developed by Stewardson and Messiter who assumed the fluid to be incompressible. For supersonic flows the same problem was studied by Daniels. The transonic range of velocities has been investigated by Bodonyi & Kluwick [14] and Korolev [15]. Their results will be exposed below.

The boundary conditions on a plate occupying the negative semi-axis $x < 0$ are of the form (5.1). On the center line of a wake they are supplemented by the constraints

$$u' = \frac{\partial u'}{\partial y'} = 0 \quad \text{at} \quad y' = 0, \; x > 0 \tag{6.1}$$

The problem is to integrate the equations (1.3a, b), (4.3a, b) and (4.4) simultaneously under the general conditions (4.5a, b) and (4.6) as well as under the specific conditions (5.1) and (6.1) for the given type of flows. The analysis of the supercritical regime admits essential simplifications because in this case the solution (1.8) of the equations (1.3a, b) is known beforehand.

The pressure distributions in the viscous sublayer are plotted in Fig. 3. In the subcritical regime with $K_\infty < 0$ the displacing effect of the trailing edge increases as the absolute value of the similarity parameter decreases. The flow accelerates to sonic velocity at some point when $K_\infty = K_\infty^{cr} = -0.43$. With K_∞ varying over a range of $K_\infty^{cr} < K_\infty < 0$ the external potential field involves a local supersonic pocket terminated in a shock. The expansive flow in the supercritical regime with $K_\infty > 0$ is given by the nonlinear eigenfunction discussed in the previous section. The following compression downstream of the trailing edge retains the external field supersonic. The influence of the trailing edge becomes especially strong as $K_\infty \to 0$ which is in line with the difference in the asymptotics (5.2) and (5.4).

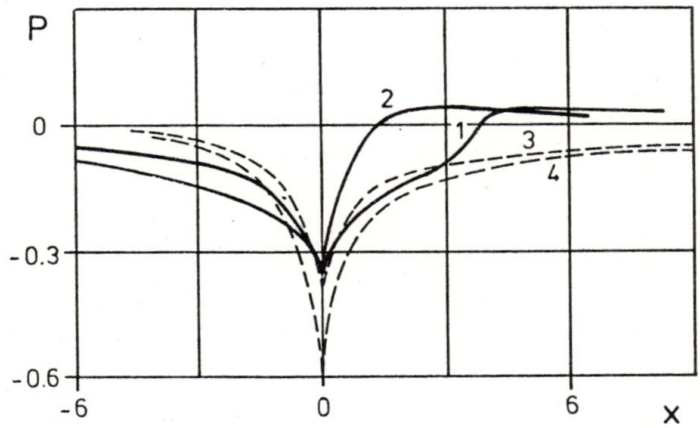

Fig. 3 The pressure distributions in the vicinity of the trailing edge of a flat plate. $1: K_\infty = -.05$, $2: K_\infty = -\infty$, $3: K_\infty = .1$, $4: K_\infty = \infty$

For the strictly sonic oncoming stream the additional drag coefficient ΔC_D caused by the trailing edge is estimated as

$$\Delta C_D \sim R^{-\frac{4}{5}} C^{\frac{4}{5}} \lambda^{-\frac{2}{5}} \left(\frac{T_w^*}{T_\infty^*} \right)^{\frac{3}{2}}$$

Flows Around Obstacles

The laminar boundary layer near convex and concave corners has been studied by Bodonyi [16] and Korolev [17], respectively, under the same assumption that the regime was supercritical. Qualitatively, their results are quite similar to those which were obtained earlier by many authors for a supersonic oncoming stream. In particular, a concave corner, if large enough, is capable of producing the adverse pressure gradient that is sufficient for causing boundary layer separation.

Of greater concern is the flow pattern around a small hump on an otherwise smooth surface. For $M_\infty = 1$ the excess pressure upstream of the obstacle decays according to the Franckl'-Guderley solution, whence

$$p' = a(-x)^{-\frac{1}{2}} , \quad a > 0 \tag{7.1}$$

The comparison of (5.4) with (7.1) shows that the hump affects the far-field disturbances more heavily than the expansion of a free-interacting boundary layer.

Calculations made by Korolev for an obstacle of the shape $y'_w = h_0(1 + \cos x)$, $-\pi \leq x \leq \pi$ are illustrated in Fig. 4 ($h_0 = 2$). Simultaneous integration of the equations (1.3a, b), (4.3a, b) and (4.4) subjected to the boundary

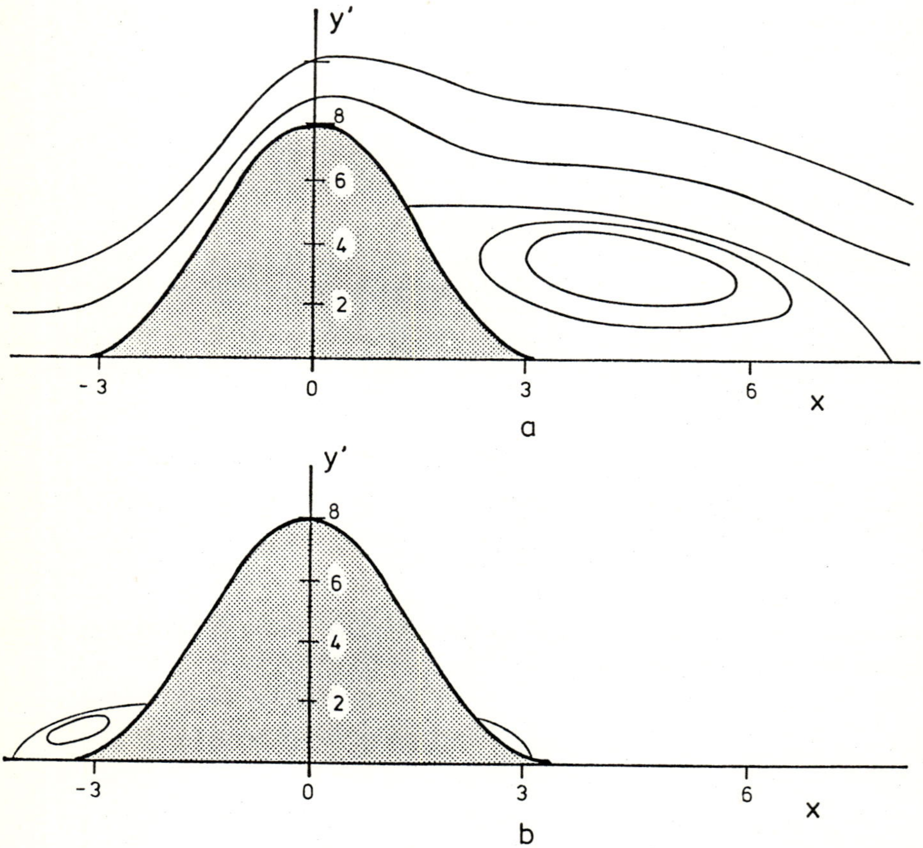

Fig. 4 The flow past a hump on an otherwise flat plate: $a: K_\infty = -3$, $b: K_\infty = 3$

conditions (4.5a), (4.6) and $u' = v' = 0$ at $y' = y'_w$ was performed for $K_\infty = \pm 3$. The flow pattern changes drastically when the supercritical regime supersedes the subcritical one. In the former case, separation occurs first on the windward side of the hump, whereas the latter case is characterized by the developing of a recirculation bubble on its leeward side. If the value of $|K_\infty|$ is supposed to be fixed, the size of the separation zone in a subcritical boundary layer will be much greater than in a supercritical one.

The following remark should be made in connection with the results under discussion. The separation bubble depicted in Fig. 4a is essentially subsonic in nature and qualitatively very similar to that emerging in the incompressible flow past the same hump, (see, for example, [18]). Meanwhile, the boundary layer on an airfoil in the subcritical regime separates, as a rule, under the influence of the pressure rise across a shock front terminating a local supersonic pocket. We need additional calculations to clear up the essence of this phenomenon. Besides, in practice the boundary layer is always turbulent. Some important results on the interaction of a weak normal shock wave with a turbulent boundary layer have been reported by Adamson, Melnik and Messiter, but they are beyond the scope of this review.

Unsteady Boundary Layer

The simultaneous integration of the equations (1.3a, b), (4.3a, b) and (4.4) subjected to the boundary conditions (4.5a, b), (4.6) and (5.1) resolves the problem of the Blasius boundary layer response to the low frequency disturbances in linear as well as in nonlinear approximations. First of all, attention should be paid to the fact that time enters the equations for the potential flow region and drops out of all the relations for the viscous sublayer [13]. For this reason the unsteady characteristics in the transonic range are determined by processes which are completely different from those inherent in flows with sub- or supersonic velocities.

Let ω and k, respectively, be the frequency wave number in the system of units specified by scalings (1.1) and (1.2) for the potential region. Then the pressure perturbations with the amplitude a tending to zero can be represented by

$$p = a \exp \left[i(\omega t + kx - \sqrt{K_\infty k^2 + \omega k}\ y) \right] \tag{8.1}$$

where by the root $\sqrt{K_\infty k^2 + \omega k}$ we mean its branch with the negative imaginary part. This excess pressure gives rise to oscillations in the boundary layer with ω and k connected by the dispersion relation

$$(K_\infty k^2 + \omega k)^{\frac{1}{2}} = k \left(\frac{ik}{k_0} \right)^{\frac{4}{3}} \tag{8.2}$$

A single valued branch of the function $k^{\frac{1}{3}}$ can be chosen by drawing a cut along the positive imaginary semi-axis in the complex k-plane. The constant k_0 is prescribed by the second of the formulae (5.2).

The above constraints on the left-hand side of (8.2) result as a consequence of the inequalities

$$\frac{-5\pi}{7} \leq \arg k \leq \frac{-2\pi}{7} \tag{8.3}$$

They imply that we cannot specify, at an initial time-step $t = 0$, the pressure at the outer edge of the viscous sublayer as a harmonic function of the longitudinal coordinate and trace the ensuing evolution of oscillations in time. This is a distinctive feature of the disturbances at hand that makes them completely dissimilar to the Tollmien-Schlichting waves of the classic hydrodynamic stability theory. The pressure in the form (8.1) necessarily decays upstream according to (8.3), regardless of the value of the parameter K_∞. Thus, the velocity of the oncoming stream can be both sub- and supersonic in our consideration.

On the other hand, the frequency

$$\omega = k\left[\left(\frac{ik}{k_0}\right)^{\frac{8}{3}} - K_\infty\right]$$

is probably characteristic of unsteady processes inherent in the beginning and development of boundary layer separation. In this case the analysis presented yields the asymptotics of disturbances as $x \to -\infty$. For revealing a tendency that properly features the changes in stability in the transonic range of velocities, one needs to turn to the triple deck theory for a subsonic boundary layer (supersonic boundary layers are, as known, stable against two-dimensional oscillations in the framework of this theory). It is impossible, however, to achieve overlapping of the results if Mach numbers are sufficiently close to 1 owing to the differences in time scalings for sub- and transonic flows.

Acknowledgement

The author is very much obliged to Mrs. E.A. Filippina and Dr. V.I. Zhuk for their kind assistance in preparing the manuscript of this lecture.

References

[1] Fal'kovich, S.V.: Prikl. Mat. Mekh. 10 (1946) 503 (in Russian).
[2] Von Kàrmàn, Th.: J. Math. Phys.26 (1947) 182.
[3] Lin, C.C., Reissner, E. & Tsien, H.S.: J. Math. Phys. 27 (1948) 220.
[4] Guderley, K.G. The Theory of Transonic Flow. London: Pergamon 1962.
[5] Ryzhov, O.S. A Study of Transonic Flows through Laval Nozzles. Moscow: Computing Center, Akad. Nauk SSSR 1965 (in Russian).
[6] Cole, J.D. & Cook, L.P. Transonic Aerodynamics. Amsterdam: North-Holland 1986.
[7] Diesperov, V.N., Lifshitz, Yu.B. & Ryzhov, O.S. In: Proc. IUTAM Symp. Transsonicum II, Gottingen 1975. Berlin: Springer 1976.
[8] Cook, L.P. & Ziegler, F.J.: SIAM J. Appl. Math. 46 (1986) 27.
[9] Gal'perin, V.G., Gorsky, I.P., Kovalev, A.P. & Christianovitch, S.A.: Uch. Zapiski TSAGI 5 (1974), No. 5, 1 (in Russian).
[10] Holder, D.W.: J. Roy. Aeronaut. Soc. 68 (1964) 501.
[11] Messiter, A.F., Feo, A. & Melnik, R.E.: AIAA J. 9 (1971) 1197.
[12] Bodonyi, R.J. & Kluwick, A.: Phys. Fluids 20 (1977) 1432.

[13] Ryzhov, O.S.: Dokl. Akad. Nauk SSSR 236 (1977) 1091 (in Russian).
[14] Bodonyi, R.J. & Kluwick, A.: Quart. J. Mech. Appl. Math. 35 (1982) 265.
[15] Korolev, G.L.: Izv. Akad. Nauk SSSR, Mekh. Zhidk. Gaza (1983) No. 3, 23 (in Russian).
[16] Bodonyi, R.J.: Quart. J. Mech. Appl. Math. 32 (1979) 63.
[17] Korolev, G.L.: Uch Zapiski TSAGI 14 (1983) No. 4, 105 (in Russian).
[18] Smith, F.T. & Merkin, J.H.: Comp. Fluids 10 (1982) 7.

Some Asymptotic Problems of Transonic Flow Theory*

J.D. Cole and L.P. Cook

Rensselaer Polytechnic Institute, Troy, NY
University of Delaware, Newark, DE, USA

Summary

The use of asymptotic methods to derive characteristic problems of transonic flow theory is illustrated in two ways. First suitable limit process expansions in angle of attack α (with Mach number $M_\infty \to 1$) are constructed in the regions near and far from pointed wings with aspect ratio $O(1)$ and thickness ratio $\delta = O(\alpha)$. The matching of these expansions defines suitable boundary value problems. Second, expansions in terms of the similarity parameter $K = \frac{1-M_\infty^2}{\alpha}$ show the generalization of the Law of Stabilization for this case.

Basic Equations and Problems

The basic configuration is shown in Fig. 1. The body shape is given by

$$B(x,y,z) = 0 = y - \alpha F_{u,\ell}(x,z) \qquad 0 < x < 1, \quad |z| < z_{LE}(x) \tag{1}$$

with the chord ℓ normalized to be unity. $F_{u,\ell}(x,z)$ is the equation of the upper and lower surfaces respectively. For simplicity wings with an untwisted mean surface are considered here. In terms of a thickness distribution $t(x,z)$ and a camber distribution $c(x)$,

$$F_{u,\ell}(x,z) = -x + c(x) \pm Tt(x,z)$$

where $T = \frac{\delta}{\alpha}$, $\delta =$ thickness ratio. To the orders considered the flow is irrotational and isentropic and it is convenient to use the potential equation to represent the Euler equations. Thus: velocity $\overrightarrow{q} = \nabla\Phi$ and $p/p_\infty = (\rho/\rho_\infty)^\gamma$, $p =$ pressure, $\rho =$ density and

$$\begin{aligned} a^2(\Phi_{yy} + \Phi_{zz}) = &-(a^2 - \Phi_x^2)\Phi_{xx} + 2\Phi_x(\Phi_y\Phi_{xy} + \Phi_z\Phi_{xz}) \\ &+ 2\Phi_y\Phi_z\Phi_{yz} + \Phi_y^2\Phi_{yy} + \Phi_z^2\Phi_{zz} \end{aligned} \tag{2}$$

where $a =$ local speed of sound:

* This work was supported by the US Air Force Office of Scientific Research under grant AFOSR 88-0037 and the National Science Foundation under grant NSF RII 8620258

J. Zierep, H. Oertel (Eds.)
Symposium Transsonicum III
IUTAM Symposium Göttingen, 24.-27. 5. 1988

$$\frac{a^2}{U^2} = \frac{1}{M_\infty^2} + \frac{\gamma-1}{2}\{1-\frac{q^2}{U^2}\}, \quad \frac{a^2}{a_\infty^2} = \left(\frac{\rho}{\rho_\infty}\right)^{\gamma-1} = \left(\frac{p}{p_\infty}\right)^{\frac{\gamma-1}{\gamma}}$$

The solution $\Phi(x,y,z)$ must be found such that the boundary condition of tangent flow

$$\overrightarrow{q} \cdot \nabla B = 0 \qquad \text{on } B = 0 \tag{3}$$

is satisfied, disturbances die out at upstream infinity and the Kutta condition is satisfied at a sharp subsonic trailing edge. For simplicity the case of a trailing edge normal to the stream is considered here, although the theory in general covers other cases.

Inner Expansion and Far Fields

The inner expansion valid initially in a region $(y,z) \cdot O(1)$ around the wing and vortex wake is associated with a limit process $\alpha \to 0, M_\infty \to 1, (x, y^*, z; K, T \text{ fixed})$ where K = transonic similarity parameter = $\frac{1-M_\infty^2}{\alpha}$ for this case; y^* is a vertical coordinate measured from the body surface or wake

$$\begin{aligned} y^* &= y - \alpha f(x) \quad 0 < x < 1 \\ &= y - \alpha f(1) \quad x > 1, f(1) = 0 \end{aligned} \tag{4}$$

and $f(x) = -x + c(x)$

The inner expansion represents small disturbances on uniform flow in the inner length scale and has the form

$$\Phi = U\left\{x + \alpha \log \frac{1}{\sqrt{\alpha}}(-S_1(x)) + \alpha\varphi_1(x,y^*,z) + \alpha^2\hat{\varphi}_2 + \cdots\right\} \tag{5}$$

where

$$\hat{\varphi}_2 = \log^2 \frac{1}{\sqrt{\alpha}}\left(-\frac{1}{2}S_{31}(x) + \varphi_{22}\right) + \log\frac{1}{\sqrt{\alpha}}(G_{31}(x) - S_3(x) + \varphi_{21}) + \varphi_2$$

$\alpha\varphi_1$ is the dominant term needed to satisfy the boundary condition (3) and the log terms are switchback terms needed for matching to the outer expansion. $S_1(x)$ is a source strength. Thus, velocity components are

$$\frac{q_x}{U} = \frac{1}{U}\frac{\partial\Phi}{\partial x} = 1 + \alpha\log\frac{1}{\sqrt{\alpha}}(-S_1') + \alpha\varphi_{1_x} + \alpha^2(\hat{\varphi}_{2_x} - \varphi_{1_{y^*}} f_x)$$
$$\frac{q_y}{U} = \frac{1}{U}\frac{\partial\Phi}{\partial y} = \alpha\varphi_{1_{y^*}} + \alpha^2\hat{\varphi}_{2_{y^*}} + \cdots$$
$$\frac{q_z}{U} = \frac{1}{U}\frac{\partial\Phi}{\partial z}\alpha\varphi_{1_z} + \alpha^2\hat{\varphi}_{2_z} + \cdots$$

From the basic equation (2) and boundary condition (3) we find a sequence of inner equations and boundary conditions on the wing $0 < x < 1$.

$$\varphi_{1_{y^*y^*}} + \varphi_{1_{zz}} = 0\,, \quad \varphi_{1_{y^*}}(x, 0\pm, z) = F_{u,\ell_x} \tag{6}$$

$$\varphi_{22_{y^*y^*}} + \varphi_{22_{zz}} = (\gamma+1)S_1'S_1'', \qquad \varphi_{22_{y^*}}(x,0,z) = 0 \tag{7}$$

$$\varphi_{21_{y^*y^*}} + \varphi_{21_{zz}} = (KS_1' - (\gamma+1)S_1'\varphi_{1_x})_x \qquad \varphi_{21_{y^*}}(x,0,z) = 0 \tag{8}$$

$$\begin{aligned}\varphi_{2_{y^*y^*}} + \varphi_{2_{zz}} &= -(K - (\gamma+1)\varphi_{1_x})\varphi_{1_{xx}} + (\varphi_{1_{y^*}}^2 + \varphi_{1_z}^2)_x + 2\varphi_{1_{y^*z}}F_z + \varphi_{1_{y^*}}F_{zz}\\ &\varphi_{2_{y^*}}(x,0\pm,z) = F_{u,\ell_x}\varphi_{1_x} + F_{u,\ell_z}\varphi_{1_z},\ y^* = 0\pm\end{aligned} \tag{9}$$

Eqn. (6) is the Laplace eqn. in a cross-plane just as in Jones' [1] original theory. However, here it is better to think of this equation as the linearized theory (Prandtl-Glauert) eqn. with Mach number effectively unity. The sequence of cross-plane problems can be solved over wing and wake with far-field conditions determined by matching with the near field behavior of the outer solutions.

The first inner problem can be split into symmetric (non-lifting) and anti-symmetric (lifting) parts $\varphi_1 = \varphi_1^S + \varphi_1^A$where $\varphi_1^S(y) = \varphi_1^S(-y), \varphi_1^A(y) = -\varphi_1^A(-y)$ with corresponding boundary conditions on the wing

$$\varphi^A_{1_{y^*}}(x,0,z) = -1 + c_x = f_x, \qquad \varphi^S_{1_{y^*}}(x,0\pm,z) = \pm Tt_x \tag{10}$$

The far-field of φ_1^A looks like a dipole with intensity equal to the scaled lift $\ell_1(x)$ up to the station x,

$$\begin{aligned}\varphi_1^A &\to \frac{\ell_1(x)}{2\pi}\frac{\cos\theta^*}{r^*} + O\left(\frac{1}{r^{*3}}\right), \quad 0 < x < 1 \ \text{(wing symmetric in } z)\\ &\to \frac{\ell_1(1)}{2\pi}\frac{\cos\theta^*}{r^*} + O\left(\frac{1}{r^{*3}}\right), \qquad x > 1 \quad r^* = \sqrt{y^{*2}+z^2}, \theta^* = \tan^{-1}\frac{z}{y^*}\end{aligned} \tag{11}$$

The explicit solution for φ^A is given by the flow around a flat plate.

$$\varphi_1^A + i\psi_1^A = -if_x\left\{\xi - \sqrt{\xi^2 - z_{LE}^2(x)}\right\}, \xi = z + iy^* = r^* e^{i(\frac{\pi}{2}-\theta^*)} \tag{12}$$

In this case, using

$$\frac{p}{p_\infty} = 1 - \delta\left\{-\alpha \log \frac{1}{\sqrt{\alpha}} S_1' + \alpha\varphi_{1_x}\right\} + \cdots \tag{13}$$

$$\ell_1(x) = \frac{L(x)}{\rho_\infty U^2 \ell^2 \alpha} = \pi(-f_x z_{LE}^2(x)); \quad L(x) = \text{lift up to station } x (\text{Jones' result}) \tag{14}$$

As consequence of the Kutta condition could be taken that the jump in pressure $[p]_{TE} = (p(0+) - p(0-))$ at the trailing edge is zero.* This is satisfied automatically if the planform $z_{LE}(x)$ and $f(x)$ are such that

$$[\varphi_{1_x}]_{TE} = -2\left\{f_{xx}\sqrt{z_{LE}^2 - z^2} + f_x \frac{z_{LE} z_{LE}'}{\sqrt{z_{LE}^2 - z^2}}\right\} = 0 \text{ at } x = 1 \tag{15}$$

Otherwise a local solution needs to be introduced in the trailing edge region. The symmetric part φ_1^S is represented by a source distribution over $y^* = 0$

$$\varphi_1^S = \frac{T}{x}\int_{-z_{LE}}^{z_{LE}} t_z(x;z') log\sqrt{y^{*2} + (z-z')^2}dz' + G_1(x) \tag{16}$$

$$\varphi_1^S \to S_1(x)\log r^* + G_1(x) + O\left(\frac{1}{r^{*2}}\right), S_1 = \frac{T}{\pi}\frac{\partial}{\partial x}\int_{-z_{LE}}^{z_{LE}} t(x;z')dz' \tag{17}$$

$G_1(x)$ is determined from the order solution.

The far-field of the second order potential φ_2 is found from the particular solution of (9) corresponding to the dominant terms on the right hand side. Splitting again into symmetric and anti-symmetric parts

$$\varphi^S_{2_{y^*y^*}} + \varphi^S_{2_{zz}} = (\gamma+1)\varphi^S_{1_x}\varphi^S_{1_{xx}} + \cdots$$

* This can also be thought of as the requirement that the pressure does not jump across the trailing edge plane $x = 1$.

or

$$\varphi^S_{2_{r^*r^*}} + \frac{1}{r^*}\varphi^S_{2_{r^*}} = (\gamma+1)S_1'S_1'' log^2 r^* - \{KS_1' - (\gamma+1)S_1'G_1'\}' \log r^* - \{K - (\gamma+1)G_1'\}\, G_1'' + O\left(\frac{\log r^*}{r^{*2}}\right) \tag{18}$$

The particular solution as $r^* \to \infty$ is

$$\begin{aligned}\varphi_2^S \to (\gamma+1)S_1'S_1''&\left\{\frac{{r^*}^2}{4}\log^2 r^* - \frac{{r^*}^2}{2}\log r^* + \frac{3}{8}{r^*}^2\right\}\\ &- \{KS_1' - (\gamma+1)S_1'G_1'\}'\left\{\frac{{r^*}^2}{4}\log r^* - \frac{{r^*}^2}{4}\right\}\\ &- (K - (\gamma+1)G_1')G_1''\frac{{r^*}^2}{4} + \cdots\end{aligned} \tag{19}$$

A comparison of (19) with (17) shows the non-uniformity of the inner expansion as $r^* \to \infty$ and the necessity of an outer expansion. The non-uniformity is produced by the typical transonic term $\varphi^S_{1_x}\varphi^S_{1_{xx}}$. The far-field of the anti-symmetric part follows similarly $\varphi^A_{2_{y^*y^*}} + \varphi^A_{2_{zz}} = (\gamma+1)(\varphi^A_{1_x}\varphi^S_x)_x + \cdots$ or

$$\varphi^A_{2_{r^*r^*}} + \frac{1}{r^*}\varphi^A_{2_{r^*}} + \frac{1}{r^{*2}}\varphi^A_{2_{\theta^*\theta^*}} = \cos\,\theta^*\left\{(\gamma+1)\frac{(S_1'\ell_1')'}{2\pi}\frac{\log r^*}{r^*} - \frac{(K\ell_1' - (\gamma+1)\ell_1'\ell_1'\prime)'}{2\pi}\frac{1}{r^*} + \cdots\right\} \tag{20}$$

$$\varphi_2^A \to \cos\,\theta^*\left\{(\gamma+1)\frac{(S_1'\ell_1')'}{2\pi}\left(\frac{r^*}{4}\log^2 r^*\right) \left(\frac{K\ell_1' - (\gamma+1)\ell_1'G_1'}{2\pi}\right)'\frac{r^*}{2}\log\, r^* + O\left(\frac{\log\, r^*}{r^*}\right) + \cdots\right\} \tag{21}$$

The coupling of lift and thickness also produces a non-uniformity as $r^* \to \infty$ but of a smaller order than for φ_2^S.

Outer Expansion and Near Field

The outer expansion is generated from a typical transonic limit to give a non-linear outer equation capable of representing mixed flows and shock waves. The limit process has $\alpha \to 0, M_\infty \to 1, (x, \tilde{y}, \tilde{z}; K$ fixed) where $(\tilde{y}, \tilde{z}) = \sqrt{\alpha}(y, z)$. The form is

$$\Phi = U\left\{x + \alpha\phi_1(x,\tilde{y},\tilde{z}) + \alpha^{3/2}\phi_2(x,\tilde{y},\tilde{z}) + \cdots\right\} \tag{22}$$

The velocity components and speed of sound are given by

$$\begin{aligned} \frac{q_x}{U} &= 1 + \alpha\phi_{1_x} + \alpha^{3/2}\phi_{2_x} + \cdots, \\ \frac{q_y}{U} &= \alpha^{3/2}\phi_{1_{\tilde{y}}} + \alpha^2\phi_{2_{\tilde{y}}} + \cdots, \\ \frac{q_z}{U} &= \alpha^{3/2}\phi_{1_{\tilde{z}}} + \alpha^2\phi_{2_{\tilde{z}}} + \cdots \end{aligned} \tag{23}$$

$$\frac{a^2}{U^2} = 1 + \alpha\left\{K - (\gamma-1)\phi_{1_x}\right\} - \alpha^{3/2}(\gamma-1)\phi_{2_x} + \cdots \tag{24}$$

From (2) we find

$$(K - (\gamma+1)\phi_{1_x})\phi_{1_{xx}} + \tilde{\nabla}^2\phi_1 = 0, \quad \tilde{\nabla}^2 = \frac{\partial^2}{\partial\tilde{y}^2} + \frac{\partial^2}{\partial\tilde{z}^2} \tag{25}$$

$$(K - (\gamma+1)\phi_{1_x})\phi_{2_{xx}} - (\gamma+1)\phi_{1_{xx}}\phi_{2_x} + \tilde{\nabla}^2\phi_2 = 0 \tag{26}$$

(25) is the transonic small-disturbance equation valid far away from the wing and (26) is the variational equation. Eqns. (25) and (26) are supplemented with shock jump conditions in the usual way. ϕ_1 has a dominant $S(x)\log\tilde{r}$, $\tilde{r} = \sqrt{\tilde{y}^2 + \tilde{z}^2}$ term representing a line source singularity, which is necessary for matching. Thus

$$\begin{aligned} \phi_{1_{\tilde{r}\tilde{r}}} + \frac{1}{\tilde{r}}\phi_{1_{\tilde{r}}} &= (\gamma+1)\phi_{1_x}\phi_{1_{xx}} - K\phi_{1_{xx}} \\ &= (\gamma+1)S_1'S_1''\log^2\tilde{r} - \{KS_1' - (\gamma+1)S_1'G_1'\}'\log\tilde{r} \\ &\quad - \{K - (\gamma+1)G_1'\}G_1'' + \cdots \end{aligned}$$

so that as $\tilde{r} \to 0$

$$\begin{aligned} \phi_1 \to & S_1(x)\log\,\tilde{r} + G_1(x) \\ & + (\gamma+1)S_1'S_1''\left\{\frac{\tilde{r}^2}{4}log^2\tilde{r} - \frac{\tilde{r}^2}{2}\log\tilde{r} + \frac{3}{8}\tilde{r}^2\right\} \\ & - \{KS_1' - (\gamma+1)S_1'G_1'\}'\left\{\frac{\tilde{r}^2}{4}\log\tilde{r} + \frac{3}{8}\tilde{r}^2\right\} \\ & - \{K - (\gamma+1)G_1'\}G_1''\frac{\tilde{r}^2}{4} + 0(\tilde{r}^4\log^3\tilde{r}) \end{aligned} \tag{27}$$

Similarly ϕ_2 has a dipole singularity to match φ_1^A, $\phi_2 \to \frac{\ell_1(x)}{2\pi}\frac{\cos\,\theta}{\tilde{r}}, + \cdots$

$$\phi_{2_{\tilde{r}\tilde{r}}} + \frac{1}{\tilde{r}}\phi_{2_{\tilde{r}}} + \frac{1}{\tilde{r}^2}\phi_{2_{\theta\theta}} = (\gamma+1)(\phi_{1_x}\phi_{2_x})_x - K\phi_{2_{xx}}$$

$$= \cos\ \theta\left\{(\gamma+1)\frac{(S_1'\ell_1')'}{2\pi}\frac{\log\tilde{r}}{\tilde{r}} + \frac{((\gamma+1)G_1'\ell_1' - K\ell_1')'}{2\pi}\frac{1}{\tilde{r}} + \cdots\right\}$$

$$\phi_2(x,\tilde{r},\theta) \to \frac{\ell_1}{2\pi}\frac{\cos\theta}{\tilde{r}} + (\gamma+1)\frac{(S_1'\ell_1')'}{2\pi}\left\{\frac{\tilde{r}}{4}\log^2\tilde{r} - \frac{\tilde{r}}{4}\log\tilde{r}\right\}\cos\ \theta$$
$$+ \frac{((\gamma+1)G_1'\ell_1' - K\ell_1')'}{2\pi}\frac{\tilde{r}}{2}\log\tilde{r}\cos\theta + 0(\tilde{r}^3\log^2\tilde{r}) \tag{28}$$

Asymptotic Matching

Asymptotic matching of inner and outer expansions is carried out in an intermediate limit which is in the overlapping region of inner and outer expansions. The representative point r_η runs to infinity but not as fast as in the outer limit.

$$r = \frac{r_\eta}{\eta(\alpha)} \to \infty, \quad \tilde{r} = \frac{\sqrt{\alpha}}{\eta(\alpha)}r_\eta \to, 0 \quad \sqrt{\alpha} << \eta << 1 \tag{29}$$

Also used are $r^* = \frac{\tilde{r}_\eta}{\eta} - \alpha f \cos\ \theta + \cdots$, $\cos\ \theta^* = \cos\theta - \alpha f\frac{\cos 2\theta}{2r} + \cdots$,and

$$\log\tilde{r} = \log\frac{\tilde{r}_\eta}{\eta} - \log\frac{1}{\sqrt{\alpha}}, \qquad \log^2\tilde{r} = \log^2\frac{r_\eta}{\eta} - 2\log\frac{1}{\sqrt{\alpha}}\log\frac{r_\eta}{\eta} + \log^2\frac{1}{\sqrt{\alpha}}$$

When inner and outer expansions are written in the intermediate variable they must match ($\longleftrightarrow$) identically. That is

$$\alpha\log\frac{1}{\sqrt{\alpha}}(-S_1(x)) + \alpha\varphi_1(x;\frac{r_\eta}{\eta},\theta) + \alpha^2\tilde{\varphi}_2(x;\frac{r_\eta}{\eta},\theta)$$
$$\leftrightarrow \alpha\phi_1(x;\frac{\sqrt{\alpha}}{\eta}r_\eta) + \alpha^{3/2}\phi_2(x,\frac{\sqrt{\alpha}}{\eta}r_\eta,\theta) + \cdots \tag{30}$$

When the matching is carried out (or verified) the first principal result is that the source strength $S_1(x)$ in the outer expansion defining the dominant equivalent body of revolution is proportional to the rate of change of cross-section area (c.f. eqn (17)) of the wing. This is the famous result of Oswatitsch and Keune [2]. Secondly, far-field boundary conditions are obtained for the switchback terms $\varphi_{22}, \varphi_{21}$. Taking into account the particular solutions of eqns (7), (8), matching shows

$$\varphi_{22} \to \frac{\gamma+1}{4} S_1' S_1'' \tilde{r}^* + \frac{\gamma+1}{4} \frac{(S_1'\ell_1')'}{2\pi} y^* + \cdots \tag{31}$$

$$\begin{aligned}\varphi_{21} \to & -\frac{\gamma+1}{2} S_1' S_1''(r^{*2}\log r^* - r^{*2}) - (\gamma+1)(S_1'\ell_1')'\frac{r}{2}\log r^* \cos\theta \\ & + (KS_1' - (\gamma+1)S_1'G_1')'\frac{r^{*2}}{4} \\ & + \left\{\frac{\gamma+1}{4}\frac{(S_1'\ell_1')'}{2\pi} + \frac{(K\ell_1' - (\gamma+1)G_1'\ell_1')'}{2(2\pi)}\right\} y^* + \cdots\end{aligned} \tag{32}$$

There is an induced upwash $\sim y^*$ produced by the coupling of thickness and lift distributions, and this is the far-field boundary condition for $\varphi_{22}, \varphi_{21}$.

In order to obtain the results of next section it is necessary to carry the inner far-fields to further terms, which are not presented here. It is also necessary to add $\alpha^2 \log \frac{1}{\sqrt{\alpha}} \phi_{31}$ and $\alpha^2 \phi_3$ in the outer expansion. The entire procedure is carried through as above.

Lift and Drag

For simple planforms the dominant term of the lift is that of Jones' theory (14) with $x = 1$

$$\ell_1(1) = \frac{L_1}{\rho_\infty U^2 \ell^2 \alpha} = \pi(-f_x)b^2, \quad z_{LE}(1) = b \tag{33}$$

and the induced drag D_i is such that $D_{i_1} = \frac{\alpha}{2} L_1$. Higher order corrections to the lift $O(\alpha^2 \log^2 \frac{1}{\sqrt{\alpha}}, \alpha^2 \log \frac{1}{\sqrt{\alpha}}, \alpha^2)$ come from $\varphi_{22}, \varphi_{21}, \varphi_2$ and these corrections show dependence on K. There is a corresponding induced drag. There is also a wave drag associated with the shock waves and this can be calculated from the entropy increase in the wave system. For small disturbances to a free stream

$$D_w = \rho_\infty T_\infty \ell^2 \int\int_{-\infty}^{\infty} [S]_s \, dy dz \tag{34}$$

where $[S]_s$ = jump increase in entropy across the shock. The integral is taken over all shocks in the system (c.f. [3] p165ff). If we combine the outer potentials

$$\hat{\phi} = \phi_1 + \sqrt{\alpha}\phi_2 + \alpha \log \frac{1}{\sqrt{\alpha}} \phi_{31} + \cdots \tag{35}$$

Then

$$D_w = -\rho_\infty U^2 \frac{\gamma+1}{12}\alpha^2\ell^2 \int_0^\infty\int_0^{2\pi} = [\phi_x]_s^3 d\theta \tilde{r} d\tilde{r} \tag{36}$$

The integral in (36) can be expressed in terms of an outward momentum flow from a control cylinder $\tilde{r}_c \to 0$ and a kinetic energy term at infinity. To show this, note that the equations for ϕ_1, ϕ_2, and ϕ_{31} combine to give

$$(K - (\gamma+1)\hat{\phi}_x)\hat{\phi}_{xx} + \hat{\nabla}^2\hat{\phi} = O(\alpha) \approx 0 \tag{37}$$

A momentum flux differential conservation form associated with (37) is

$$\begin{aligned} \tilde{r}\left(K\frac{\hat{\phi}_x^2}{2} - \frac{\delta+1}{3}\hat{\phi}_x^3 - \frac{1}{2}\hat{\phi}_{\tilde{r}}\hat{\phi}_x\right)_x &+ (\tilde{r}\hat{\phi}_{\tilde{r}}\hat{\phi}_x)_{\tilde{r}} \\ &+ \frac{1}{\tilde{r}}(\hat{\phi}_x\hat{\phi}_\theta)_\theta = 0 \end{aligned} \tag{38}$$

By integrating over all space outside a control cylinder and upstream of $x = x_2$ and taking account of shock jumps, it can be shown that

$$\begin{aligned} \frac{D_w}{\rho_\infty U^2\ell^2\alpha^2} = &\lim_{\tilde{r}_c\to 0}(-\tilde{r}_c)\int_{-\infty}^{\infty} dx \int_0^{2\pi} \hat{\phi}_x(x_1\tilde{r}_c,\theta)\hat{\phi}_{\tilde{r}}(x,\tilde{r}_c,\theta)d\theta \\ &- \lim_{x_2\to\infty}\int_{\tilde{r}_c}^{\infty}\tilde{r}d\tilde{r}\int_0^{2\pi} d\theta\frac{(\tilde{\nabla}\hat{\phi})^2}{2}|_{x=x_2} \end{aligned} \tag{39}$$

From (27) the dominant terms are

$$\frac{D_w}{\rho_\infty U^2\ell^2\alpha^2} = \int_0^1\left(S_1(x)G_1'(x;K) + \alpha\log\frac{1}{\sqrt{\alpha}}(S_{31}G_1' + S_1G_{31})\right)dx \tag{40}$$

The first term is the usual formula for the drag of an equivalent body of revolution. In the second term S_{31} is the source strength of ϕ_{31} and can be shown to be

$$S_{31} = 2\left(S_1S_1' + \frac{\gamma+1}{2}\frac{\ell_1'\ell_1''}{(2\pi)^2}\right).$$

and

$$\phi_{31} = S_{31}(x)\log\tilde{r} + G_{31}(x) \qquad \text{as} \qquad \tilde{r}\to 0.$$

Stabilization

Stabilization expresses the slow change of the outer flow field with M_∞ for M_∞ close to one. This can be expressed by asymptotic expansion in K. The original outer region (α expansion) is divided in two regions I, II. Region I is the inner part where the flow is represented as sonic flow ($K = 0$) plus corrections. II is the outer part, of the scale of the sonic bubble, where the equivalent body appears as a point singularity. The limit process for region I has $K \to 0, (x, \tilde{y}, \tilde{z}$ fixed). The axisymmetric solution of (25) turns out to have the expansion [4].

$$(\gamma+1)\phi_1(x,\tilde{r};K) = \psi_0(x,\tilde{r}) + K^{1/3}C_{1/2} + Kx + K^{5/3}\psi_1(x,\tilde{r}) + \cdots \tag{41}$$

$\psi_{0,1}$ satisfy

$$\psi_{0_x}\psi_{0_{xx}} + \psi_{0_{\tilde{r}\tilde{r}}} + \frac{1}{\tilde{r}}\psi_{0_{\tilde{r}}} = 0 \tag{42}$$

with the boundary conditions

$$\text{as } \tilde{r} \to 0, \qquad \psi_0 \to (\gamma+1)\left\{S_1(x)\log\tilde{r} + G_1(x)\right\} \tag{43}$$

$$\text{as } \tilde{r} \to \infty, \qquad \psi_0 \to \tilde{r}^{-2/7}\frac{f(a\xi)}{a^3} + \tilde{r}^{-6/7}\frac{f_1(a\xi)}{a^3} + \cdots \text{ahead of the shock} \tag{44}$$

$$\psi_0 \to \tilde{r}^{-2/7}\frac{f(a\xi)}{a^3} + \tilde{r}^{-4/7}\frac{f_1(a\xi)}{a^3} + \cdots \text{behind the shock} \tag{45}$$

The higher order terms in (44) and (45) come from [5],[6]. Here $\xi = \frac{x}{\tilde{r}^{4/7}}$ is the similarity coordinate for sonic flow, and a is a scale factor depending on the equivalent body.

$$-(\psi_{0_x}\psi_{1_x})_x + \psi_{1_{\tilde{r}\tilde{r}}} + \frac{1}{\tilde{r}}\psi_{1_{\tilde{r}}} = 0 \tag{46}$$

with the conditions

$$\text{as } \tilde{r} \to 0, \qquad \psi_{1_{\tilde{r}}} \to 0 \tag{47}$$

$$\text{as } \tilde{r} \to \infty, \qquad \psi_1 \to \tilde{r}^{8/7} \frac{F(a\xi)}{a^3} + \cdots \tag{48}$$

Region II has an expansion in rescaled variables to preserve the mixed subsonic-supersonic structure. The limit has $K \to 0$ $(\bar{x}, \bar{r})$ fixed where $\bar{x} = K^{2/3}x, \bar{r} = K^{7/6}\tilde{r}$. (see Fig. 2) The expansion is

$$(\gamma + 1)\phi_1(x, \tilde{r}; K) = K^{1/3}\bar{\psi}_0(\bar{x}, \bar{r}) + \cdots \tag{49}$$

$\bar{\psi}_0$ is then determined from

$$(1 - \bar{\psi}_{0_{\bar{x}}})\bar{\psi}_{0_{\bar{x}\bar{x}}} + \bar{\psi}_{0_{rr}} + \frac{1}{\bar{r}}\bar{\psi}_{0_{\bar{r}}} = 0 \tag{50}$$

with the condition

$$\begin{aligned} &\text{as } \bar{x} \to -\infty \qquad \bar{\psi}_{0_{\bar{x}}}, \bar{\psi}_{0_{\bar{r}}} \to 0 \\ &\text{as } \bar{r} \to 0, \psi_0 \to \bar{r}^{-2/7}\frac{f(a\xi)}{a^3} + C_{1/2} + \bar{x} + \frac{\bar{r}^{8/7}F(a\xi)}{a^3} + \cdots \end{aligned} \tag{51}$$

both ahead and behind the shock.

Note that $\xi = \frac{x}{\tilde{r}^{4/7}} = \frac{\bar{x}}{\bar{r}^{4/7}}$ is invariant under rescaling. Matching from Region I to Region II of (51) has produced the far field of Eqn (48). A similar procedure applies to the non-axisymmetric part of ϕ_2 which has the behavior in Region I, determined after matching

$$\phi_2(x, \tilde{r}, \theta; K) = \chi_0(x, \tilde{r}, \theta) + K^{5/3}\chi_1(x, \tilde{r}, \theta) + \cdots \tag{55}$$

Here

$$-(\psi_{0_x}\chi_{0_x})_x + X_{0_{\tilde{r}\tilde{r}}} + \frac{1}{\tilde{r}^2}X_{0_{\theta\theta}} = 0 \tag{56}$$

and

$$-(\psi_{0_x}\chi_{1_x})_x + \chi_{1_{\tilde{r}\tilde{r}}} + \frac{1}{\tilde{r}}\chi_{1_{\tilde{r}}} + \frac{1}{\tilde{r}^2}\chi_{1_{\theta\theta}} = (\psi_{1_x}\chi_{0_x})_x \tag{57}$$

Corresponding boundary and matching conditions are

$$\text{as} \quad \tilde{r} \to 0, \qquad \chi_0 \to \frac{\ell_1(x)}{2\pi\tilde{r}}\cos\theta, \quad \tilde{r}\chi_1 \to 0 \tag{58}$$

$$\text{as} \quad \tilde{r} \to \infty, \qquad \chi_0 \to \tilde{r}^{-1} \cos\theta \frac{h(a\xi)}{a^3} + O(\tilde{r}^{-9/7}) \tag{59}$$

This last result follows from Euvrard [5]. Matching with the representation in Region II shows

$$\text{as} \quad \tilde{r} \to \infty, \quad \chi_1 = \tilde{r}^{3/7} \cos\theta \frac{H_f(a\xi)}{a^3} + \cdots \tag{60}$$

H_f is a forced solution which is known explicitly in region II

$$\phi_2(x, \tilde{r}, \theta; K) = K^{7/6} \bar{\chi}(x, \tilde{r}, \theta) + \cdots \tag{61}$$

where

$$((1 - \bar{\psi}_{0_{\bar{x}}})\bar{\chi}_{\bar{x}}) + \bar{\chi}_{\bar{r}\bar{r}} + \frac{1}{\bar{r}}\bar{\chi}_{\bar{r}} + \frac{1}{\bar{r}^2}\bar{\chi}_{\theta\theta} = 0 \tag{62}$$

with

$$\bar{x} \to -\infty, \quad \bar{\chi}_{\bar{x}} \to 0 \quad \text{and} \quad \bar{r} \to 0, \quad \bar{\chi} = \frac{1}{\bar{r}} \cos\theta \frac{h(a\xi)}{a^3} + \bar{r}^{3/7} \cos\theta H_f \frac{(a\xi)}{a^3} + \cdots \tag{63}$$

Remarks

This study complements the results of Cheng and Hafez [7] and earlier workers for the case when $\alpha >> \delta$. A recent contribution to that case appears in Ref. 8. Application of the stabilization law to the equivalent body shows how $G_1(x; K)$ depends on K. The stabilization result agrees with that of Diesperov and Lifshits [9] ahead of the tail shock.

Thus,

$$\frac{D_w}{\rho_\infty U^2 \ell^2 \alpha^2} = D_w^* + K^{5/3} D_{Wc}; \quad \begin{array}{l} D_w^* = \text{wave drag at sonic} \\ D_{wc} = \text{correction} \end{array} \tag{64}$$

Corrections to lift and induced drag due to coupling of equivalent body lift S_1, and ℓ_1 appear in ϕ_2 and are $O(\alpha^2 \log^2 \frac{1}{\sqrt{\alpha}}, \quad \alpha^2 \log \frac{1}{\sqrt{\alpha}}, \alpha^2)$. The stabilization for ϕ_2 has the same form as that for ϕ_1.

References

[1] Jones, R.T.: Properties of low aspect ratio pointed wings at speeds below and above the speed of sound, NACA TR 835, 1946.

[2] Oswatitsch, K. and Keune, F.: Ein Äquivalenzsatz für nichtangestellte Flügel kleiner Spannweite in schallnaher Strömung. Zeit. f. Flugwiss. v3, no 2 Feb. 1955 p 29-46.

[3] Cole, J.D. and Cook, L.P.: *Transonic Aerodynamics* , North Holland 1986.

[4] Cole, J.D. and Cook, L.P.: Transonic choking and stabilization for flows about slender bodies. *Mech Res. Comm.* to appear.

[5] Euvrard, D.: Etude asymptotique de l'écoulement a grande distance d'un obstacle se déplaçant à la vitesse du son. (III). Ecoulement tridimensionnel en amont des ondes de choc. *J. de Mécanique* v7, no3 Sept 1968.

[6] Tournemine, G.: Comportement asymptotique de l'écoulement sonique autour d'un corps de révolution de dimensions finies, en aval de l'onde de choc. *J. de Mécanique* v7, no 3 Sept 1968.

[7] Cheng, H.K. and Hafez, M.M.: Transonic equivalence rule: a nonlinear problem involving lift. *J. Fluid Mech.* v72 pt. 1 1975, p 161-187.

[8] Cole, J.D. and Malmuth, N.: Wave drag due to lift for transonic airplanes, Proc. NASA Langley Transonic Symposium, Apr. 1988, to appear.

[9] Diesperov, V.N. and Lifshits, Iu. B.: On the drag of bodies of revolution at transonic speeds, PMM v39, no 2, 1975 p 290-297.

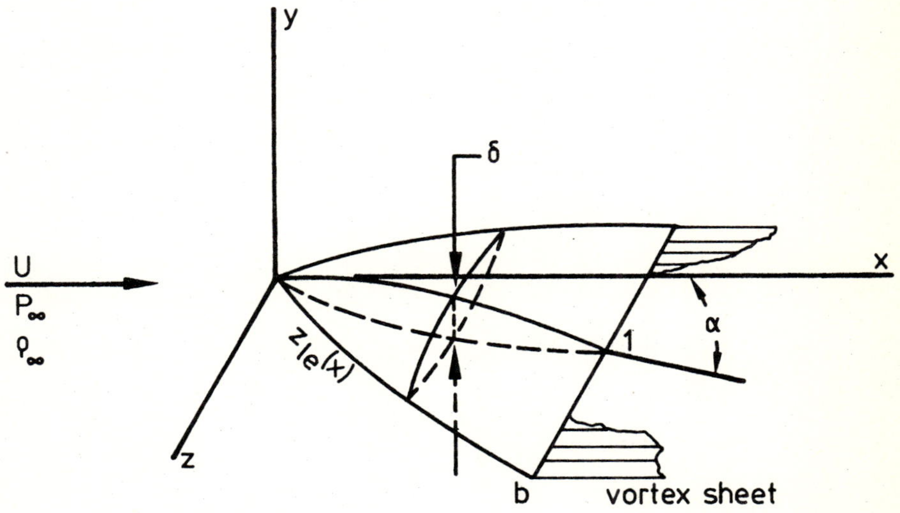

Fig. 1 Pointed wing

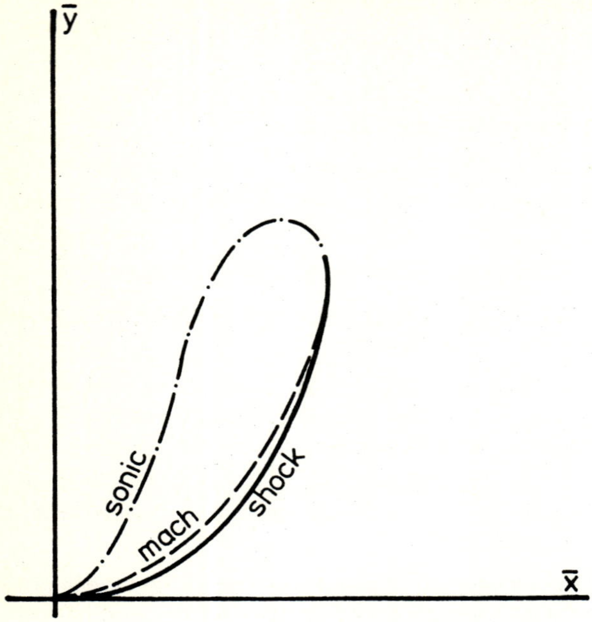

Fig. 2 Rescaled coordinates

Transonic Degeneracy in Systems of Conservation Laws

J.-P. Guiraud

Laboratoire de Modélisation en Mécanique (CNRS N° 2290)
Université P&M. Curie, Tour 66, Place Jussieu,
75 230 Paris Cedex 05, FRANCE

ONERA Direction de l'Aérodynamique
BP 72 - 92322 Chatillon Cedex, FRANCE

Summary

We examine the way in which the system of two equations for two unknown functions of two variables which occurs in transonic aerodynamics may be derived asymptotically for systems of conservation laws. Some already known applications are resumed in a unified way.

The simplest transonic degeneracy

Starting from a constant steady state, with pressure p_0, density ρ_0, and velocity equal to the speed of sound a_0, aligned with the x axis, we consider a steady two-dimensional flow which is a small purturbation from it, and we set the pressure p, the density ρ, and the two components u and v of the velocity

$$\begin{aligned} p &= p_0(1+\delta\bar{p}) \ , \quad \rho = \rho_0(1+\delta\bar{\rho}) \ , \\ v &= a_0\zeta\bar{v} \ , \quad u = a_0(1\pm\epsilon+\eta\bar{u}) \ , \end{aligned} \tag{1}$$

assuming that the quantities denoted with a bar are functions of the two variables x and $y = \lambda\bar{y}$. In these formulae, δ, ϵ, and ζ are small parameters while λ is the inverse of a small parameter. The $\pm$ sign in front of ϵ is introduced so that ϵ be always positive. In order to recover the transonic equations we simply need to set the following relations between the parameters

$$\delta = \eta \ , \quad \epsilon = K\eta \ , \quad \lambda = \eta^{-\frac{1}{2}} \ , \quad \zeta = \eta^{\frac{3}{2}} \ , \tag{2}$$

and a straightforward limiting process leads to

$$(\pm K + 2u)\frac{\partial\bar{u}}{\partial x} = \frac{\partial\bar{v}}{\partial\bar{y}} \ , \qquad \frac{\partial\bar{v}}{\partial x} = \frac{\partial\bar{u}}{\partial\bar{y}} \ , \tag{3}$$

and, of course, we have the complementary relations

$$\gamma^{-1}\bar{p} = \bar{\rho} = -\bar{u} \ . \tag{4}$$

Systems of two independent variables

We raise the degree of generality by assuming a system of quasi-linear equations of two independent variables,

J. Zierep, H. Oertel (Eds.)
Symposium Transsonicum III
IUTAM Symposium Göttingen, 24.-27. 5. 1988

$$A(U)\,\frac{\partial U}{\partial x} + B(U)\,\frac{\partial U}{\partial y} = 0 \quad , \tag{5}$$

where U stands for an N element column matrix, function of x and y, while $A(U)$ and $B(U)$ are NxN square matrices, functions of U. We assume that U is close to some constant value U_0 and we expand it by setting

$$U = U_0 + \epsilon_1 U_1\,(x,\ \bar{y}) + \epsilon_2 U_2\,(x,\ \bar{y}) + \epsilon_3 U_3\,(x,\ \bar{y}) + \cdots , \tag{6}$$

where $\epsilon_1 > > \epsilon_2 > > \epsilon_3$ ... is an ordered sequence of gauge small parameters. We notice that the coordinate y has been rescaled in the way suggested by the simplest example (1), by

$$y = \lambda\,\bar{y} \tag{7}$$

with $\lambda > > 1$. We assume that A and B may be expanded by

$$\begin{aligned} A(U) &= A_0 + \epsilon_1 A_1\,(U_1) + O(\epsilon_2) + O(\epsilon_1^2) \ , \\ B(U) &= B_0 + O(\epsilon_1) \ . \end{aligned} \tag{8}$$

Substituting (6) (7) (8) into (5), we find, in zero order,

$$A_0\,\frac{\partial U_1}{\partial x} = 0 \quad , \tag{9}$$

which is easily solvable, provided det $(A_0) = 0$, by

$$U_1 = \bar{u}\,(x,\ \bar{y})\,R \quad , \tag{10}$$

where $u\,(x,\ \bar{y})$ is some scalar function of x and $\bar{y}$, unknown until now, while R is a right eigenvector (L being a left one), such that

$$A_0 \cdot R = L \cdot A_0 = 0 \quad . \tag{11}$$

We assume that R and L are well-defined, each up to a constant factor, and that this arbitrariness has been removed by some normalization. For the next order we choose

$$\epsilon_2 = \epsilon_1\,\lambda^{-1} > > \epsilon_1^2 \ , \tag{12}$$

and we find

$$A_0\,\frac{\partial U}{\partial x} + B_0\,R\,\frac{\partial u}{\partial \bar{y}} = 0 \quad . \tag{13}$$

This equation is not solvable unless

$$L \cdot B_0 \cdot R = 0 \quad , \tag{14}$$

and we assume that this relation holds. As a matter of fact, one may check that (14) holds when (5) is the system of steady two-dimensional Euler equations of gasdynamics and when U is the column matrix describing a uniform sonic flow parallel to the x axis. Equation (13) may be solved by setting

$$U_2 = \bar{v}\,(x,\ \bar{y})\,\Delta + u_2\,(x,\ \bar{y})\,R \quad , \tag{15}$$

where u_2 is an arbitrary scalar function of x and $\bar{y}$ (to be determined at a later stage in the process), while Δ is a column matrix solution of the following equation:

$$A_0 \Delta + B_0 R = 0 \quad , \tag{16}$$

and v is related to u through the equation

$$\frac{\partial \bar{v}}{\partial x} = \frac{\partial \bar{u}}{\partial \bar{y}} \quad . \tag{17}$$

Going to the next order in the process of expanding equation (5), we find that the proper choice of gauges is

$$\epsilon_3 = \epsilon_1^2 = \frac{\epsilon_2}{\lambda} \quad , \tag{18}$$

and we get the following equation:

$$A_0 \frac{\partial U_3}{\partial x} + A_1(R) \cdot R\,\bar{u}\,\frac{\partial \bar{u}}{\partial x} + B_0 \Delta \frac{\partial \bar{v}}{\partial \bar{y}} + B_0 R \frac{\partial u_2}{\partial y} = 0 \tag{19}$$

We do not consider this equation as one giving U_3 (this occurs at a later stage that we do not consider here). We consider it just to the extent that it must be solvable with respect to U_3. This requires that a compatibility condition be satisfied. This condition is simply obtained by multiplying (19) to the left by L. The unknown u_2 disappears from the resulting equation thanks to the following relation,

$$L \cdot B_0 \cdot R = - L \cdot A_0 \cdot \Delta = 0 \quad , \tag{20}$$

which is an obvious consequence of (11). We choose some constant a, depending on the system under consideration, such that

$$L \cdot A_1(R) \cdot R + a\, L \cdot B_0 \cdot \Delta = 0 \quad . \tag{21}$$

Multiplying (19) to the left by L and using (20) and (21), we find the following equation:

$$a\,u \frac{\partial \bar{u}}{\partial x} = \frac{\partial \bar{v}}{\partial \bar{y}} \quad . \tag{22}$$

We find, at the end of this process, that the simplest transonic degeneracy, so well-known in gasdynamics since von Kármán [1], may be obtained from a quite general system of conservation laws in two dimensions. It includes the following features. Firstly, the two independent variables must be scaled with lengths having differing orders. This feature is well-known from transonic aerodynamics. The second feature is that the lines corresponding to a constant value of the undistorted coordinate should be characteristic for the linear system associated with (5), obtained by replacing the matrices A and B by their values A_0 and B_0 corresponding to U_0. The last feature that we wish to mention is the condition (14). We observe that two scalar unknown functions $\bar{u}(x, \bar{y})$ and $\bar{v}(x, \bar{y})$ dominate. The first one is involved in the leading approximation $\epsilon_1 \bar{u} R$, and the second one in the second approximation $\epsilon_2 \bar{v} \Delta$.

Some minor modifications lead to known results. We consider two of them. The first one is when det (A_0) is not exactly zero but one may write

$$A_0 = A_{0,0} + K\epsilon_1 A_{0,1} + O(\epsilon_1) \ , \tag{23}$$

and det $(A_{0,0}) = 0$. We replace $A_0$by $A_{0,1}$ throughout the preceding analysis and the relation (21) by

$$L \cdot A_{0,1} \cdot R + a\, L \cdot B_0 \Delta = 0 \ , \tag{24}$$

and we get the following system:

$$(K + a\,\bar{u}) \frac{\partial \bar{u}}{\partial x} - \frac{\partial \bar{v}}{\partial \bar{y}} = 0 \ , \quad \frac{\partial \bar{v}}{\partial x} - \frac{\partial \bar{u}}{\partial \bar{y}} = 0 \ . \tag{25}$$

The second modification that we consider concerns the case when $A = A(U, x, y)$ depends on independent variables as well as on U, the same being true for B. On the other hand we assume that such a dependency is weak, being felt through δx and δy, δ being some small parameter. Again we set $A(0, x, y) = A_0(\delta x, \delta y)$ and we assume that

$$A_0 = A_{00} + \delta A_{001} x + \delta A_{001} y + O(\delta) \ . \tag{26}$$

The preceding analysis is valid up to and including (17), provided we use A_{00} instead of A_0 throughout. Corresponding to (18) we choose

$$\epsilon_3 = \frac{\epsilon_2}{\lambda} = \mathrm{SUP}_{\mathrm{ord}}\,(\epsilon_1^2, \ \epsilon_1 \frac{\delta}{\lambda}) \ , \tag{27}$$

where $\mathrm{SUP}_{\mathrm{ord}}(\eta, \zeta)$ means η when $\eta >> \zeta$ and ζ when $\eta << \zeta$, respectively. According to (27), we set

$$\epsilon_1^2 = K_1\, \epsilon_3 \qquad \epsilon_1 \frac{\delta}{\lambda} = K_2\, \epsilon_3 \ , \tag{28}$$

and thus, instead of (19), we obtain the following equation,

$$\begin{aligned} A_{00} \frac{\partial U}{\partial x} + K\, y\, A_{001} \frac{\partial \bar{u}}{\partial x} + K\, A_{10}(R) \cdot R\, \bar{u}\, \frac{\partial \bar{u}}{\partial x} + \\ + B_{00}\, \Delta \frac{\partial \bar{v}}{\partial \bar{y}} + B_{00}\, R \frac{\partial u_2}{\partial \bar{y}} = 0 \ , \end{aligned} \tag{29}$$

where B_{00} stands for $B(0,0,0)$. The remaining part of the analysis is unchanged, and we end with the system

$$(K_1\, \bar{y} + a\, K_2\, \bar{u}) \frac{\partial \bar{u}}{\partial x} - \frac{\partial \bar{v}}{\partial y} = 0 \ , \quad \frac{\partial \bar{v}}{\partial x} = \frac{\partial \bar{u}}{\partial \bar{y}} \ . \tag{30}$$

More general systems of conservation laws

From now on, we consider systems of conservation laws in three dimensions, namely

$$A_a(U) \frac{\partial U}{\partial x_a} = 0 \ , \quad \frac{\partial U}{\partial t} + A_k(U) \frac{\partial U}{\partial x_k} = 0 \ . \tag{31}$$

We use either x_0 or t for time and we assume that repeated indexes imply summation, from one to three for Latin ones, and from zero to three for Greek ones.

We assume that we know one solution for (31), depending weakly on x_α, and we write for this solution

$$U_0 = U_0(\delta x_\alpha) \quad , \tag{32}$$

where δ is some small parameter. We look for a slightly perturbed solution to (31) by setting for it

$$U_0 = U_0(\delta x_\alpha) + \epsilon_1 U^*[\xi(x_\alpha, \lambda), \eta(x_\alpha, \mu), x_\alpha] \quad , \tag{33}$$

where ϵ_1 is a small parameter, while λ^{-1} and μ^{-1} are large ones ($\lambda << \mu << 1$). We assume that

$$\frac{\partial \xi}{\partial x_\alpha} = \frac{1}{\lambda} p_\alpha(x_\alpha) \quad , \quad \frac{\partial \eta}{\partial x_\alpha} = \frac{1}{\mu} q_\alpha(x_\alpha) \quad , \tag{34}$$

and we write

$$A_\alpha(U) = A_\alpha(U_0) + \epsilon_1 A_\alpha^1(U_0 | U^*) + O(\epsilon_1) \quad . \tag{35}$$

We assume that, owing to the structure of U, as indicated by (32), we may expand $A_\alpha(U_0)$ by

$$A_\alpha(U_0) = A_\alpha^{00} + \delta x_\beta A_{\alpha\beta}^{01} + O(\delta) \quad , \tag{36}$$

but we may re-express x_β through ξ, η, and two other variables in such a way that one has

$$x_\beta = \mu b_\beta \eta + O(\lambda) \tag{37}$$

and (36) may be rewritten as

$$A_\alpha(U_0) = A_\alpha^{00} + \mu\delta\eta b_\beta A_{\alpha\beta}^{01} + O(\mu\delta) \quad , \tag{38}$$

finally we set $A_0(U_0 | U^*) = A_\alpha(U^*)$. We look for an expansion

$$U^* = \epsilon_1 U^*{}_1 + \epsilon_2 U^*{}_2 + \epsilon_3 U^*{}_3 + O(\epsilon_3) \quad , \tag{39}$$

where ϵ_1, ϵ_2, ϵ_3 is an ordered sequence of gauge parameters all of which are small. In zero order we find

$$p_\alpha A_\alpha^{00} \frac{\partial U^*}{\partial \xi} = 0 \quad , \tag{40}$$

so that it is required that

$$\det(p_\alpha A_\alpha^{00}) = 0 \quad . \tag{41}$$

We introduce the right R and left L eigenvectors, such that

$$p_\alpha A_\alpha^{00} \cdot R = p_\alpha L \cdot A_\alpha^{00} = 0 \quad , \tag{42}$$

and we assume that they are unique once normalized. We find that

$$U^*{}_1 = u^*(\xi, \eta, x_\alpha) R(\delta x_\alpha)$$

and we go to the next order by choosing

$$\frac{\epsilon_2}{\lambda} = \frac{\epsilon_1}{\mu} > > \mathrm{SUP}_{\mathrm{ord}} \left(\frac{\mu \delta \epsilon_1}{\lambda} , \epsilon_1^2 \right) , \tag{43}$$

and so the analysis continues as before; we look for Δ, such that

$$p_a A_a^{00} \cdot \Delta + q_a A_a^{00} \cdot R = 0 , \tag{44}$$

but this is not solvable unless

$$q_a L \cdot A_a^{00} \cdot R = 0 , \tag{45}$$

which is the analogue of (14), but with a quite different interpretation. As a matter of fact, in the case of (14) the condition might be unsatisfied, while, on the contrary, here, by a proper choice of the function η, (45) may always be forced to hold. One should consider that (45) is a partial differential equation for the function η, in the same way as (40) has to be considered a partial differential equation for ξ. We may look for U^* as follows:

$$U^*_2 = v^* \Delta + u^*_2 R , \tag{46}$$

and we find the second of equations (51) below. For the next order, the proper choice of gauges is

$$\frac{\epsilon_3}{\lambda} = \frac{\epsilon_2}{\mu} = \mathrm{SUP}_{\mathrm{ord}} \left(\frac{\epsilon_1^2}{\lambda} , \frac{\epsilon_1 \delta \mu}{\lambda} \right) , \tag{47}$$

and we write, accordingly,

$$\epsilon_1^2 = K_1 \epsilon_3 , \quad \epsilon_1 \delta \mu = K_2 \epsilon_3 , \tag{48}$$

then we get the following equation:

$$\begin{aligned} & p_a A_a^{00} \frac{\partial U^*_3}{\partial \xi} + q_a A_a^{00} \frac{\partial v^*}{\partial \eta} + q_a A_a^{00} R \frac{\partial u^*_2}{\partial \eta} + \\ & \quad + \{ K_1 p_a A_a^{10}(R) \cdot R u^* + K_2 \eta p_a b_\beta A_{a\beta}^{01} R \} \frac{\partial u^*}{\partial \xi} = 0 , \end{aligned} \tag{49}$$

and we observe that a term $\epsilon_1 A_a^{00} (\partial / \partial x_a) (R u^*)$ which might be considered at this stage is, indeed, of a higher order. As a matter of fact, $R = R(\delta x_a)$, and the term under consideration has to be written $\epsilon_1 \delta A_a (\partial R / \partial y_a) u^*$ so that it is δ times smaller than the term proportional to K_2 in (49). Again, we do not solve (49) as an equation for U^*_3 but, instead, as one giving a compatibility condition by multiplying it to the left by L. We observe that, through this operation, the terms in U^*_3 and u^*_2 disappear owing to $L \cdot A_a^{00} = 0$. We write

$$\begin{aligned} & p_a L \cdot A_a^{00} (R) \cdot R + a q_a L \cdot A_a^{00} \cdot \Delta = 0 , \\ & p_a b_\beta L \cdot A_{a\beta}^{01} \cdot R + b q_a L \cdot A_a^{00} \cdot \Delta = 0 , \end{aligned} \tag{50}$$

and we get the first equation in the following transonic system (the second has already been obtained from just after (46)):

$$\frac{\partial v^*}{\partial \eta} = (K_1 a\, u^* + K_2 b\, \eta)\, \frac{\partial u^*}{\partial \xi} \quad ,$$
$$\frac{\partial v^*}{\partial \xi} = \frac{\partial u^*}{\partial \eta} \quad . \tag{51}$$

Discussion

The system (51) may be reduced to the linear Tricomi equation when $K_1 = 0$ or to the non-linear one when $K_2 = 0$. The relevant cases depend on the gauges. When $\delta\mu >> \epsilon_1$, the relevant degeneracy is the linear Tricomi equation, while it is the nonlinear one when $\delta\mu << \epsilon_1$. On the other hand, when $\delta\mu = \epsilon_1$ (equality as orders of magnitudes), the full system has to be retained. The gauge ϵ_1 scales the dominant term of the perturbations, while δ^{-1} scales the space displacement or the time shift which gives an order one variation of the basic solution U_0. Finally, the gauge μ^{-1} scales the space displacement or time shift (directed in the proper direction) which gives an order one variation of η. The condition $\epsilon_1 = \delta\mu$ states that an order one variation of η is accompanied by an order ϵ_1 variation in U_0, and, more specifically, variations in the eigenvalues and eigenvectors which give rise to the degeneracy. Translated into transonic aerodynamic language, the condition $\delta\mu = \epsilon_1$ states that the variation of the speed of sound of the basic undisturbed (but non-homogeneous) state accompanying an order one transverse (corresponding to the y direction in the example following (1)) reduced displacement is of the same order as the dominant order in the perturbed velocity. This result seems to be quite good and is found to be rather universal. Universal again appears to be the transonic system. We summarize the features which appear to be of importance in the derivation. First, there must exist a wave front (ξ = Const) and there should occur rapid ($\lambda << 1$) variations across it. But this is not by itself sufficient. There would simply be accumulation near some wave front, a situation which is known to be described by a Burgers equation. A second feature has to be fulfilled in order to obtain the transonic degeneracy. One needs to have rapid but slightly less ($\lambda << \mu << 1$) rapid variations across transverse fronts (η = Const). This means that phenomena to which the transonic system is relevant are basically two-dimensional. One of the two dominant variables has to be constant along waves and it carries the steepest variations. The other dominant variable is not arbitrary and it has to be in agreement with the condition (45) which does not appear to have, at first sight, a simply stated physical significance. We observe that the ratio of steepnesses is $\mu/\lambda = \varphi$ and we find from (43) and (47) that

$$\epsilon_1 = \varphi^{-2} \quad , \quad \epsilon_2 = \varphi^{-3} \quad , \tag{52}$$

which may be used in (39) to rewrite it as

$$U^* = \epsilon_1\, u^* R + \epsilon_1^{\frac{3}{2}}\, (v^* \Delta + u^* R) + 0\{\mathrm{SUP}_{\mathrm{ord}}\, (\epsilon_1^2,\ \epsilon_1\, \delta\mu)\} \ . \tag{53}$$

Concerning the term containing $\mathrm{SUP}_{\mathrm{ord}}$, we observe that it corresponds to the statement that the third term in the expansion (53) is dominated either by the square of the perturbations, that is by the linearities in the equations - and this occurs when $\epsilon_1^2 >> \epsilon_1 \delta\mu$ - or by the variations in the basic state, a situation which is relevant when $\epsilon_1^2 << \epsilon_1 \delta\mu$. Here, we make an observation concerning the change from (36) to (38). The term $x_\beta A^{01}_{\alpha\beta}$ had to be written as a combination of ξ, η and two other variables independent of ξ and η. These last ones are not considered in the result because they occur in the equations just as parameters. Concerning the resulting expression, linear in ξ and η, we observe that the part corresponding to η dominates the one corresponding to ξ owing to $\lambda << \mu$. The transonic system has been found in a number of situations and we emphasize here the reflection of short waves on caustics. Guiraud [2] investigated the reflection of a sonic boom on a caustic, a situation which has been further investigated by a number of authors, namely Obermeier [3], Eremenko [4], Seebass, Murman and Krupp [5], Pechuzal and Kevorkian [6], among others. Some other focussing effects in aerodynamics have been studied by Fung [7] and Cramer [8].

References

[1] Von Kármán, T. The similarity law for transonic flow. Jal Math.Phys. 26 (1947) 182-190.

[2] Guiraud, J.-P. Acoustique géométrique, bruit balistique des avions supersoniques et focalisation. Jal. de Mécanique 4 (1965) 215-267.

[3] Obermeier, F. Some comments on sonic boom focussing. Z.A.M.M. 58 (1975) T295-T297.

[4] Eremenko, V.A. On the structure of the velocity field in the vicinity of a caustic. P.M.M. 41 (1977) 1147-1151.

[5] Seebass, R., Murman, E. and Krupp, J.A. Finite difference calculation of a discontinuous signal at a caustic. Third conference on sonic boom research; NASA SP 255.

[6] Pechuzal, G. and Kevorkian, J. Supersonic-transonic flow generated by a thin airfoil in a stratified atmosphere. SIAM Jal. Appl. Math. 38 (1977) 8-33.

[7] Fung, Y. Shock wave formation at a caustic. SIAM Jal. Appl. Math. 39 (1980) 335-371.

[8] Cramer, M.S. The focussing of weak shock waves at an axisymmetric arête; J.F.M. 110 (1981) 249-253.

Transonic Equivalence Rule: Unresolved Aspects of Asymptotic and CFD Studies

H.K. Cheng

Department of Aerospace Engineering
University of Southern California
Los Angeles, California 90089-1191, USA

Summary

Theoretical and experimental studies of the transonic equivalence rule with lift are discussed with a view for shedding light on several unresolved aspects of the related asymptotic, CFD and wind-tunnel investigations.

Introduction

The transonic area rule has been an important aerodynamic concept and has played a major role in high-speed aircraft design since the late fifties. Oswatitsch and Keune [1] perceived that the cross flow next to a slender body is nearly two-dimensional as in the slender-body theory for subsonic and supersonic flows[2,3]; accordingly, the flow far away from the body is controlled by a line source determined by the axial rate of change of the body cross-section area. Hence, bodies with the same cross-section area distribution share the same transonic outer flow, and, in this sense, they are "equivalent". In an extensive study on wave pattern and drag in a transonic wind tunnel, Whitcomb [4] arrived at the same rule independently. A comprehensive review on equivalence rules in transonic and supersonic flows was given by Oswatitsch [5].

The original versions of Refs. 1, 4 and subsequent studies [6–9] were restricted to a very slight degree of asymmetry associated with the lift, i.e., $\alpha << \tau$, where τ is a thickness ratio and α is an angle of attack or a parameter controlling the wing camber. This restriction was relaxed in a formulation by Ashley and Landahl [10] to $\alpha/\tau = 0(1)$. An important extension was made later by Spreiter and Stahara [11] who recognized the applicability of the transonic area rule to pointed non-slender wings with a (non-vanishing) apex angle of the unit order. Underlying the aforementioned works is an axisymmetric (outer) flow far from the body which is consistent (matched) with the cross flow next to the slender wing/body.

When α or the lift becomes much larger, in particular, when α is comparable to $\sqrt{\tau}$, Cheng and Hafez [12] found that the outer flow can no longer remain axisymmetric

J. Zierep, H. Oertel (Eds.)
Symposium Transsonicum III
IUTAM Symposium Göttingen, 24.-27. 5. 1988

and is controlled by both a line source and a line doublet. It was noted that many modern transport and similar aircraft designs fall in this domain of $\alpha/\sqrt{\tau} = O(1)$. In this paper, we shall discuss the theoretical development which led to an extension of the equivalence rule to this domain as well as a domain dominated completely by the lift. Attention will be directed to the requirements of the analysis, issues on the flow model and formulation. An experiment by Y.Y. Chan on the equivalence rule with lift [29] represents an important recent development and deserves greater attention.

Remarks on the Theoretical Development

A feature overlooked in Ref. 12 was the additional source strength, which could have been anticipated from an examination of the nonlinear lift contribution to the cross flow. This important correction was noted by Barnwell [13] and will be explained further later. Subsequent formulation by Cheng and Hafez [14] has uncovered several additional source-like corrections of comparable magnitude. A later report by Barnwell [15] recovered essentially the same results, using body-oriented coordinates and exploring also the effect of the separation vortices. Cramer [16] studied a limit of Cheng and Hafez's theory [14] corresponding to the case of a zero thickness and confirmed most of the results of Ref. 14; his subsequent analysis [17] brought complete agreement. The nonlinear-lift correction was used in an early experimental study of transonic transport design [18], but the amount of area needed to compensate for the lift effect was arrived at empirically. The importance of the effect in engineering applications has been studied in an aerodynamic-design study of transport aircraft [19] and in drag-rise estimates for fighter-type configurations [20].

Inner Solution: Nonlinear Corrections and Significance

We consider a winged configuration with swept leading edges. Let the half span be b and its streamwise length scale be ℓ. The ratio $\lambda \equiv b/\ell$ controls the leading-edge sweep angle and, unlike in the slender wing theory, will be assumed to be of unit order or even slightly larger (see below). The key to the theoretical development lies in the nonlinear corrections to the lift-dominated inner (Jones') solutions, which determine the region of nonuniformity and establish the scales for the nonlinear outer flow. Interestingly, these second-order corrections are not completely accounted for by the standard transonic small-disturbance (TSD) equation [21, 23] but produce effects of first-order importance in the outer flow. The reason can be seen more clearly from the familiar partial differential equation written for the perturbation potential

$$\phi_{yy} + \phi_{zz} = (M_\infty^2 - 1)\phi_{xx} + M_\infty^2 U_\infty^{-1}\left[(1+\frac{\gamma-1}{2}M_\infty^2)\phi_x^2 + \phi_y^2 + \phi_z^2\right] + \cdots \quad (1)$$

omitting the triple products. The second-order terms on the right correct Jones' approximation for the nonlinear compressibility effect and contribute as a distributed source in the cross-flow plane. At large distance $r = (y^2 + z^2)^{1/2}$ from the wind/body axis, the integrable part of the distribution produces an effect equivalent to a line source, the strength of which does not vanish with the thickness ratio, as long as the perturbation velocity does not vanish. Its magnitude relative to that from the geometric cross section $S_c(x)$, or more precisely to dS_c/dx, is apparently of the order α^2/τ (for a unit-order λ). The nonintegrable part of the distribution arises from the dominant line doublet behavior in ϕ, which results in a behavior proportional to $\alpha^2(\ell nr)^2$ for the correction to ϕ. From a viewpoint of the theoretical development, this was an important term, for matching it with the corresponding (inner) expansion of the outer solution yields a term $\alpha^2|\ell n\varepsilon|\ell nr$ for ϕ in the outer flow. This signifies an addition to the cross-section area of the equivalent body noted first by Barnwell [13]. Returning to the distributed source effect, the last two square terms in Eq. (1) signify the rate of change in the cross-flow kinetic-energy flux, absent from the standard TSD equation. The line source contributed by these two terms leads to a nonvanishing total volume flux, and hence an equivalent afterbody with a cross-section area proportional to the vortex drag.

The inner solutions carried out for the nearly planar wing [14], though complicated in appearance for the higher approximations, are straightforward. From their analytical structure at large r/b, of which the details are omitted, one may infer not only the region of its nonuniformity, identifying thereby the scale for the outer region $b' \equiv b/\varepsilon$, but also three basic parameters controlling the relative importance of three distinct nonlinear effects in and on the outer flow. Namely,

$$K=(1-M_\infty^2)\lambda^2\varepsilon^{-2}, \quad \sigma_* \equiv \sqrt{(\gamma+1)|2n\varepsilon|}M_\infty\lambda^{2/3}\alpha\tau^{-1/2}, \quad \Gamma_* \equiv \frac{8}{(\gamma+1)|2n\varepsilon|\lambda^2}. \quad (2,a,b,c)$$

Here, K is the similarity parameter in the TSD equation, σ_* is a doublet strength, the square of which controls the nonlinear lift contribution to the equivalent source, and Γ_* measures the part of the lift-dependent equivalent source unaccounted for by the TSD theory. For finite K, σ_*, and Γ_*, the nonuniformity determines b', hence the scale factor

$$\varepsilon \equiv \sqrt{\gamma+1}M_\infty\tau^{1/2}\lambda^{3/2}, \quad (3)$$

while ϕ in the nonuniform and outer regions can be scaled by τUb. For an unbounded σ_*, which corresponds to a thoroughly lift-dominated flow and includes the zero-thickness limit, another set of scales appears to be more appropriate. The two scale sets are completely equivalent, although use of (3) for a numerically large σ_* may result in unduly large or small magnitudes in certain variables. For the present purpose, normalization based on (3) will suffice and is adopted below.

Control of the Nonlinear Outer Flow

The major objective of the inner solution developed and substantiated in Refs. [14,15,16] is to arrive at an expression for ϕ far from the body, with which an equivalence rule governing the nonlinear outer flow can be inferred. Expressed in the outer variable $\eta = \varepsilon r/b$, the form of ϕ in question can be arranged into

$$\begin{aligned}\phi/\tau Ub \sim\ & (2\pi)^{-1}S'_e(x)\ln\eta + (2\pi)^{-1}D_o(x)\eta^{-1}\sin\omega + \beta_o(x) \\ & + \varepsilon(2\pi)^{-1}[\tilde{D}_1(x)\eta^{-1}\cos\omega + \sigma_*|\ln\varepsilon|^{-1/2}m_{32}(x)\eta^{-2}\sin 2\omega] \\ & + (8\pi)^{-2}|\ln\varepsilon|^{-1}(F'^{2})'(2\ln^2\eta + \cos^2\omega) + \cdots \end{aligned} \tag{4}$$

In above, ω is the azimuthal angle, F(x) is the total lift up to the x-station (normalized by αUb^2); m_{32} and $\tilde{D}_1$ are of unit order and are functionals of the chordwise and spanwise loadings at x-station, but vanish under bilateral symmetry; and

$$D_o(x) = \sigma_*|\ln\varepsilon|^{-1/2}F(x) = \sigma_1 F(x) \tag{5a}$$

$$S'_e(x) = \frac{d}{dx}\{S_c(x)+\sigma_*^2[(8\pi)^{-1}(1+\frac{1}{2}|\ln\varepsilon|^{-1})F_x^{\,2}+\frac{1}{2}|\ln\varepsilon|^{-1}T(x)+\frac{1}{8}\Gamma E(x)]\} \tag{5b}$$

where E(x) is a function of the differential side wash across the wing and vortex wake recognizable as the kinetic energy of the cross flow, and T(x) assumes the same form of E(x) with the side wash ϕ_y replaced by ϕ_x (normalized by $\rho_\infty U^2\alpha^2\lambda^2 b$). The matching with the outer solution provides the function β(x) which determines the average pressure at the x-station.

The Transonic Equivalence Rule for the Outer Flow

In terms of η and ω, the TSD equation with the similarity parameter K is recovered in the outer flow for $\Phi \equiv \phi/\tau Ub$. With a vanishing disturbance required in the far field

($\eta \to \infty$), the outer solution is completely determined by the line source with S'_e, line doublet with D_o or $\sigma_1 F(x)$, and to a lesser extent by the line quadropoles, according to Eq. (4). The equivalence rule that follows is: the outer-flow structure at a given K, including shocks and their associated drag rise, is identical and can be correlated in η and ω, as long as the axial distributions of the lift and the cross-section of the equivalent body, i.e. $\sigma_1 F(x)$ and $S_e(x)$, remain the same, subject to a relative error of the order $\varepsilon|\ell n \varepsilon|$ in general, and of the order $\varepsilon^2|\ell n \varepsilon|^2$ for configurations with bilateral symmetry. Since neither the details of load and thickness distributions, nor σ_* and Γ_* appear explicitly, the equivalence principle in question is more general than the classical transonic similitude.

Limitations and Untreated Aspects

Implicit in deriving the expressions for E(x) and T(x) is the assumption of a finite pressure difference at the leading edge, referred to as a "shock-free entry" and implies a "droop leading edge" [14,19]. Otherwise, not only the validity of the E and T integrals but the flow physics are questionable because of flow separations. The latter is likely to occur in the lift-dominated regime. The expressions given for E and T in Ref. 14 have also relied on the assumption of a nearly planar, pointed wing with a sweep angle corresponding to λ = 0(1) or slightly larger. A nonplanar example may promise a high degree of control on the lift-dependent S_e but requires a new derivation for E and T expressions. The case corresponding to $\mathcal{R} \equiv 4b^2/S_w >> 1$ and a non-vanishing λ would represent the problem of a high-aspect-ratio swept wing of considerable interest, if only the expressions for the E and T [14] could yield meaningful values in the limit corresponding to a (swept) lifting line. One may, nevertheless, introduce subregions around the local wing chord to allow a lifting-line treatment of the inner-flow region as in Ref. 23, with which new forms for E and T functions have to be determined. This approach may also allow nonlinearity to enter into the component mixed flow on the swept wing, as in Ref. 24. The interesting prospect, wherein the theory of transonic lifting line [24, 25] (for the inner region) and the transonic equivalence rule (for the outer region) coexist, can be anticipated by comparing the transonic similarity parameter K of Eq. (2a) with that for the swept-wing [24, 25]

$$K_n = (1-M_n^2)\cdot M_n^{-1}\alpha_n^{-2/3} \tag{6}$$

where M_n and α_n are the reference Mach number and incidence of the normal-component flow, respectively. An examination discloses that K and K_n can both belong to order unity and both theories may apply provided M_∞ is close enough to one.

Mach 0.98 Transport Design Study

The drag rise associated with shock loss in the outer flow for a prescribed $D_o(x)$ and $S'_e(x)$ may be correlated according to the equivalence rule as

$$C_{D_w} = D_w/\tfrac{1}{2}\rho_\infty U^2 S_w = \frac{1}{2}\tau^2 M_\infty^2 Æ\, f(K) \tag{7}$$

where f is of unit-order and function of K for a fully developed shock, except at a large σ_*, for which f is of the order σ_*^2. This represents the part of drag that cannot be controlled by the knowledge of supercritical wing sections; whether the effect in question is important in practice may nevertheless be inferred from the magnitude of $\frac{1}{2}\tau^2 M_\infty^2 Æ$ which gauges the order of C_{D_w}. A rise of 0.002 in C_D has been used in aerodynamic design to define the drag-rise Mach number [26]. For the example studied by Goodmanson [26] with M_∞ = 0.98, τ = 0.0263 and $Æ$ = 6.3, this gauge factor is found to be very close to 0.002, indicating the engineering relevance of the theoretical description. It is therefore worthwhile to examine how much the nonlinear lift effect may contribute to the equivalent body in this specific example, and how alternative lifting-surface arrangements can affect the magnitude and distribution of

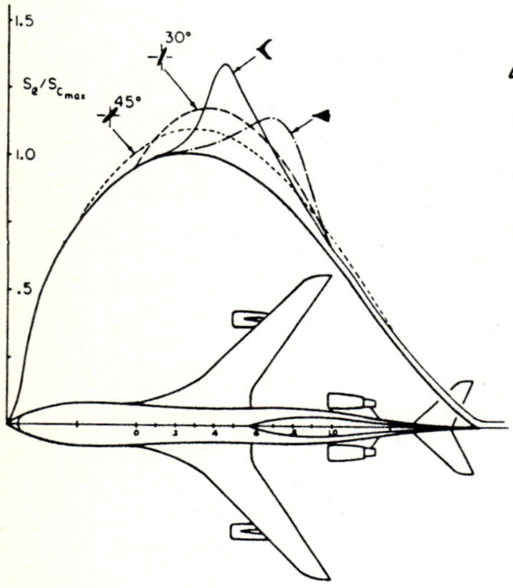

Fig. 1 Equiv. body cross-sec. area of Mach 0.98 design with four lift-arrangement options.

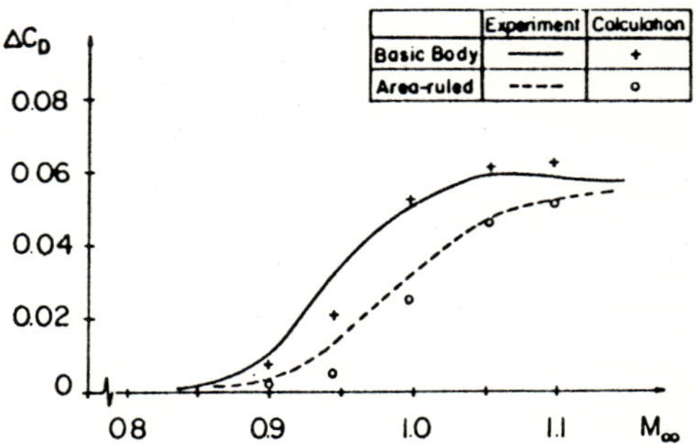

Fig. 2 Predicted and measured drag-rise characteristics of a NACA config. at zero. lift.

$S'_e(x)$. Figure 1 reproduces the result of such a study [19]. The basic design assumes a gross weight 287,300 lb., an altitude 39,000 ft., a half span 60 ft., a root chord 80 ft., an overall length 171 ft. and a maximum cross-section area 270 ft^2. From these, the scale ratio $\varepsilon = b/b'$ of the theory was calculated to be $\varepsilon = 0.0512$ with which σ_* and Γ_*, hence the S'_e of Eq. (5b), can be evaluated with functions F', T and E, which were determined for a variety of load distributions. The results of $S_e - S_c$ obtained for four representative planforms are shown and added to the smooth geometrical cross-section area $S_c(x)$ (in solid curve) orginally used in Goodmanson's study. The particularly rapid rise of 35% over a distance comparable to the root chord found with the swept-wing design indicates clearly the need for compensating the lift effect in this case.

CFD and Experimental Studies

The shocks occurring in the vicinity of a swept wing are known to be much stronger than those in the outer flow. In view of the above observation and the overwhelming success of modern CFD codes in predicting and controlling shocks on wings and bodies the question on the relevance of the equivalence rule arises. This issues becomes more critical to the study of high aspect ratio swept wings, for which most earlier wing-tunnel measurements [4] and flow-field calculations [11,17] are not too helpful. Whereas the earlier experiment by J.B. McDevitt [27] for the combinations of a body with a swept wing with an aspect ratio of 6 did show the merit of the area-ruling, it is not altogether clear if this benefit could have resulted primarily from a favorable interaction of the shock and boundary layer on the wing. Thus an inviscid flow-field analysis of the same problem and its comparison with experiments should provide a more concrete verification of the equivalence rule. This analysis was made by Caughy and Jameson [28] as a part of their work to illustrate the ability of a 3-D full-potential program to model fuselage geometry effects. In their analysis, the flows about two wing-body configurations with the same wing tested by McDevitt were studied; a Sears-Haak body was used in the fuselage of the first configuration, and area-ruling was applied to the second, slightly enlarged configuration which retained the same total volume and the Sears-Haak distribution. The predicted drag-rise calculated from surface pressure at zero-lift is reproduced from Ref. [28] in Fig. 2 for the first model (crosses) and for the second, area-ruled model (open circles). The measured data from Ref. [27] are in solid curve and in dashes for the first and second models, respectively. The comparison shows not only encouraging agreement between experimental and CFD studies, but indicates relevance and credence of the equivalence rule in aerodynamic-design study. With the relative importance of the lift contribution brought out earlier, CFD studies of the incidence effect on these and other configurations of interest can and should be made.

Experimental Study of the Equivalence Rule with Lift

A transonic-flow experiment designed to study the nonlinear lift effect on the equivalence rule was carried out by Y.Y. Chan [29]. Two models of wing-bodies having the same geometrical cross-section area distribution $S_c(x)$ but with (delta) wings of different thicknesses and aspect ratios (and spans) were used. Force measurements, surface and Schlieren flow visualizations were made at zero lift and with lift. The angle of attack of the two configurations was so chosen that they had, in addition to the same S_c, the same lift distribution $\sigma_1 F(x)$, and hence the same $S_e(x)$. Two equivalent bodies of revolution have also been tested; one corresponds to the zero lift and the other with lift. Figure 3a reproduces from Ref. 29 the drag rise characteristics of the two wing-bodies at zero lift, along with that of the body of revolution having the same $S_c(x)$, in the form of a rescaled drag coefficient and a transonic similarity parameter. The projected model geometries are shown on the left of Fig. 3a. The data for the wing-body having a smaller wing (WB1) are labelled with open circles, data for the wing-body having the larger wing (WB2) with triangles, and data for the body of revolution with squares. The discrepancies between WB1 and WB2 at large drag rise revealed in Fig. 3a indicate a departure from the equivalence rules. On the other hand, data for WB2 and for the equivalent body of revolution are

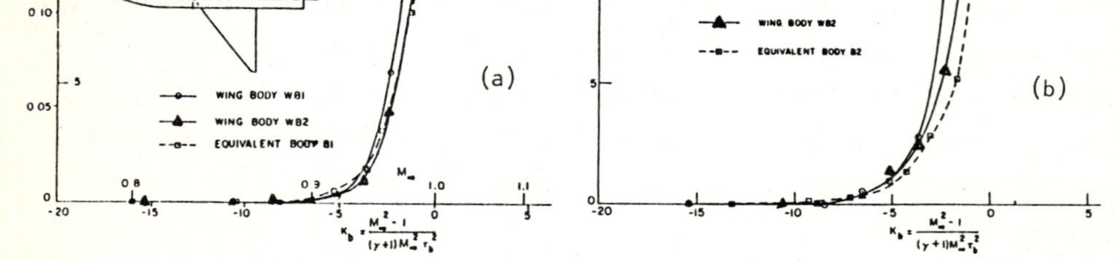

Fig. 3 Drag rise of wing-bodies WB1, WB2 and equiv. b. revolution: (a) zero lift (b) design lift condition.

in excellent agreement, suggesting that the boundary layer may have separated partly on WB1. The three data sets nevertheless agree reasonably well with one another at low ΔC_{D_o}-value, and predict almost the same drag-rise Mach number. Figure 3b compares the corresponding data at lift, where the data marked in squares were those for the equivalent body of revolution with the cross-section area taken to be the same as the $S_e(x)$ computed for WB1 and WB2. Here, the departure in data of WB1 and WB2, hence the equivalence rule, again occurs at higher drag rise as in Fig. 3a. The drag is also seen to be noticeably lower on the equivalent body of revolution, which was determined to be caused also by flow separation [29]. At low drag-rise values, however, the three data sets correlate extremely well and indicate also a decrease in the drag-rise Mach number from that at zero lift. In spite of discrepancies due to flow separation, the general trend of the dragcharacteristics is described rather well by the equivalence rule with lift, even though the designed magnitude of $S_e(x)$ due to lift in the experiment is rather small, amounting to no more than 15% of the $S_c(x)$.

Whereas continuation of Chan's experiment to a higher lift is of great interest, several requirements in the theory not met by the present experiment must be noted: (1) the wings with symmetrical profiles used do not satisfy strictly the shock-free entry requirement of the theory, (2) the triangular planform does not meet the requirements of a vanishing $F'(x)$ at the trailing edge, (3) the equivalent body of revolution at angle of attack cannot produce the correct doublet strength to fully simulate the $S_e(x)$ and $D_o(x)$ distributions. In spite of these shortcomings, the experiment has succeeded in substantiating the equivalence rule and the lift effects.

Closing Remarks

Experimental and CFD studies [27, 28, 29] confirm that the transonic equivalence rule can be extended to study and control drag rise of transonic swept-wing aircraft models. Apart from the several untreated and unresolved issues noted earlier, there are a few fundamental aspects in need of clarification. Most outstanding among these is perhaps the transonic vortex lift and its relationship to the equivalence rule. In the context of the asymptotic theory, the nonuniformity brought about by a shock occurring in both inner and outer regions has yet to be treated.

Acknowledgement

The author would like to thank Drs. S. Berndt, Y.Y. Chan, D.A. Caughey and M.M. Hafez for their suggestions on the material discussed. An expression of appreciation must be given to the U.S. Office of Naval Research, Fluid Dynamics Branch, for the support the author received during the course of his research in the transonic wing theory.

References

1. Oswatitsch, K., Keune, F.Z.: Flugwiss 3 (1954) 29-46.
2. Jones, R.T.: NACA Rept. 835.
3. Ward, N.G.: Quart. J. Mech. Appl. Math. 11 (1949) 75-97.
4. Whitcomb, R.T.: NACA Rept. 1273 (1956).
5. Oswatitsch, K.: Appl. Mech. Rev. 10 (1957) 543-545.
6. Cole, J.D, Messiter, A.F.: ZAMP 1 (1957).
7. Berndt, S.B.: ZAMP 9 (1955).
8. Berndt, S.B.: Aero. Res. Inst. Sweden, FAA Rept. 70 (1956).
9. Drougge, G.: Aero. Res. Inst. Sweden, FAA Rept. 83 (1959).
10. Ashley, H., Landahl, M.T.: Aerodynamics of Wings and Bodies, Addison-Wesley (1965).
11. Spreiter, J.R., Stahara, S.S.: AIAA J. 9 (1971) 1784-1791.
12. Cheng, H.K., Hafez, M.M.: AIAA J. 10 (1972) 1115-1116.13.
13. Barnwell, R.W.: AIAA 11 (1973) 764-765.
14. Cheng, H.K., Hafez, M.M.: J. Fluid Mech. 72 (1975) 161-187.
15. Barnwell, R.W.: NASA TR R-440 (1975).
16. Cramer, M.S.: J. Fluid Mech. 95 (1979) 223-
17. Cramer, M.S.: J. Fluid Mech. 109 (1981) 257-258.
18. Langhans, R.A., Flechner, S.G.: NASA TMX-2622 (1971).
19. Cheng, H.K.: AIAA J. 15, (1977) 366-373.
20. Sedin, Y.C.J.: Proc. 11th Congr. Int. Counc. Aero. Sci. 2 (1978) 71-84.
21. Guderley, K.G.: The Theory of Transonic Flow, Pergamon (1962).
22. Cole, J.D., Cook, L.P.: Transonic Aerodynamics, North-Holland (1986).
23. Cheng, H.K.: AIAA J. 16 (1978) 1211-1212.
24. Cheng, H.K., Meng, S.Y.: J. Fluid Mech. 97 (1980) 531-556.
25. Cook, L.P.: Quart Appl. Math. (1979) 178-202.
26. Goodmanson, L.T.: Aeronautics & Astronautics 46 (1971).
27. McDevitt, J.B.: NACA RM A 55B21 (1955).
28. Caughey, D.A., Jameson, A.: AIAA J. 17 (1979) 175-181.
29. Chan, Y.Y.: Nat. Res. Coun. Canada, Nat. Aero. Est. Aeron. Rept. LR-609/NRC 20225 (1982).

Perturbation Analysis of Unsteady Transonic Flow through Cascades

A. Kluwick, F. Koller

Institut für Strömungslehre und Wärmeübertragung
Technische Universität Wien, ÖSTERREICH

1. Introduction

The flow characteristics of cascades operating in the transonic regime have become increasingly important for the design of modern jet engines. However, in this regime even small amplitude oscillations of the blades may cause large variations of the magnitude of the aerodynamic forces and moments. Therefore, an inexpensive method for the prediction of steady and unsteady pressure distributions on oscillating blades in cascades would be particularly helpful when the flow is transonic.

2. Problem formulation

In this study we consider disturbances caused by an oscillating two-dimensional cascade which is placed into a uniform stream of a perfect gas with constant ratio γ of the specific heats at slightly subsonic speed, Fig. 1. It will be assumed that the thickness δ, the camber δ_c of the blades, the amplitude ν_p of the plunging motion nondimensional with the chord length $\tilde{c}$, the angle of incidence α and the amplitude ν_t of the pitching motion are small compared to one. If viscous effects are neglected, the field quantities can be expressed in terms of the velocity potential $\tilde{\Phi}$ to the order considered. Introducing the nondimensional variables

$$x=\frac{\tilde{x}}{\tilde{c}}, \quad y=\frac{\tilde{y}}{\tilde{c}}, \quad \tau=\tilde{\omega}\tilde{t}$$

$$u=\frac{\tilde{u}}{\tilde{a}^*}, \quad v=\frac{\tilde{v}}{\tilde{a}^*}, \quad \phi=\frac{1}{\tilde{a}^*\tilde{c}}\tilde{\phi}, \quad k=\frac{\tilde{\omega}\tilde{c}}{\tilde{a}^*} \tag{1}$$

where $\tilde{x}$, $\tilde{y}$, $\tilde{t}$, $\tilde{u}$, $\tilde{v}$, $\tilde{a}^*$, $\tilde{\omega}$ and k denote Cartesian coordinates, the time, the velocity components, the critical speed of

J. Zierep, H. Oertel (Eds.)
Symposium Transsonicum III
IUTAM Symposium Göttingen, 24.-27. 5. 1988

sound, a characteristic frequency and the reduced frequency, respectively, the gasdynamic equation can be written in the form:

$$k^2\phi_{\tau\tau} + \text{grad}\phi \cdot \text{grad}\left[2k\phi_\tau + \frac{1}{2}(\text{grad}\phi)^2\right] - \\ -a^2\Delta\phi = 0 \qquad (2)$$

Here the speed of sound is given by

$$a^2 = (\gamma - 1)\left[\frac{\gamma+1}{2(\gamma-1)} - k\phi_\tau - \frac{1}{2}(\text{grad}\phi)^2\right] \qquad (3)$$

The boundary condition at the blade surface $y=h^*(x,\tau)$ (Fig.1) requires:

$$\phi_y(x,h^*,\tau) = k\frac{\partial h^*(x,\tau)}{\partial \tau} + \phi_x(x,h^*,\tau)\frac{\partial h^*(x,\tau)}{\partial x} \qquad (4)$$

Vortex sheets caused by the time dependence of the circulation around the blades are assumed to emanate from the trailing edges. Hence, the difference in pressure across a blade vanishes as the trailing edge is approached (unsteady Kutta condition). Furthermore, appropriate jump conditions across moving shocks and vortex sheets have to be satisfied.

Approximate solutions to the set of equations (2) to (4) can be obtained in various ways. Linearization of the governing equations leads to the complete decoupling of the effects of airfoil shape and airfoil motion. Accordingly, the unsteady part of the problem describes the flow through a cascade of flat plates oscillating about zero incidence. Solutions for the case of subsonic flow were given by Whitehead [1] and Smith [2], for the case of supersonic flow by Kurosaka [3] and Adamczyk & Goldstein [4]. Surampudi & Adamczyk [5] applied the linearization concept to transonic flows.
The interaction between effects due to thickness and motion of the airfoils can be treated approximately if the airfoil oscillations cause only small perturbations of the steady flow field. As in linear theory, the governing equations for the unsteady part of the problem are linear (time linearization) but the coefficients now vary with the spatial

coordinates. Consequently, these equations have to be solved numerically in general. Numerical solutions for subsonic flow were presented by Verdon & Caspar [6], and for supersonic flow by Vogeler [7]. Verdon & Caspar [8] employed this method for transonic flow.
Asymptotic analysis, however, shows that the equations obtained by means of time linearization can be solved analytically for a certain regime of the relevant parameters. Moreover, asymptotic methods offer the possibility of treating strong interactions, that is when steady and unsteady effects can no longer be separated. This effect has not been studied so far.

3. Flow structure

Before turning to a more detailed outline of the analysis it is useful first to compare briefly the flow fields associated with a single airfoil and a cascade at transonic speeds.

In the case of a single airfoil in steady flow nonlinear theory comes into operation if the transonic similarity parameter $\chi^*=|1-M_\infty^{*2}|/\delta^{2/3}=O(1)$. Here $\varepsilon=(1-M_\infty^{*2})/K$ and M_∞^* denotes the free stream value of the critical Mach number. K is a constant of order one. The same result holds for a cascade if the spacing between the blades is $l=O(\varepsilon^{-1/2})$. If, however, $l=O(1)$, as in the case considered here, the flow properties of the regions between the blades resemble those through a channel rather than those past a single airfoil. As a consequence, nonlinear effects will be of importance if $\delta=O(\varepsilon^2)$ leading to velocity disturbances $u=O(\varepsilon)$, $v=O(\varepsilon^2)$. Assuming that $v=O(\varepsilon^2)$ also outside the channel region then yields the estimate $u=O(\varepsilon^{3/2})$ indicating that the regions upstream and downstream of the cascade are governed by linear equations, Messiter & Adamson [9].

The structure of transonic flow past oscillating airfoils has been elucidated by Lin, Reissner and Tsien [10]. Their results show that nonlinear theory applies if $\chi^*=O(1)$ and if the value of the reduced frequency is small: $k=O(\delta^{2/3})$. In order to study the related unsteady cascade problem, the perturbation potential $\varphi=\phi-x$ in the channel region is expanded in the form

$$\varphi = \varepsilon\varphi_1 + \varepsilon^2\varphi_2 + \dots \tag{5}$$

Insertion into (2) to (4) and equating terms of equal order of magnitude leads to sets of equations for $\varphi_1, \varphi_2, \dots$. Some conclusions which can be drawn from those equations assuming pitching and plunging oscillations of the same frequency are summarized in Table 1.

parameter range	flow characterization
$\delta = O(\varepsilon^2)$, $k = O(\varepsilon)$, $\nu = O(\varepsilon^2)$	nonlinear
$\delta = O(\varepsilon^2)$, $k = O(\varepsilon)$, $\nu = o(\varepsilon^2)$	time linearization
$\delta = O(\varepsilon^2)$, $k = o(\varepsilon)$, $\nu = o(\varepsilon^2)$	quasisteady
$\delta = o(\varepsilon^2)$, $k = o(\varepsilon)$, $\nu = o(\varepsilon^2)$	linear

Table 1: Properties of the flow inside the channel region for various parameter ranges.
δ... thickness or angle of incidence
ν... amplitude of pitching oscillation ν_t or $k\cdot\nu_p$ of the plunging oscillation

As in the studies of unsteady channel flows by Messiter & Adamson [11] and Kluwick [12], the nonlinear case $\delta=O(\varepsilon^2)$, $k=O(\varepsilon)$, $\nu=O(\varepsilon^2)$ listed in table 1 includes flows accelerating from subsonic to supersonic speed and shocks moving distances of $O(1)$.

According to the transonic small disturbance theory of steady flows through lightly loaded cascades, [9], solutions describing the flow properties in the channel region cannot be matched directly to those holding outside. Of course, a similar difficulty arises in the investigation of unsteady flows. Consequently, a more complicated flow structure, sketched in Fig. 1, has to be considered. Orders of magnitude of the velocity components and suitably stretched coordinates for the various regions are indicated in Table 2 for the nonlinear parameter regime. The same form of the stretched variables, however, applies in the case of time linearization which will be outlined in the next section.

regions	velocity perturbations	stretched coordinates
channel	(1): $u = O(\varepsilon)$ $v = O(\varepsilon^2)$	x y
entrance	front (5): $u = O(\varepsilon^{3/2})$ $v = O(\varepsilon^2)$ rear (2): $u = O(\varepsilon)$ $v = O(\varepsilon^2)$	$x^* = \pi(x-d)/[l(\varepsilon K^*)^{1/2}]$ $y^* = \pi(y-l)/l;\ K^*=O(1)$ $x^* = \pi(x-l)/[l(\varepsilon K_r^*)^{1/2}]$ $y^* = \pi y/l;\ K_r^*=O(1)$
near field	front (6): $u = O(\varepsilon^{3/2})$ $v = O(\varepsilon^2)$ rear (3): $u = O(\varepsilon)$ $v = O(\varepsilon^2)$	x y x y
far field	front (7): $u = O(\varepsilon^{3/2})$ $v = O(\varepsilon^2)$ rear (4): $u = O(\varepsilon)$ $v = O(\varepsilon^2)$	$\bar{x} = x$ $\bar{y} = (\varepsilon K^*)^{1/2}(y-lx/d)$ $\bar{x} = x-l$ $\bar{y} = (\varepsilon K_r^*)^{1/2}[y-l(x-l)/d]$

Table 2: Velocity perturbations and coordinates for various regions indicated in Fig. 1.

4. Time linearization for a flat plate cascade

As pointed out already, the concept of time linearization applies if $\delta=O(\varepsilon^2)$, $k=O(\varepsilon)$ and $\nu=O(\varepsilon^3)$. Specifically, it will be assumed that $k=k^*\varepsilon$, $\nu_t=\nu_t^*\varepsilon^3$ with constants k^* and ν_t^* of order one. Furthermore, it will be assumed for simplicity that the blades are flat plates which perform oscillations about their mid chord points. The mean angle of attack is zero. A real turbomachine-cascade would be composed of N blades. Hence, in the two-dimensional model the flow repeats itself after N blades.

The leading order term of the perturbation potential $O(\varepsilon)$ accounts for the difference between the actual (subsonic) value of M_∞^* and its value 1 at critical flow conditions. In the channel region the lowest order unsteady term is $O(\varepsilon^2)$ and the expansion proceeds in powers of ε. In contrast, unsteady effects enter the flow description at $O(\varepsilon^{5/2})$ and $O(\varepsilon^2)$ in

regions upstream and downstream of the channel, respectively. The expansions proceed in powers of $\varepsilon^{1/2}$.
Insertion of these expansions into (2), (3) and (4) then yields the isolated problems for the various regions, the solutions of which have to be matched asymptotically to obtain the description of the whole flow field. Perturbations of the streamwise velocity components up to order $\varepsilon^{5/2}$ have been calculated.
In the channel region the unsteady correction term $O(\varepsilon^2)$ to the steady part of the perturbation potential satisfies a linear inhomogeneous differential equation with constant coefficients. Application of the appropriate boundary and matching conditions leaves a set of constants c^n, $n=1,....,N$, undetermined.

The first two terms of the expansion valid in the channels match the first three terms of the solutions holding in the entrance regions. In regions (2) the first two terms of velocity perturbations are governed by the homogeneous and an inhomogeneous Laplace equation, respectively. As shown in [9], solutions to the homogeneous equation can be found using a Schwarz - Christoffel mapping. Again, a set of constants c^n remains undetermined.
The flow in the rear far-field (4) is described by a linear homogeneous and a linear inhomogeneous wave equation. A solution to the homogeneous equation can be derived from that of the classical linear wave equation

$$(1-M_\infty^2)\psi_{xx} + \psi_{yy} - 2kM_\infty\psi_{x\tau} - k^2\psi_{\tau\tau} = 0$$

where M_∞ denotes the free-stream Mach number, Fung [13]. A particular solution to the inhomogeneous equation can be expressed in terms of the lower order solution. Matching of these solutions to those valid in (2) finally provides the condition by which the set of constants c^n can be calculated, thus yielding the solution to the problem. In the case of constant phase shift $\sigma=2\pi/N$ between adjacent blades the evaluation of this condition reduces to the solution of a single equation for c^1.

Solutions describing flow properties upstream of the cascade can be found in a similar manner. Since there is flow around the leading edges of the blades, the solutions valid in the entrance regions (5) contain terms not present in regions (2), however. In the far-field (7) the calculations lead to a homogeneous wave equation in leading order. Its solution consists of two parts; one represents pressure waves originating from the blade surfaces outside the channels while the other one characterizes excess and deficit in mass flow through the channels as compared to steady flow.

Results for a flat-plate cascade are presented in Fig. 2. The difference in pressure Δp between the upper and lower side of a blade, nondimensionalized with $\rho^* \tilde{a}^{*2}/2$, compares qualitatively well with Verdon's numerical results [14], although the channel region is considerably shorter than wide.

5. Concluding remarks

It has been shown that asymptotic methods can be applied successfully to study the flow through lightly loaded oscillating cascades in the transonic regime. Numerical results have been obtained so far for the case of time linearization but work on the fully nonlinear case is in progress, Koller [15].

One of the authors, (A.K.) would like to thank Professors T.C.Adamson and A.F.Messiter for many helpful discussions during his visits to the University of Michigan.

References

[1] Whitehead, D.S., Vibration and Sound generation in a Cascade of Flat Plates in Subsonic Flow, ARC R&M 3685.

[2] Smith, S.N., Discrete Frequency Sound Generation in Axial Flow Turbomachines, Cambridge University Report CUED/A-Turbo/TR29, (1971) or ARC R&M 3709.

[3] Kurosaka, M., On the unsteady Supersonic Cascade with a Subsonic Leading Edge - An Exact First Order Theory - Part 1, Journal of Engineering for Power, Transactions of ASME, Ser. A, Vol. _96_, Jan. 1974, 13-22.

[4] Adamczyk, J.J., Goldstein, M.E., Unsteady Flow in a Supersonic Cascade with Subsonic Leading Edge Locus, AIAA Journal _16_, No 12, 1248-1254 (1978).

[5] Surampudi, S.P., Adamczyk, J.J., Unsteady Transonic Flow over Cascade Blades, AIAA Journal, _24_, No. 2, 293-302 (1986).

[6] Verdon, J.M. and Caspar, J.R., Development of a Linear Unsteady Aerodynamic Analysis for Finite-Deflection Subsonic Cascades, AIAA Journal, _20_, No. 9, 1259-1267, (1982).

[7] Vogeler, F.K., Gitter mit harmonisch schwingenden Schaufeln in Überschallströmung, Dissertation an der RWTH Aachen, (1983).

[8] Verdon, J.M. and Caspar, J.R., A Linearized Unsteady Aerodynamic Analysis for Transonic Cascades, Journal of Fluid Mechanics, _149_, 403-429, (1984).

[9] Messiter, A.F., Adamson Jr., T.C., Transonic Small-Disturbance Theory for Lightly Loaded Cascades, AIAA Journal 19, No. 8, 1047-1054 (1981).

[10] Lin, C.C., Reissner, E. and Tsien, H.S., On two-dimensional non-steady motion of a slender body in a compressible fluid, J.Math.Phys., _27_, 220-231 (1948).

[11] Messiter, A.F., Adamson Jr., T.C., Forced Oscillations of Transonic Channel and Inlet Flows with Shock Waves, AIAA Journal, 22, No. 11, 1590-1599, (1984).

[12] Kluwick, A., Schallnahe Wellenausbreitungsvorgänge in schlanken Düsen, Acta Mechanica, 15, 105-119 (1972).

[13] Fung, Y.C., An Introduction to the Theory of Aeroelasticity, J.Wiley & Sons, Inc., New York, (1955).

[14] Verdon, J.M., Unsteady Aerodynamics of Blade Rows, Paper presented at the Tenth U.S. National Congress of Applied Mechanics, June 16-20, 1986.

[15] Koller, F., Theorie kleiner Störungen für schallnahe instationäre, ebene Strömung durch Schaufelgitter, Dissertation, Techn. Univ. Vienna, in preparation (1988).

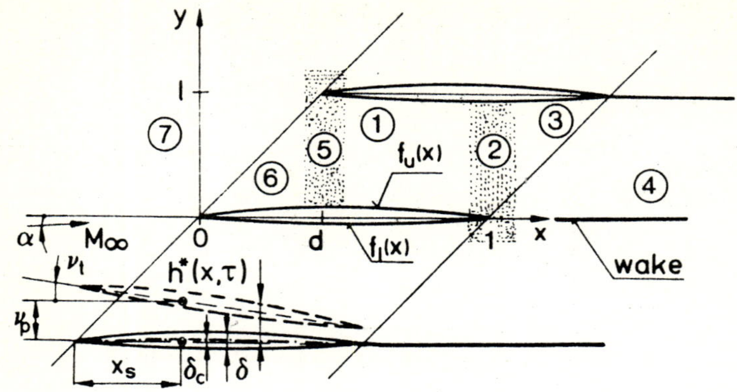

Fig.1: Two-dimensional cascade, regions for calculation.

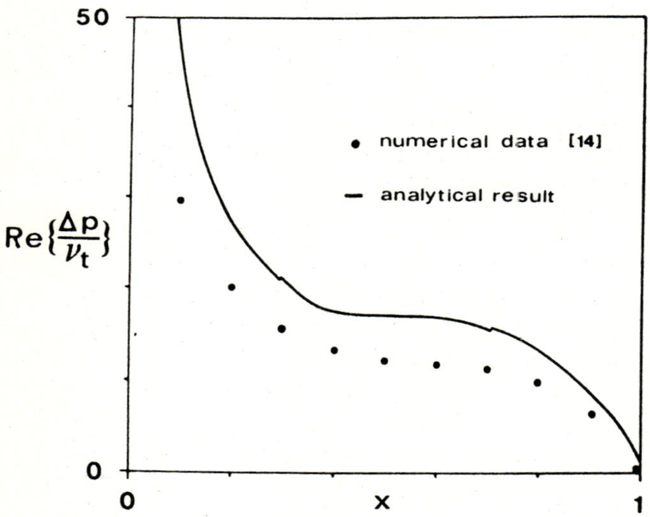

Fig.2: Real part of the unsteady pressure difference between upper and lower surfaces of the blades of a flat-plate cascade; l=d=0.707 (Fig.1), M_∞=0.9, k=0.092, phase shift 180°.

Viscous Transonic Flow in the Vicinity of a Corner Point

V.N. Diesperov

USSR Academy of Sciences
Computing Center
Moscow 117333, USSR

Summary

The asymptotic analysis is done for the Navier-Stokes system for transonic flow at high Reynolds numbers near a vortex of a convex corner. The presence of a sonic line is assumed. A local potential Vaglio-Laurin flow type and a flow with a free stream line are considered as limiting ones. It was shown that the character of the flow in incident laminar boundary layers depends on the pressure gradient that is established from the solution of a global problem.

External Potential Flows

A transonic flow is considered in the vicinity of the corner point of the profile break formed as a result of the intersection of two smooth curves, AO and OD, as shown in Figs. 1, 2 . We take Cartesian coordinates whose axis x coincides with the tangent at curve AO. Let φ, v_x, v_y be the perturbation potential and its associated velocities. The equation of the curve AO is set as

$$y = -k(n)(-x)^{4-3/n} + \cdots, x < 0; \quad \frac{3}{2} \leq n < 2, \quad k(n) > 0 \tag{1}$$

For certain values of $n \notin (3/2, 2)$ the coefficient k can be equal to zero. In this case the curve AO becomes a straight line. The incident subsonic flow along AO is supposed to reach sonic velocity at point O, and so a flow with a sonic line starting from point O is generated. The potential flows in the vicinity of the corner point are described by the similarity solutions of the Falcovich-Kármán-Guderley equation

$$\varphi = y^{3n-2} f(\xi), \quad \xi = \beta x / y^n, \quad \beta = (1+\kappa)^{-1/3} \tag{2}$$

For $n = 6/5$ [1] we have $k(6/5) = 0$ and the flow at a free streamline OB (see Fig. 1) described by the solution (2) in parametric form [2]

$$\xi = d_{6/5} \frac{2\omega + 3}{3\omega^{3/5}}, \quad f = d_{6/5}^3 \frac{16\omega + 9}{3\omega^{4/5}}, \quad -\infty < \omega < 0, d_{6/5} > 0 \tag{3}$$

For $n = 5/4$ we have $k(5/4) = 0$ and obtain the Vaglio-Laurin flow [3] (see Fig. 2). In this case the solution (2) can also be presented in a closed parametric form [4]

$$\xi = d_{5/4} \frac{t - 8/5}{(t-1)^{5/8}}, \quad f(\xi) = d_{5/4}^3 \frac{7t^2 - 140t + 160}{21(t-1)^{7/8}}, \quad 1 < t < \infty \tag{4}$$

J. Zierep, H. Oertel (Eds.)
Symposium Transsonicum III
IUTAM Symposium Göttingen, 24.-27. 5. 1988

For $n = 6/5$, $5/4$ the curvature AO is taken into account by means of the higher order terms of the solution expansion. However, for $3/2 \leq n < 2$ there is one more class of solutions (2) with $k \neq 0$ of a Vaglio-Laurin type. In the vicinity of the limiting hodograph characteristic these solutions for the stream function ψ can be written as [5]

$$\psi = D_j v_y^{2j} (1 - \zeta)^{2j + 1/6} F(j + 2/3,\ j + 1/6;\ 2j + 7/6;\ 1 - \zeta), \qquad j = \frac{1}{6(n - 1)} \tag{5}$$

Here $\zeta = 4\beta^{-3} v_x^3/(9v_y)^2$. The solution for $n = 3/2$, when AO has a constant curvature, is given in [6].

Thus, we have a non-unique solution of Prandtl-Meyer type for the local transonic problem of an ideal gas flow in the vicinity of the corner point.

The solution (3)-(5) possesses a favourable pressure gradient tending to $-\infty$, when $(-x) \to 0$. As a result we have

$$\frac{dp}{dx} = -\kappa \gamma \epsilon \tilde{d}_n (-x)^{\gamma - 1} + \cdots, \quad 1/3 \leq \gamma < 1 \ (\gamma = 2 - 2/n) \tag{6}$$

Boundary Layer under the Effect of an Infinite Favourable Pressure Gradient

As experimental investigations [7,8] show, the viscosity has a great influence upon the flow in the vicinity of a corner point when an expansion wave is formed in the transonic flow near the vertex corner. Owing to the presence of the boundary layer, the expansion wave ceases to be centered and does not start exactly at the corner point. Incompressible boundary layer interaction in the preliminary separation region was studied in [9-12]

Let us assume that the surface, which has a break, is thermally insulated; viscosity and thermal conductivity coefficients are proportional to the temperature; the Prandtl number equals unity; the Reynolds number Re is based on the characteristic size of the streamlined body and the critical values of the flow parameters and that the thermodynamic state of the gas under consideration is described by the Clapeyron equation.

The whole region of the boundary layer in front of the corner is subdivided into two parts under the effect of the pressure gradient. The flow in the lower viscous sublayer is incompressible in the first order approximation owing to small values of the velocities and the absence of a large temperature gradient. The vortex flow in the main part of the boundary layer is governed by the Euler equations in the first approximation. In the case of the formation of a nonlinear velocity profile under the effect of the pressure gradient near the corner point, the stream function Ψ in viscous sublayer is given as follows: $\Psi = (-x)^{(\gamma+2)/4} F_0(\eta)$, $\eta = Y/(-x)^{(2-\gamma)/4}$. The boundary problem can be reduced by means of some affine transformation to the following

$$\frac{(\gamma + 2)}{2\gamma} \Phi_0 \Phi''_0 - \Phi'^2_0 - \Phi'''_0 = 1\,, \tag{7}$$

$$\Phi_0(0) = \Phi'_0(0) = 0\,;\quad \Phi_0 = b_0\,\zeta^{(2+\gamma)/(2-\gamma)} + \cdots,\ \zeta \to \infty,\ b_0 > 0 \tag{8}$$

Under condition (8), equation (7) can only be solved if $0 < \gamma < 2/3$ [137]. In the case of $\gamma = 2/3$ ($n = 3/2$) there is no solution [13]. We can see that under conditions $2/3 < \gamma \leq 1$ and $\Phi''_0(0) > 0$, there are no physically available solutions. From the properties of the homogeneous equation (7) it follows [14] that the function $|\Phi_0(\zeta)|$ may tend to infinity in the limit $\zeta \to \infty$ only in accordance with (8). By multiplying equation (7) and its first derivative on $\exp[-(\gamma+2)\cdot\int_0^\zeta \Phi_0 d\zeta]$ and then integrating from 0 to ζ, it is easy to show that under the condition that $\Phi''(0) > 0$ we have $\Phi_0 \geq 0$, $\Phi'_0 \geq 0$, $\Phi''_0 > 0$ and $\Phi'''_0 < 0$ at $\zeta \geq 0$. However, the last inequality is inconsistent with (8) at $2/3 < \gamma \leq 1$.

Let us consider now the solution in the viscous sublayer as follows:

$$\Psi = R(0)\left\{\frac{\lambda}{2}\,Y^2 = \cdots + \epsilon\,\frac{\gamma\tilde{d}_n}{\lambda R(0)}\,(-x)^\gamma \Phi(\zeta)\right\},\ \zeta = \lambda^{1/3} R^{2/3}(0)\,\frac{Y}{(-x)^{2/3}}$$

Here, $R(0)$ is the density in the corner point O, and λ is the skin-friction coefficient. We assume that $\lambda \gtrsim 1$. The solution of a linear equation for Φ, which satisfies the no-slip conditions and the condition of an algebraic behaviour at infinity reads

$$\Phi = \zeta\int_0^\zeta [K_1 \Psi(-\gamma,\ 1/3;\ \xi) - \frac{1}{\gamma}]\,\zeta^{-2} d\zeta,\ \ K = \frac{\Gamma(2/3-\gamma)}{\gamma\Gamma(2/3)} \tag{9}$$

Here, $\Psi(.,.;\zeta)$ is the Tricomi function, $1/3 \leq \gamma < 1$, $\gamma \neq 2/3$.

The skin-friction is given by

$$\tau_W = \lambda + [2\cdot 3^{-4/3}\gamma R^{1/3}(0)\,K_1\,\frac{\Gamma(-2/3)}{\Gamma(-\gamma)}\,\tilde{d}_n]\,\epsilon\lambda^{-1/3}(-x)^{\gamma-2/3} + \cdots \tag{10}$$

The solution for $\gamma = 2/3$, which satifies the no-slip condition, has the exponential behaviour at $\zeta \to \infty$. Thus, the solution (5) with $j = 1/3$ ($n = 3/2$) cannot be realized in the viscous flow near the corner point O .

From the solution in the main part of the boundary layer adjoined to a viscous sublayer for $Y \to \infty$, we have for the transverse velocity

$$V = \epsilon\,[C_{11}(-x)^{\gamma-4/3} + C_{12}(-x)^{\gamma-1} + \cdots],$$

$$C_{11} = (\gamma - 1/3)\,\gamma R^{-1/3}(0)\,\tilde{d}_n D_1 \lambda^{-5/3},$$

$$C_{12} = -\gamma\tilde{d}_n\int_0^\infty \left[\frac{1-v^2R}{v^2R} - \frac{1}{\lambda^2 R(0)\,Y^2}\right] dY \tag{11}$$

$$D_1 = -3^{1/3}\,\frac{(1-\gamma)\,\Gamma(4/3-\gamma)}{(3\gamma-1)\,\Gamma(2-\gamma)}\,\Gamma(1/3),\quad 1/3 < \gamma < 1,\ \gamma \neq 2/3$$

The displacement effect of the boundary layer is described by formula (11). The first term is due to the inertia force, and the second one is the pressure gradient and the behaviour of the flow parameters in the incident boundary layer. The expansion (11) makes sense if $(-x) << \lambda^{-5}$. On the other hand, the skin-friction expansion is valid if

$$(-x) >> \epsilon^{3/(2-3\gamma)} \lambda^{-4/(2-3\gamma)} \tag{12}$$

As a result, we have that the obtained expansions are valid under the condition that $\epsilon << \lambda^{5(\gamma-2/5)}$ $(\lambda \gtrsim 1)$. From the last condition we get for $\gamma = 2/5$ an $\epsilon << 1$.

Interaction of the Boundary Layer with the Vaglio-Laurin Flow

For $\gamma = 2/5$ the external potential flow is described by the Vaglio-Laurin solution (4) with $d_{5/4} = 3^{-3/8}\, 5^{-1/4}\, \beta^{-7/8}\, \tilde{d}_{5/4}^{5/8}\, \epsilon^{5/8}$. For $\epsilon = O(1)$ the assumption about the linear velocity profile in the oncoming boundary layer becomes incorrect for $Y \to 0$.

At first we consider the case $\epsilon = O(1)$. For $\eta \to \infty$ the function F_0 behaves in the following way:

$$F_0 = B_0\eta^{3/2} + B_{00}\eta^{1/2}\ln\eta + B_{01}\eta^{1/2} + \cdots \tag{13}$$

The appearance of the logarithmic term is the result of the fact that the pressure and inertial forces on the external edge of a viscous sublayer become of the same order and resonance arises. The matching of the expansion (13) with the solution in the main part of the boundary layer yields: $v(Y) = 3B_0Y^{1/2}/2 + \cdots$, $Y \to 0$. From the solution in the main part of the boundary layer it follows that for $Y \to \infty$ we have

$$V = [A_0 ln(-x) + A_1](-x)^{-3/5} + \cdots, \quad (A_{0,1} = const) \tag{14}$$

Additional disturbances proportional to $Re^{-1/2}$ are present in the external flow as a result of the displacement effect of the boundary layer. The perturbation potential owing to (14) is taken in the form

$$\varphi = y^{7/4}f(\xi) + Re^{-1/2}y^{1/4}[\chi_{1/4} ln y + f_{1/4}] + \cdots,$$

$$\chi_{1/4} = B_{1/4}^{(S)}(t-1)^{-1/8} + B_{1/4}^{(A)}(t-1)^{3/8},$$

$$f_{1/4} = \left[\frac{1}{6}B_{1/4}^{(S)}\ln\frac{t^4}{(t-1)^3} + B_{1/4}^{(A)}\,arctg\sqrt{t-1} - B_{1/4}^{(S)} + D_{1/4}^{(S)}\right](t-1)^{-1/8} + \tag{15}$$

$$+\left[\frac{1}{2}B_{1/4}^{(A)}\ln\frac{t^4}{t-1} - \frac{4}{3}B_{1/4}^{(S)}\,arctg\sqrt{t-1} - 3B_{1/4}^{(A)} + D_{1/4}^{(A)}\right](t-1)^{3/8}$$

As we can see from (15), solutions $\chi_{1/4}$ and $f_{1/4}$ are stratified into the direct sum of two fundamental ones. For this reason the constants $B_{1/4}^{(S)}$ and $D_{1/4}^{(S)}$ remain arbitrary. We shall give them the same order of magnitude as $B_{1/4}^{(A)}$ and $D_{1/4}^{(A)}$, respectively. The latter parameters are expressed in terms of A_0 and A_1 by matching (14) and (15). The comparison of the terms in (15) shows that they become of the same order of magnitude at distance

$$(-x)_* = O[Re^{-5/12}\ln^{5/6}Re], \quad y_* = O[(-x)_*^{4/5}]$$

Over the supersonic region $z = x/y = O(1)$, $y \to 0$ the expansion of the total potential Φ takes the form

$$\tilde{\Phi} = yg_0(z) + y^{5/3}g_1(z) + Re^{-1/2}[W(z) + y^{1/3}(P_{1/4}(z)\ln y + C_{1/4}(z) + \cdots)] \quad (16)$$

The functions $g_0(z)$, $g_1(z)$ are defined in [15,16]. The function $W(z)$ is arbitrary and corresponds to the solution of the simple (noncentered) expansion-wave type. For $z \to 0$ it can be written as follows:

$$W(z) = \delta[4B_{1/4}^{(A)}\ln\delta - \frac{2}{3}\pi B_{1/4}^{(S)} - 3B_{1/4}^{(A)} + D_{1/4}^{(A)}] + \cdots, \quad \delta = \beta z/d_{5/4}$$

The expansion (16) ceases to be valid for $x_* = O(Re^{-3/10})$ $y_* = O(Re^{-3/10})$. The skin-friction coefficient is equal to $\tau_w = O[Re^{1/12}\ln^{1/6}Re]$ for a distance of the order of $(-x)_*$.

The uncertainty of the constants $B_{1/4}^{(S)}$, $D_{1/4}^{(S)}$ and of the function W are due to the fact that the behaviour of the boundary layer flow at the vicinity of the corner point is unknown. Such strong round off of the corner may be connected in particular with the formation of a separation bubble in the vicinity of a corner point [8].

We can introduce the term $y^{1/4}\ln^2 y\, \Phi_{1/4}(\xi)$ into the expansion (15) and so satisfy the limiting condition of the Prandtl-Meyer solution for $x > 0$, $y \to 0$. However, in this case the disturbances caused by the changes of the boundary conditions in the supersonic vicinity of the corner point have a very strong influence on the solution over the subsonic region.

Now, we consider the case $\epsilon << 1$. The boundary layer generates the disturbances in the external potential region according to (11). In contrast to the case $\epsilon = O(1)$, the displacement thickness is formed by the intertial forces at a first approximation. Now the subsonic flow depends only weakly on the displacement thickness which is due to the pressure gradient and corresponds to the second term in (11). As a result of the first and second terms in (11), the displacement thickness changes the curvature in the vicinity of point O and generates the disturbances in the external flow. Accordingly, the perturbation potential takes the form

$$\begin{aligned}
\tilde{\varphi} &= Re^{-1/2}[y^{-1/6}f_{-1/6}(\xi) + y^{1/4}f_{1/4}(\xi) + \cdots], \\
f_{-1/6} &= D_{-1/6}^{(S)}(t-1)^{1/12}F\left(\frac{5}{18}, \frac{1}{3}; \frac{1}{2}; 1-t\right) + \\
&+ D_{-1/6}^{(A)}(t-1)^{7/12}F\left(\frac{5}{6}, \frac{7}{9}; \frac{3}{2}; 1-t\right), \\
f_{1/4} &= D_{1/4}^{(S)}(t-1)^{-1/8} + D_{1/4}^{(A)}(t-1)^{3/8}
\end{aligned} \quad (17)$$

The constants $D_{-1/6}^{(S)}$ and $D_{-1/6}^{(A)}$ are assumed to be of the same order. The solution $f_{1/4}$ like $\psi_{1/4}$ in (15), is stratified into the direct sum of the fundamental ones. Matching (11) and (17) gives

$$D_{-1/6}^{(A)} = \frac{2\beta^{7/4}\tilde{d}_{5/4}^{5/12}}{3^{19/12}d^{5/6}R^{1/3}(0)} D_1\lambda^{-5/3}\epsilon^{5/12}, \quad D_{1/4}^{(A)} = \frac{5^{3/4}\beta^{9/8}}{3^{3/8}\tilde{d}_{5/4}^{-3/8}} C_{12}\epsilon^{5/8}$$

The disturbances in the external flow induced by the boundary layer become of the same order with the base solution at distances

$$(-x)_* = O(\epsilon^{-15/46}\lambda^{-25/23}Re^{-15/46}) \tag{18}$$

We presuppose that the expansions in the boundary layer and the expansions in the external potential region are valid at the values of x and are of the same order. The comparison between (12) and (18) shows that this condition is satisfied if λ, ϵ and Re are connected by the relation

$$\lambda\epsilon^{-25/24}Re^{-1/12} = K, \quad K = O(1)$$

This relation is similar to that obtained for the incompressible corner flow in [11,12].

In the expansion of the total potential $\tilde{\Phi}$, as well as in (16), the arbitrary function $W(z)$ reaches the supersonic region $x > 0$, $y \to 0$. For $z \to 0$ it has the form

$$W(z) = \left[D^{(S)}_{-1/6}\alpha_{-1/6} + D^{(A)}_{-1/6}\beta_{-1/6}\right] d^{2/3}_{5/4} z^{-2/3} + D^{(A)}_{1/4} d^{-1}_{5/4} z + \cdots,$$

$$\alpha_{-1/6}/\beta_{-1/6} = 2\Gamma(7/9)\,\Gamma(2/3)\,\Gamma^{-1}(5/18)\,\Gamma^{-1}(1/6)$$

The expansion $\tilde{\Phi}$ in the supersonic region $z = O(1)$, $y \to 0$ loses its meaning when $y_* = O(\epsilon^{-1}Re^{-3/10})$, $x_* = O(y_*)$. The influence of the second term in (11) at $\epsilon \to 1$ owing to the pressure gradient on the external flow becomes of the same order because of the influence of inertial forces.

At $x > 0$, $y \to 0$ we can require the passing of the total potential into the solution of the Prandtl-Meyer type. In this case it is necessary to introduce the term $y^{1/4}\ln y\, \Phi_{1/4}(\xi)$ into (17). The connection between the constants is as follows: $\beta_{-1/6}D^{(S)}_{-1/6} = -\alpha_{-1/6}D^{(A)}_{-2/6}$. As a result the expansion $\tilde{\Phi}$ ceases to be valid when $y_* = O(\epsilon^{-25/16}Re^{-3/8})$. Thus, for $\epsilon << 1$ the boundary layer in the supersonic vicinity of the corner vertex weakly influences the subsonic flow.

These solutions show that the solution near the corner point depends on the pressure gradient of the whole problem.

In the case of solution 6 the quantity $\epsilon\tilde{d}_n$ is expressed in terms of the coefficient k. The distinguishing feature of this flow consists of a weak perturbation of the skin-friction (10) by the pressure gradient (6) $(2/3 < \gamma < 1)$. Unlike the Vaglio-Laurin flow, the profiles in the boundary layer are less filled in the limit $Y \to 0$. This is due to the fact that function Φ becomes negative at $2/3 < \gamma < 1$. The first and second terms in the expansion of the perturbation potential are similar to (17).

Flow Near Corner Point with Free Streamline

Consider the case $\epsilon = O(1)$. The velocity profile at $Y \to 0$ is as follows: $v(Y) = 7B_0Y^{2/5}/5 + \cdots$.

The perturbation potential φ may easily be continued in the vicinity of a free streamline [17]. The solution in the mixing layer exists and is unique [14]. Expansion φ ceases to be correct at $(-x)_* = O(Re^{-3/7})$. If the integral $\int_0^\infty (1 - v^2R)/(v^2R)\,dY$ equals zero, the displacement thickness is formed by inertial forces for $Y \to \infty$. Then, the characteristic size of the disturbed region becomes equal to $(-x)_* = O(Re^{-6/13})$. In the linear case $(\epsilon << 1)$ the quantities λ, ϵ and Re are connected by the relation $\lambda\epsilon^{-15/13}Re^{-3/26} = K$.

References

[1] Diesperov, V.N.: Uch. Zapiski TSAGI 13 (1982), No.2, 15 (in Russian).4 (1984), p. 275.
[2] Diesperov, V.N.: Transonic Flows over Profile Break Points: Computing Center, Akad. Nauk SSSR 1987 (in Russian).
[3] Vaglio-Laurin, R.: Fluid Mech. 9 (1960), No.1, 81.
[4] Fal'kovich, S.V., Cernov, I.A.: Prikl. Math. Mekh. 28 (1964) 280 (in Russian).
[5] Diesperov, V.N.: Prikl. Mat. Mekh. 45 (1981) 652.
[6] Lifshitz, Yu.B., Shifrin, E.G.: Izv. Akad. Nauk SSSR, Mekh. Zhidk. Gaza (1971), No.2, 67 (in Russian).
[7] General Theory of High Speed Aerodynamics. Sears, W.R. (ed.). Princeton: Princeton University Press 1954.
[8] Bryson, A.E.: N.A.C.A. Rep. (1952), No., 1094.
[9] Ackerberg, R.C.: Fluid Mech. 44 (1970), No. 2, 211.
[10] Sychev, V.V. (ed.), Ruban, A.I., Sychev, Vic. V., Korolev, G.H.: Asymptotic Theory of Separated Flows. Moscow: Nauka 1987 (in Russian).
[11] Ryzhov, O.S.: Proc. IUTAM Symp. on Boundary-Layer Separation, London 1986. Berlin: Springer 1987.
[12] Bogdanova, E.V., Ryzhov, O.S.: Prikl. Mat. Mekh. 50 (1986) 394.
[13] Mc.Leod, J.B.: Quart. J. Math. 23 (1972), No. 89, 63.
[14] Diesperov, V.N.: Prikl. Mat. Mekh. 50 (1986) 403.
[15] Boichenko, V.S., Lifshitz, Yu.B.: Uch. Zapiski TSAGI 7 (1976), No. 2, 8 (in Russian).
[16] Esin, A.I., Chernov, I.A.: Prikl. Mat. Mekh. 41 (1977) 292.
[17] Diesperov, V.N.: Izv. Akad. Nauk SSSR, Mekh. Zidk Gaza (1983), No.5, 130 (in Russian).

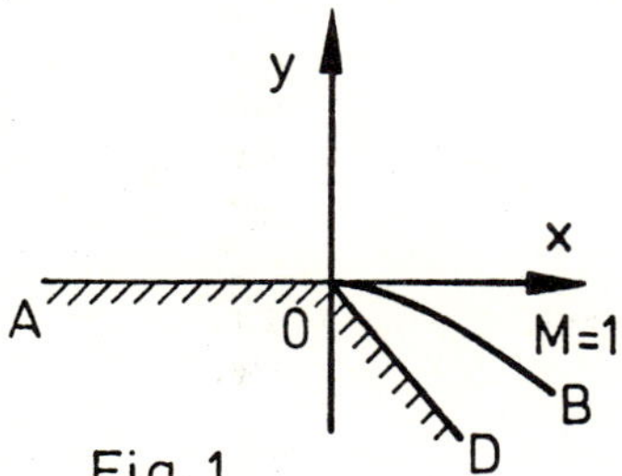

Fig. 1

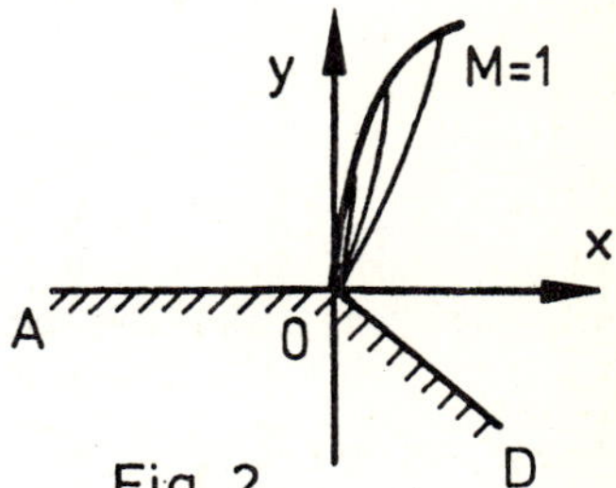

Fig. 2

Stability and Transition

Stability of Three-Dimensional Boundary Layers on Swept Wings at Transonic Speeds

L.M. Mack

Jet Propulsion Laboratory
California Institute of Technology
Pasadena, California 91109, USA

Summary

Linear, compressible, quasi-parallel, spatial stability theory is applied to the boundary layer on the upper surface of an infinite-span transonic swept wing. The stability theory is formulated for a class of standing and travelling waves that is suitable for comparison with experiment. Results of new calculations are presented for the wing without suction to demonstrate the principal instabilities. There is standing and travelling-wave crossflow instability, viscous instability, inflectional streamwise instability and a combination of the latter with crossflow instability. A discussion is given of the relation of these instabilities to the crossflow and streamwise boundary-layer profiles.

1. Introduction

The technical importance of maintaining laminar flow on the wings of large transonic aircraft has led to the study of the instability of three-dimensional (3D) boundary layers. The stability problems unique to 3D boundary layers were first encountered in the early 1950's. The most important reference from this period concerning stability theory is the classic paper of Gregory, Stuart and Walker [1] in which the phenomenon of crossflow instability was identified and explained.

The first viscous stability calculations of 3D boundary layers were carried out by Brown [2], but the subject then remained dormant until the middle 1970's when there was a rebirth of interest in maintaining laminar flow on large aircraft. By this time computer-aided stability calculations were common. The first program intended for general use was the SALLY code of Srokowski and Orszag [3]. This code was used by Hefner and Bushnell [4] to analyze low-speed swept-wing transition data. The SALLY code was later superseded by the COSAL code of Malik [5] which uses compressible stability theory. Both of these codes applied a form of the well-known e^N or N-factor method to indicate when transition may be expected. The COSAL code has found wide use in aircraft design studies. An alternative method of calculation was developed by Rozendaal [6] based on an early version of the 3D compressible code of Mack.

Different methods of applying linear stability theory to 3D boundary layers were developed at about the same time by Cebeci and Stewartson [7], Lekoudis [8,9,10], Mack

J. Zierep, H. Oertel (Eds.)
Symposium Transsonicum III
IUTAM Symposium Göttingen, 24.-27. 5. 1988

[11,12,13] and Nayfeh [14,15]. Later work is by Nayfeh and Reed [16], Arnal et al. [17] and Dallman and Bieler [18]. References [7, 17, 18] use incompressible stability theory; the others use compressible theory. All use the quasi-parallel theory except Nayfeh and Nayfeh-Reed. It was shown by Mack [12] that it is essential to use compressible stability theory at transonic speeds, except possibly for crossflow instability in the region of the wing before the direction of the maximum crossflow reverses.

An important effect that is not considered in any of the cited references is the effect of curvature. It was shown by Malik et al. [19] that in-plane streamline curvature cannot be neglected for stationary waves in the rotating-disk boundary layer, and by Malik and Poll [20] that both streamline and surface curvature are important for a yawed circular cylinder. A single calculation by Collier and Malik [21] at a location on the lower surface of a swept wing indicates that curvature may also have to be included for stationary crossflow waves on a swept wing where there is appreciable curvature. Unfortunately, at present there are not enough calculations to quantify the meaning of the word "appreciable".

All of the calculations in the present paper were done for an infinite-span swept wing using linear quasi-parallel stability theory with curvature neglected. Recent work on the subsequent development of the instability waves beyond the primary linear instability, such as wave interactions and secondary instabilities, will not be considered. The new calculations are for the same 35° swept wing as used in [8-10,12,13], but without suction and at a lower freestream temperature.

2. Formulation of stability theory

The normal-mode stability equations used here are for a perfect gas and may be found elsewhere [12,22]. In their complete form, these equations constitute an eighth-order system of ordinary differential equations with constant specific heat ($\gamma = 1.4$), the Sutherland viscosity law, a constant bulk viscosity of 0.8, and a Keyes formula for the thermal conductivity coefficient. However, in order to speed up the time-consuming eigenvalue calculations, the simplified sixth-order system has been used for all calculations. This system results from dropping one of the dissipation terms in the energy equation, and has been shown in the past [12,22] to give maximum amplification rates within about 2-3% of the eighth-order results.

As shown in Fig. 1, the x coordinate of the stability equations is aligned with the local potential velocity U_p^*. The angle between the x axis and the direction of the undisturbed freestream is the flow angle ψ_p. With a positive sweepback angle of ψ_{sw}, positive crossflow is to the right (z direction). The wave motion shown in the figure with the wavenumber vector $\vec{k}$ at the angle ψ with respect to the x axis, and with growth only in the x_s direction, is given below [eq.(4)].

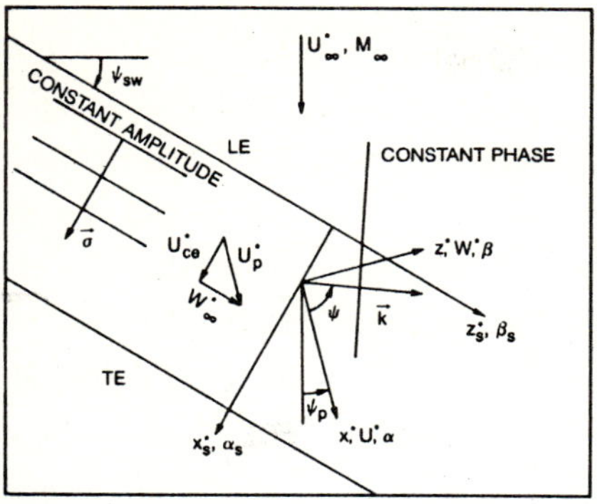

Fig. 1 Notation and coordinate systems of infinite-span swept wing.

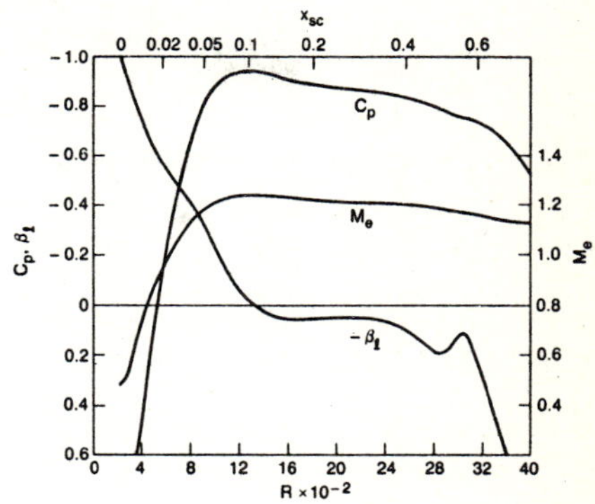

Fig. 2 Pressure distribution, local Mach number and pressure-gradient parameter of 35° swept wing.

The equations are dimensionless. The velocity U_p^* is the reference velocity, and the reference length is

$$L^* = (\nu_e^* x_s^* / U_{se}^*)^{1/2}, \tag{1}$$

where ν_e^* is the edge kinematic viscosity, and the subscript s refers to the coordinate system with x_s^* in the chordwise direction and z_s^* in the spanwise direction. Asterisks refer to dimensional quantities. The local Reynolds number is

$$R = U_p^* L^* / \nu_e^*. \tag{2}$$

The definition of the length scale in eq. (1) has been chosen so that R will have a nonzero value at the attachment line. The remaining freestream parameter that appears explicitly in the equations is the edge Mach number M_e.

The final stability equations are for a single Fourier component, or normal mode. A typical normal mode is

$$u(x,y,z,t) = \hat{u}(y) \exp\left[i(\alpha x + \beta z - \omega t)\right], \tag{3}$$

where the wavenumber components α and β and the frequency ω are complex. Thus, there are six real wave parameters. Two of the parameters are calculated as eigenvalues for a specified boundary-layer profile; the other four must be specified in some way. As there are no experiments with controlled disturbances with which to compare, it is not clear as to just what sort of calculations are the most useful. The approach taken here is to assume the spatial wave motion of Fig. 1, which has the form

$$u(x,y,z,t) = A\hat{u}(y)\exp\left[\left(\int_{(x_s^*)_{in}}^{x_s^*} \alpha_s^* dx_s^* + \beta_s^* z_s^* - \omega^* t^*\right)\right]. \quad (4)$$

Both β_s^*, the spanwise wavenumber at the initial position $(x_s^*)_{in}$, and ω^*, the constant frequency, are real and specified. Consequently, the wave grows only in the chordwise direction. The complex chordwise wavenumber component α_s is obtained as an eigenvalue from the stability equations along with the eigenfunction $\hat{u}(y)$. The initial amplitude A is a constant. Locally, such a wave in the x, z coordinates is a normal mode with both α and β complex. A single travelling wave of this type could be generated at $(x_s^*)_{in}$ in an experiment. Although there seems to be no way to produce a single standing wave of this type, a small roughness element would generate a continuous spectrum of standing waves that can be decomposed into waves of this type as was done by Mack [23] for the rotating-disk experiment of Wilkinson and Malik [24].

One more condition is needed to determine the downstream values of β_s from the assigned initial value. As eq. (4) represents a physical wave, the irrotationality, or phase continuity, condition is applied to the complex wavenumber vector. With the simplification of an infinite-span wing, this condition reduces to keeping the dimensional spanwise wavenumber constant as the wave moves downstream, or, in dimensionless terms,

$$\beta_s = (\beta_s)_{in}(L^*/L_{in}^*). \quad (5)$$

The stability problem is now completely specified, and spatial stability theory can be used to find the complex α_s as an eigenvalue at each chord position for the specified β_s^* and ω^*.

The spatial amplification rate is $\sigma = -\alpha_{si}$. The overall wave growth can be characterized by the N factor, defined to be

$$\ln N = Re_c \int_{(x_{sc})_o}^{x_{sc}} (\sigma/R)(U_p^*/U_\infty^*) dx_{sc}, \quad (6)$$

where x_{sc} is the surface distance referenced to the wing chord ch, and Re_c is the undisturbed freestream Reynolds number based on ch. The integral is evaluated for constant real dimensional frequency and spanwise wavenumber. The stability code uses the dimensionless frequency $F_\infty = \omega^* \nu_\infty^* / U_\infty^{*2}$. At each chordwise station, ω, which is real and defined in terms of the local length and velocity scales, has to be recalculated from F_∞. It is customary to start the integration at $(x_{sc})_c$, the first x_{sc} where the wave component becomes unstable, or, if the wave is already unstable at the initial position, at $(x_s)_{in}$. Obviously, if the N factor were to be related to the actual amplitude of the wave motion given by eq. (4), the integration would be required to start at $(x_s)_{in}$.

In previous papers by the author [11-13], the wave growth was taken to be either in the group-velocity direction or in the potential-flow direction. These two directions rarely differ by more than a few degrees. The amplification rate $\sigma(\overline{\psi})$ for growth in the arbitrary direction $\overline{\psi}$ is approximately related to $\sigma(\phi)$, the amplification rate for growth in the group-velocity direction, by

$$\sigma(\phi) \;=\; \sigma(\overline{\psi})\,\cos(\overline{\psi}\,-\,\phi), \tag{7}$$

where $\overline{\psi}$ and ϕ, the real part of the group-velocity angle, are both measured with respect to the x direction. The rate $\sigma(\overline{\psi})$ is not to be confused with the component of $\sigma(\phi)$ in an arbitrary direction. In the formulation of eq. (4), $\overline{\psi} = -(\psi_{sw} + \psi_p)$. It is to be noted that the N factor calculated from $\sigma(\phi)$ and the path length along the group-velocity trajectory is the same as given by eq. (6) to within the accuracy of eq. (7), because for growths in different directions the path length varies inversely as the amplification rate.

Cebeci and Stewartson [7] and Nayfeh [14] applied the harmonic point-source saddle-point condition of real group-velocity angle to the eigenvalue calculation, but did not evaluate the amplitude by this method. COSAL [5] calculates the maximum temporal amplification rate of an assigned frequency, and the group velocity of the most unstable wave. From the Gaster transformation, it obtains the approximate maximum spatial amplification rate in the group-velocity direction, and calculates the N factor by integrating the succession of maximum amplification rates along the group-velocity trajectory. This procedure automatically satisfies the saddle-point condition of Cebeci-Stewartson and Nayfeh. It may be pointed out that in the point-source calculation of [23], the author found that the proper saddle-point condition to use is the more complicated one of Gaster [25] for a growing boundary layer. Also, in that case the ray of maximum amplitude was not the one with the spatial amplification-rate vector in the direction of the group velocity.

3. Classification of mean velocity profiles

Instability in the quasi-parallel theory depends on the local boundary-layer profiles of velocity and temperature. Although viscous instability is still important at transonic Mach numbers, crossflow instability is an inviscid, or inflectional, instability. In addition, an adverse pressure gradient in the absence of suction produces a streamwise inflection point and a strong inviscid instability. Hence, it is useful to relate the calculated instabilities to the forms of the streamwise and crossflow velocity profiles. The only change from incompressible stability theory is that an inflection point is replaced by a generalized inflection point, i.e. the point where

$$\frac{d}{dy}\left(\rho \frac{dQ}{dy}\right) = 0. \tag{8}$$

In this equation, ρ is the density and Q is the mean velocity in some particular direction.

For the 35° swept wing, the chordwise pressure coefficient, the Levy-Lees pressure-gradient parameter β_l, and the local Mach number are plotted against both R and x_{sc} in Fig. 2. The undisturbed freestream conditions are: Mach number, 0.8912; temperature, 268.3° K (483° R); pressure, 0.3066 atm. The wing chord normal to the leading edge is 2.0 m, and the chord Reynolds number is $Re_c = 13.9 \times 10^6$. The flow angle ψ_p and the maximum positive and negative crossflow velocities are given in Fig. 3 for the wing with no suction. Reverse crossflow starts at about $x_{sc} = 0.25$ and laminar separation takes place at about $x_{sc} = 0.59$.

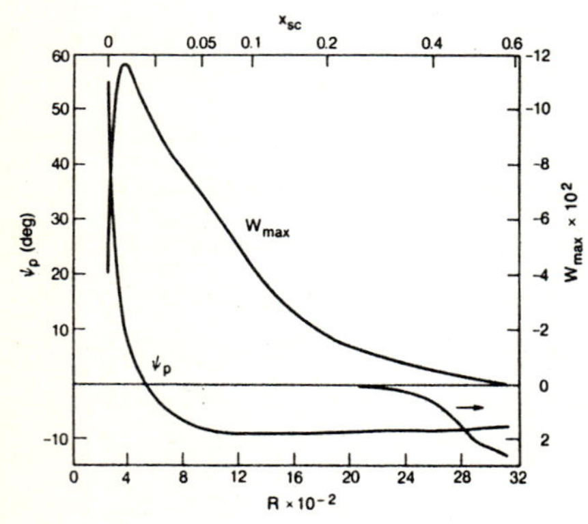

Fig. 3 Maximum negative and positive crossflow velocity and flow angle of 35° swept wing with no suction.

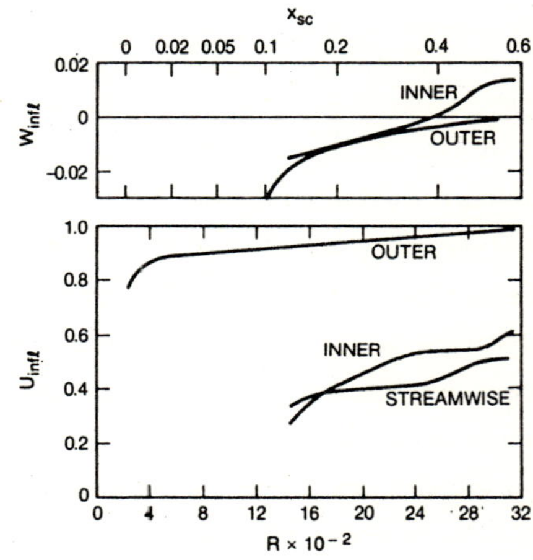

Fig. 4 Generalized inflection points of crossflow and streamwise profiles.

The generalized inflection points (referred to henceforth simply as inflection points) of the streamwise and crossflow profiles are as shown in Fig. 4. In the favorable pressure-gradient region, the crossflow is negative and the crossflow profile has one inflection point in the outer part of the boundary layer. The streamwise profile has no inflection point, as expected. As soon as the adverse pressure gradient starts, at $x_{sc} = 0.11$, the streamwise profile develops an inflection point and a second inflection point appears in the inner portion of the crossflow profile. The inner inflection point moves from negative to positive crossflow at $x_{sc} = 0.39$, well downstream of the start of reverse crossflow. The negative crossflow decreases steadily from this station on, but there is still a small amount at separation. The main influence of suction, besides a decrease in the amount of crossflow, is that the streamwise profile can be kept without an inflection point over the entire chord. The crossflow profile still goes through the evolution just described.

In the incompressible temporal stability theory, the governing velocity profile is the one in the direction of the wavenumber vector. This profile may be called the directional velocity profile. In the spatial theory, the stability equations can still be put in 2D form, but the directional profile and Reynolds number are both complex. In the compressible spatial theory, only the simplified sixth-order equations can be put in 2D form, but with complex Mach number and the temperature profile of the 3D flow. However, the velocity profile occurs mostly as $\alpha U + \beta W$ and it is still helpful to think of the instability in terms of the directional profiles even though numerical results can no longer be calculated from this profile alone.

4. Stability results - Standing waves

4.1 Negative crossflow region

An essential result of Stuart [1] was that an inviscid neutral standing wave exists when the inflection point of the directional velocity profile occurs at zero mean velocity. This same result carries over to compressible flow when the inflection point is the generalized inflection point. Thus, a neutral standing wave can be expected at finite Reynolds number in the negative crossflow region with a wave angle ψ near $90^\circ - \epsilon_{ip}$, where ϵ_{ip} is the deflection angle of the velocity at the inflection point. For example, at $x_{sc} = 0.0059$, $\epsilon_{ip} = 5.01^\circ$, and there is a neutral point at $k = 0.871$ with $\psi = 85.09^\circ$ (k is the magnitude of the wavenumber vector). The unstable region extends from the neutral point down to a completely viscous second neutral point at $k = 0.186$. At smaller k, the wave is damped with a complicated behavior of the damping rate. Because of the irrotationality condition, it is the wave components with small wavenumbers near the attachment line that are amplified further downstream. For example, at $x_{sc} = 0.310$, the most amplified wavenumber is $k = 0.524$, but for this same wave component at $x_{sc} = 0.0033$, near where crossflow instability starts, $k = 0.093$ and $\sigma = -0.016$. The critical Reynolds number for this wave component is 329 ($x_{sc} = 0.058$).

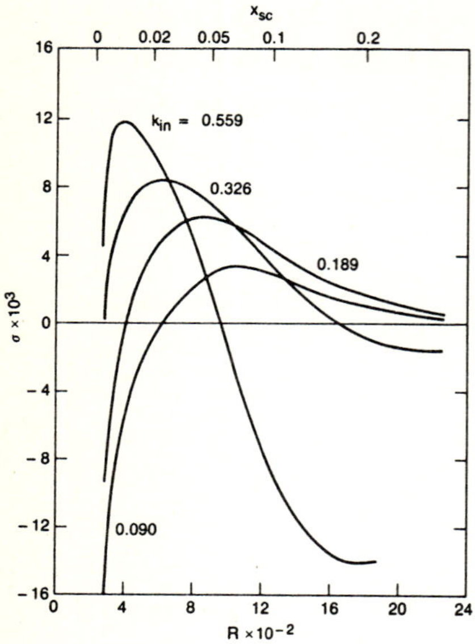

Fig. 5 Amplification rate of four stationary wave components in negative crossflow region.

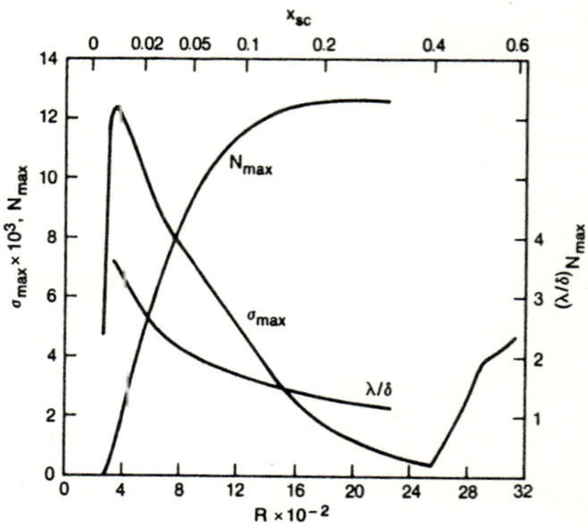

Fig. 6 Maximum amplification rate, maximum N factor and ratio of wave length to boundary-layer thickness at N_{max}; stationary waves.

The JPL stability code calculates the eigenvalues and N factors as a function of R for a band of spanwise wavenumbers with the same frequency. The wave components are given by eq. (4) and are subject to the irrotationality condition. If the wavenumber band has been properly chosen, the code yields the maximum amplification rate, σ_{max}, and the maximum N factor, N_{max}. The amplification rates of four wave components with zero frequency and identified by their initial wavenumber magnitudes at x_{sc} = 0.0033 are given in Fig. 5. The highest wavenumbers become unstable first. As the crossflow diminishes, all amplification rates approach zero. In Fig. 6, σ_{max} and N_{max} are given as functions of R and x_{sc} along with λ/δ, the ratio of the wave length to the boundary-layer thickness, corresponding to N_{max}. The curve for N_{max} ends at $x_{sc} = 0.31$ ($R = 2270$), where $W_{max} = -0.0095$, because numerical difficulties prevented the band of wave components from being carried past this point. However, calculations were still possible at individual stations, so that σ_{max} could be calculated all the way to laminar separation.

The maximum amplification rate varies with R as does W_{max}. As the latter becomes small, the wave angles are concentrated more and more in a narrow band near 90°. For a given wave component, N of course increases as long as there is instability. However, N_{max} does not continue to increase; as the amplification rates decline with W_{max}, N_{max} will eventually reach a maximum and also decline. This behavior is in contrast to the N factor of the COSAL program, which must of necessity continue to increase as long as there is instability (see [12] for a comparison with the similar behavior of the SALLY code).

4.2 Reverse crossflow region

When there is reverse crossflow, i.e., both negative and positive crossflow, the wave angle of σ_{max} is still related to the deflection angle of the outer inflection point until the inner inflection point moves to positive W and the maximum crossflow is positive. There are not separate solutions for the two inflection points. With a single inflection point, the wave angle is less than 90° with negative maximum crossflow, and greater than 90° with positive maximum crossflow. With two inflection points, normally $\psi <$ 90° as long as W at the inner inflection point is negative, and $\psi >$ 90° when W at the inner inflection point is positive. However, there is a transition zone. At the last station before both W at the inner inflection point and the maximum crossflow velocity become positive, ψ was found to be slightly greater than 90° for small wavenumbers and slightly less than 90° for all larger wavenumbers.

4.3 Positive crossflow region

When the first station is reached with the maximum crossflow positive ($x_{sc} = 0.419$), the inner inflection point is also at $W > 0$ and all wave angles are greater than 90°. Although there is also negative crossflow for some distance downstream, the positive crossflow region can be considered to start at this station, because from this point on it is the inner inflection point which governs the instability.

The question of whether all wave components pass continuously through 90° and thus belong to the same family of solutions is an important one. Although, the boundary-layer stations are widely spaced and the entire band of wavenumbers could not be carried through this region, a single wave component could be, albeit with numerous failures of the eigenvalue search procedure. The code operates so that when there is a failure of the search procedure at a station, interpolated profiles are formed to calculate an eigenvalue closer to the previous station. In this way, repeated failures give a detailed history of the wave. The results of several calculations with single wave components show that in spite of an almost discontinuous change in the slope of σ vs R where the maximum crossflow velocity changes sign, the wave angle passes continuously through 90°. Therefore, the σ_{max} curve of Fig. 6 represents the same family of solutions, and to calculate N factors as defined by eq. (6) in the positive crossflow region it is necessary to start the calculation before the critical Reynolds number in the region near the attachment line.

Starting at the first station with the maximum crossflow positive, a band of wave components was carried to the separation point. The maximum amplification rate increases with W_{max}, but as this remains less than 3%, σ_{max} just before separation is less than half of the peak value in Fig. 6. The incremental N factor to be added to N_{max} is less than three.

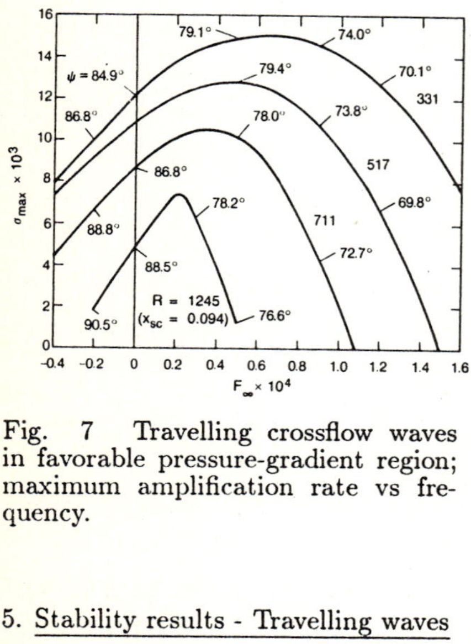

Fig. 7 Travelling crossflow waves in favorable pressure-gradient region; maximum amplification rate vs frequency.

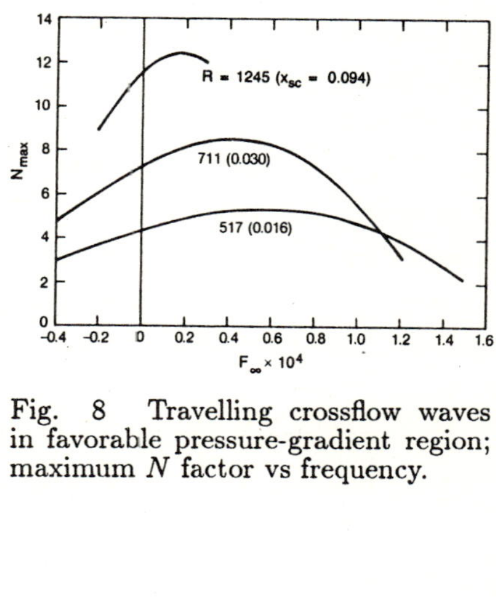

Fig. 8 Travelling crossflow waves in favorable pressure-gradient region; maximum N factor vs frequency.

5. Stability results - Travelling waves

5.1 Crossflow waves

Crossflow instability is not limited to standing waves. Indeed, it has long been known that the peak amplification rate occurs for a travelling, not a standing, wave. The maximum amplification rate, where the maximum is with respect to wavenumber, is given in Fig. 7 as a function of frequency for four Reynolds numbers in the region of favorable pressure gradient. There are two families of unstable frequencies. The first has the wavenumber vector in the direction opposite to the crossflow and includes the peak amplification rate. The second family is shown for convenience in Fig. 7 with negative frequencies. With positive frequencies, this family has the wavenumber vector in the same direction as the crossflow velocity. It is the first family that includes the peak amplification rate, has the widest band of unstable frequencies, and has the largest variation in wave angle.

The maximum N factors of the two families of waves are given in Fig. 8 at three Reynolds numbers. As the Reynolds number increases, the amplification rates decrease, the unstable frequency bandwidth decreases, the wave angles of the unstable frequencies deviate less from 90°, and both the peak amplification rate and the peak N factor occur closer to zero frequency. All of the directional profiles over the unstable angular range have a single inflection point well away from the wall. The narrowing of the unstable frequency bandwidth is mainly a consequence of the thickening of the boundary layer.

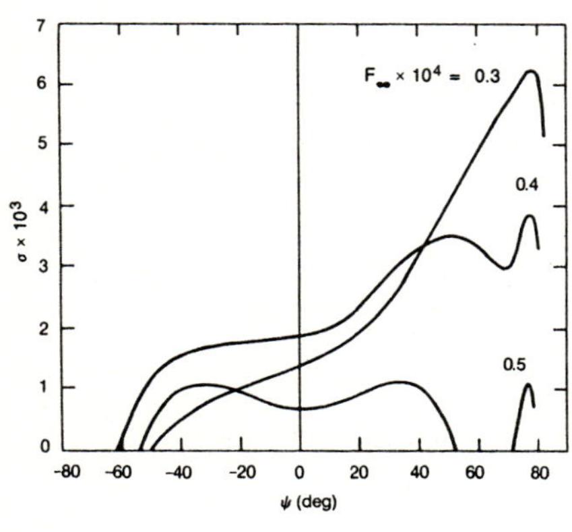

Fig. 9 Amplification rate vs wave angle for three travelling waves at $x_{sc} = 0.094 (R = 1245)$.

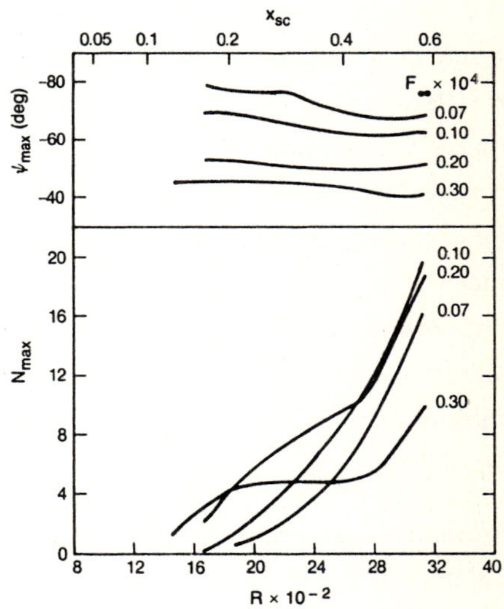

Fig. 10 Maximum N factor and wave angle at N_{max} for four frequencies in adverse pressure-gradient region; negative wave angles only.

5.2 Viscous streamwise instability

Instability is not limited to crossflow instability. In order to get a complete picture of the instability it is necessary to examine the full range of wave angles between −90° and +90° for each frequency. Only a limited number of calculations of this type have been carried out. The results for the amplification rates of three frequencies are given in Fig. 9 at the most downstream station of Figs. 7 and 8. This station is located just before the start of the adverse pressure gradient. The extent of the crossflow instability region is clear for the two larger frequencies, but not for the lowest frequency. The main question to answer concerns the nature of the instability for $\psi < 70°$. The directional profiles have a single inflection point for $\psi > 64°$, two inflection points for $64° > \psi > 45°$ and no inflection points for $\psi < 45°$. It is concluded that the instability for $\psi < 70°$ is primarily viscous, but with an influence of crossflow for the lower frequencies. This conclusion is supported by the distribution of σ with wave angle for $F_\infty = 0.5 \times 10^{-4}$ where there are maxima at about $\pm 35°$. Such maxima are characteristic of viscous instability at supersonic Mach numbers.

5.3 Inflectional streamwise and crossflow instability

The inflectional streamwise profiles that develop for $R > 1245$ in the absence of suction result in a stronger instability than exists on the same wing with suction, where there can be only viscous instability aside from the influence of crossflow. The latter influence makes the complete picture of instability in the adverse pressure-gradient region very complicated. More calculations than have been carried out for this paper are needed for a complete clarification. The results in [12] for the same wing with suction show that in the adverse pressure-gradient region with negative maximum crossflow, a weak viscous streamwise instability exists. When the maximum crossflow reverses, a stronger inflectional instability develops at large negative wave angles. In the present case without suction, the inflectional streamwise profile has an important influence on the instability.

A stronger instability than found in [12] develops for negative wave angles even before the maximum crossflow reverses. Figure 10 shows the maximum N factors and corresponding wave angles for four frequencies. The spanwise wavenumber band of this calculation included only negative wave angles. These wave angles are unstable only in the adverse pressure-gradient region. The fact that the maximum amplification rates and N factors occur at large wave angles, and with the wave angle of the maxima increasing with decreasing frequency, is not particular to the present boundary layer. Indeed, the same behavior is found in a Blasius boundary layer and in the nonsimilar 2D boundary layer with the pressure distribution of Fig. 2. However, in those flows the distribution of amplification rate with wave angle is symmetric about $\psi = 0$, whereas here the crossflow, even when small, results in a marked asymmetry. An examination of the directional velocity profiles shows that there is a single inflection point at about the same distance from the wall over almost the whole range of wave angles, but there are marked differences in the curvature of the velocity profile at the wall. The curvature increases with decreasing wave angle, so that small curvature and low amplification rates occur for $\psi > 0$, and large curvature and high amplification rates for $\psi < 0$. There are no maxima in the distribution of amplification rate with wave angle for $\psi > 0$. In view of the strong influence of crossflow, it does not seem proper to refer to this instability as streamwise instability. Indeed, the influence of crossflow is destabilizing in that the amplification rates are larger than for the 2D boundary layer with the same pressure distribution. Consequently, the instability of Fig. 10 is an example of streamwise instability modified by crossflow.

6. Concluding remarks

The complete linear instability characteristics of the boundary layer on a swept wing require extensive calculations to show in their entirety. The limited number of calculations presented here show several different regions of instability. In the favorable pressure-gradient region near the leading edge, there is standing and travelling-wave crossflow instability. A weak viscous instability develops towards the end of this region, and there is streamwise instability in the adverse pressure-gradient region. This instability is greatly influenced by even weak crossflow, and is strongest for waves with large negative wave angles.

Suction reduces crossflow instability and can eliminate the streamwise inflectional instability. There are still inflectional profiles at large negative wave angles in the positive crossflow region, but the amplification rates are less than when the streamwise profile is also inflectional. With suction the boundary layer can remain attached over the entire chord. A detailed study of the aft region with large positive crossflow and non-inflectional streamwise profiles remains to be carried out.

Acknowledgement

The work described in this paper was carried out at the Jet Propulsion Laboratory, California Institute of Technology, under contract with the National Aeronautics and Space Administration (NASA). Support from the Aerodynamics Division of the Office of Aeronautics and Space Technology, NASA, is gratefully acknowledged.

References

1. Gregory, N., Stuart, J.T. and Walker, W.S., On the Stability of Three-Dimensional Boundary Layers with Application to the Flow due to a Rotating Disk, *Phil. Trans. R. Soc. London*, Vol. A 248, pp.155-199, 1955.
2. Brown, W.B., A Stability Criterion for Three-Dimensional Laminar Boundary Layers, in "Boundary Layer and Flow Control" (G.V. Lachmann, ed.), Vol.2, pp.913-923, Pergamon Press, London, 1961.
3. Srokowski, A.J. and Orszag, S.A., Mass Flow Requirements for LFC Wing Design, AIAA Paper 77-1222, 1977.

4. Hefner, J.N. and Bushnell, D.M., Application of Stability Theory to Laminar Flow Control, AIAA Paper 79-1493, 1979.

5. Malik, M.R., COSAL - A Black-Box Compressible Stability Analysis Code for Transition Prediction in Three-Dimensional Boundary Layers, NASA CR-165925, 1982.

6. Rozendaal, R.A, Variable Sweep Transition Flight Experiment (VSTFE) - Stability Code Development and Clean-up Glove Data Analysis, NASA Conference Publication 2487, Part 3, pp. 845-860, 1987.

7. Cebeci, T. and Stewartson, K., Stability and Transition in Three-Dimensional Flows, *AIAA J.*, Vol.18, pp.398-405, 1980.

8. Lekoudis, S.G., Stability of Three-Dimensional Boundary Layers over Wings with Suction, AIAA Paper 79-0265, 1979.

9. Lekoudis, S.G., Stability of the Boundary Layer on a Swept Wing with Cooling, *AIAA J.*, Vol.18, pp.1029-1035, 1980.

10. Lekoudis, S.G., The Linear Instability due to the Compressible Crossflow on a Swept Wing, *ASME J. of Fluids Engineering*, Vol.102, pp.502-509, 1980.

11. Mack, L.M., Three-Dimensional Effects in Boundary-Layer Stability, in "Twelfth Symposium on Naval Hydrodynamics", pp.63-76, National Academy of Sciences, Washington, D.C., 1979.

12. Mack, L.M., On the Stability of the Boundary Layer on a Transonic Swept Wing, AIAA Paper 79-0264, 1979.

13. Mack, L.M., Compressible Boundary Layer Stability Calculations for Sweptback Wings with Suction, *AIAA J.*, Vol. 20, pp.363-369, 1982.

14. Nayfeh, A.H., Three-Dimensional Stability of Growing Boundary Layers, in "Laminar-Turbulent Transition" (R. Eppler and H. Fasel, eds.), pp.223-238, Springer-Verlag, Berlin, 1980.

15. Nayfeh, A.H., Stability of Three-Dimensional Boundary Layers, *AIAA J.*, Vol.18, pp.406-416, 1980.

16. Nayfeh, A.H. and Reed, H.L., Stability of Compressible Three-Dimensional Boundary Layers, AIAA Paper 82-1009, 1982.

17. Arnal, D., Habiballah, M. and Coustols, E., Laminar Instability Theory and Transition Criteria in Two and Three-Dimensional Flow, La Recherche Aerospatiale 1984-2, 1984.

18. Dallmann, U. and Bieler, H., Analysis and Simplified Prediction of Primary Instability of Three-Dimensional Boundary Layer Flows, AIAA Paper 87-1337, 1987.

19. Malik, M.R., Wilkinson, S.F. and Orszag, S.A., Instability and Transition in Rotating Disk Flow, *AIAA J.*, Vol.19, pp.1131-1138, 1981.

20. Malik, M.R. and Poll, D.I.A., Effect of Curvature on Three-Dimensional Boundary Layer Stability, *AIAA J.*, Vol.23, p.1362, 1985.

21. Collier, F.S., Jr. and Malik, M.R., Stationary Disturbances in Three Dimensional Boundary Layers over Concave Surfaces, AIAA Paper 87-1412, 1987.

22. Mack, L.M., Boundary Layer Linear Stability Theory, in "Special Course on Stability and Transition of Laminar Flow", AGARD Report 709, 1984.

23. Mack, L.M., The Stationary Wave Pattern Produced by a Point Source on a Rotating Disk, AIAA Paper 85-0490, 1985.

24. Wilkinson, S.P. and Malik, M.R., Stability Experiments in Rotating-Disk Flow, *AIAA J.*, Vol. 23, pp. 588-628, 1985.

25. Gaster, M., The Development of a Two-Dimensional Wave Packet in a Growing Boundary Layer, *Proc. Roy. Soc A.*, Vol. 384, pp. 317-332, 1982.

Transition Prediction in External Flows via Linear Stability Theory

D.M. Bushnell, M.R. Malik, and W.D. Harvey

NASA
Langley Research Center
Hampton, Virginia, USA

Summary

Paper summarizes progress in relating linear stability theory to transition location. Linear stability theory computations for both flight and low-disturbance wind tunnels, at low and high speeds, and for T-S, Görtler, and cross-flow modes indicate that, in the absence of Morkovin "bypasses," transition at low background disturbance levels corresponds to N-factors the order of 9 to 11 (in agreement with the early two-dimensional low speed flow research of A. M. O. Smith and van Ingen) provided that the appropriate physics, such as curvature effects, are included. This work indicates a much wider range of applicability for the e^N method than previously conjectured. Therefore, if roughness and waviness and background disturbance fields are small, the e^N method can evidently be utilized to correlate transition location and parameterize for such effects as (1) Mach number, (2) pressure gradient, (3) wall temperature, (4) angle of attack, (5) wall mass transfer, (6) sweep, (7) flow history, (8) geometry/curvature, (9) body rotation, (10) flow chemistry, and (11) bluntness. The next stage of research in the relationship between stability theory and transition will prove extraordinarily difficult, as exquisite and often immeasurable details of the environmental disturbance fields will be required. Transonic specific and accentuated background disturbances and their influences upon transition are also discussed.

Introduction and Discussion

Increased knowledge concerning the processes and predictability of transition is of current interest for a multitude of applications, including (a) improved accuracy and interpretation of conventional wind tunnel aerodynamic data and corresponding flight design, (b) laminar flow control, both natural and "unnatural" across the speed range from sailplanes to SST's, (c) sensor performance and observables, (d) dynamic surface loading, (e) flow separation control for "super maneuverability" and RPVS and, more recently, (f) hypersonic vehicles including "aero-space planes." In general, transition is important for lower Reynolds numbers whether attained through small vehicles, high altitude, or short distances on large bodies although for hypersonic flows Reynolds numbers of interest

J. Zierep, H. Oertel (Eds.)
Symposium Transsonicum III
IUTAM Symposium Göttingen, 24.-27. 5. 1988

for transition may range up to 100 × 10^6. The historical and continuing confusion regarding transition prediction and "correlation" is probably mainly attributable to, on the one hand, the penchant of man (even scientific man) for simple explanations and, on the other hand, the extraordinary richness, sensitivity, and multiplicity of physics which can lead to turbulence (e.g., Refs. 1-4). The present paper will attempt to indicate under what circumstances transition can be "predicted" (estimated) using linear stability theory.

For low-level background disturbances and in the absence of Morkovin bypasses (Ref. 1) the events leading to transition can be conveniently divided into six approximately serially occurring processes (a) generation and occurrence of ambient and body-produced disturbance fields, (b) modification of these initial disturbance fields by the body flow field, (c) interaction and internalization of this modified disturbance field by the body viscous flow ("receptivity" problem, Ref. 1), (d) amplification of this received/internalized disturbance field via normal modes as described by linear stability theory (possibly accompanied by continued forcing/receptivity), (e) finite amplitude wave-wave interaction/spectral broadening (the non-linear "end game") and, finally, (f) "transition", the first occurrence of Emmons spots or at least a significant deviation from laminar-like heat transfer and shear behavior. Although what is sought in most applications is the actual location of transition (the stage denoted (f) above), the situation is more akin to a long lingering illness than a sudden stroke, i.e., the transition location sought is the final stage of a process that can (1) involve a multitude of stages, (2) extend for extraordinary distances, (3) involve amplifications the order of 20,000 and, (4) be influenced by an immense parameter space.

Basically, transition occurs when disturbance amplitudes are the order of 5 percent (usual order of the fluctuation level in turbulent flows) with the non-linear end game beginning an order of magnitude or more earlier. Since, for the case of low ambient/vehicle initial disturbance fields the total amplification can be the order of 20,000 and the non-linear portion involves only the last factor of 10 or so, most of the transition processes can be approximated by linear theory. In tracking through the various steps leading to transition it becomes obvious that, while much of the amplification occurs in the linear instability region (for low level initial disturbance fields) the actual transition location can be dictated as much by the processes occurring **upstream** of normal mode amplification as by the normal mode amplification process itself. Since transition occurs at a disturbance **amplitude**, the **initial** disturbance **modes, levels, spectra** and **orientations**, etc., and how these are **processed** by the body **inviscid** and **viscous** flows are required, along with linear and non-linear stability theory, to predict transition. Unfortunately, very little information is available concerning the first three stages of the transition process. The initial ambient and vehicle disturbance levels are hardly, if ever, measured, and never the complete three-dimensional spectra actually required for all modes. In fact, it is usually not even clear what constitutes the background or initial disturbance fields, e.g., the dominant **sources** of these initial disturbances are still not identified in many cases, much less quantified. Therefore, "the rational prediction of transition Reynolds numbers in free flight borders on the impossible because of the lack of information on the disturbance environment" (Ref. 2). This state of affairs is probably due to (a) difficulties in measuring the often minute initial disturbance fields, particularly in the flight case (b) the great difference in the nature of the initial disturbance fields between flight and ground facilities and (c) the historical propensity to focus upon the non-linear "end game" where amplitudes are high and hence easily measurable and which has been perceived as "where the action is." In actuality the linear theory and non-linear end game portions are, thanks to large computational efforts, insightful experiments and clever theories, rapidly becoming understood and in the opinion of the present authors there is a need to shift emphasis to the first three stages (identification/quantification of initial and vehicle-induced disturbances, and processing of these by the vehicle inviscid and viscous flow fields) which define the initial disturbance fields upon which the linear and non-linear machinery ultimately operates.

Obviously, as the initial disturbance levels (as processed by the body inviscid and viscous flows)increase, less linear amplification is required to reach non-linear levels/transition and, in the limit, the linear processes can be "bypassed" completely (Ref. 5) for sufficiently large ambient/vehicle disturbances with efficient internalization. Many of these cases are describable, to first order, by spectrally averaged turbulence closures based upon energy methods (Ref. 6). Such large initial amplitudes can alter not only the location of transition but the parametric trends as well (Refs. 7-9). **Hence, linear stability theory can only apply to the transition process/prediction in the presence of low background**

disturbances where it describes the rate dominating physics. Details of the linear theories have been experimentally confirmed in many instances and even in high disturbance environments and at non-linear levels the most unstable modes often correspond to the frequency/wavelength indicated by linear theory. Therefore, a zeroth order application of linear theory is the identification of dangerous portions of the background disturbance spectrum and data evaluation in general (Ref. 10). As will be shown herein, the linear stability approach to correlation of transition information termed the e^N method appears to have (with essentially the same N values) a broad range of applicability (with success bordering on the phenomenal, considering the disparate physics and disturbance conditions involved). **Therefore, by inference, in many cases of engineering interest the internalized disturbance levels are apparently the same order of magnitude,** which is an extraordinarily fortunate happenstance. The linear stability theory/e^N approach also allows parameterization of transition information for parameters which affect the mean flow such as (1) Mach number, (2) pressure gradient, (3) wall temperature, (4) angle of attack, (5) wall mass transfer, (6) sweep, (7) flow history, (8) body rotation, (9) chemistry, (10) curvature, and (11) bluntness. However, constant vigilance is required to either avoid or recognize and account for bypasses. This presupposes extensive knowledge of both bypass physics and initial (internalized) disturbance characteristics and amplitude, information which is often non-existent. Also, the actual "most excited" instability mode in a particular flow may not be the most unstable, but rather the one whose product of initial disturbance level and integrated amplification is the greatest, i.e., **some appropriate initial disturbance is necessary to excite a particular mode.**

Three general classes of experimental data can be identified vis-a-vis the e^N method, based upon general levels of ambient disturbance fields. In free flight with nearly "unlimited visibility" (very few particles) and on sailplanes (no engine noise/vibration) with smooth "wave-free" surfaces and in the absence of clear air turbulence the N factors can be the order of 15 or greater. In the usual case of atmospheric powered flight and in low turbulence tunnels with "good surfaces" the N factors are the order of 10. For conventional tunnels, where $\frac{u'_\infty}{U_\infty}$ can be the order of 1 percent or greater and in gas turbine engines where stream disturbances can approach 10 percent (Ref. 11) the N factors are generally 5 or less. **It should be noted that the usual excursions about the N = 10 level for the intermediate**

(and more usual class of data) correspond, according to References 12-14, to excursions in transition location the order of 20 percent, which is reasonable considering that correlations of various "transition Reynolds numbers" usually exhibit "scatter" the order of factors of 5 or more.

e^N Method

The usual types of linear instabilities in boundary layers are of three types; Tollmien-Schlichting (T-S), Görtler, and cross-flow instability. The equations governing these instabilities are derived by superposing a small disturbance on a (known) basic state and then linearizing the Navier-Stokes equations with respect to the imposed disturbance. The solution of the linear stability equations yields the disturbance growth rate (σ) which may be integrated to compute the exponent N or the N factor as

$$N = \ln (A/A_o) = \int_{s_o}^{s} \sigma (s)\, ds \tag{1}$$

where s is the distance along the body to the transition location. Note that this is an **amplitude ratio and not an amplitude level**, hence, the expected dependence upon initial (internalized) disturbance levels.

It is a simple matter to compute the N factor using Eq. (1) for two-dimensional basic flow. For three-dimensional boundary layers, however, the procedure is somewhat involved and requires the determination of group velocity along with satisfying certain other constraints (Refs. 15-19).

e^N at Subsonic Speeds

Various investigators, including the present authors, have analyzed subsonic experimental transition data from both nominally low-disturbance wind tunnels and flight using the e^N method. The bulk of these results is summarized in Table 1. The test cases considered cover all three types of instability mechanisms and even include cases with wall bulges and small separation bubbles. The result of these studies indicates that the e^N method correlates the transition location remarkably well, e.g., a value of N = 10 constitutes a reasonable representation of the bulk of the results. However, one must compute the basic flow correctly and include the dominant physics in the stability equations. For example, the flow over a hump had to be computed using interacting boundary-layer theory. In the study of stability of the flow past a swept cylinder, the curvature effects had to be accounted for in the stability equations. If curvature is neglected, then the computed N factors are around 16-18 at transition on a swept cylinder. This indicates the importance of curvature for the design of leading edges of swept wings and more recent unpublished computations confirm this point.

The swept cylinder calculation also highlighted the importance of unsteady cross-flow waves which had previously been ignored in the calculations. However, they have been observed in recent experiments (Ref. 20) confirming the predictions of Malik and Poll. In general, abnormally high or low N values are traceable to extraordinarily low or high ambient disturbance fields, respectively.

e^N at Supersonic and Hypersonic Speeds

All three types of instability mechanisms persist into the supersonic regime. While compressibility has a stabilizing effect on Görtler and cross-flow modes (Refs. 40 and 41), the nature of the instability essentially remains unchanged. The growth rate of the T-S mode also decreases with Mach number but more rapidly and the most amplified waves are now oblique. At Mach numbers above 4, however, there is an additional unstable mode, known as the second mode, which dominates transition in high Mach number flows (Ref. 42). The second mode is most unstable when two-dimensional. This mode is destabilized with cooling but may be stabilized with suction and favorable pressure gradient (Ref. 43).

Recent computations indicate that the e^N method may also be used to predict the onset of transition in supersonic and hypersonic boundary layers. Table 2 summarizes the N values obtained from supersonic low-disturbance tunnels (of which there is only one, the Pilot Quiet Tunnel at NASA Langley) and flight using a compressible stability analysis computer code (Ref. 44). The list again contains all three types of instability mechanisms including the 'second mode' at high Mach numbers. The correlation with the experimental data is extremely good thus far. Of particular interest are the flat-plate experiments at Mach 3.5. Supersonic data from conventional wind tunnels indicated that the transition Reynolds number on a flat plate is lower than on a cone. e^N calculations gave completely opposite trends which motivated an experiment that was performed in the NASA Langley Low-Disturbance Tunnel. This experiment (Ref. 9) confirmed the predictions of the e^N method and also demonstrated the differential effect of noise, present in conventional supersonic tunnels, upon various model geometries.

e^N at Transonic Speeds

Initial Disturbances - The well-known propensity of disturbance fields to focus at transonic speeds implies that special care is needed in this speed range, even for the flight case. Conceptually, disturbances generated in flight by engines, juncture vortex flows, flow separations and radiated noise from attached turbulent flows on adjacent surfaces can influence transition, but applicable detailed studies are unavailable except for the X-21 project (Refs. 48-51) and the F-15 cone experiment (Ref. 52). In the former case the most prominent disturbance affecting the wing boundary layer at transonic speeds was adjudged to be sound radiated from the turbulent fuselage boundary layer. In the F-15 cone experiment the model was mounted in front of the fuselage and the most dominant (non-supersonic) disturbance mode was evidently the engine noise radiated forward onto the model.

In the case of ground facilities at transonic conditions multitudinous disturbance sources exist (Fig. 1) which typically create tremendous acoustic fields (Fig. 2). The usual dominant acoustic sources are disturbances associated with test section wall openings (slots, perforations) or upstream traveling waves generated by flow separation dynamics associated with model support struts and tunnel diffusers.

Transonic N Factors - The occurrence of sizable initial disturbance levels in many transonic flows suggests an examination of the limited transonic data as a function of corresponding turbulence level (deduced from the sound field) for the T-S mode (Fig. 3). The Mach number is indicated by each symbol. Also shown for comparison is a correlation derived by Mack based upon low-speed flat-plate experiments (Ref. 39, verified for other cases in Refs. 6 and 24) where the dominant stream disturbance mode was vorticity rather than sound. The transonic N values are in the same range as both lower and higher speed results discussed previously when corrected for the elevated transonic disturbance levels (see also Ref. 53). The lower levels from the LFC airfoil study can be at least partly attributed to observed unsteady model shock waves and instrumentation location (transition occurred downstream of the location for which N was determined but upstream of the next instrumented point, N was computed to increase rapidly between these locations).

Concluding Remarks

Stability theory computations for both flight and low disturbance tunnels at high and low speeds and for T-S, Görtler and cross flow modes indicate that in the absence of Morkovin "bypasses" and in the presence of stream disturbance levels the order of .05 percent transition corresponds to N-factors in the nominal range from 9 to 11, in agreement with the work of Smith et al. for two-dimensional low-speed wing flows. The present work indicates a much wider range of applicability for the e^N method than previously conjectured. The e^N approach can be, and indeed has already been, used to parameterize transition prediction as a function of parameters which affect the mean flow profiles for such purposes as LFC (suction, pressure gradient, wall temperature, compliant walls) and minimum drag designs. Moreover, the e^N approach provides an "expected transition Reynolds number" which can be utilized to infer the product of background disturbance field, body flow-field alteration thereof and receptivity, i.e., if the number deviates significantly from this expected value, then something untoward is occurring which requires further study or perhaps the calculation was carried out for the incorrect disturbance mode.

As a cautionary note, the actual N value corresponding to a particular experimental transition location is a function of the protocol utilized in applying the linear theory, especially in the three-dimensional case, as well as the fidelity and accuracy of the mean field calculations and the inclusion (or absence) of relevant physics which can influence the linear waves such as curvature. In addition, several applications of the method have carried the computation to the "middle" or even the end of the actual transitional flow region, thereby adding insult (trying to predict the occurrence of fully turbulent flow with a linear theory) to injury (applying the linear theory beyond the linear region to the position of initial Emmons spot production).

What is beyond e^N for transition prediction? Obviously some type of amplitude method. Unfortunately, this will be paced by increase in knowledge of the complete ambient and vehicle-induced initial disturbance field and this may be a very slow pace indeed (at least at the current rate of progress).

Acknowledgment

The authors wish to emphasize their indebtedness to the works of M. V. Morkovin who, since the 1950's, has provided the conceptual forefront of transition research.

References

1. Morkovin, Mark V.: Instability, Transition to Turbulence and Predictability. AGARDograph No. 236, presented at the AGARD Fluid Dynamics Panel Symposium on "Laminar-Turbulent Transition, Lyngby, Denmark, May 1977.

2. Reshotko, Eli: Boundary-Layer Stability and Transition. Annual Review of Fluid Mechanics, Vol. 8, 1976, pp. 311-349.

3. Kachanov, Yu. S.; Kozlov, V. V.; and Levchenko, V. Ya: Emergence of Turbulence in Boundary Layers. Vozniknoveniye Turbulentnosti v Progranichnom Sloye, Publishing House "Nauka," Novosibirsk, 1982, pp. 1-151.

4. Morkovin, Mark V.: Understanding Transition to Turbulence in Shear Layers. AFOSR-TR 83-0931, May 1983.

5. Morkovin, Mark V.: Bypass Transition to Turbulence and Research Desiderata. Symposium on "Transition in Turbines," NASA Lewis Research Center, Cleveland, OH, May 1984. CP-2386.

6. Arnal, Daniel: Description and Prediction of Transition in Two-Dimensional, Incompressible Flow. AGARD Report R-709, "Special Course on Stability and Transition of Laminar Flow," Mar. 1984, pp. 2-1-2-71.

7. Potter, J. Leith: Review of the Influence of Cooled Walls on Boundary-Layer Transition. AIAA Journal, Vo. 18, No. 8, August 1980, pp. 1010-1012.

8. Lysenko, V. I.; and Maslov, A. A.: The Effect of Cooling on Supersonic Boundary-Layer Stability. J. Fluid Mech., Vol. 147, 1984, pp. 39-52.

9. Chen, F.-J.; Malik, M. R.; and Beckwith, I. E.: Comparison of Boundary-Layer Transition on a Cone and Flat Plate at Mach 3.5. Presented at the AIAA 26th Aerospace Sciences Meeting, Reno, NV, January 11-14, 1988. Paper No. AIAA 88-0411.

10. Reshotko, Eli: Stability Theory as a Guide to the Evaluation of Transition Data. AIAA Journal, Vol. 7, No. 6, June 1969, pp. 1086-1091.

11. Transition in Turbines. NASA CP 2386, Proceedings of a symposium sponsored by the NASA Lewis Research, NASA Lewis, Cleveland, OH, May 15-16, 1984.

12. Morkovin, Mark V.: Guide to Experiments on Instability and Laminar-Turbulent Transition in Shear Layers. Illinois Institute of Technology, Chicago, Illinois, partially supported under AFOSR Contract F-49620-77-0013.

13. Smith, A. M. O.; and Gamberoni, Nathalie: Transition, Pressure Gradient, and Stability Theory. Douglas Aircraft Company, Inc., Report ES 26388, August 1956.

14. Kaups, K.: Transition Prediction on Bodies of Revolution. McDonnell Douglas Report MDC J6530, April 1974.

15. Srokowski, A. J.; and Orszag, S. A.: Mass Flow Requirements for LFC Wing Design. AIAA Paper No. 77-1222, 1977.

16. Cebeci, T.; and Stewartson, K.: On Stability and Transition of Three-Dimensional Flows. AIAA Journal, Vol. 18, 1980, pp. 398-405.

17. Nayfeh, A.: Stability of Three-Dimensional Boundary Layers. AIAA Journal, Vol. 18, 1980, pp. 406-416.

18. Mack, L. M.: On the Stability of the Boundary Layer on a Transonic Swept Wing. AIAA Paper No. 79-0264.

19. Malik, M. R.; and Orszag, S. A.: Comparison of Methods for Prediction of Transition by Stability Analysis. AIAA J., Vol. 18, 1980, pp. 1485-1489.

20. Bippes, H.; and Nitschke-Kowsky, P.: Experimental Study of Instability Modes in a Three-Dimensional Boundary Layer. AIAA Paper No. 87-1336.

21. Jaffe, N. A.; Okamura, T. T.; and Smith, A. M. O.: Determination of Spatial Amplification Factors and Their Application to Predicting Transition. AIAA Journal, Vol. 8, 1970, pp. 301-308.

22. Van Ingen, J. L.: A Method of Calculating Transition Region for Two-Dimensional Boundary Layers with Distributed Suction. Netherlands Wissenschaftliche Gesellschaft für Luft- und Raumfahrt, Europäischer Luftfahrtkongress, 6th, Munich, West Germany, September 1-4, 1965.

23. Huang, T. T.; and Hannan, D. E.: Pressure Fluctuations in the Regions of Flow Transition. David Taylor Naval Ship R&D Center Report No. 4723, 1975.

24. Lauchle, G. C.; Eisenhuth, J. J.; and Gurney, G. B.: Boundary-Layer Transition on a Body of Revolution. J. Hydronautics, Vol. 14, 1980, pp. 117-121.

25. Winter, K. G.; and Maskell, E. C.: The Reynolds Number for Transition on a Flat Plate in the RAE 4-ft. × 3-ft. Low Turbulence Wind Tunnel. RAE TM Aero-1854, 1980.

26. Malik, M. R.: Stability Theory Applications to Laminar-Flow Control. Research in Natural Laminar Flow and Laminar-Flow Control. NASA CP-2487, 1987. (Eds.: J. N. Hefner and F. E. Sabo).

27. Nayfeh, A. H.; Ragab, S. A.; and Al-Maaitah, Ayman: Effect of Bulges on the Stability of Boundary Layers. To appear in Physics of Fluids, 1988.

28. Cebeci, T.; and Egan, D. A.: The Effect of Wave-Like Roughness on Transition. AIAA Paper No. 88-0139.

29. Saric, W. S.; Takagi, S.; and Mousseux, M.: The ASU Unsteady Wind Tunnel and Fundamental Requirements for Freestream Turbulence Measurements. AIAA Paper No. 88-0053.

30. Ishida, Y.; and Itoh, N.: Application of the e^N Method to Calculations of Laminar Flow Control. Proceedings of the Third NAL Symposium on Aircraft Computational Aerodynamics, June 13-14, 1985, NAL, Tokyo, Japan, pp. 131-138.

31. Redeker, G.; Horstmann, K. H.; Koster, H.; and Quast, A.: Investigations on High Reynolds Number Laminar Flow Airfoils. ICAS-86-1.1.3, 15th Congress of the ICAS, September 7-12, 1986, London, UK. (Eds.: P. Santini and R. Staufenbiel).

32. Power, L.: A Comparison Between Measured and Computed Locations of Transition on Nine Forebodies of Revolution. Proceedings of Low-Speed Boundary Layer Transition Workshop, II, Rand Corporation Report R-P-6119, 1978. (Eds.: W. S. King and M. Yokota).

33. Runyan, L. J.; and George-Falvy, D.: Amplification Factors at Transition on an Unswept Wing in Free Flight and on a Swept Wing in Wind Tunnel. 17th Aerospace Sciences Meeting, New Orleans, Louisiana, January 15-17, 1979, AIAA Paper No. 79-0267.

34. Hefner, J. N.; and Bushnell, D. M.: Application of Stability Theory to Laminar Flow Control. AIAA Paper No. 79-1493, 1979.

35. Malik, M. R.; Wilkinson, S. P.; and Orszag, S. A.: Instability and Transition in Rotating Disk Flow. AIAA J., Vol. 19, 1981, pp. 1131-1138.

36. Malik, M. R.; and Poll, D. I. A.: Effect of Curvature on Three-Dimensional Boundary Layer Stability. AIAA J., Vol. 23, 1985, pp. 1362-1369.

37. Smith, A. M. O.: On the Growth of Taylor-Görtler Vortices Along Highly Curved Walls. Quart. Appl. Math., Vol. 13, 1955, pp. 233-262.

38. Wazzan, A. R.; Okamura, T. T.; and Smith, A. M. O.: Spatial and Temporal Stability Charts for the Falkner-Skan Boundary-Layer Profiles. DAC 67086, 1968, McDonnell Douglas Corp.

39. Mack, L. M.: Transition and Laminar Instability. Jet Propulsion Laboratory Publication 77-15, Pasadena, CA, 1977.

40. El-Hady, N. M.; and Verma, A. K.: Growth of Görtler Vortices in Compressible Boundary Layers Along Curved Surfaces. AIAA Paper No. 81-1278, presented at the 14th Fluid and Plasma Dynamics Conference, Palo Alto, CA, June 23-25, 1981.

41. Bushnell, D. M.; and Malik, M. R.: Supersonic Laminar Flow Control. Research in Natural Laminar Flow and Laminar Flow Control, NASA CP-2487, 1987, (Eds.: J. N. Hefner and F. E. Sabo).

42. Mack, L. M.: Boundary-Layer Linear Stability Theory. Special Course on Stability and Transition of Laminar Flow, AGARD-R-709, 1984.

43. Malik, M. R.: Prediction and Control of Transition in Hypersonic Boundary Layers. AIAA Paper No. 87-1414.

44. Malik, M. R.: COSAL - A Black-Box Compressible Stability Analysis Code for Transition Prediction in Three-Dimensional Boundary Layers. NASA CR-165925, 1982.

45. Malik, M. R.: Instability and Transition in Supersonic Boundary Layers. Laminar Turbulent Boundary Layers. (Eds.: E. M. Uram and H. E. Weber), Energy Resources Technology Conference, New Orleans, Louisiana, Feb. 12-16, 1984.

46. Chen, F.-J.; Malik, M. R.; and Beckwith, I. E.: Instabilities and Transition in the Wall Boundary Layers of Low-Disturbance Supersonic Nozzles. AIAA Paper No. 85-1573.

47. Collier, F. S.; and Johnson, J. B.: Supersonic Boundary-Layer Transition on the LaRC F-106 and DFRF F-15 Aircraft - Part I: Transition Measurements and Stability Analysis. Research in Natural Laminar Flow and Laminar Flow Control. NASA CP-2487, 1987 (Eds.: J. N. Hefner and F. E. Sabo).

48. Charmichael, R. F.; and Pelke, D. E.: In-Flight Noise Measurements Performed on the X-21A Laminar Flow Aircraft. Northrop Corporation, NOR-64-81, April 1964.

49. Rogers, O. R.; Curran, E. T.; and Wrobel, N. A.: Aero-Acoustic Aspects of Laminar Flow Control. ASD-TDR-62-718, Vol. 1.

50. Mangiarotty, R. A.: Effects of Engine Noise on Aircraft Wing Laminar Boundary-Layer Stability. J. Acoust, Soc. A. 70 (1), July 1981, pp. 98-109.

51. Rooney, T. R.; Charmichael, R. F.; and Eldred, K. M.: Investigation of Noise with Respect to the LFC, NB-66 Aircraft. Northrop Corporation, NOR-61-010, April 1961.

52. Fisher, David F.; and Dougherty, N. Sam, Jr.: In-Flight Transition Measurement on a 10° Cone at Mach Numbers from 0.5 to 2.0. NASA TP-1971, June 1982.

53. Vijgen, P. M. H. W.; Dodbele, S. S.; Pfenninger, W.; and Holmes, B. J.: Analysis of Wind-Tunnel Boundary-Layer Transition Experiments on Axisymmetric Bodies at Transonic Speeds Using Compressible Boundary-Layer Stability Theory. AIAA Paper No. 88-0008.

Table 1

Correlation of e^N Method With Experimental Data in Subsonic Boundary Layers

Ref.	Geometry	Test Conditions	Instability Mechanism	N Values	Comments
13	Flat/curved plates, airfoils, bodies of revolution	Wind tunnels/flight (both favorable and adverse pressure gradient)	T-S	4.2 - 20.9	Approximate methods used (similarity profiles and Pretsch stability charts). Most data could be correlated with N = 9 with less than 20% error in transition Re No. (Re_t) (hence the "e^9" method). The case with N = 20.9 used an estimated and probably inaccurate pressure distribution.
21	Flat plate, airfoils, bodies of revolution	Mostly wind tunnels; one flight test	T-S	6.8 - 12.1	Solved for nonsimilar boundary layer profiles and actually solved Orr-Sommerfeld equation. N = 10 was found to correlate data with less than 16.5% error in Re_t.
22	Flat plate	Wind tunnel	T-S	9.2	Approximate methods
14	Bodies of revolution	Flight/wind tunnels	T-S	6.9 - 12.9	Nonsimilar boundary layer and solution of Orr-Sommerfeld Eq.
23	Bodies of revolution	Anechoic wind tunnel	T-S	9 - 10	Uses stability charts of Wazzan, Okamura and Smith (38) for similar boundary layers. Transition location not well defined.

24	Body of revolution	Water tunnel with varying free-stream turbulence intensity	T-S	9	N was 9 when freestream turbulence intensity was on the order of .1%. Variation of N with freestream turbulence was in agreement with Mack's correlation (39).
25	Flat plate	RAE wind tunnel	T-S	10 - 13	Turbulence level less than .02%
26	Airfoil	Flight	T-S	12.8	Flow approaching separation. N = 10 will give 2% error in Re_t.
27	Flat plate with a hump	Wind tunnel	T-S	7 - 10	Basic flow computed with interactive boundary-layer theory. Stability analysis using Orr-Sommerfeld Eq. Computes through separation bubble.
28	Flat plate with a hump	Wind tunnel	T-S	~ 8.4	Interactive boundary layer.
29	Flat plate	ASU wind tunnel	T-S	9	
30	NACA-0012 Airfoil		T-S	9.5	
31	NACA 65 215-114 Airfoil	Low-Turbulence Wind Tunnel	T-S	10 - 13	SALLY Code
32	Bodies of revolution	Towing tank	T-S	9 - 10	Intermittent flow begins at e^{11}. Only the forebodies for which flow remained attached are quoted here.

33	Swept wing	Wind Tunnel	Crossflow	12	
	Unswept wing	Flight	T-S	15	Sail plane (no engine noise).
15	Swept wing	Wind Tunnel	T-S/Crossflow	7 - 11	Uses SALLY Code
34	Swept wing	Wind tunnels/flight	T-S/Crossflow	7 - 11	Uses SALLY Code
35	Rotating disk	Laboratory	Crossflow	11	Effect of rotation and curvature included in the stability equation.
36	Swept cylinder	Wind Tunnel	Crossflow	~ 11	Unsteady disturbances much more unstable than the steady cross-flow vortices, curvature terms included.
37	Curved plates and airfoils with concave curvature	Wind Tunnel	Görtler	~ 10	In one case N was lower, around 6. In this case, however, flow was separated and T-S instability may have been a factor but was not considered.

Table 2

Correlation of e^N Method With Experimental Transition Data in Supersonic Boundary Layers

Ref.	Body Geometry	Test Conditions	Mechanism	Instability N Value	Comments
45	Sharp cone	LaRC Quiet Tunnel and flight	1st T-S Mode	9 - 11	
46	Nozzle wall	LaRC Quiet Tunnel	Görtler	9 - 11	In the concave region, Mach number varies about 2 - 3.5. Favorable pressure gradient.
47	Swept wing	Flight	Crossflow	6 - 11	Mach numbers up to 1.76. Only zero frequency considered. Unsteady disturbances and curvature effect analysis still required.
9	Flat plate	Quiet Tunnel	1st T-S Mode	~ 10	This Quiet Tunnel experiment colloborated predictions of e^N method which had been in contradiction with the experiments performed in conventional supersonic wind tunnels.
43	Sharp cones	Flight	1st and 2nd T-S Mode	~ 10	Trend of the data up to Mach 7 and effect of cooling on transition predicted well with e^{10} method. (Further unpublished comparisons at cold wall 2nd mode conditions indicates e^{10} still works in hypersonic flows.)

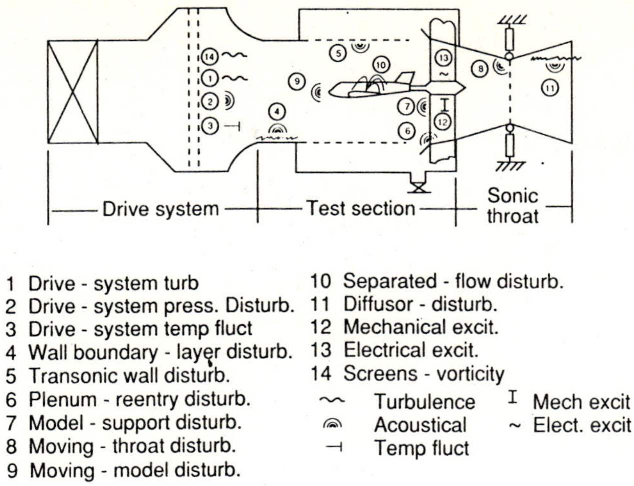

Figure 1 - Sources of flow disturbances in transonic wind tunnels.

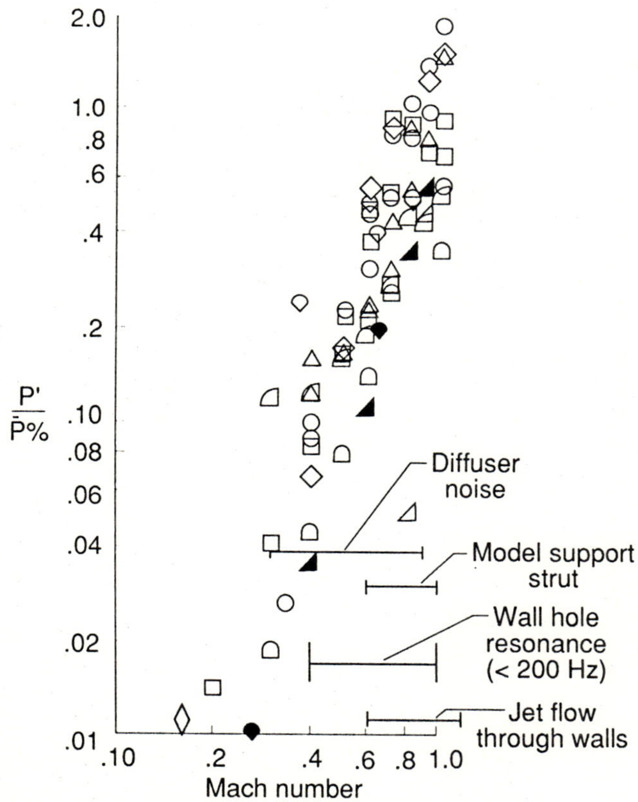

Figure 2 - Freestream pressure disturbances in subsonic and transonic wind tunnels.

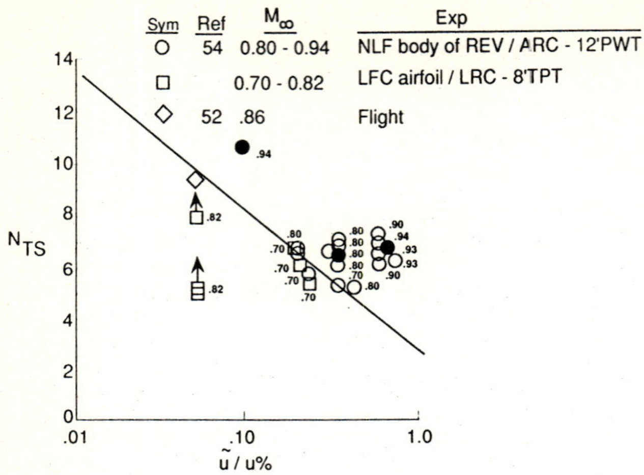

Figure 3 - Variation of transonic N-factor at transition with disturbance level.

Theoretical Investigation of Instability and Transition in the DFVLR-F5 Swept-Wing Flow

T.M. Fischer, U. Ehrenstein, F. Meyer

DFVLR
Institute for Theoretical Fluid Mechanics
D-3400 Göttingen, FR Germany

Summary

Primary and secondary stability analysis is applied to the flow over the swept DFVLR-F5 wing. The crossflow as well as the Tollmien-Schlichting instability is investigated. The influence of stationary crossflow vortices of finite amplitude on the transition process is analysed. In addition, the nonlinear development of the crossflow instability is demonstrated by numerical simulation.

1. Introduction

The design of modern aircraft wings requires the prediction of laminar flow stability and instability as well as suitable criteria for defining regions of laminar, transitional and turbulent flow. For the transonic transport airplanes under consideration, a wing sweep is included, and the flow over the wing must be regarded as three-dimensional. Linear stability theory provides a tool for determining instability and the type of small amplified disturbances. Based on the linear stability results and on additional experimental data, the so-called e^N method and its modifications are widely used for transition prediction [8, 17]. However, real disturbances of finite amplitude which are necessary to set the transition process going cannot be treated properly by a linear stability theory. Unfortunately, there is a dearth of experimental investigations in order to obtain reliable information on the validity and universality of the e^N method especially in its application to three-dimensional flows.

Before the three-dimensional laminar flow breaks down, stationary streamwise vortices – the so-called crossflow vortices –, but also travelling waves, are observed [1, 2, 16]. The onset of growing stationary and unsteady disturbances within a three-dimensional boundary layer can be described successfully by a linear (primary) stability theory [1, 3, 14]. In the present paper, the e^N method based on integration of the linear amplification rates is applied to the flow over the swept DFVLR-F5 model wing. However, the presence of stationary crossflow vortices will change the stability characteristics of the boundary-layer flow. Therefore, the classical e^N method is extended by taking finite amplitudes of the stationary crossflow vortices into account. The interaction of stationary vortices with small unsteady crossflow disturbances in the front region of the wing can then be analysed in the framework of a theory of secondary instability. In addition, the linear (primary) stability results show a strong amplification of obliquely travelling waves in the mid-chord region of the wing. Our aim is to study the effect of the different

J. Zierep, H. Oertel (Eds.)
Symposium Transsonicum III
IUTAM Symposium Göttingen, 24.-27. 5. 1988

types of instability on the transition process and to incorporate nonlinear interactions between stationary vortex structures and travelling disturbance waves into a transition model.

The specification of disturbance amplitudes requires, for instance, the solution of nonlinear eigenvalue problems [11]. In the present investigation, the time-dependent growth of crossflow disturbances at finite amplitudes is described by a numerical simulation of the Navier-Stokes equations. In particular, numerical simulation yields the disturbance amplitudes, which can be used for the investigation of secondary instability.

2. The e^N Method

The DFVLR-F5 wing has been defined in [23] as a model wing for transonic flow with a leading edge sweep angle of 21°. It has been tested in the transonic wind tunnel of the DFVLR in Göttingen [24], and we take the free-stream Mach and Reynolds number as $M_\infty = 0.82$ and $Re_\infty = 2 \cdot 10^6$ according to the experiment, using the total chord length $L = 0.2\,m$ as reference length. The following stability analysis, which is based on a compressible boundary-layer calculation [22], has been performed for successive chord locations x_c at a fixed spanwise position $y_c = 50\%$. We remark that it is conjectured from acenaphtene visualizations [24] that, for this spanwise position, transition occurs at about $x_c = 70\%$.

For the purpose of stability and transition investigation, the swept-wing boundary layer is assumed to be infinite in the spanwise direction, and parallel:

$$Q_0(z) = (U_0(z),\ V_0(z),\ 0)\ , \tag{1}$$

where the z-axis is defined normal to the wall in a Cartesian coordinate system (x, y, z). This means that, in the framework of a local stability analysis, the influence of non-parallel flow and all curvature effects will be neglected. The local Reynolds number Re_δ is based on the velocity U_{ext} of the external inviscid flow and on a boundary-layer thickness δ. For a small disturbance q of the basic flow Q_0, the normal-mode ansatz

$$q(x_s, y_s, z, t) = \hat{q}(z)\, e^{i(\alpha_s x_s + \beta_s y_s - \omega t)}, \quad q = (u_s, v_s, w)\,, \tag{2}$$

leads to an eighth-order primary stability system in the compressible case and to the well-known Orr-Sommerfeld eigenvalue problem for a fourth-order system in the incompressible case [13]. Here α_s and β_s are the (real) wavenumbers in the streamwise and crossflow direction, x_s and y_s, and ω is the complex frequency in a temporal growth concept. For the numerical solution of these linear stability equations, the computer codes LISA [4] and COSTA [5] have been developed on the basis of Chebyshev collocation for boundary value problem formulations.

For computation of the primary N factor, we use the so-called envelope method [25], where, for a given physical frequency $f = (\omega_r/(2\pi)) \cdot (U_{ext}/\delta)$, the temporal amplification ω_i is maximized with respect to α_s and β_s. At this maximum, the group velocity ratio can be shown to be real. The spatial amplification γ in the direction of the real group velocity $c_g = (\partial\omega_r/\partial\alpha_s\,,\ \partial\omega_r/\partial\beta_s)$ is then obtained by using the Gaster transformation, $\gamma = \omega_i/|c_g|$. Integration of the spatial amplification rates along a curve which is everywhere tangential to c_g yields the primary N factor

$$N(x_{c1}) = \ell(y_c) \cdot \int_{x_{c0}}^{x_{c1}} \frac{\gamma}{\delta \cdot \cos\Theta}\, dx_c\,, \qquad \gamma(x_{c0}) = 0\,, \tag{3}$$

where $\ell(y_c)$ is the total chord length at the span location y_c and $\Theta = \sphericalangle(x_c, c_g)$.

Fig. 1 shows the dependence of the primary N factor upon the chord location x_c for various frequencies f. The N factor based on the compressible stability calculation is about 15% lower than that obtained from the incompressible solution (given by the dashed lines). This means that compressibility has a stabilizing influence, which is well-known [13]. The qualitative behaviour of the N factor with respect to

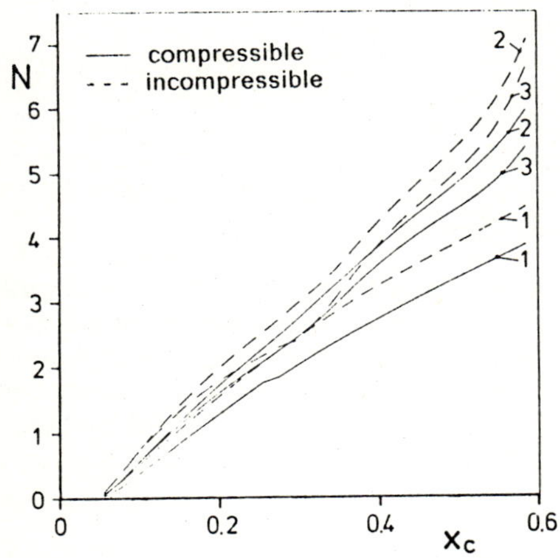

Fig.1 Primary N factor, for frequencies (1) f= 0 Hz, (2) 20 kHz, (3) 25 kHz.

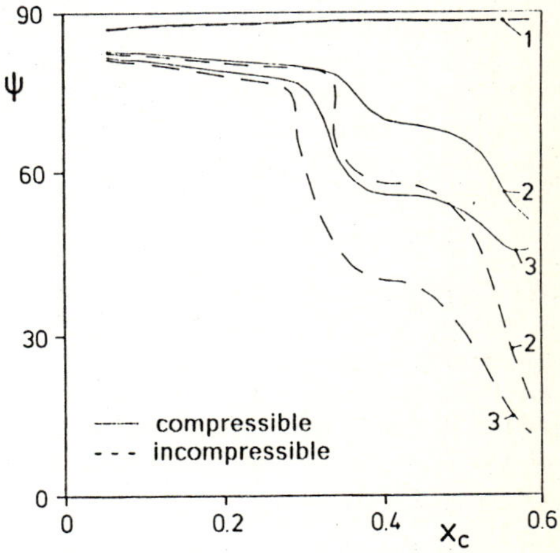

Fig.2 Waveangle ψ, for frequencies (1) f= 0 Hz, (2) 20 kHz, (3) 25 kHz.

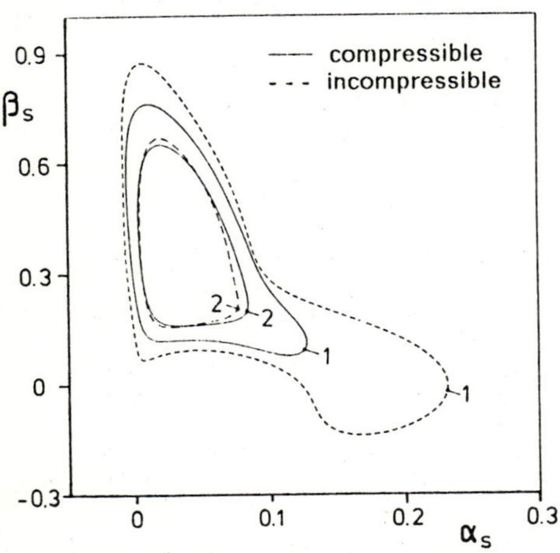

Fig.3 Lines of constant temporal amplification (1) ω_i= 0.001, (2) 0.002, for chord location x_c= 29.5%.

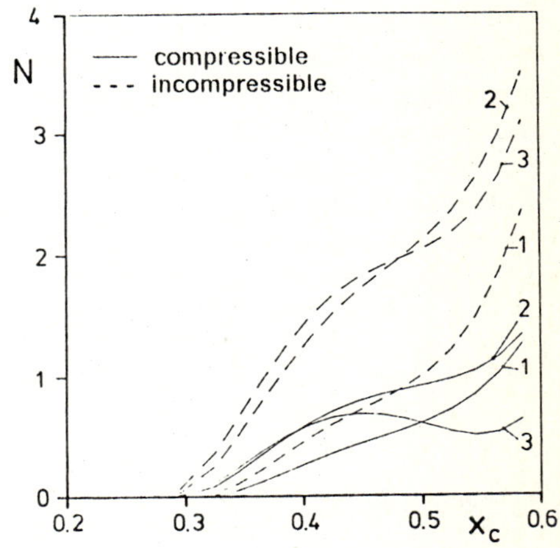

Fig.4 Tollmien-Schlichting N factor, for frequencies (1) f= 20 kHz, (2) 30 kHz, (3) 35 kHz.

the frequency f in the compressible case is quite similar to that in the incompressible case. In particular, the maximum N factor is obtained for $f = 20kHz$. The waveangle ψ (defined as $\tan\psi = \beta_S/\alpha_S$) which belongs to the maximum of the temporal amplification according to the envelope method is shown in fig. 2. Up to a chordwise position of $x_c = 30\%$, the compressible and the incompressible stability analysis lead to nearly identical results. In this region, the crossflow (CF) disturbances with $\alpha_S \ll \beta_S$ are most amplified. Further downstream, however, obliquely travelling waves become dominant, and there are considerable differences between the compressible and the incompressible results except for $f = 0Hz$.

In fig. 3, lines of constant temporal amplification ω_i are plotted on the α_S, β_S-wave number plane for the chord location $x_c = 29.5\%$. In the compressible as well as in the incompressible case, we find that the CF disturbances (including the zero-frequency disturbances) are most amplified. At further downstream chord locations, there exist, in addition, amplified disturbance waves travelling in the streamwise direction (i.e. with $\beta_S = 0$). The N factor for these so-called Tollmien-Schlichting (TS) waves is shown in fig. 4. Here the compressible stability calculation leads to a drastic reduction of the amplification rates.

We conclude that the flow region up to a chordwise position of $x_c = 30\%$ is governed by the CF instability. In this region, an incompressible analysis of the disturbances seems to be justified for investigating the compressible boundary-layer flow. Downstream of $x_c = 30\%$, the most amplified disturbances are obliquely travelling waves, and compressibility effects must not be neglected in a detailed analysis. Note, however, that compressibility does not affect the zero-frequency disturbances in a significant manner.

The different behaviour of the amplified disturbances in the front and in the midchord region of the wing suggests to define separate N factors for CF disturbances and for particular obliquely travelling waves. The (most amplified) zero-frequency disturbances, which correspond to the stationary CF vortices observed in three-dimensional boundary-layer experiments [2, 21], can be chosen to define a crossflow N factor N_{CF}. On the other hand, the amplified disturbance waves with CF wavenumber $\beta_s = 0$, for instance, can be used to compute a Tollmien-Schlichting N factor N_{TS} in dependence upon the frequency f. In practice, crossflow and Tollmien-Schlichting N factor are given different weights to obtain a unique transition criterion for varying Reynolds number, sweep angle and pressure distribution (cf. [20]). However, a convincing relationship between N_{CF} and N_{TS} has not been found yet. This might be due to the fact that real disturbances of finite amplitude which are necessary to set the transition process going cannot be treated properly by a linear stability theory. In the following section, we try to extend the classical e^N method by taking finite amplitudes (of the stationary CF vortices) into account and also, in a sense, to give estimates for the validity of the linear primary stability calculations.

3. The Extended e^N Method

An extension of the classical e^N method will be given based on a secondary stability analysis but will be considered only for the flow region (up to 30% chord length), which is governed solely by travelling and zero-frequency crossflow disturbances. The idea is that an increase of the disturbance amplitudes in the front region of the wing will cause premature transition due to CF instability. Thus, we introduce the amplitude of the stationary CF vortices at an anterior wing position as an additional parameter. The amplitudes of the initial disturbances are determined by free-stream disturbances as well as roughness on the wing surface. The CF vortices seem to be favoured by roughness elements.

In general, nonlinear interactions between stationary vortex structures and travelling disturbance waves must be expected. If, for a fixed wing location within a short distance downstream of the onset of instability, the influence of the non-stationary disturbances was negligible, the amplitude of the stationary CF disturbance would be allowed to reach a saturation state. Whereas the linear primary stability theory predicts exponential growth of the CF vortices, the solution of an initial boundary value problem for the Navier-Stokes equations by a quasi two-dimensional simulation yields a nonlinear saturation of this growth at relatively large disturbance amplitudes [15]. Fig. 5 shows the dependence of the saturation amplitude

$$\varepsilon = \max_{0 \leq z < \infty} \left(\frac{\beta_0}{2\pi} \cdot \int_0^{2\pi/\beta_0} |u_s(y_v, z)|^2 dy_v \right)^{1/2} \tag{4}$$

upon the chord location x_c, where the wavelength $2\pi/\beta_0$ and the angle between the streamwise direction x_s and the direction y_v normal to the vortex axes are taken from the linearly most amplified zero-frequency disturbance according to section 2. – Recent investigations indicate that there is a considerable influence of the stationary disturbances on the basic flow, which leads to a reduction of the saturation amplitudes. – In an actually convective flow, the amplitude of the stationary CF vortices will not reach this saturation state. In the present work, the spatial growth of the stationary CF vortices is described by linear theory. However, for larger amplitudes of the zero-frequency disturbances, nonlinear damping of the amplitude growth becomes important and this would require further investigations.

The presence of stationary crossflow vortices will change the stability characteristics of the boundary-layer flow. Thus, we introduce a new basic flow

$$\tilde{Q} = Q_0 + \varepsilon_0 q_0 , \quad q_0(y_v, z) = \hat{q}_0(z)\, e^{i\beta_0 y_v} + c.c. , \tag{5}$$

where $\varepsilon_0 q_0$ denotes the primary zero-frequency disturbance with a finite amplitude ε_0 according to definition (4), and investigate this primarily disturbed basic flow for instability in the framework of a secondary stability analysis [7]. By using the shape assumption [9], the primary disturbance will be chosen strictly periodic, where $\hat{q}_0(z)$ is the complex eigenfunction from the linear stability theory, β_0 is the real wavenumber normal to the vortex axes – the spatial or temporal amplification will be neglected – and c.c. means the complex conjugate. From the numerical simulation of the CF vortices at finite amplitude ε_0, we find that the nonlinear distortion of the amplitude coefficient $\hat{q}_0$ is not significant but that the amplitudes of the higher harmonics are of about one half of ε_0. The influence of these nonlinear effects on the secondary stability results will be discussed later.

For a small disturbance q_1 of the new basic flow $\tilde{Q}$, we again obtain a linear stability system. To these equations, Floquet-theoretical methods can be applied and the secondary disturbance q_1 admits the following representation [10]:

$$q_1(x_v, y_v, z, t) = \hat{q}_1(y_v, z)\, e^{i(\alpha_1 x_v - \omega_1 t)}, \quad \hat{q}_1(y_v, z) = e^{-i\sigma\beta_0 y_v} \sum_{k=-K}^{K} \hat{q}_{1,k}(z)\, e^{ik\beta_0 y_v}. \quad (6)$$

Here α_1 and ω_1 are the (complex) wavenumber in the direction x_v of the stationary vortex axes and circular frequency, respectively, and σ is a detuning parameter, which describes harmonic ($\sigma = 0$), combination ($0 < \sigma < 1/2$ or $1/2 < \sigma < 1$) and subharmonic ($\sigma = 1/2$) resonances. The resulting eigenvalue problem determines

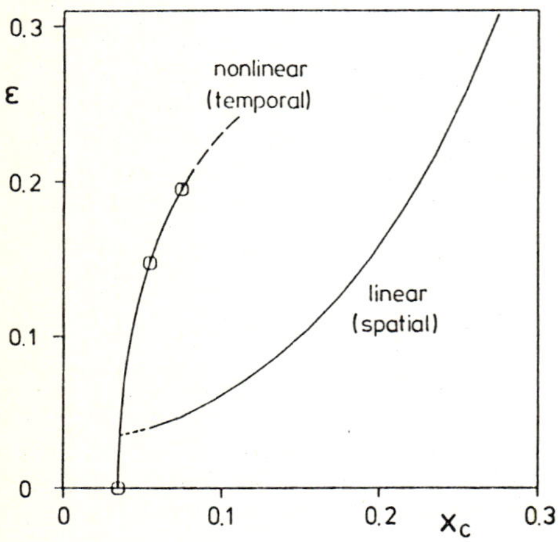

Fig.5 Amplitude ε of the stationary crossflow vortices (relative to the velocity of the external inviscid flow).

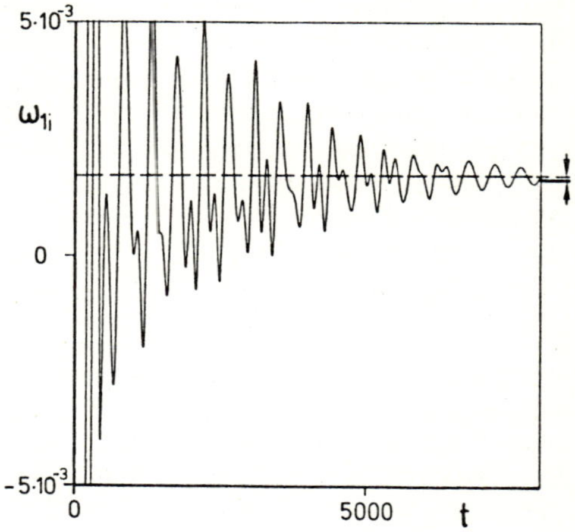

Fig.6 Effect of the shape assumption for the primary crossflow vortices on the secondary amplification rate ω_{1i}.

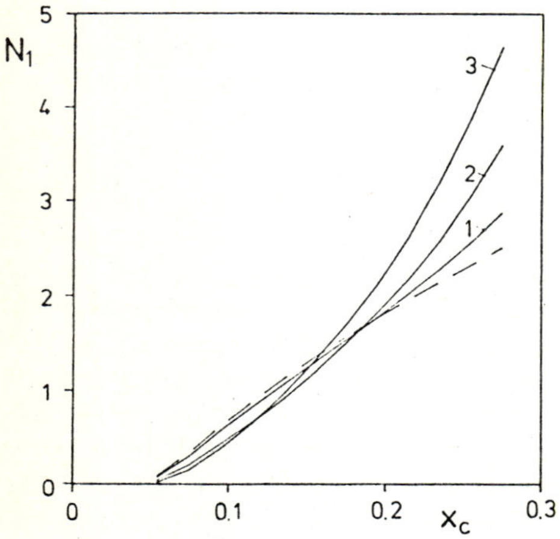

Fig.7 Secondary N factor N_1, for frequency $f_1 = 20$ kHz and primary initial amplitudes (1) $\varepsilon_0 = 2\%$, (2) 4%, (3) 6% taken at $x_c = 5.5\%$.

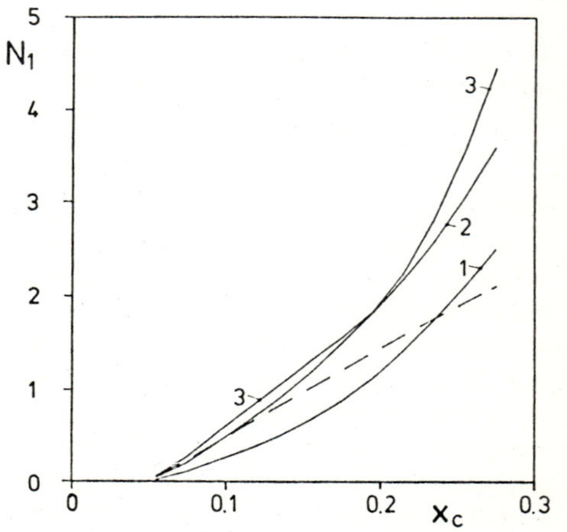

Fig.8 Secondary N factor N_1, for frequencies (1) $f_1 = 10$ kHz, (2) 20 kHz, (3) 30 kHz and primary initial amplitude $\varepsilon_0 = 4\%$ taken at $x_c = 5.5\%$.

α_1 or ω_1 and the amplitude functions $\hat{q}_{1,k}(z)$ within a spatial or temporal growth concept and is solved numerically by means of a Galerkin approximation [6].

The truncation of the Fourier series (6) must be done carefully. We have found a great influence of the higher harmonics of the fundamental secondary waves $\hat{q}_{1,0}$, $\hat{q}_{1,1}$ on the numerical eigensolutions, in particular for increasing amplitude ε_0 of the primary disturbance. Whereas the stability analysis in [18] is based solely on the use of the fundamental secondary waves (see also the criticism in [19], where $\hat{q}_{1,1}$, $\hat{q}_{1,2}$ are used instead), our secondary stability calculations have been performed initially with $\hat{q}_{1,0}$, $\hat{q}_{1,1}$, $\hat{q}_{1,2}$ and, for increasing chord location x_c, $\hat{q}_{1,-1}$ and $\hat{q}_{1,-2}$ have been added successively.

A justification of the shape assumption (5) can now be given by a comparison of the secondary amplification rates ω_{1i} with those for which the primary zero-frequency disturbance has been defined properly by the nonlinear saturation state of a Navier-Stokes simulation. Here, the temporal growth of small secondary disturbance amplitudes will be described again by the solution of an initial boundary value problem for the Navier-Stokes equations — with the steady bifurcating solution kept fixed —, which yields the time-dependent secondary amplification,

$$\omega_{1i}(t) = \frac{d}{dt}\left(\max_{0 \le z < \infty} \ell n\,|\hat{u}_{1,1}(z,t)|\right). \tag{7}$$

The limit of this temporal amplification rate for large t then corresponds to the imaginary part of the secondary eigenvalue. In fig. 6, the results of a calculation of $\omega_{1i}(t)$ in the subharmonic resonance case $\sigma = 1/2$ at an anterior chordwise position are shown. We find that the limiting value of $\omega_{1i}(t)$ for large t is quite well approximated by the eigenvalue (given by a side-mark on the right-hand side in fig. 6) from a secondary stability analysis, which is based on the shape assumption with ε_0 equal to the saturation amplitude ($\varepsilon_0 = 12\%$).

For a given physical frequency $f_1 = (\omega_{1r}/(2\pi)) \cdot (U_{ext}/\delta)$, the secondary N factor has been computed by using the spatial growth concept with $\omega_{1i} = 0$ and α_1 complex. We note that the resulting values for the real wavenumber part α_{1r} are of one order less than β_0, so that the secondary disturbances are composed of waves travelling preferably oblique to the potential flow direction. Integration of the spatial amplification rates $-\alpha_{1i}$ yields the secondary N factor

$$N_1(x_{c1}) = -\,\ell(y_c) \cdot \int_{x_{c0}}^{x_{c1}} \frac{\alpha_{1i}}{\delta \cdot \cos\Theta'}\,dx_c\,, \qquad \alpha_{1i}(x_{c0}) = 0\,, \tag{8}$$

where $\ell(y_c)$ is the total chord length at the span location y_c and $\Theta' = \sphericalangle(x_c, x_v)$. The increase of the secondary N factor with the chordlength x_c for $f_1 = 20kHz$ and for various initial amplitudes ε_0 of the primary zero-frequency disturbance is shown in fig. 7. Since, at least for small values of x_c, the maximum amplification rate is obtained in the combination resonance case rather than in the harmonic or subharmonic case [7], we have chosen $\sigma = 1/4$ in the series representation (6) of the secondary disturbances. The primary N factor according to section 2 is given by the dashed line. If the initial amplitude ε_0 of the stationary CF vortices taken at $x_c = 5.5\%$ is less than 2%, the secondary N factor is comparable to that obtained from the primary stability theory. However, for increasing initial amplitude ε_0, the secondary crossflow-disturbance waves become more and more important. In addition, the frequency f_1 of the most amplified disturbances increases. For an initial amplitude of $\varepsilon_0 = 4\%$ at $x_c = 5.5\%$, the maximum secondary N factor is obtained at about $f_1 = 30kHz$ as is indicated in fig. 8. (Here the dashed line corresponds to the N factor of the primary zero-frequency disturbance.)

Let us conclude that the method of secondary instability provides a powerful concept for modelling transition of crossflow type in dependence upon the amplitudes of the stationary CF vortices which are produced initially by external disturbances. If these initial amplitudes are sufficiently low, the results of a secondary stability analysis are similar to those of the classical (primary) stability theory and, relatively far downstream, the transition process will be forced by growing obliquely travelling waves. Else secondary crossflow-disturbance waves are strongly amplified and, possibly, will lead to premature transition in the front region of the wing. The secondary N factor of an extended e^N method should be correlated with experimental transition data. However, for this purpose, the amplitudes of the stationary CF vortices have to be determined or at least to be estimated. A comparison of the present theory with the results of the DFVLR-transition swept-plate experiment [2] is under preparation.

4. Nonlinear Development of the Crossflow Instability

In this section, we consider the more general case of an arbitrary set of nonlinear interacting wavenumbers. To this aim, we solve the three-dimensional time-dependent incompressible Navier-Stokes equations by numerical simulation in a spatial domain which is bounded by the smallest wavelengths L_x, L_y in the wall-parallel directions [12, 15]. The spatial discretization is based on a spectral method with Fourier expansions in the horizontal, vortex-oriented coordinates x_v, y_v,

$$q\,(x_v, y_v, z, t) = \sum_{k_x} \sum_{k_y} \hat{q}\,(k_x, k_y, z, t)\, \exp(ik_x\alpha_v x_v + ik_y\beta_v y_v)\,, \quad q = (u_v, v_v, w) \qquad (9)$$

($\alpha_v = 2\pi/L_x$, $\beta_v = 2\pi/L_y$), and on Chebyshev collocation in the normal coordinate z. The sum (9) extends over the wavenumber range $0 < |k_{x(y)}| < N_{x(y)}/2$. The total flow field then consists of the boundary-layer profile Q_0 plus the disturbance q. In the following calculations, we consider the subharmonic resonance case (i.e. $\beta_v = \beta_0/2$) at a chord location x_c within a short distance downstream of the onset of instability. Initially, q is composed of a zero-frequency disturbance of relatively small amplitude ($\varepsilon = 4.24\%$) and a single wave propagating oblique to the direction x_v of the stationary vortex axes, with max $|\hat{u}_s(1,1,z,0)| = 0.02$. The wavenumber of the travelling wave in the x_v-direction is chosen as $\alpha_v = \alpha_1$, where the secondary stability analysis predicts strong amplification. The numerical resolution varies from $N_x \times N_y \times N_z = 16 \times 32 \times 60$ initially to $32 \times 64 \times 60$ in the later stages.

In order to check the adequacy of the horizontal resolution of the numerical simulation, we show in fig. 9 the energy $E\,(0,0)$ of the basic flow and that of the Fourier expansion (9), defined by

$$E\,(k_x, k_y, t) = \frac{1}{2} \int_0^{\infty} |\hat{q}\,(k_x, k_y, z, t)|^2 dz\,. \qquad (10)$$

at time $t = 1800$. We observe a monotonic decay with increasing wavenumbers. It is noticeable that a large number of modes in the y_v-direction is amplified, while the modes decay very rapidly in the vortex-axes direction x_v. We remark that the energy of the highest wave numbers is very small and, accordingly, the resolution of our simulation is sufficient.

In fig. 10, the distribution of rms fluctuations at t = 1800 is given. The root mean square value of the streamwise variation around the average of u_v,

$$(u_v)_{rms} = [\overline{(u_v - \bar{u}_v)^2}]^{1/2}, \qquad \bar{u}_v = \frac{1}{L_x}\int_0^{L_x} u_v dx_v, \tag{11}$$

describes the strength of fluctuations in the x_v-direction. We relate this rms value with the time-averaged rms value in the experiment. The rms profiles $(u_v)_{rms}(z)$ at different y_v-positions show two relative maxima in the z-direction and a strong dependence on y_v. The maximum value of $(u_v)_{rms}$ varies between 1.8% and 9%. In the perspective view of the rms distribution, two wavelengths of the crossflow vortex are shown. This pattern is different from the experimental observation in

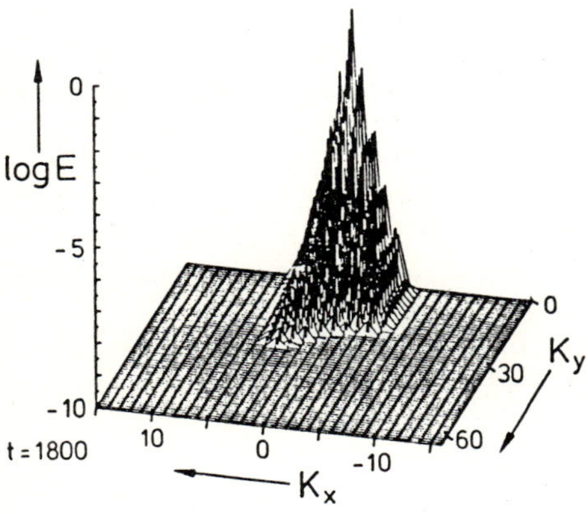

Fig.9 Energy $E(k_x, k_y)$ of horizontal Fourier modes.

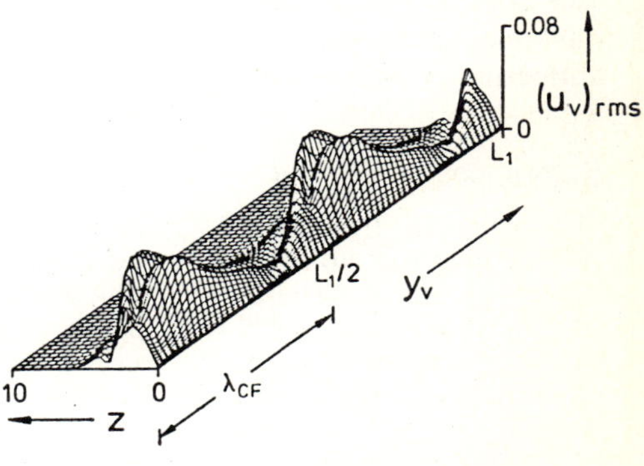

Fig.10 Streamwise rms fluctuations $(u_v)_{rms}(y_v, z)$ at t = 1800.

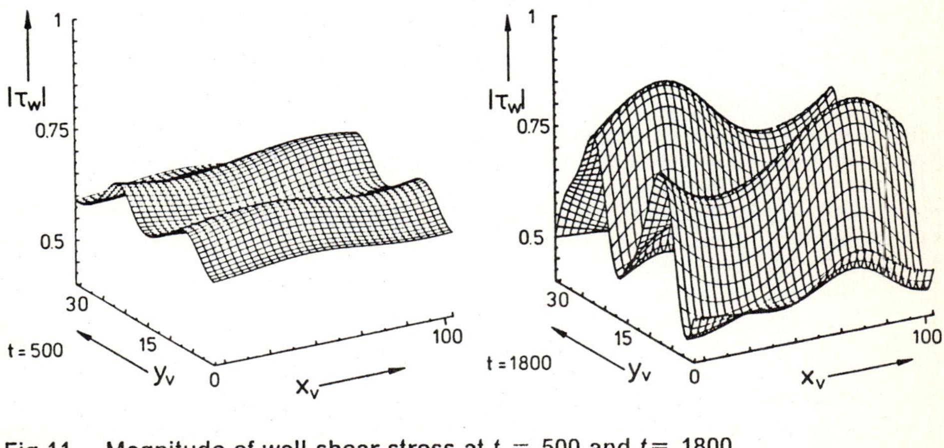

Fig.11 Magnitude of wall shear stress at t = 500 and t = 1800.

[1], where two local maxima in y_v-direction have been found. However, the secondary stability calculations of the previous section performed at later chord locations x_c yield rms fluctuations of the experimentally observed type.

Finally, in fig. 11, we show the wall shear stress distribution, defined by

$$|\tau_W| = \left[\left(\frac{\partial u_v}{\partial z}\right)^2 + \left(\frac{\partial v_v}{\partial z}\right)^2\right]^{1/2} . \tag{12}$$

For the early stages (t = 500), the variation of the wall shear stress around the value for the mean flow, $|\tau_W| = 0.63$, is rather weak. The staggered pattern of the maxima is caused by the subharmonic nature of the initial disturbance. For t = 1800, the wall shear stress varies in the region of $0.45 \leq |\tau_W| \leq 0.85$. The pattern of $|\tau_W|$ is moving in the x_v-direction while the streamwise average $\overline{|\tau_W|}$ remains stationary (except for an increase in amplitude). We identify $\overline{|\tau_W|}$ with the streaks observed in oil flow visualization experiments.

Acknowledgements

The authors are indebted to Dr. U. Dallmann and Dr. L. Kleiser for helpful suggestions, discussions and comments and to Dr. L.M. Mack for valuable remarks concerning the compressible stability characteristics.

References

1. Arnal, D.; Coustols, E.; Juillen, J.C.; Rech. Aérosp. No. 1984-4 (1984), p. 275.
2. Bippes, H.; Nitschke-Kowsky, P.; AIAA Paper No. 87-1336 (1987).
3. Dallmann, U.; Bieler, H.; AIAA Paper No. 87-1337 (1987).
4. Ehrenstein, U.; DFVLR-IB 221-88 A 09 (1988).
5. Ehrenstein, U.; computer code COSTA.
6. Fischer, T.M.; DFVLR-IB 221-88 A 10 (1988).
7. Fischer, T.M.; Dallmann, U.; DFVLR-FB 87-44 (1987); see also, AIAA Paper No. 87-1338 (1987).
8. Hefner, J.N.; Bushnell, D.M.; AIAA Paper No. 79-1493 (1979).
9. Herbert, T.; AIAA Paper No. 83-1759 (1983).
10. Herbert, T.; Bertolotti, F.P.; Santos, G.R.; In: Proc. ICASE/NASA Workshop "Stability of Time-Dependent & Spatially Varying Flows", 1985, Hampton, Va.
11. Koch, W.; In: Proc. Workshop "Propagation in Nonequilibrium Systems", 1987, Les Houches, France. Springer Proceedings in Physics.
12. Laurien, E.; DFVLR-FB 86-05 (1986).
13. Mack, L.M.; In: AGARD Report No. 709 (1984).
14. Malik, M.R.; Poll, D.I.A.; AIAA Paper No. 84-1672 (1984).
15. Meyer, F.; Kleiser, L.; DFVLR-IB 221-87 A 05 (1987).
16. Poll, D.I.A.; J. Fluid Mech. 150 (1985), pp. 329-356.
17. Redeker, G.; Horstmann, K.-H.; In: DGLR-Report 86-03 (1986), p. 317.
18. Reed, H.L.; AIAA Paper No. 84-1678 (1984).
19. Reed, H.L.; Phys. Fluids 30 (1987), pp. 3419-3426.
20. Rozendaal, R.A.; NASA CR-3975 (1986).
21. Saric, W.S.; Yeates, L.G.; In: "Laminar-Turbulent Transition" (V.V. Kozlov ed.), p. 429. Berlin: Springer 1985.
22. Schwamborn, D.; In: Notes on Num. Fluid Mech., Vol.7 (M. Pandolfi and R. Piva eds.), p. 315. Braunschweig: Vieweg 1984.
23. Sobieczky, H.; DFVLR-IB 221-86 A 18 (1986).
24. Sobieczky, H.; Hefer, G.; Tusche, S.; AIAA Paper No. 87-2485 (1987).
25. Srokowski, A.J.; Orszag, S.A.; AIAA Paper No. 77-1222 (1977).

Transition Prediction in Transonic Flow

D. Arnal

ONERA/CERT
Department of Aerothermodynamics
Toulouse, FRANCE

Summary

The paper presents stability computations for two-dimensional, transonic flows on adiabatic walls. The effects of streamwise pressure gradients are discussed. An analytical representation of the stability diagrams is then used in order to develop a simplified e^n method. Some examples of application are given.

1. Linear stability computations

1.1. Problem formulation

The principle is to introduce sinusoïdal small disturbances into the linearized Navier-Stokes equations in order to compute the range of unstable frequencies. It is assumed that any fluctuation r' (velocity, pressure, density or temperature) is expressed by :

$$r' = r(y)\exp\left[i(\alpha x + \beta z - \omega t)\right] \qquad (1)$$

The amplitude function r depends on y only (parallel flow approximation). In the general case, α, β, ω are complex numbers.

This leads to a system of 6 ordinary differential equations in y for the amplitude functions r(y). Due to the homogeneous boundary conditions, the problem is an eigenvalue one : when the mean flow is specified, nontrivial solutions exist only for certain combinations of the parameters α, β, ω and R, where R is the Reynolds number.

There are several numerical methods available for the solution of compressible linear eigenvalue problems. The results presented in the next sections were obtained using a "shooting method". The

J. Zierep, H. Oertel (Eds.)
Symposium Transsonicum III
IUTAM Symposium Göttingen, 24.-27. 5. 1988

stability equations are transformed in a system of 8 first-order differential equations ; analytical solutions are computed in the free stream, and the integration is performed up to the wall with a fourth-order Runge-Kutta technique. The eigenvalues are then found using a Newton-Raphson iteration scheme. The numerical technique is similar to that described by Mack /1/ or Gasperas and Antar /2/, for instance.

1.2. Basic flow

The basic flow (mean velocity and mean temperature profiles) is obtained by solving similarity equations with a numerical method developed by Gleyzes and Cousteix /3/. The similarity equations are obtained by assuming that :

$$\frac{U}{Ue} = f(\eta) \quad , \quad \frac{hi}{hie} = g'(\eta), \text{ with } \quad \eta = y/\delta \qquad (2)$$

where δ is the boundary layer thickness.

This leads to a system of two ordinary differential equations which differ little from those obtained with the Levy-Lees transformation. Assumption (2) implies that the wall heat flux, the free-stream Mach number M_e, the pressure gradient parameter $\Lambda\delta = \frac{\delta^2}{\nu}\frac{dU_e}{dx}$ and the total temperature Tie in the free-stream are constant in the streamwise direction. This is verified in some particular cases only (for example for zero pressure gradient flows), and self-similar solutions can be computed. In the cases where the previous requirements are not met (negative or positive pressure gradients), local similarity solutions are obtained.

Stability diagrams were computed for 30 basic flows in the transonic range. The characteristics of the mean profiles and the detailed results can be found in Jelliti's thesis /4/. Five values of the free-stream Mach number were considered (M_e = 0, 0.6, 0.9, 1.1 and 1.3). For each M_e, eight values of $\Lambda\delta$ (positive and negative) were used in order to study the effects of streamwise pressure gradients. The calculations covered a range of incompressible shape factors H_i between 2.4 and 3.4. In all cases, the wall is adiabatic and the stagnation temperature is 300 K.

The results which will be presented below were obtained by using the spatial theory, i.e. ω is real, α and β are complex. It is also assumed that the amplification vector $(-\alpha_i, -\beta_i)$ is aligned with the mean flow direction ($\beta_i = 0$), so that relation (1) takes the form :

$$r' = r(y)\exp(-\alpha_i x)\exp\left[i(\alpha_r x + \beta_r z - \omega t)\right] \qquad (3)$$

The angle $\psi = \tan^{-1}(\beta_r/\alpha_r)$ represents the wavenumber direction.

1.3. Examples of results

In our calculations, the Reynolds number is computed by using the displacement thickness δ_1, the free-stream velocity U_e and the kinematic viscosity ν_e at the outer edge of the boundary layer. α_i, α_r and ω are equal to $\alpha_i^*\delta_1$, $\alpha_r^*\delta_1$ and $\omega^*\delta_1/U_e$, where the stars denote dimensional quantities.

a - Flat plate results. In order to check the validity of the numerical code, stability results obtained for zero pressure gradient flows were compared with those published by other authors. Figure 1 shows a comparison between the stability diagram computed for $M_e = 0.9$ and the neutral curve obtained by Boehman and Mariscalco /5/ at the same Mach number. The stability diagram corresponding to the Blasius profile is also plotted in this figure. The first observation is that the shape of the curves of equal amplification rates do not change very much when the Mach number increases. However, the amplification rates of the unstable waves are clearly smaller in transonic flow, even if the critical Reynolds number remains of the same order.

It is well known that an important aspect of the instability in compressible boundary layers is the effect of the wavenumber angle ψ. Up to $M_e \cong 1$, the most unstable waves are two-dimensional ($\psi = 0°$), but at higher Mach numbers, the maximum value of $-\alpha_i$ at a given Reynolds number is obtained for a non-zero value ψ_{max} of the wavenumber angle. ψ_{max} increases generally as the Reynolds number increases, but this dependence is rather weak. A comparison between our results and those obtained by Mack /6/ for $M_e = 1.3$ is shown in Figure 2. As Mack used the temporal theory, the spatial amplification rates were converted into temporal amplification rates by using the Gaster transformation /7/. The evolution of $\omega_{i max}$ (maximum value of ω_i at a given $R\delta_1$) is plotted as a function of $R\delta_1$ for two-dimensional waves and for the most unstable oblique waves. It is clear that the effect of the wave orientation becomes increasingly important at high Reynolds numbers.

b - Results with pressure gradients. As for the flat plate case, the wavenumber direction ψ remains a significant parameter for supersonic Mach numbers. The maximum values of $-\alpha_i$ computed for $M_e = 1.3$ and $R\delta_1 = 5\,000$ are plotted in Figure 3 as a function of ψ and H_i, which can be considered as a pressure gradient parameter. The value of ψ_{max} does not depend very much on H_i, except for strong negative pressure gradients.

Stability diagrams computed for $M_e = 0$ and 0.9 at nearly the same value of H_i ($H_i \cong 3.05$, decelerated flow) are compared in Figure 4. The observations are the same as those already noticed for the flat plate case : there is no change in the shape of the curves of equal amplification rate, but compressibility is stabilizing.

2. A simplified method for stability computations

At the present time, to predict transition, one must place reliance on empirical procedures. In the e^n method, the logarithm of the amplitude ratio A/A_0 is computed for various frequencies by integrating $-\alpha_i$ along the streamwise distance. (A is the wave amplitude, and the index 0 refers to the point where the wave becomes unstable). It is assumed that transition occurs when the ratio A/A_0 of the locally most amplified frequency reaches a critical value e^n, with $n \sim 7$ to 10. In fact, the value of n at transition onset depends on the disturbance environment. For low speed flows, a major parameter is the free-stream turbulence level Tu, and Mack suggested the following relationship /8/ :

$$n = -8.43 - 2.4 \ln Tu \qquad (4)$$

Although the validity of the e^n method is not firmly established in compressible flows, promising results were obtained, for instance by Malik /9/. Unfortunately, this method is rather time-consuming and expensive, so that the development of simpler calculation techniques presents an unquestionable interest. This section describes a simplified method for stability computations and some applications to two-dimensional flows.

2.1. Principle

Stability diagrams are often plotted in the $(F, R\delta_1)$ coordinates, where F is a dimensionless frequency defined by :

$$F = \frac{2\pi f \nu_e}{U_e^2} = \omega / R\delta_1 \text{, where f is a physical frequency} \quad (5)$$

The basic assumption is that, for a given mean flow and for a given F, the curve $\alpha_i(R\delta_1)$ can be represented by two half-parabolas, the equations of which are written in the form :

$$\frac{\alpha_i}{\alpha_{iM}} = 1 - \left(\frac{R\delta_1 - R_M}{R_i - R_M}\right)^2 \quad (6)$$

with $R_i = R_0$ if $R\delta_1 < R_M$ and $R_i = R_1$ if $R\delta_1 > R_M$

The parameters α_{iM}, R_M, R_0 and R_1 are defined in Figure 5. It was also observed that the variations of these parameters with F could be approximated by the following relationships :

$$\begin{aligned} \alpha_{iM} &= B(F_0 - F) \\ R_M &= K F^E \\ |R_i - R_M| &= R_M C_i (F_{Ci} - F) \quad , \quad i = 0,1 \end{aligned} \quad (7)$$

B, F_0, K, E, C_0, C_1, F_{C0}, F_{C1} are tabulated functions of the free-stream Mach number and of a pressure gradient parameter, for instance the incompressible shape factor H_i. For M_e = 1.1 and 1.3, only the values corresponding to ψ_{max} are taken into account.

Figure 6 shows the "exact" and analytical evolutions of α_i as a function of $R\delta_1$ for various frequencies. Two basic flows are considered ; in both cases, the free-stream Mach number is 1.3, and the pressure gradient is either negative (Figure 6a) or positive (Figure 6b).

2.2. Examples of application

a - Incompressible flow. At first, the simplified method was applied to a low speed configuration, for which detailed measurements /10/ and stability computations /11/ were available. In this case, transition occurs in a positive pressure gradient, since the shape factor at transition onset is close to 3. Figure 7a shows some curves of integrated amplification rates. Using relation (4) with the experimental value of Tu (0.15%) gives x_T = 0.41 m, in good agreement with the measurements. Comparison between Figure 7a (analytical method) and Figure 7b (exact stability computations) for f = 340, 400 and 500 Hz illustrates the accuracy of the simplified method.

b - Theoretical case. The main interest of this method is that the CPU time is from 10^4 to 10^6 lower than that which is required for solving the stability equations. As a consequence, it becomes possible to perform parametric studies such as that which is presented in Figure 8. The free-stream Mach number increases from 0 to 1.1 between x/c = 0 and 0.01 (c= 0.5 m, Ti= 300 K, Pi= 1atm) and then, between x/c = 0.01 and 1, M_e increases linearly up to 1.2, remains constant or decreases linearly up to 1 or 0.9. A value of Tu = 0.1% was assumed in the computations. For each case, 4 results are given:

. Curves 1 and 2 represent the results obtained by applying a criterion previously developed at ONERA/CERT for incompressible flows. The criterion is a correlation between the momentum thickness Reynolds number at transition, the free-stream turbulence level and a mean Pohlhausen parameter (extension of Granville's criterion, /12/). We used either compressible quantities (curve 1) or incompressible quantities (curve 2).

. Curve 3 is deduced from the analytical method previously described. In principle, it gives the most rigorous results.

. Curve 4 was also obtained by using the analytical method, but the compressibility effects were neglected : instead of the real value of the free-stream Mach number, M_e = 0 was imposed in the simplified stability computations.

Curves 1, 2 and 4 are close together, but the disagreement with curve 3 becomes more and more pronounced for zero and negative pressure gradients.

c - Transonic flow on a CAST10 airfoil. The simplified method was also used in the case of a wing profile in transonic flow. The experiments were performed in the induction driven tunnel T2 at ONERA/CERT on a supercritical CAST10 airfoil /13/. Curves of total amplification rate are presented in Figure 9 for an angle of attack equal to $-0.25°$. Theoretical transition location obtained with $n = 8.15$ ($Tu = 0.1\%$) is not far from that given by the experiments. Figure 10 shows a comparison between measured and predicted transition locations, as a function of the angle of attack α ; the agreement is satisfactory. The dotted line corresponds to numerical results obtained when $M_e = 0$ is used in the stability computations.

Conclusions

The first part of the paper presented stability computations obtained by a direct numerical solution of the stability equations. The stabilizing effect of compressibility as well as the influence of the wave orientation were demonstrated. The trends remain unchanged for negative, zero or positive pressure gradients.

A simplified calculation method was then developed. It is of great practical interest because it allows to perform rapid and systematic computations. The agreement with experiments was good for the cases where transition onset was measured. On the other hand, it was shown that simple transition criteria developed for low speed flows cannot be used with confidence in configurations where compressibility effects become significant, i.e. in zero and negative pressure gradients. The deficiencies of the criteria are not so important in decelerated flows, where pressure gradient effects play a dominant role. Obviously, these conclusions assume that the e^n method is valid in transonic flow for all pressure gradients.

References

1. Mack, L.M. : Methods of Computational Physics. Vol. 4. New York : Academic Press 1965.
2. Gasperas, G ; Antar, B. : AIAA Paper 85-1699 (1985).
3. Gleyzes, C ; Cousteix, J. : ONERA Technical Report n° 8/5050 (1974).
4. Jelliti, M. : Thesis, ENSAE Toulouse (1986).
5. Boehman ; Mariscalco. : AFFDL -TR - 76 - 148 (1976)

6. Mack, L.M. : AGARD Report n° 709 (1984)

7. Gaster, M. : J. Fluid Mech., Vol. 14 (1962).

8. Mack, L.M. : AGARD Report n° 224 (1977).

9. Malik : Energy Resources Technology Conference, New Orleans (1984).

10. Arnal, D. ; Juillen, J.C. : ONERA Technical Report n° 17/5018 (1981)

11. Habiballah, M. : Thesis, ENSAE Toulouse (1981)

12. Arnal, D ; Habiballah, M ; Coustols, E. : La Recherche Aérospatiale n° 1984-2 (1984).

13. Blanchard, A. ; Mignosi, A ; Dor, J.B. ; Seraudie, A. : 22ème Colloque d'Aérodynamique Appliquée, Lille (1985)

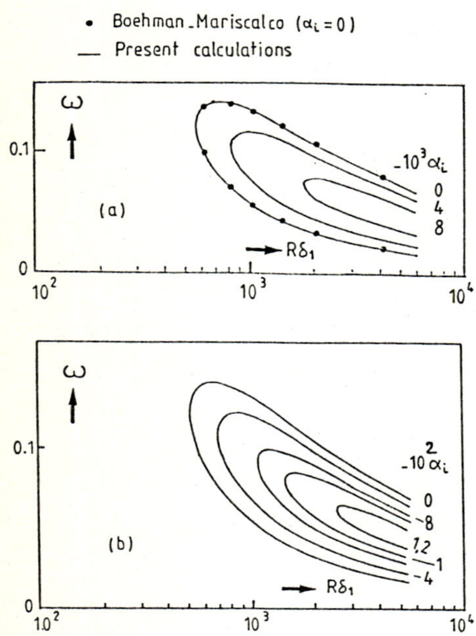

Fig. 1 - Stability diagrams for flat plate flow

a - Me = 0.9 ,
b - Me = 0

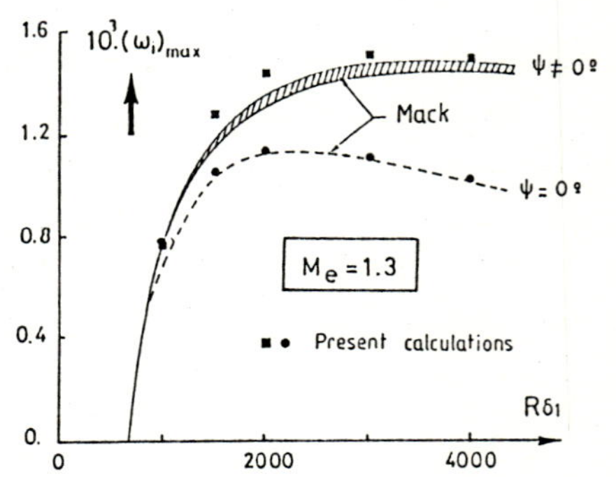

Fig. 2 - Maximum amplification rate as a function of the Reynolds number.

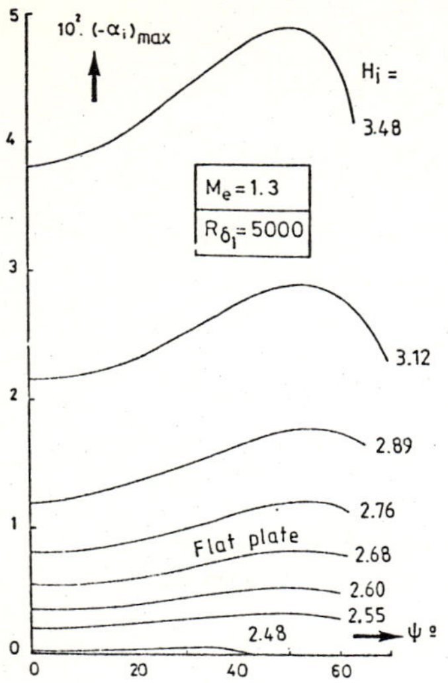

Fig. 3 - Effect of ψ and Hi on the maximum amplification rate

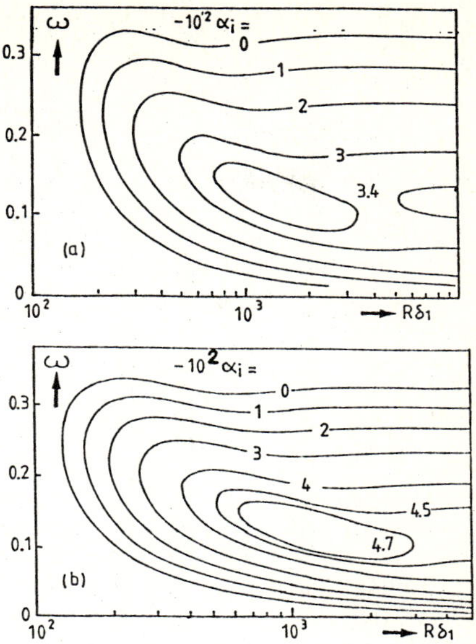

Fig. 4 - Stability diagram with positive pressure gradients
a - Me = 0.9 ,
b - Me = 0

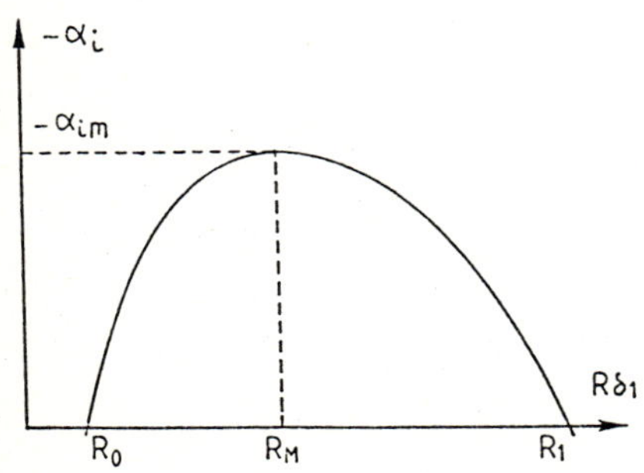

Fig. 5 - Notations for the simplified method

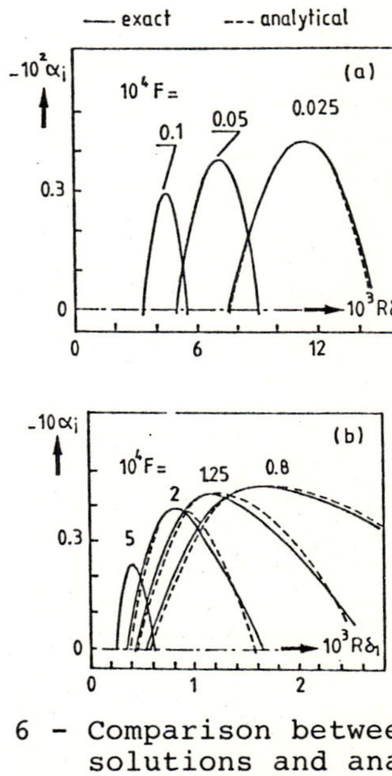

Fig. 6 - Comparison between "exact" solutions and analytical representations. Me = 1.3
a) Hi = 2.55, b) Hi = 3.48

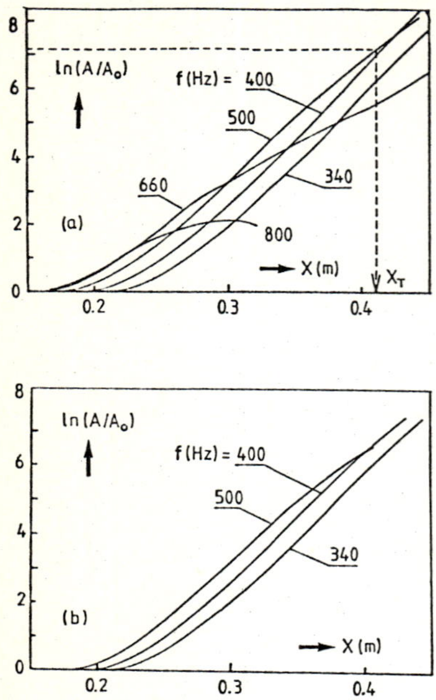

Fig. 7 - Incompressible flow
a) Simplifed method
b) Exact stability computations

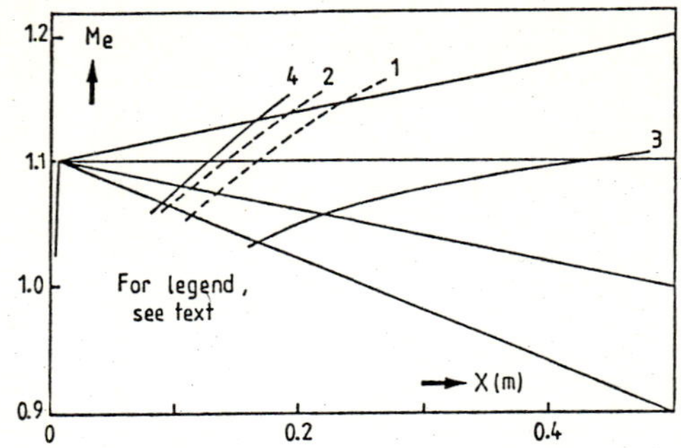

Fig. 8 - Theoretical case
Curves 1 and 2 : transition criterion.
Curve 3 : e^n method with $M_e \neq 0$
Curve 4 : e^n method with $M_e = 0$

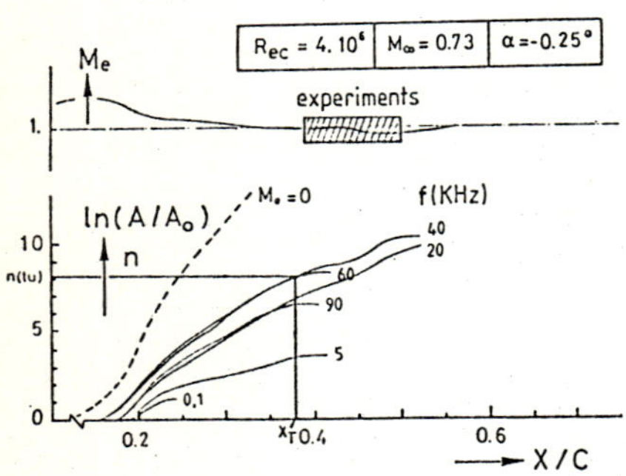

Fig. 9 - CAST10 airfoil : application of the simplified method

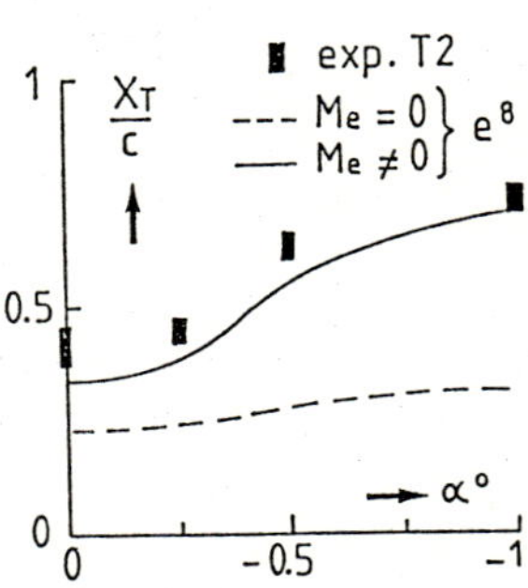

Fig. 10 - CAST10 airfoil : comparison between theory and experiments

Vortical Flows

Transonic Vortical Flow

E.M. Murman, A.M. Goodsell[1], R.E. Malecki[2]

Computational Fluid Dynamics Laboratory
Department of Aeronautics and Astronautics
Massachusetts Institute of Technology
Cambridge, Massachusetts 02139, USA

Introduction

Transonic flow is generally associated with the inviscid fluid effects when a flow with a freestream Mach number $M_\infty \simeq 1$ accelerates to locally supersonic velocities, or decelerates to locally subsonic velocities, as it moves streamwise past a body. However, transonic flows can occur in many other circumstances such as on high lift devices at low M_∞, at the tips of helicopter blades, and around corners in ducts. In this paper, a class of transonic flows for slender, highly swept wings at high angle of attack ($\alpha \geq 10°$) will be considered. Under these conditions, leading edge vortices generally form as shown schematically in Figure 1. If $M_\infty \sim 1$, the streamwise flow can be transonic in the classical sense, leading to mixed subsonic-supersonic flow and shock waves on the wing. However, the leading edge vortices create additional transonic cross flow effects due to the high swirl velocities associated with them.

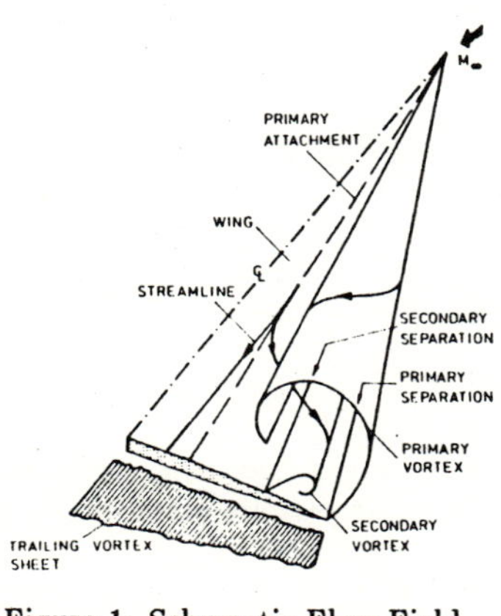

Figure 1: Schematic Flow Field

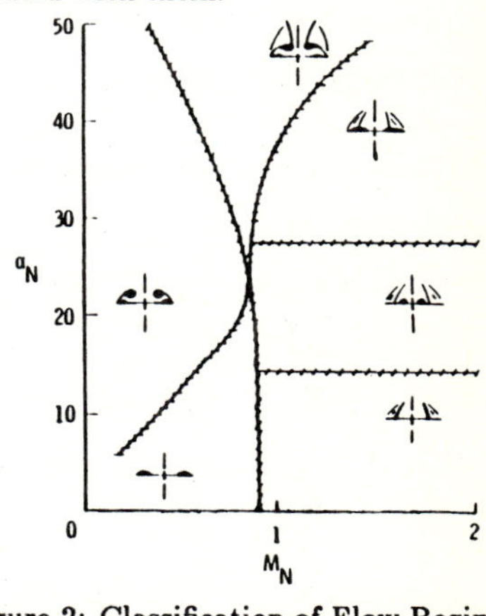

Figure 2: Classification of Flow Regimes

The classes of vortical transonic crossflows are generally correlated in terms of the angle of attack and Mach number normal to the wing leading edge given by the formulas [1]

$$\alpha_n = \tan^{-1}(\tan\alpha \sec\Lambda) \tag{1}$$

$$M_n = M_\infty \sin\alpha \csc\alpha_n \tag{2}$$

where Λ is the wing leading edge sweep angle. Figure 2 shows a correlation from experimental data presented and discussed by Miller and Wood [2,3] for sharp leading edge wings. Other flow

[1]Present address, NASA Ames Research Center, Moffett Field, CA.
[2]Present Address, Pratt & Whitney Aircraft Group, Hartford, Conn.

J. Zierep, H. Oertel (Eds.)
Symposium Transsonicum III
IUTAM Symposium Göttingen, 24.-27. 5. 1988

patterns have also been discovered experimentally by Vorropoulos and Wendt [4] and Szodruch and Peake [5]. A paper in this conference by Narayan and Seshadri [6] further discusses these flow regimes and also presents results for round leading edge wings. For the latter case, a Reynolds number based on some leading edge length scale must also be an important parameter.

Transonic vortical flows are a complex mixture of streamwise and crossflow effects. The paper by Elsenaar and Bütefisch [7] presents results of an extensive experimental investigation for a 65° delta wing which serve to illustrate this point. A complete model which can handle the interacting inviscid and viscous effects undoubtedly requires Navier-Stokes equations, and one such approach is given by Rizzi, Drougge, and Müller [8]. For sharp leading edge wings, the separation of the flow to form a primary vortex is fixed by the geometry via a Kutta condition, and an Euler equation model should be adequate if vorticity is introduced. However, important viscous effects such as shock wave-boundary layer interaction, secondary vortex formation, and boundary layer transition clearly won't be represented. Although the question of whether such models can represent vortex breakdown is not settled, numerical solutions seem to indicate that bursting of the primary vortex is captured if the mesh is fine enough. In principle, a potential flow with added vortex sheets or filaments as suggested by Desai et al [9] could be used. However, the direct solution of the Euler and/or Navier-Stokes equations seems more promising, although more expensive.

In this paper, Euler equation models are considered for sharp edge geometries. The conical Euler equations are considered first to investigate the crossflow phenomena. Following this is a discussion of the origin of the vorticity in Euler equation calculations. To conclude the paper, three-dimensional solutions are presented which combine both streamwise and crossflow transonic effects. The geometry considered is that reported by Elsenaar and Bütefisch [7].

Governing Equations

The three-dimensional unsteady Euler equations in conservation form are:

$$\frac{\partial}{\partial t}\begin{bmatrix}\rho\\ \rho u\\ \rho v\\ \rho w\\ \rho E\end{bmatrix}+\frac{\partial}{\partial x}\begin{bmatrix}\rho u\\ \rho u^2+p\\ \rho uv\\ \rho uw\\ u(\rho E+p)\end{bmatrix}+\frac{\partial}{\partial y}\begin{bmatrix}\rho v\\ \rho uv\\ \rho v^2+p\\ \rho vw\\ v(\rho E+p)\end{bmatrix}+\frac{\partial}{\partial z}\begin{bmatrix}\rho w\\ \rho uw\\ \rho vw\\ \rho w^2+p\\ w(\rho E+p)\end{bmatrix}=0. \tag{3}$$

$(u,v,w)=\vec{u}$ are the Cartesian velocity components in the x,y,z directions, respectively, ρ is the density, E the total energy per unit mass, and p the static pressure given by the equation of state

$$\frac{p}{\rho}=(\gamma-1)\left[E-\frac{u^2+v^2+w^2}{2}\right]. \tag{4}$$

Unsteady terms have been retained for the iterative procedure used to reach the steady solution. The tangential flow boundary condition is enforced on the solid body while the Kutta condition will be enforced implicitly by the numerical method. Freestream conditions are applied far away from the body.

Introducing the conical variables

$$\eta=\frac{y}{x}\qquad \xi=\frac{z}{x}\qquad r=\sqrt{x^2+y^2+z^2} \tag{5}$$

and assuming conical self-similarity (the solution is independent of r), the Euler equations become

$$\frac{r}{\sqrt{1+\eta^2+\xi^2}}\frac{\partial}{\partial t}\begin{bmatrix}\rho\\ \rho u\\ \rho v\\ \rho w\\ \rho E\end{bmatrix}+\frac{\partial}{\partial \eta}\begin{bmatrix}(v-\eta u)\rho\\ (v-\eta u)\rho u-\eta p\\ (v-\eta u)\rho v+p\\ (v-\eta u)\rho w\\ (v-\eta u)(\rho E+p)\end{bmatrix}+\frac{\partial}{\partial \xi}\begin{bmatrix}(w-\xi u)\rho\\ (w-\xi u)\rho u-\xi p\\ (w-\xi u)\rho v\\ (w-\xi u)\rho w+p\\ (w-\xi u)(\rho E+p)\end{bmatrix}+2\begin{bmatrix}\rho u\\ \rho u^2+p\\ \rho uv\\ \rho uw\\ u(\rho E+p)\end{bmatrix}=0. \tag{6}$$

These equations are valid for wing geometries generated by rays passing through the origin and for $M_\infty > 1$ so that the outer boundary condition is also conical. Again, the unsteady terms are retained only for the iterative solution process. The equations are solved on the unit sphere by setting r = 1. Boundary conditions are the same as the Euler equations except that freestream conditions are applied ahead of the conical bow shock.

Solution Procedure

The basic solution scheme is a finite volume spatial discretization with a multi-stage integration in time, as proposed by Jameson *et al.*[10]. It will be outlined for the conical Euler equations. The solution of the three-dimensional Euler equations, which is very similar, is fully explained by Roberts [11] who implemented the version used in this paper.

The conical Euler equations (6) written in vector form

$$\frac{\partial}{\partial t}\mathbf{U}+\frac{\partial}{\partial \eta}\mathbf{F}+\frac{\partial}{\partial \xi}\mathbf{G}+\mathbf{H}=0 \tag{7}$$

may be transformed using the divergence theorem into an integral equation and discretized as

$$\frac{d\mathbf{U}_{ij}}{dt}A_{ij}+\sum_{\ell=1}^{4}[\mathbf{F}_\ell n_{\xi_\ell}+\mathbf{G}_\ell n_{\eta_\ell}]+\mathbf{H}_{ij}A_{ij}=0. \tag{8}$$

A_{ij} is the area of a computational cell and the summation represents the fluxes through the four cell sides with normal vectors $\vec{n}_\ell = (n_{\xi_\ell}, n_{\eta_\ell})$. The flux vectors $\mathbf{F}_\ell, \mathbf{G}_\ell$ on the cell sides are evaluated from simple averages of flux vectors at the cell centers $\mathbf{F}_{ij}, \mathbf{G}_{ij}$ and $\mathbf{F}_{i\pm1,j\pm1}, \mathbf{G}_{i\pm1,j\pm1}$.

A multistage integration scheme is used to solve the equations in time to reach a steady state. In order to accelerate the iterative solution, the time step in each cell is determined by the linear stability criteria for that cell. Since the time steps are different for each cell, the solution is not time accurate. The tangential flow boundary condition on the body is enforced by zeroing the convective fluxes in (6) and extrapolating the pressure from the adjacent cell center, $p_{j_{body}} = p_{j_{i=1}}$.

This scheme requires added second and fourth order damping to capture shocks and damp unwanted high frequency modes of the discrete solutions. The damping formulation follows Rizzi and Eriksson [12]. The second order damping is pressure-weighted, and is of the form

$$\mathbf{D}_2(\mathbf{U})=\nu_2\left[\delta_i\left(\frac{\delta_i^2 p}{max|\delta_i^2 p|}\delta_i\mathbf{U}\right)+\delta_j\left(\frac{\delta_j^2 p}{max|\delta_j^2 p|}\delta_j\mathbf{U}\right)\right] \tag{9}$$

where δ_i and δ_j are undivided central difference operators defined by

$$\delta_i(\mathbf{U}_{i,j})=\mathbf{U}_{i+\frac{1}{2},j}-\mathbf{U}_{i-\frac{1}{2},j} \qquad \delta_j(\mathbf{U}_{i,j})=\mathbf{U}_{i,j+\frac{1}{2}}-\mathbf{U}_{i,j-\frac{1}{2}}. \tag{10}$$

The fourth-order damping is unweighted, and of the form

$$\mathbf{D}_4(\mathbf{U})=\nu_4\left[\delta_i^4\mathbf{U}+\delta_j^4\mathbf{U}\right]. \tag{11}$$

The pressure switch in the second order damping is normalized to vary from $O(\Delta^2)$ to 1 throughout the flow field, where Δ is a measure of the mesh spacing. It is designed to turn on at embedded shock waves. In these calculations it was found that the second difference damping needed to be turned on at the wing leading edge. This was done by setting the pressure switch to 1 in a few cells in this region. Values of $\nu_2 = .05$ and $\nu_4 = .01$ were used for the calculations unless noted.

The added damping terms (9) and (11) together with the basic discretization error from (8) implies that the numerical calculations represent the solution of a modified differential equation

$$\frac{\partial}{\partial \eta}\mathbf{F} + \frac{\partial}{\partial \xi}\mathbf{G} + \mathbf{H} \simeq \nu_2 \tilde{\lambda} \Delta \left[\frac{\partial}{\partial \xi}\left(S_\xi \frac{\partial \mathbf{U}}{\partial \xi}\right) + \frac{\partial}{\partial \eta}\left(S_\eta \frac{\partial \mathbf{U}}{\partial \eta}\right)\right] + O\left(\Delta^2\right) - \nu_4 \tilde{\lambda} \Delta^3 \left[\frac{\partial^4 \mathbf{U}}{\partial \xi^4} + \frac{\partial^4 \mathbf{U}}{\partial \eta^4}\right] \tag{12}$$

where S_ξ, S_η are the pressure switches in (9) and $\tilde{\lambda}$ the maximum characteristic speed of propagation of a disturbance $\simeq |\vec{u}| + c$. The first group of terms on the right hand side represents the second order damping, the second group the truncation errors, and the third group the fourth order damping. In the neighborhood of the tip and embedded shock waves, $S_\xi, S_\eta \sim 1$ and the second difference damping is proportional to the first power of the mesh spacing. Away from these regions, $S_\xi, S_\eta \sim \Delta^2$ and the second difference damping is proportional to the cube of the mesh spacing. The fourth difference damping is everywhere proportional to the cube of the mesh spacing. The solutions automatically satisfy the Kutta condition, and the present authors believe this is due to the added numerical viscosity.

Conical Euler Solutions

Numerous solutions have been computed for the conical Euler equations for sharp edge geometries using the method as outlined above or in a somewhat modified form. All the primary vortex and shock wave structures illustrated in Figure 2 have been reproduced, as well as other shock structures reported in [4,5]. Secondary vortices due to boundary layer separation under the primary vortex and vortices created by shock wave-boundary layer interactions are not modeled by the Euler equations. The most comprehensive presentation of the variety of solutions which have been obtained may be found in the PhD thesis of Powell [13]. A comparison of the conical and a three-dimensional Euler solution are given in [14], while comparisons between the conical Euler and conical Navier-Stokes solutions with data are given in [15]. In this section, two representative examples will be presented.

An interesting example illustrating the nature of the transonic crossflow effects is a flat plate delta wing with a leading edge sweep $\Lambda = 75°$ in a $M_\infty = 1.95$ flow at an angle of attack $\alpha = 25°$. The above conditions correspond to $M_n = .94, \alpha_n = 61°$ and fall in the middle top region of Figure 2. Experimental results for this condition are reported by Monnerie and Werlé [16]. The calculations were done on an O-type grid which had 128 cells around the wing, windward to leeward symmetry plane, and 128 cells between the wing and the farfield boundary outside the bow shock.

The general topology of the flow field is illustrated in Figure 3 by the conical streamlines which are everywhere tangential to the crossflow velocity vectors. The view corresponds to looking upstream from the trailing edge with only the right hand side of the wing shown. The dominant feature is the sheet emanating from the wing leading edge and spiraling into the vortex core[3]. On the windward side, the attachment line is almost halfway between the centerline and leading edge of the wing, while on the leeward side it is at the wing centerline.

[3]The blank area in the middle of the core is due to an inaccurate streamline integration [17] and not an apparent source as suggested in [18].

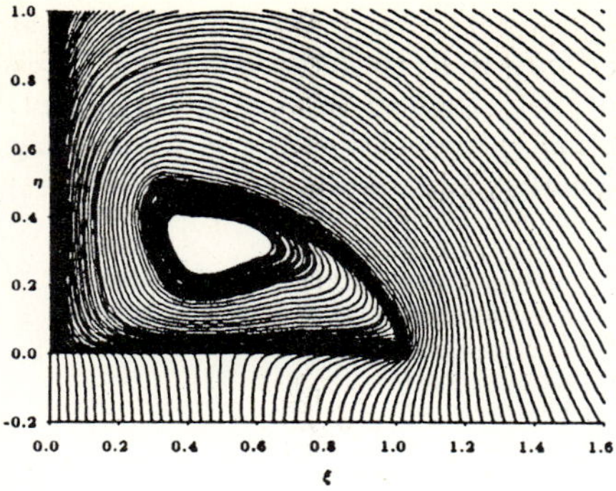

Figure 3: Conical streamlines.

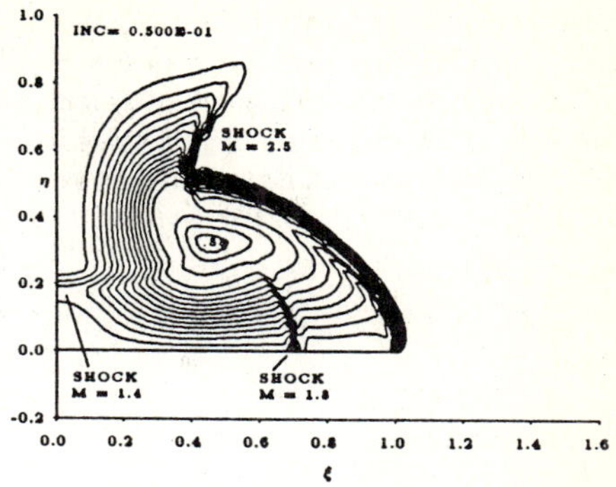

Figure 4: Total Pressure Loss.

Contours of constant total pressure loss $1-(P_t/P_{t_\infty})$ are shown in Figure 4. There are large losses in the shear layer from the leading edge, and in the center of the vortex the total pressure loss is 88 %. It can be seen that cross flow shocks are present above and below the primary vortex and that there is also a shock between the two primary vortices. This latter shock is observed in the shadowgraph picture from the experiment [16]. The swirling flow accelerates to cross flow Mach number $M_c = 2.5$ above the vortex before passing through the first shock. It then turns downwards towards the wing reaching $M_c = 1.4$ ahead of a second shock. The flow then accelerates to $M_c = 1.8$ on the wing surface under the vortex where it experiences the third shock. A comparison of the cross flow streamlines and the shock locations will reveal a sort of transonic nozzle effect for the last two shocks. There is a minimum spacing between the vortex and the symmetry plane and the vortex and the wing. Examination of the crossflow Mach number contours which aren't shown here reveals a converging-diverging nozzlelike pattern.

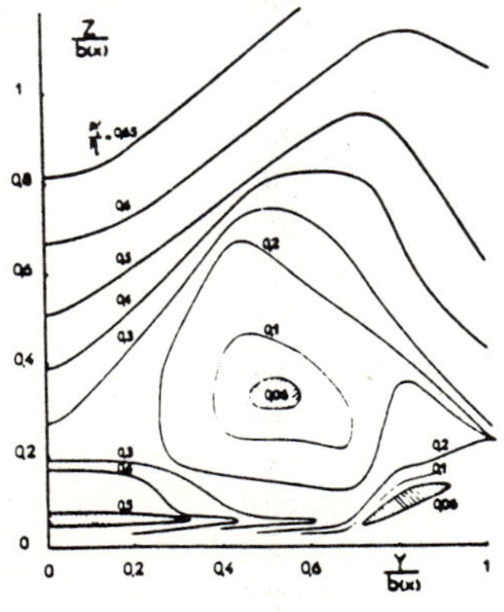

Figure 5: Experimental pitot pressure.

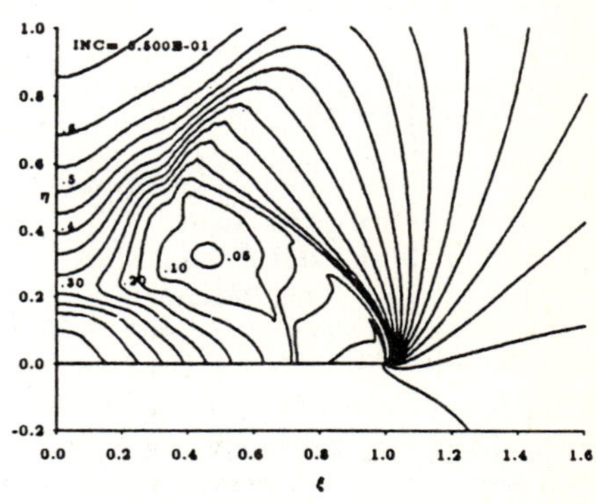

Figure 6: Computed pitot pressures.

Monnerie and Werlé reported measured pitot pressure (normalized by the freestream total pressure) values on the leeside of the wing which are shown in Figure 5. Pitot pressures computed from the Euler calculations using the Rayleigh pitot formula [19] are shown in Figure 6. The computed values would correspond to a probe which is everywhere aligned with the flow, while in the experiments the probe was aligned with the wing centerline. Vortical flows have large flow angularity as illustrated by the swirl angle (ratio of cross flow velocity magnitude to radial velocity) contours of Figure 7. Generally, pitot pressure measurements are accurate to within 1% if the flow angularity is less than 10° [20]. This is really only true right in the vortex core region, and the computed (.05) and measured (.06) values are in excellent agreement. Throughout the rest of the region there is good qualitative agreement except where the wing boundary layer and secondary vortex formation dominate.

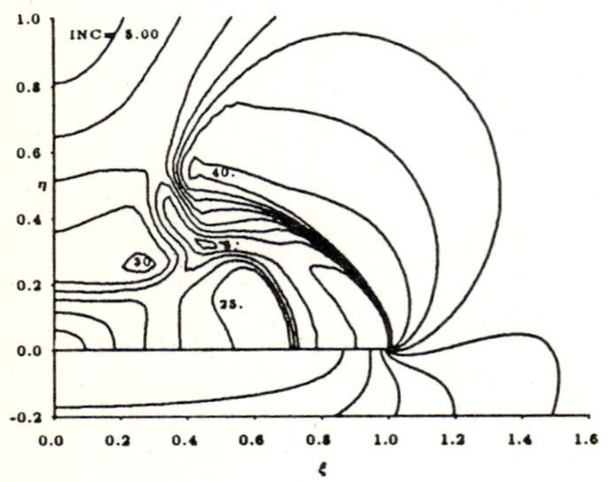

Figure 7: Computed swirl angles.

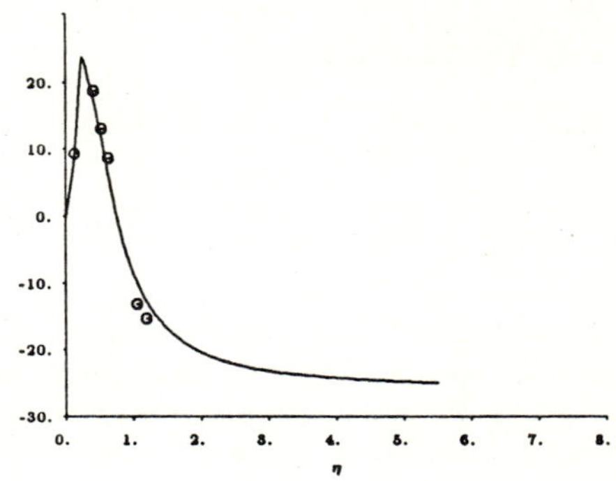

Figure 8: Symmetry plane flow angles.

Measurements were also made of the flow angle in the leeside symmetry plane. A small wedge in the flow was adjusted until the top and bottom shock waves viewed with a shadowgraph were symmetric. The estimated accuracy[4] is ±2°. Figure 8 compares the measured and computed values which are in good agreement. The change in flow angularity across the shock spanning, the port and starboard vortices is clear.

Calculations have been done on coarser grids to investigate the sensitivity of the solution to grid resolution. The following table displays some of these values.

Grid	32 × 32	64 × 64	128 × 128
η *vortex*	.287	.307	.320
ξ *vortex*	.445	.470	.438
Normal Force Coefficient	.729	.739	.742
Max Swirl Angle	38°	40°	40°
Max Total Pressure loss	.872	.882	.880
Min Pitot Pressure	.075	.052	.048

It can be seen that the position of the vortex, the normal force coefficient, the maximum total pressure loss, and maximum swirl angle are all insensitive to mesh spacing. The pitot pressure is somewhat sensitive to grid refinement as the streamwise Mach number in the vortex core increases with finer meshes. This is consistent with the computed vortex core size becoming smaller with mesh refinement. In general, the solution on the 128 × 128 grid can be considered converged except for the size and details of the inner vortex core region[13]. However, its position and maximum total pressure loss are converged as illustrated in the table. The conclusions are representative of all our conical Euler calculations for supersonic freestream flows.

The second example of a conical Euler solution is taken from [15] which presents six different comparisons between conical Euler, conical Navier-Stokes, and experimental results. Details of the calculations and comparisons may be found in that paper. The particular example presented here is for a delta wing with a leading edge sweep $\Lambda = 75°$ in a $M_\infty = 2.8$ flow at an angle of

[4]Personal communication between authors.

attack $\alpha = 8°$. These parameters give $M_n = .82, \alpha_n = 28.5°$ which lie in the left middle region of Figure 2.

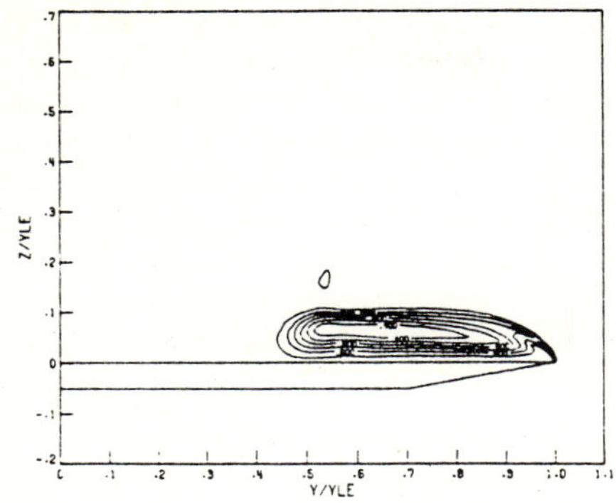

Figure 9: Total pressure loss, Euler.

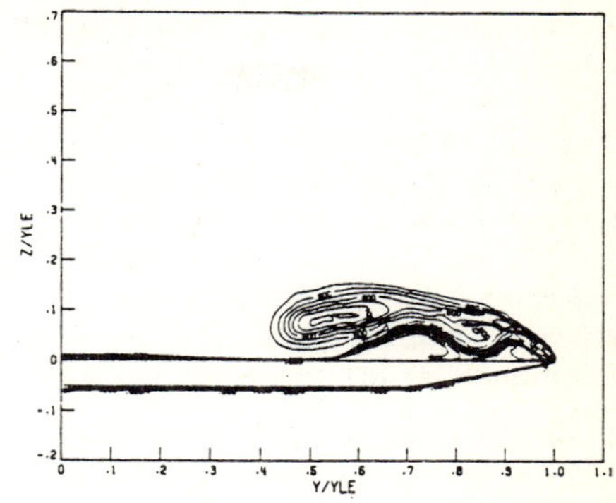

Figure 10: Total pressure loss, Navier-Stokes.

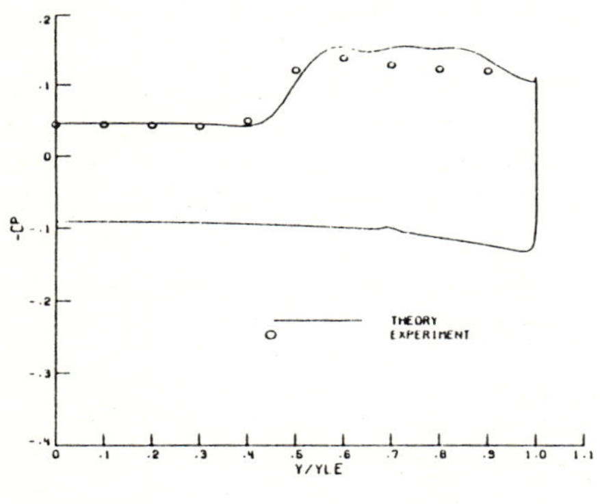

Figure 11: C_p, Euler and data.

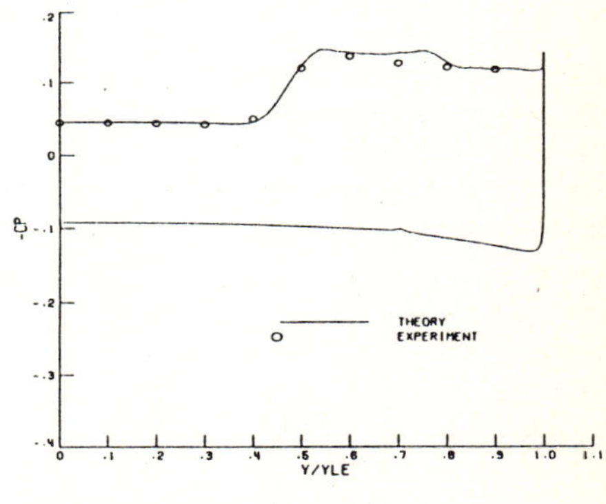

Figure 12: C_p, Navier-Stokes and data.

A comparison of the total pressure loss contours between the conical Euler and conical Navier-Stokes (laminar) is given by Figures 9 and 10. The Navier-Stokes model clearly shows the viscous boundary layer, the secondary separation, and for this case a tertiary separation. The latter affect the position and shape of the primary vortex by their displacement effects. However, both the Euler and Navier-Stokes models yield a primary vortex with essentially the same total pressure loss in the core. A comparison of the surface pressures given in Figures 11 and 12 shows that both models are in reasonable agreement with the data. The effects of the secondary and tertiary vortex (including their displacement effects on the primary vortex) are small, but well represented by the Navier-Stokes solution.

Total Pressure Loss

The origin of the total pressure loss in the Euler equation solutions is not immediately apparent and has been the source of considerable controversy. The total pressure loss can equivalently be represented as entropy (S) production, and by Crocco's equation

$$T\nabla S + \vec{u} \times \vec{\omega} = \nabla h_{\circ} \tag{13}$$

can be related to a vorticity ($\vec{\omega}$) distribution since the total enthalpy ($h_{\circ}$) is constant for this class of flows. One source of total pressure loss is the conical bow shock wave. However, these losses are two to three orders of magnitude smaller than the values in the vortical regions. The results shown in Figures 3 to 8 contain strong embedded shocks which clearly have an associated total pressure loss. Comparing Figures 3 and 4 it can be seen that these losses are approximately convected along streamlines. There are additional total pressure losses in the shear layer from the wing tip and vortex core. The latter is clearly not convected along streamlines. Many examples have been computed which have either no embedded shocks or very weak ones, but which have total pressure losses of 30% or more. It is clear that some mechanism other than shock waves is involved.

A possible explanation could be that the losses represent numerical error introduced by the discretization process or added damping terms as represented by Equation (12). In fact this seems to be true, but the interpretation is a bit surprising. Careful numerical studies have been done [13,21] in which grid density and artificial viscosity constants ν_2 and ν_4 were varied by an order of magnitude. The total pressure loss is extremely insensitive to these parameters. See, for example, the table in the previous section. The total pressure loss was also found to be insensitive to grid distribution, initial conditions, and convergence tolerance.

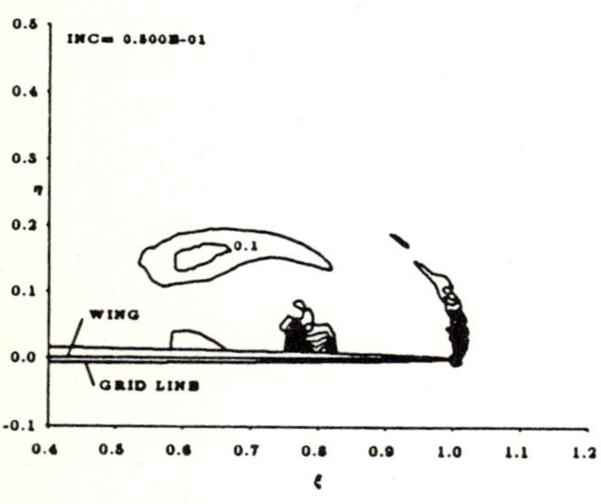

Figure 13: Second order damping magnitude

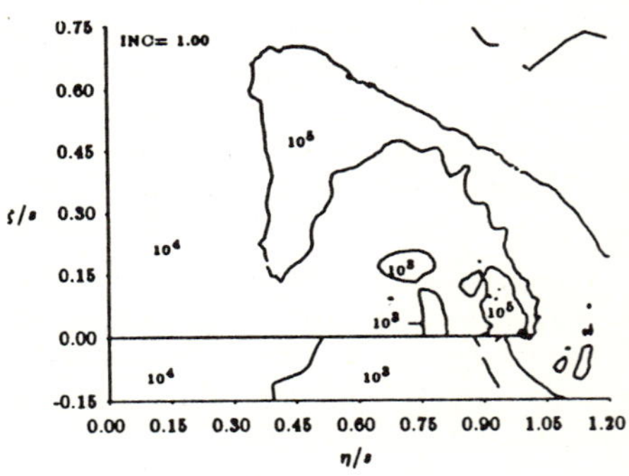

Figure 14: Equivalent Reynolds number

A clue to the source of the losses can be found by comparing the magnitude of the second order damping terms with the sink term **H** in Equation 12 as shown in Figure 13 taken from [18]. The results are for a $\Lambda = 75°$ flat plate delta wing at $M_\infty = 1.3, \alpha = 10°$ with a 128×128 cell grid. It can be seen that in the feeding sheet region, the vortex core, and around an embedded shock under the vortex, the damping terms are significant.

Further insight can be obtained by looking at a simplified equation set corresponding to the one-dimensional continuity and momentum equations [13]

$$\frac{\partial}{\partial x}(\rho u) = \tilde{\epsilon}_\rho \frac{\partial^2}{\partial x^2}\rho \tag{14}$$

$$\frac{\partial}{\partial x}\left(\rho u^2 + p\right) = \tilde{\epsilon}_{\rho u} \frac{\partial^2}{\partial x^2}(\rho u) \tag{15}$$

where the numerical diffusion coefficients have been denoted by $\tilde{\epsilon}$. Differentiating momentum and using continuity yields

$$\rho u \frac{\partial u}{\partial x} + \frac{\partial p}{\partial x} = \rho \tilde{\epsilon}_{\rho u} \frac{\partial^2 u}{\partial x^2} + u \left(\tilde{\epsilon}_{\rho u} - \tilde{\epsilon}_{\rho} \right) \frac{\partial^2 \rho}{\partial x^2} + 2 \tilde{\epsilon}_{\rho u} \frac{\partial \rho}{\partial x} \frac{\partial u}{\partial x}. \tag{16}$$

Now, if $\tilde{\epsilon}_{\rho u} = \tilde{\epsilon}_{\rho}$, i.e. the same artificial viscosity formulation is used for both continuity and momentum equations, which is the usual procedure, then there are only two added terms. The first is a diffusion term, while the last one is a convective term modifying the coefficient of $\frac{\partial u}{\partial x}$. Powell [13] used this idea to define an effective numerical Reynolds number $\sim \tilde{\epsilon}_{\rho u}^{-1}$ which varies throughout the flow field due to mesh resolution, the magnitude of the pressure switch, and the value of $\tilde{\lambda}$. Figure 14 taken from [13] shows the magnitude of the numerical Reynolds number for the same wing as Figure 13 but at $M_\infty = 1.1, \alpha = 10°$. An embedded grid was used such that in the region of interest the grid density is 256×256. Even for such a fine grid, it can be seen that the effective Reynolds number in the primary vortex core is about 10^3!

It is worth keeping in mind that most Navier-Stokes algorithms contain artificial diffusion effects (whether added explicitly as herein or included implicitly as with upwind methods) which are needed for the Euler part of the algorithm. Unless the grid is extremely fine in the vortex region, the artificial diffusion terms probably dominate the real viscous terms in such calculations.

Another interesting numerical result can be understood from Equation 16. A calculation was done in which the second order damping was suppressed from the continuity equation ($\tilde{\epsilon}_{\rho} = 0$) but retained in all the other equations. This had a strong influence on the maximum total pressure loss. It can be seen from (16) that the coefficient $\tilde{\epsilon}_{\rho u} - \tilde{\epsilon}_{\rho}$ does not zero out and an added diffusion term results. By explicitly correcting for this, the calculations with suppressed second differencing damping in the continuity equation gave the same total pressure loss as the original computations. So the magnitude of the total pressure loss is sensitive to the form of the added viscosity, but not to its magnitude. Comparisons between the present finite volume algorithm and results from other algorithms have not shown significant differences, except for the results of Marconi [22] discussed in the last paragraph of this section.

Computational results indicate that the maximum total pressure loss is directly related to the aerodynamic parameters. Variations of Mach number, angles of attack, sweep, and yaw, as well as body geometry all affect the total computed pressure loss in a consistent way. That is, increasing the angle of attack produces stronger vortices and greater total pressure loss as is illustrated in the next section.

The above evidence can be summarized as follows. In the primary vortex core and the shear layer emanating from the wing tip, the added artificial viscosity terms are significant and produce a total pressure loss in a manner similar to what the actual fluid viscosity would produce. The intriguing question is why the total pressure loss is insensitive to the magnitude of the numerical diffusion coefficient. There are different explanations for the core and shear layers. For the core region, the circumferential velocity is brought to zero (see Figure 7) producing a loss. Powell's [13,23] analytical model for a high Reynolds number conical incompressible viscous vortex shows that the only parameter which sets the maximum total pressure loss is the swirl velocity at the edge of the core. As can be seen from the table in the previous section, this parameter is also independent of mesh spacing for the supersonic conical flows. For the shear layer [21], the added viscous terms together with the discrete nature of the solution produce a vortex "sheet" with a thickness related to the mesh spacing. The velocity magnitude should be the same on either side of the sheet, but the direction changes as dictated by the strength of the sheet. Traversing across the sheet, the velocity magnitude will have a minimum in the center leading to a total pressure loss. The reader is referred to [13,23,21,24] for the detailed discussion of these arguments.

The question remains as to whether the artificially introduced total pressure loss is related to the real total pressure loss. Comparisons with measured pitot pressure as given in the previous section and total pressure as given in the next section indicate that although the numerical losses may not be exact, they are a good approximation for compressible flows. For incompressible flows, the agreement is not as good [25], but the studies are not as thorough. The artificial diffusion terms are similar to, but different than, the real diffusion terms even for laminar flows. For Reynolds numbers characteristic of aeronautical applications, the vortex cores will be turbulent. A more physically based numerical viscosity model would be helpful in establishing the accuracy of the predicted total pressure losses.

The final remark of this section addresses the issue of isentropic vortex cores. Marconi [22] has suggested that the correct solution of the Euler equations should only have shock induced total pressure losses. He has used the λ-scheme which integrates the equations in a characteristic form. Entropy production must be introduced explicitly, which only occurs at shock waves. Vortical flows have been computed with zero total pressure losses. By Crocco's equation (13) this means the vorticity and velocity vectors are aligned. The results of these calculations show that the vortex position, surface pressures, and other global features of the flow are in close agreement with solutions which have total pressure losses [22]. Powell [13] modified the present solution algorithm by overriding the streamwise momentum equation with the constant total pressure requirement. Although the iterative convergence rate is poorer, solutions can be obtained and compared with the "lossy" ones. In agreement with Marconi, the global features are the same, but the values of parameters in the vortex cores are different. In particular, the streamwise velocity and Mach number are greater. Comparison with the pitot pressure measurements of Monnerie and Werlé [16] shows much poorer agreement. The present authors feel that the desirable solution of the Euler equations is that which models a viscous flow in the limit of $Re \to \infty$. The solutions with total pressure losses due to artificial viscosity are more realistic than those with no total pressure loss. Either one seems to be acceptable for prediction of surface pressures and hence forces. However, the two approaches may give different results regarding the prediction of vortex bursting in three-dimensional calculations as total pressure losses can play a role in this phenomenon [26].

Three-Dimensional Solutions

Although the conical Euler model is suitable for studying some of the basic features of leading edge vortex flows, a fully three-dimensional model is needed for realistic wing geometries as well as for subsonic freestream Mach number conditions. In this section, some results using two different grid topologies are presented and compared to experimental data for $M_\infty = .85$. The grids are rather coarse and, therefore, the results should be considered as exploratory. The calculations are for the $\Lambda = 65°$ sharp edge cropped delta wing geometry discussed by Elsenaar and Bütefisch [7]. Much more refined grid solutions for this wing are presented by Wagner and Hitzel [27] and Kumar and Das [28].

The first set of calculations were done by Malecki [29] using an O-O grid supplied by Rizzi and his co-workers (e.g. [8]). A total of 37,768 computational cells were in the field with 2048 cells adjacent to the wing surface. Although 1/16 of the total cells are on the wing, there are only 16 cells between wing and far field. From the conical Euler studies, this is not adequate to give good resolution of the vortex. Also the O-O grid has significant skewness near the cropped tip. Convergence difficulties from this were solved by raising ν_2 to .075 in this area. Only 10° angle of attack was reported by Malecki.

The second set of calculations were done by Goodsell [25] who modified Roberts' program to handle an O-H grid. In planes perpendicular to the flow, an O grid is generated similar to that used for the conical Euler calculations. A total of 73,728 cells with 768 on the wing surface were used. There are only 12 chordwise stations on the wing, but there are 48 cells between the wing and the far field. Since the vortical flow does not vary rapidly in the streamwise direction, this grid provides better resolution of the crossflow effects than the O-O grid. However, the streamwise transonic effects such as shocks will be poorly resolved. The O-H grid has less skewness than the O-O grid, and has a topology which is more flexible for complicated geometries. Goodsell reported results for $\alpha = 10°, 20°, 24°$, and the former two are discussed here.

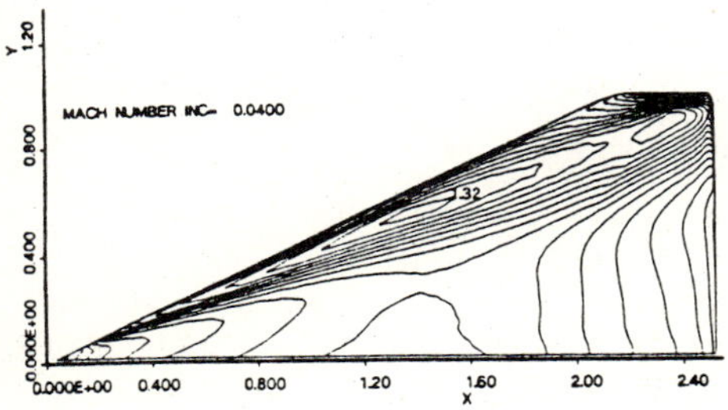

Figure 15: Surface Mach Number, O-O grid. $\alpha = 10°, M_\infty = .85$.

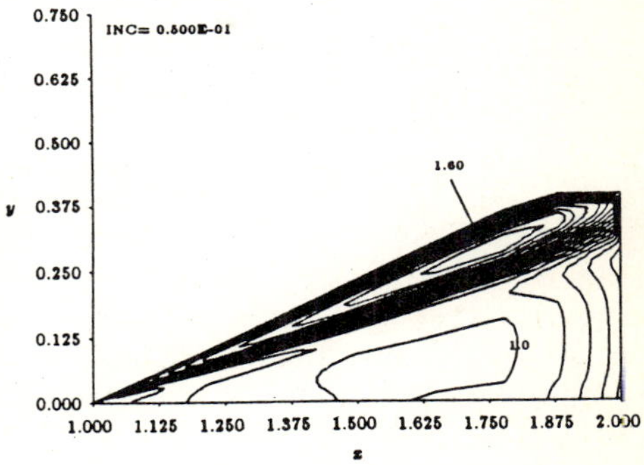

Figure 16: Surface Mach Number, O-H grid. $\alpha = 10°, M_\infty = .85$.

The results for $\alpha = 10°$ correspond to $M_n = .38, \alpha_n = 22.6°$ placing them well within the left hand middle region of Figure 2. Leeside surface Mach number contours for the two topologies are shown in Figures 15 and 16. The footprint of the vortex is evident, and the flow underneath it is supersonic. The O-O grid gives a maximum surface Mach number of 1.32 while the O-H grid with better resolution off the wing surface reaches 1.60. The flow is just supersonic over the mid portion of the wing demonstrating slight streamwise transonic effects for this lowest angle of attack. Finally, both solutions show a shock wave at the tapered tip which is aligned with the streamwise direction. The supersonic outboard flow under the vortex passes through this shock to turn downstream.

Figures 17-19 compare the computed surface pressure coefficients for the O-H grid (lines) with the measurements (symbols) at 30%, 60%, and 80% chord locations. The windward side agreement is good except at 30% chord where the centerbody may have an effect. On the leeward side, it is seen that the calculations predict significantly more suction and that the vortex is located further outboard compared to the experiments. In general the agreement is not too good. Figure 20 shows similar results for the O-O grid at 80% chord location. These calculations show less of a suction peak than the O-H grid results and in general have poorer resolution. The O-O grid calculations have been compared with Rizzi's code using exactly the same grid and are in good agreement [29].

Total pressure loss contours for the O-H grid calculations at 78% chord are shown in Figure 21. Experimental [30] total pressure contours at the 60% chord location are shown in Figure 22 for a wing with the same planform but flat upper surface. The computed results show a loss of 38% while the measurements indicate a loss of 30-40% which is in qualitative agreement. Malecki's calculations with the O-O grid predicted 44% losses at about the 90% chord location.

The results for $\alpha = 20°$ correspond to $M_n = .45, \alpha_n = 40.7°$, again in the left hand middle region of Figure 2. All results are from the O-H grid study [25], and a value of $\nu_2 = .1$ was

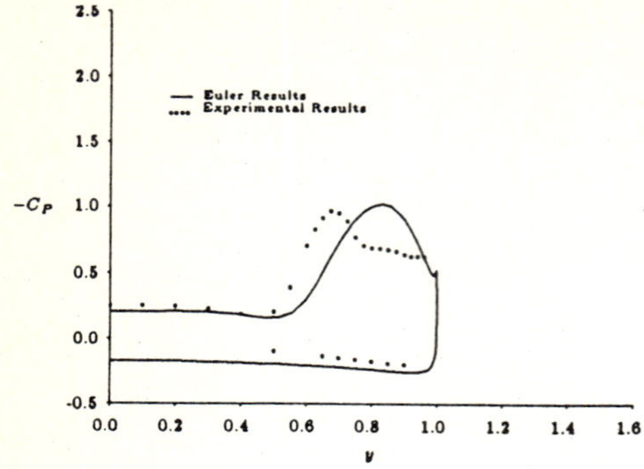

Figure 17: Surface C_p at 30% chord, O-H grid. $\alpha = 10°, M_\infty = .85$.

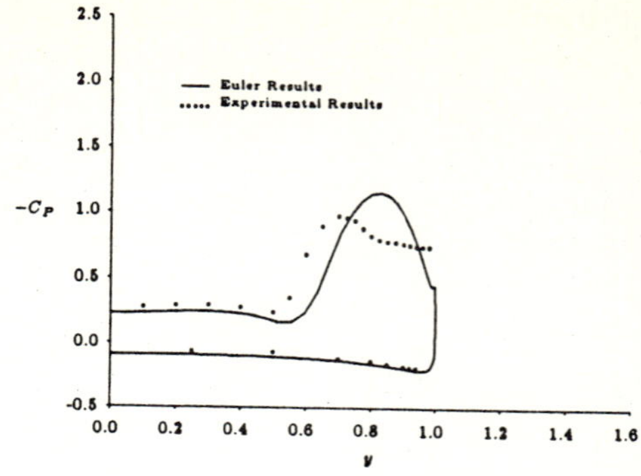

Figure 18: Surface C_p at 60% chord, O-H grid. $\alpha = 10°, M_\infty = .85$.

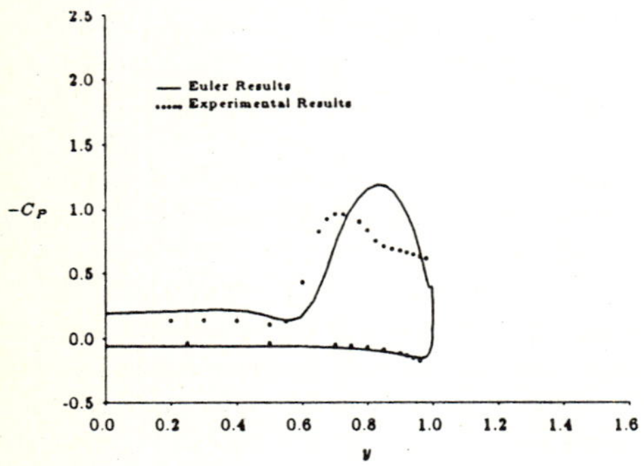

Figure 19: Surface C_p at 80% chord, O-H grid. $\alpha = 10°, M_\infty = .85$.

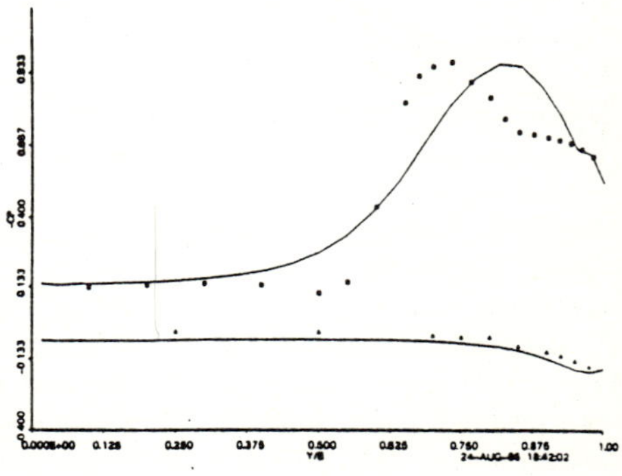

Figure 20: Surface C_p at 80% chord, O-O grid. $\alpha = 10°, M_\infty = .85$.

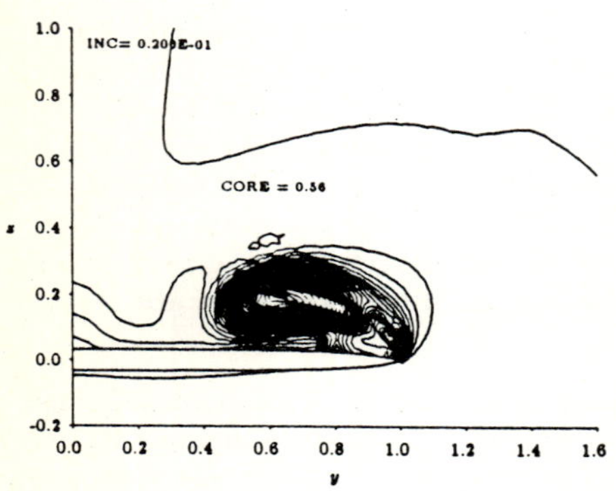

Figure 21: $(1 - P_t/P_{t\infty})$ at 78% chord, O-H grid. $\alpha = 10°, M_\infty = .85$.

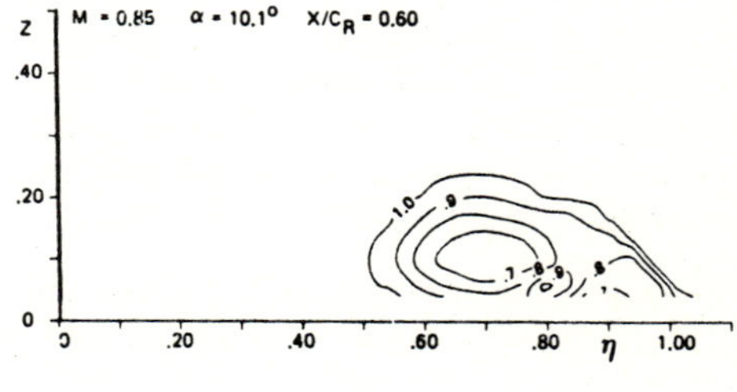

Figure 22: $(P_t/P_{t\infty})$ at 60% chord, measured. $\alpha = 10.1°, M_\infty = .85$.

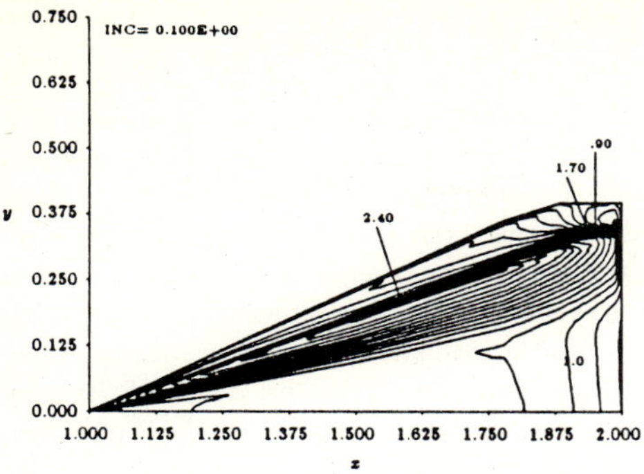

Figure 23: Surface Mach number, O-H grid. $\alpha = 20°, M_\infty = .85$.

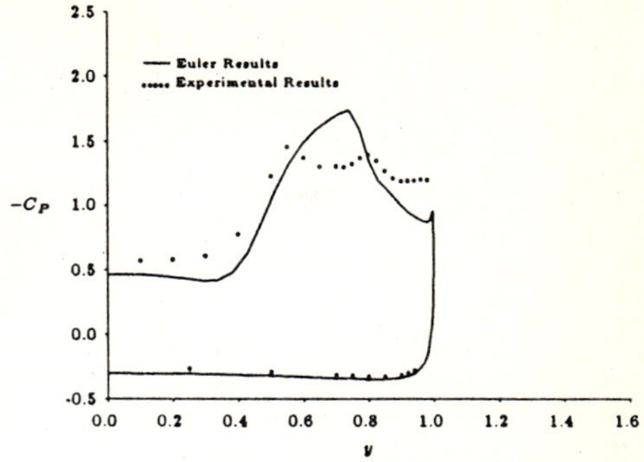

Figure 24: Surface C_p at 60% chord, O-H grid. $\alpha = 20°, M_\infty = .85$.

needed for convergence. The surface Mach number contours of Figure 23 show that the flow is highly transonic. A Mach number of 2.4 is reached under the vortex, and the streamwise flow in the center of the wing is mostly supersonic. The more refined calculations of Wagner and Hitzel [27] show that this supersonic region is terminated by a crossflow shock which interacts with the vortex. The present grid is too coarse to resolve this. The clustering of the contours indicates that there is a shock wave under the vortex. The surface C_p results at the 60% chord, Figure 24, location likewise indicate a strong suction peak terminated by a shock wave. The agreement with the measured C_p data is poor on the leeside due to the viscous effects. The secondary vortex and shock wave-boundary layer interactions significantly reduce the suction peak. A Navier-Stokes model is needed for accurate surface pressure predicitions in this speed regime. The computed and measured total pressure contours are shown in Figures 25 and 26, respectively. Again, there is reasonable agreement between the computed loss of 56% and the measured value of $\simeq 60\%$.

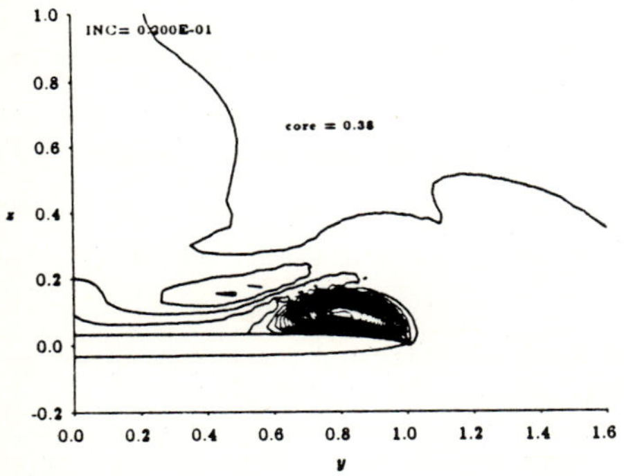

Figure 25: $(1 - P_t/P_{t\infty})$ at 78% chord, O-H grid. $\alpha = 20°, M_\infty = .85$.

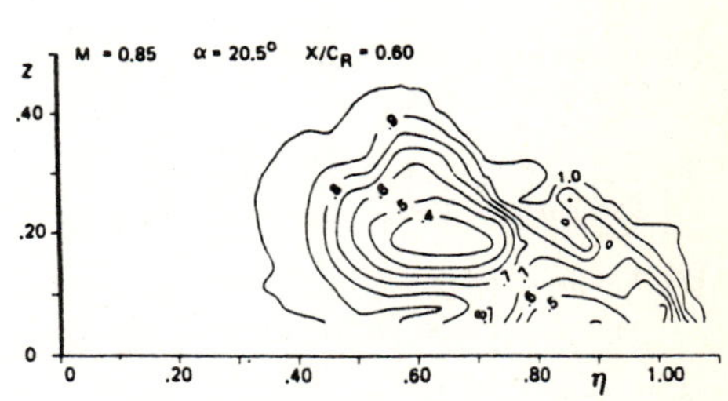

Figure 26: $(P_t/P_{t\infty})$ at 60% chord, measured. $\alpha = 20.5°, M_\infty = .85$.

Conclusions

Leading edge vortices introduce transonic crossflow effects due to their accompanying high swirl velocities. Based upon both calculations and experiments, embedded transonic zones and shock waves are the norm rather than the exception. These effects are in addition to the classical transonic flow features of lifting wings. Numerical solutions of the Euler equations for sharp edge wings appear to model the main effects of the primary vortex. Both analytical and numerical evidence point to the artificial viscosity terms as the source of vorticity. The numerical damping models routinely used by most algorithms seem to lead to realistic total pressure losses. More physically based artificial viscosity models would be desirable. Total pressure deficits may be important in the prediction of vortex bursting. For transonic freestream Mach numbers, the influences of secondary vortex formation and shock wave-boundary layer interaction are important and outside the capabilities of an Euler equation model. For supersonic freestream conditions, the secondary vortex effects have less influence on the leeside pressure distribution.

Acknowledgements

This research was supported by Office of Naval Research Grant N00014-86-K-0288 monitored by Dr. Spyridon Lekoudis.

References

[1] A. Stanbrook and L. C. Squire. Possible Types of Flow at Swept Leading Edges. *Aeronautical Quarterly*, 15(1):72–82, February 1964.

[2] D. S. Miller and R. M. Wood. Leeside Flows over Delta Wings at Supersonic Speeds. *Journal of Aircraft*, 21(9):680–686, September 1984.

[3] D. S. Miller and R. M. Wood. *Lee-Side Flow Over Delta Wings at Supersonic Speeds.* NASA TP 2430, June 1985.

[4] G. Vorropoulos and J. F. Wendt. Laser Velocimetry Study of Compressibility Effects on the Flow Field of a Delta Wing. In *AGARD-CP-342*, April 1983. Paper 9.

[5] J. Szodruch and D. Peake. *Leeward Flow over Delta Wings at Supersonic Speeds.* NASA TM 81187, 1981.

[6] K.Y. Narayan and S.N. Seshadri. Vortical Flows on the Lee-Surface of Delta-Wings. In *Symposium Transsonicum III*, May 1988.

[7] A. Elsenaar and K.A. Bütefisch. Experimental Study of Vortex and Shock Wave Development on the 65-Delta Wing. In *Symposium Transsonicum III*, May 1988.

[8] A. Rizzi, G. Drougge, and B. Müller. Navier-Stokes and Euler Solutions for the Vortex Flow Delta Wing. In *Symposium Transsonicum III*, May 1988.

[9] S.S. Desai, R. Rangarajan, U.N. Sinha, and K.S. Ravichandran. On the Development and Application of Software based on the Framework of Cartesian Coordinates. In *Symposium Transsonicum III*, May 1988.

[10] A. Jameson, W. Schmidt, and E. Turkel. *Numerical Solutions of the Euler Equations by a Finite Volume Method Using Runge-Kutta Time-Stepping Schemes.* AIAA-81-1259, June 1981.

[11] T.W. Roberts. *Euler Equation Computation for the Flow Over a Hovering Helicopter Rotor.* PhD thesis, Massachusetts Institute of Technology, November 1986.

[12] A. Rizzi and L. E. Eriksson. Computation of Flow around Wings Based on the Euler Equations. *Journal of Fluid Mechanics*, 148, November 1984.

[13] K.G. Powell. *Vortical Solutions of the Conical Euler Equations.* PhD thesis, Massachusetts Institute of Technology, June 1987. To be published by Vieweg as a Volume of Notes On Numerical Fluid Mechanics.

[14] E. Murman, A. Rizzi, and K. Powell. High Resolution Solutions of the Euler Equations for Vortex Flows. In *Progress and Supercomputing in Computational Fluid Dynamics*, pages 93–113, Birkhauser-Boston, 1985.

[15] S.N. McMIllin, J.L. Thomas, and E.M. Murman. *Euler and Navier-Stokes Solutions for the Leeside Flow Over Delta Wings at Supersonic Speeds.* AIAA-87-2270, August 1987.

[16] B. Monnerie and H. Werlé. Étude de l'Écoulement Supersonique & Hypersonique autour d'une Aile Élancée en Incidence. In *AGARD-CP-30*, 1968. Paper 23.

[17] E.M. Murman and K.G. Powell. Trajectory Integration in Vortical Flows. Technical note to appear in AIAA J.

[18] E.M. Murman, K.G. Powell, A. M. Goodsell, and M. Landahl. *Leading Edge Vortex Solutions With Large Total Pressure Losses.* AIAA-87-0039, January 1987.

[19] *Equations, Tables, and Charts for Compressible Flow.* NACA TR 1135.

[20] R. W. Ladenburg, B. Lewis, R. N. Pease, and H. S. Taylor, editors. *Physical Measurements in Gas Dynamics and Combustion.* Princeton University Press, 1964.

[21] K. Powell, E. Murman, E. Perez, and J. Baron. *Total Pressure Loss in Vortical Solutions of the Conical Euler Equations.* AIAA-85-1701, July 1985.

[22] F. Marconi. *Flat Plate Delta Wing Separated Flows with Zero Total Pressure Losses.* AIAA-87-0038, January 1987.

[23] K.G. Powell and E.M. Murman. *A Model for the Core of a Slender Viscous Vortex.* AIAA-88-0503, January 1988.

[24] K.G. Powell and E.M. Murman. An Embedded Mesh Procedure for Leading-Edge Vortex Flows. In *Proceedings of NASA Langley Transonic Symposium*, Springer-Verlag, 1988.

[25] Goodsell A.M. *3-D Euler Calculations of Vortex Flows Over Delta Wings.* Master's thesis, Massachusetts Institute of Technology, July 1987.

[26] M. G. Hall. Vortex Breakdown. In *Annual Review of Fluid Mechanics - Vol 4*, pages 195–217, Annual Reviews, Inc., 1972.

[27] B. Wagner and S.M. Hitzel. Computations Relating to the Vortex Flow Experiment. In *Symposium Transsonicum III*, May 1988.

[28] A. Kumar and A. Das. Computations of Vortex Flow on a Delta Wing at Transonic Speed. In *Symposium Transsonicum III*, May 1988.

[29] R.E. Malecki. *Euler Equation Calculations for a Cropped Delta Wing Using the CYBER 205.* Master's thesis, Massachusetts Institute of Technology, August 1986.

[30] A. Elsenaar, L. Hjelmberg, K.A. Bütefisch, and Bannink W.J. The International Vortex Flow Experiment. In *Lisbon Conference*, AGARD, May 1988.

Experimental Study on Vortex and Shock Wave Development on a 65° Delta Wing

A. Elsenaar* and K.A. Bütefisch**

* Nationaal Lucht- en Ruimtevaartlaboratorium NLR,
Anthony Fokkerweg 2, Amsterdam, NETHERLANDS

** DFVLR - Institut für Experimentelle Strömungsmechanik
D-3400 Göttingen, FRG

Summary

An analysis is presented of the transonic flow over a 65° delta wing at Mach=.85 using pressure data, flow visualizations and LDA flow field measurements. It is argued that shock waves and shock-induced separations can induce primary and secondary vortices. Near the symmetry plane another shock is observed that seems to interact in a very abrupt way with vortex breakdown .

Introduction

The flow over a delta wing at transonic conditions represents a very interesting flow field that is still far from being understood. Due to the recent advancements in CFD, this complex flow with vortices and shocks can now be modelled on the computer, at least in principle. However, these numerical simulations still require experimental data for guidance and verification.

New measuring techniques like three-component Laser Doppler Anemometry (LDA) allow detailed, non-intrusive measurements. For these reasons the time was ripe in 1983 for FFA, AFWAL, DFVLR, NLR, MBB, Dornier and the Technical Universities of Delft and Braunschweig to start a joint experimental and theoretical program known as "The International Vortex Flow Experiment on Euler Code Validation". The experimental part of this program concentrated on the transonic flow over a 65° cropped delta wing with a rounded leading edge; other configurations (different leading edge shapes and sweep angles, the addition of a canard wing) were also investigated. The Mach number in the tests varied between 0.4 and 4. The measurements comprise force and pressure data, flow visualizations (with oil and the laser sheet technique) and flow field surveys (with LDA and 7-hole pressure probes).

In 1986 a Symposium was held in Stockholm [1] to disclose the first results of this joint program. A more recent and complete account of the

J. Zierep, H. Oertel (Eds.)
Symposium Transsonicum III
IUTAM Symposium Göttingen, 24.-27.5.1988

experimental work can be found in [2] where a description of the models, the experimental facilities and the test conditions is given, followed by an analysis of some of the observed flow phenomena. In view of the theme of this conference, we will limit ourselves to some aspects of the transonic flow over a 65° delta wing with a rounded leading edge at a free-stream Mach number of 0.85. The results have been taken from force and pressure measurements in the 1.6x2 m^2 transonic wind tunnel HST of the NLR together with LDA and laser sheet measurements in the 1x1 m^2 transonic wind tunnel TWG of the DFVLR in Göttingen.

A tentative condition for shock wave formation

Transonic flow will be defined here as flow with a free-stream Mach number less than one, but high enough to induce locally supersonic flow velocities. The flow field on the leeside of a delta wing is highly three-dimensional. When supersonic pockets are formed in this flow field such that the flow has to decelerate locally to subsonic conditions, shock waves might be formed. Intuitively [3] it is often assumed that shock wave formation is only possible when the Mach number component in a direction perpendicular to the local isobar surface exceeds a value of one (to the authors' knowledge there is no proof for this proposition). This condition is generally used in swept wing design. It will be applied here as a "working hypothesis". On the basis of this hypothesis some possible shock wave locations are indicated schematically in figure 1. These shock waves and their interaction with the vortex flow, will be the topic of this paper. In the discussion we made use of the so-called x-alpha diagram (Fig. 2) that indicates the streamwise location of a particular flow phenomenon (expressed as a fraction of the root chord) as a function of the angle of attack.

Shock waves and primary vortex formation

In figure 2 the first appearance of supersonic flow along the leading edge is indicated. Due to three-dimensional effects, the highest suction peaks are observed near the wing tip for an untwisted delta wing with a constant airfoil section. In spite of this, the pressure variation in a direction perpendicular to the leading edge is still an order of magnitude larger than its variation along the leading edge (from tip to apex). The "infinite swept wing analogy" can then be used to study the leading edge flow in detail (Fig. 3), provided that the flow is attached. The analogy is in fact a special case of the adopted "working hypothesis". In figure

4 the experimental pressure distribution at the 60% root chord station has been compared with theoretical results from a three-dimensional subsonic panel method and a two-dimensional transonic full potential inviscid method (applied as indicated in figure 3). For subsonic conditions (Fig. 4a) the good agreement between the three results supports the validity of the leading edge analogy. Note, however, that due to three-dimensional effects, the angle of attack for the two-dimensional calculations is not known a priori so that results for a range of incidences have been presented. At 12° the experimentally observed pressure distribution deviates substantially from the calculations. This is not surprising: the flow starts to separate and the assumption of attached flow is violated. When the free stream Mach number is increased to a value of 0.85 (Fig. 4b) the two-dimensional calculations indicate that the local Mach number component in a direction normal to the leading edge (M_{shock}) can exceed a value of one. In that case shock formation is possible as is in fact shown by the two-dimensional calculations. When this shock is sufficiently strong (a shock Mach number of 1.3 is often taken as a critical value [4]) the flow will separate at the foot of the shock. These simple arguments indicate that shock-induced separation can be the cause of primary vortex formation at transonic flow conditions. Note that such a shock is not observed in the experiment. This is partly caused by a lack of pressure resolution near the leading edge [2] and partly due to the fact that as soon as separation starts, the single shock is replaced by a number of "shock-lets" on top of the separated flow region [5]. With increasing incidence, the local shock-induced separation will turn into an open separation with a leading edge vortex. The newly formed separation line gives rise to a second attachment line more inboard on the wing upper surface such that part of the flow moves away from the attachment line, under the vortex and in the direction of the leading edge.

Shock waves and secondary vortex formation

The boundary layer on the wing surface under the vortex experiences an unfavourable pressure gradient. When the vortex is sufficiently strong, the adverse pressure gradient can initiate separation. The secondary separation position determines to some extent the strength and location of the secondary vortex [6].

In transonic flow the absolute flow velocities close to the vortex can be very high. The LDA measurements reveal at Mach=0.85, α=20° local values in excess of 2. For the same condition a value of 2 can be deduced from the pressure measurements on the wing surface. When the LDA results are presented in a cross-flow plane (Fig. 5a), the cross-flow Mach number (the component in a direction perpendicular to an array through the wing apex) can locally exceed one. Such a "cross-flow supersonic pocket" is observed above the vortex and it has been suggested [7] that this region is terminated by a shock wave, similar to what has been observed at supersonic free-stream conditions [8]. According to the "working hypothesis" on the likelyhood of shock wave formation, this is not to be expected. In figure 6 a conjectural isobar pattern is sketched. It illustrates that the velocity vectors are almost aligned with the isobar surfaces above the vortex.

Underneath the vortex the cross-flow Mach number is also supersonic. The LDA measurements give some indication of this (Fig. 5a). Moreover, an estimate of the cross-flow Mach number on the wing surface can be obtained from an integration of the measured pressure distribution (similar to the ones shown in Fig. 7) in conical representation starting from the attachment line. The results of such a calculation [9], presented in figure 8, indicate supersonic cross flow at about 12°. Shock wave formation is then possible near the wing surface since the isobar surfaces are approximately perpendicular to the wing surface (Fig. 6). This shows that on the wing surface the cross-flow Mach number is identical to the Mach number component in a direction perpendicular to the local isobar surface. When this component exceeds a value of 1.3, shock-induced separation is likely, followed by the formation of a secondary vortex. Consequently, there are arguments to support the view that secondary vortex formation can be shock-induced at transonic conditions. A similar suggestion has been made in [10] for a highly swept wing.

Shock waves and vortex breakdown

Once primary separation has started all along the leading edge, the vortex dominated part of the flow field is fairly conical (Fig. 2). This is also indicated by the pressure distributions for the three wing stations (Fig. 7). At subsonic conditions the flow has to adjust itself to the downstream trailing edge condition and this introduces non-conical flow features as also illustrated with the (conjectured) isobar pattern (Fig. 6). In the symmetry plane the highest velocity is reached near the wing apex or somewhat further downstream (Fig. 2). From thereon the flow decelerates to trailing edge conditions. When the free-stream Mach number is sufficiently high, the flow can become locally supersonic. Unfortunately, the available pressure information is very limited in the symmetry plane: the onset of supersonic velocities can only be determined at the three pressure stations (Fig. 2). However, the LDA measurements at 20° indicate a supersonic region (Fig. 5b) that appears to be terminated by a shock wave. This shock wave was also visible on the Schlieren picture of the HST tests (Fig. 9). Such a shock wave is compatible with the "working hypothesis": for symmetry reasons the isobar surfaces near the trailing edge are perpendicular to the symmetry plane and to the wing surface. It is of interest to note here that near the trailing edge in the plane of symmetry, the local flow velocities are barely supersonic but almost normal to the isobar surfaces, whereas the flow close to the (conical) vortex is highly supersonic but almost aligned with the isobar surfaces. For this reason it is not very well possible at present to speculate about the flow structure in the intermediate region between symmetry plane and vortex but it is clear that shock wave formation in that region will be the result of a very subtle process between pressures and flow directions.

When the incidence is increased, a very interesting phenomenon has been observed in the wind tunnel at 21°. The shock wave near the 80% wing root location suddenly jumps to a much more upstream position, though followed by a second shock still near 80% chord (Fig. 2, 9). The laser sheet measurements (Fig. 10) indicate an increase in cross-sectional area of the vortex at the approximate shock locations. The pressure distributions show a sudden loss in suction peak level for the most downstream wing sections when the incidence is increased from 21° to 22° (Fig. 7). The evidence suggests a particular form of vortex breakdown. However, it is not clear at present whether the shock wave causes the vortex breakdown or the vortex breakdown initiates the sudden shock wave movement. The results do indicate a very complex coupling between shock waves and vortex breakdown at transonic conditions.

References

1. Elsenaar, A.; Eriksson, G.: Proceedings of the Symposium on the "International Vortex Flow Experiment on Euler Code Validation", Stockholm, October 1-3, 1986 (published by FFA).

2. Elsenaar, A.; Hjelmberg, L.; Bütefisch, K.A.; Bannink, W.J.: The International Vortex Flow Experiment, AGARD Symposium on "Validation of Computational Fluid Dynamics"; Lisbon, May 2-5, 1988.

3. Bickley, W.G.: Critical conditions for compressible flow, ARC R&M No 2330 (1950).

4. Délery, J.; Marvin, J.G.: Turbulent Shock Wave Boundary Layer Interaction, AGARDograph No 280 (1985).

5. Henshall, B.D.; Cash, R.F.: An experimental Investigation of Leading Edge Flow Separation from a 4 per cent Thick Two-Dimensional Biconvex Aerofoil. R&M No. 3091, 1958.

6. de Bruin, A.C.; Hoeijmakers, H.W.M.: Computation of the three-dimensional boundary layer transition and separation on a 65° swept delta wing at 20° angle of attack, in [1].

7. Schrader, K.F.; Reynolds, G.A.; Novak, C.J.: Effects of Mach Number and Reynolds Number on Leading-Edge Vortices at High Angle of Attack, AIAA paper 88-0122.

8. Miller, D.S.; Wood, R.M.: Lee-side flow over delta wings at supersonic speeds, NASA TR 2430, 1985.

9. Houtman, E.M.: Private communication.

10. Brocard, Y.; Schmitt, V.: Interaction aerodynamique entre un canard proche et une aile en flêche en écoulement transonique, ONERA TP 1980-34 (1980).

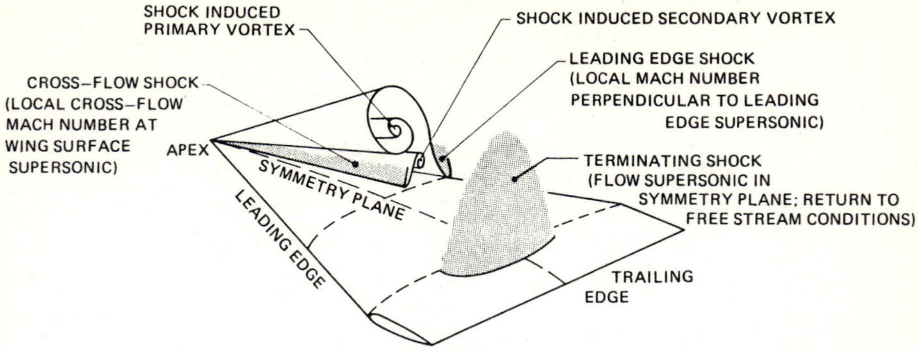

Fig. 1 Shock waves on a delta wing at transonic conditions (tentative)

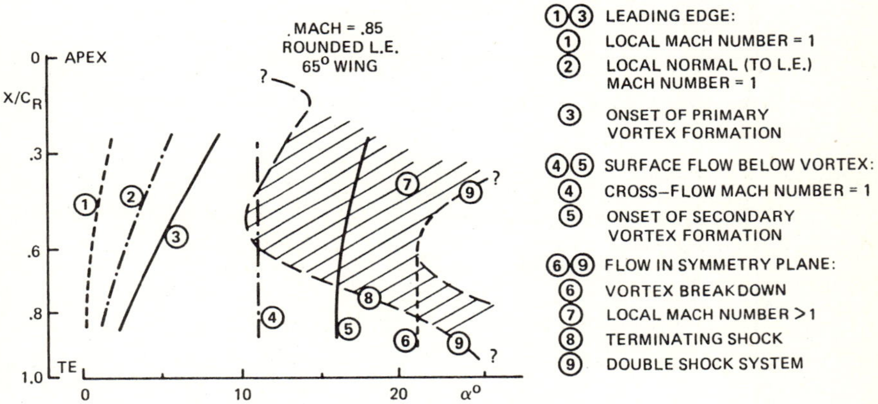

Fig. 2 Tentative view of flow development in x–α plane

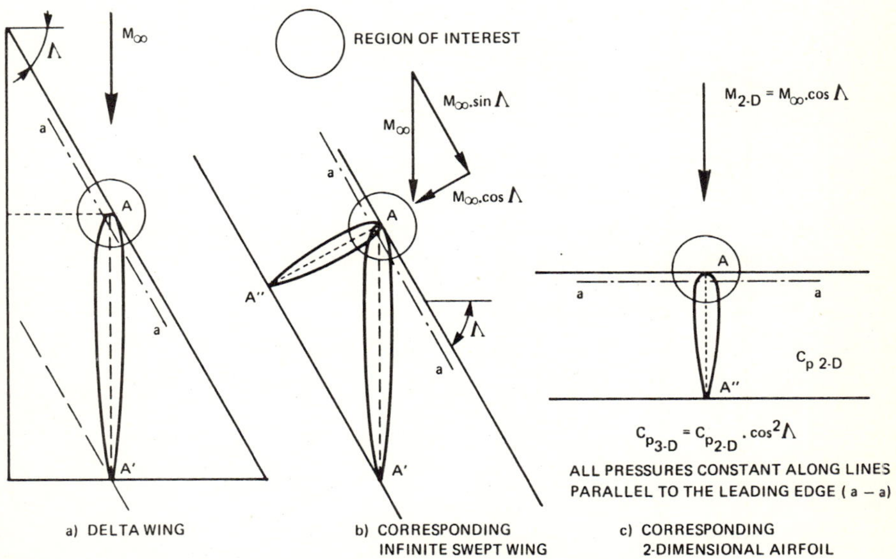

Fig. 3 Infinite swept wing analogy for leading edge flow

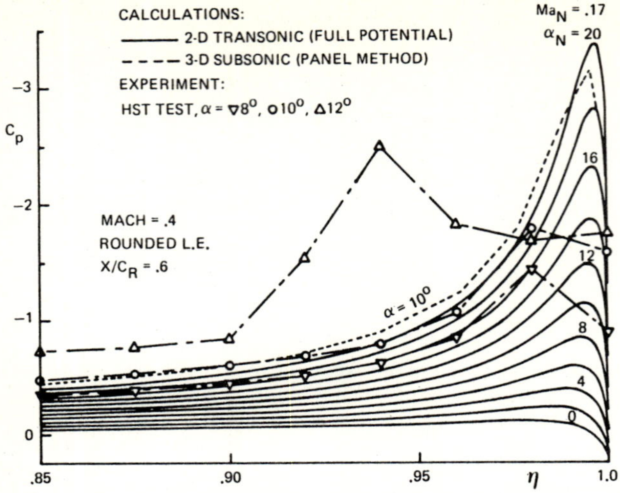

Fig. 4a Infinite swept wing analogy in leading edge region; subsonic condition

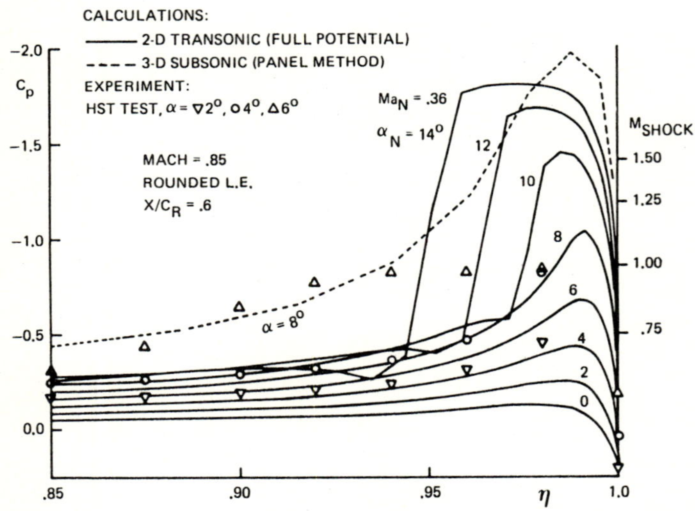

Fig. 4b Infinite swept wing analogy in leading edge region; transonic condition illustrating shock wave formation

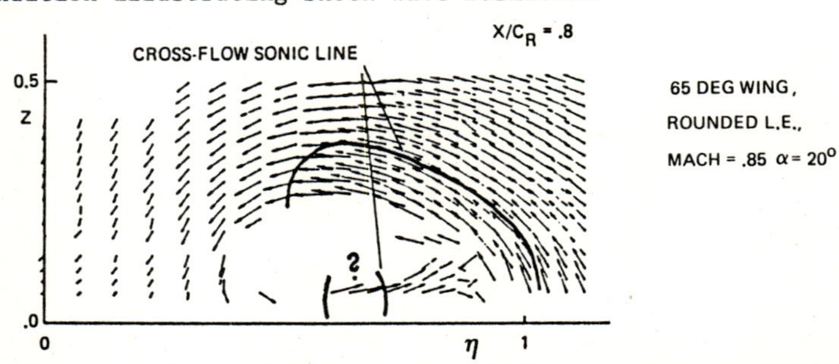

Fig. 5a Cross-flow velocity vectors (LDA measurements, DFVLR)

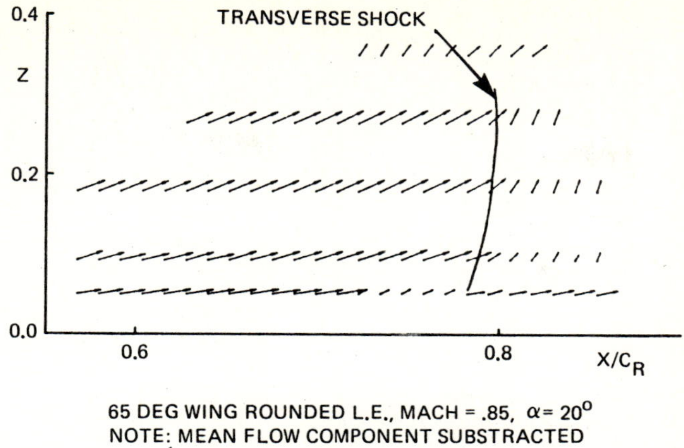

Fig. 5b Velocity vectors in plane of symmetry indicating terminating shock (LDA measurements, DFVLR)

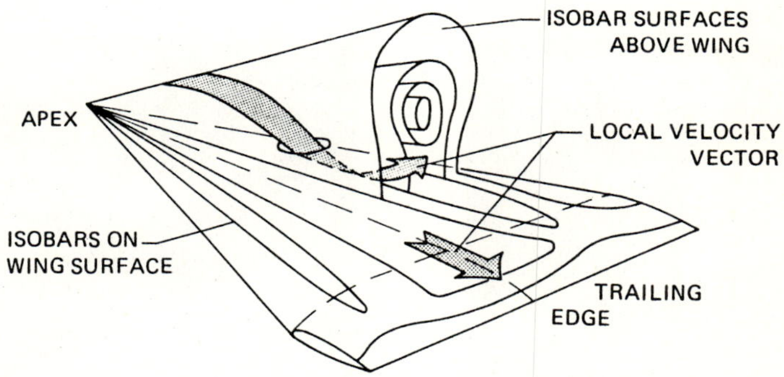

Fig. 6 Isobar surfaces on a delta wing at transonic conditions (tentative)

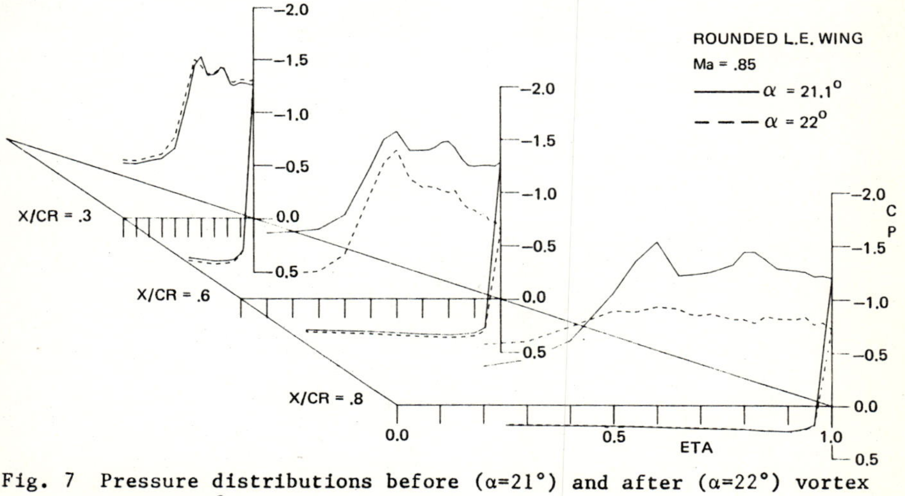

Fig. 7 Pressure distributions before (α=21°) and after (α=22°) vortex breakdown

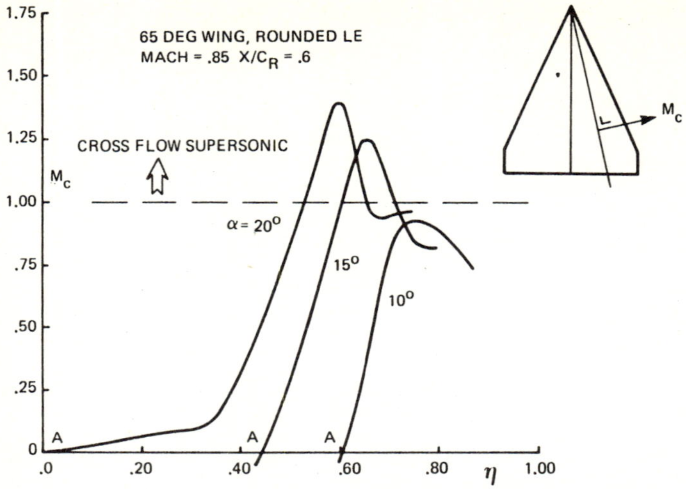

Fig. 8 Calculated surface cross-flow Mach numbers from measured pressure distributions (according to Ref. 9)

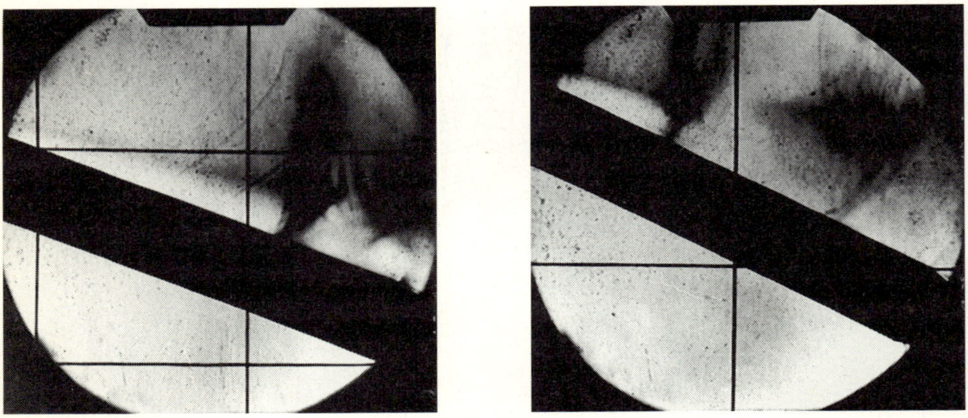

Fig. 9 Schlieren view showing shock wave displacement related to vortex breakdown (NLR tests)

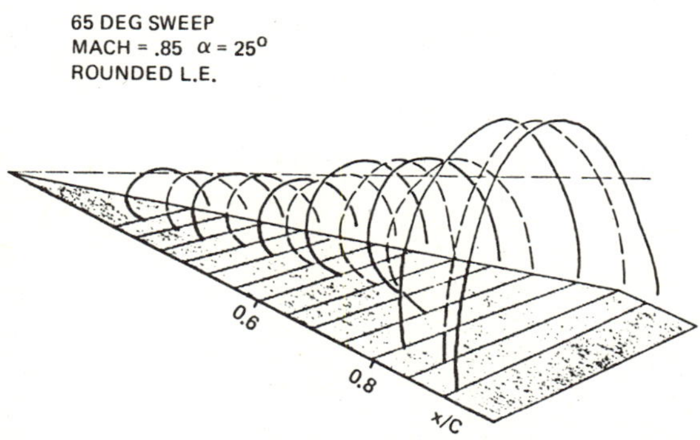

Fig. 10 Cross-sectional area of the vortex indicating vortex breakdown (laser sheet measurements, DFVLR)

Computations Relating to the Vortex Flow Experiment

S.M. Hitzel, B. Wagner

DORNIER GmbH
Numerische Strömungsmechanik
D-7990 Friedrichshafen, FRG

Summary

For slender wings the lift generated by vortex flows emerging from the leading edges may be essential for the aircraft performance and maneuverability of modern aircraft. The wing leading-edge sweep and shape control the vortex formation and therefore the aerodynamic characteristics. Depending on the sweep, vortex-breakdown as a limiter of concentrated vortices may destroy most of the vortices' advantages at high angle-of-attack by severe changes in the aerodynamic forces and moments. In the early eighties Euler methods became available for predicting these effects in the transonic and supersonic speed ranges. The "International Vortex Flow Experiment" has been set up to provide experimental data for validating Euler-codes, since the previous data basis was very poor for high speeds. This paper contributes some Euler-results obtained by the solution of time-dependent Euler-equations with a finite-volume time-stepping approach for the sharp-edged delta wing of the Vortex Flow Experiment. Especially transonic effects are discussed including their influences on vortex-breakdown. Also some attention is given to the problem of high pressure losses in the vortex cores.

Wing Vortex Flows and the Euler Simulation

To meet the demands of performance and maneuverability on modern flying vehicles a slender wing design is preferred [1]. The wing leading edges selected are often quite sharp, causing flow separation there even at low incidence. This separation forms longitudinal vortices which dominate the lift-distribution by distinct suction-peaks on the wing upper surface. The wing sweep and the leading-edge shape control the vortex configuration and therefore the wing's aerodynamic behaviour. High angle-of-attack flight of slender wing configurations may be influenced severely by the so-called vortex-breakdown phenomenon [2, 3, 4]. Vortex-breakdown means rapid change of the vortex structure and diminishing of the induced suction forces.When occurring above the wing, it affects lift and drag as well as the pitching moment of the aircraft [5]. The leading-edge vortices also cause viscous secondary vortex separation underneath themselves due to adverse pressure gradients between primary vortex and leading edge. They have a displacement effect on the leading-edge vortex depending on the Reynolds number. Behind the trailing edge a trailing-edge vortex is formed by a roll up of the trailing vortex sheet induced by the primary vortices. Hence, the formation of this vortex is a nonviscous effect [6].

Since the leading-edge vortex phenomenon had been detected [7] many trials were made to understand the vortex flowfields either experimentally or by theoretical approaches. An overview on early computational methods is given in reference [8]. These approaches are characterized by modelling the vortices and the vorticity layers using the formulation of the subsonic potential flow theory and searching for

J. Zierep, H. Oertel (Eds.)
Symposium Transsonicum III
IUTAM Symposium Göttingen, 24.-27. 5. 1988

equilibrium positions of the vortex system. First some methods for conical flow were developed, later the corresponding ideas were transferred successfully to 3-D panel-methods including those of higher order type [9].

A great success were the first Euler solutions in the early eighties which exhibited leading-edge vortex flows automatically without any modelling conditions for separation. According to HELMHOLTZ [10] such type of separation may occur at any sharp-edged body without the need of viscous effects. As described also by VON KARMAN [11],separation and the shedding of vortex-sheets may be caused by viscous effects and/or by discontinuities such as aerodynamically sharp edges. Euler-solutions using a mesh trimmed to a very high resolution of such a "round edge" possibly may not show any separation at sufficiently low subsonic speeds except a Kutta-condition is introduced explicitly. However, it is known from experiments [e.g. 12, 13, 14] that leading-edge vortices at slender wings hardly depend on Reynolds number. Thus, on engineering and economic reasoning a simulation of the primary leading-edge vortex flow with Euler-methods is worthwhile. Usually, the leading-edge roundness of existing designs [15] is small compared to the chord-length, making the assumption of sharp edges reasonable for a first guess of the vortex flow configurations.

Secondary separation naturally may never be shown in Euler-results since it depends on viscous effects. Since those secondary vortices may have some importance on the primary leading-edge vortex position especially at laminar conditions, they only may be neglected at high Reynolds numbers and at turbulent conditions where the influence of secondary separations may become small. In case of drooped leading edges or vortex-flaps the secondary vortices may generally be of more importance [15, 16].

According to experiments [17] the increase of the Mach number even lowers the Reynolds number effects on primary separation. At transonic and supersonic speeds an additional effect may cause vorticity due to CROCCO's law by shocks of variable strength. Any shock providing a gradient of entropy will result in vorticity.

The total-pressure losses found especially in the vortex cores for the Euler-results even if not connected with a shock are not yet fully understood. BROWN [18] developed the idea of a compressible slender conical vortex with constant entropy neglecting all viscous effects. In this 3-D flow the vorticity vector everywhere is parallel to the velocity-vector providing a flow which will not show total-pressure losses. This vortex core flowfield is compatible to the inviscid, incompressible outer vortex solutions of HALL [19].

However, Euler-simulated vortices usually exhibit very high total-pressure losses in the vortex core. SMITH [20] concluded that these losses must have been introduced by numerical errors caused by discretization and artifical dissipation. Keeping in mind BROWN's vortex model this effect is not very surprising. In her model finite flow values but infinite gradients are found on the vortex axis which may not be simulated due to the insufficient geometrical resolution of the computational grids. In reference 20 this effect can be seen by comparing the results for a delta wing in a coarse grid with those in a fine one. The strength of the total-pressure losses in the core essentially depends on the wing geometry and the flow conditions but not on details of the numerical code employed [21]. Thus,SMITH

[22] summarized that the total-pressure losses have only limited influence on the validity of the computational results.

Vortex-breakdown primarily depends on pressure gradients and therefore it is also almost independent of viscous effects [23]. Hence,it is not surprising that it can be simulated by Euler-methods [e.g. 24, 25, 26]. The breakdown positions as well as the basic flowfield and its influence on the aerodynamic forces and moments qualitatively agree with experiments.

Euler-methods may simulate some of the most important leading-edge vortex features discussed including all compressible effects such as shocks and shock vortex flow interactions. Although Navier-Stokes methods [e.g. 27, 28] now offer the calculation of the natural evolution of leading-edge vortices, they are not yet optimally used for the leading-edge vortex problems. Apart from the surface boundary layer also the shear layer and the vortex core have to be resolved using mesh adaptation and/or local grid refinement to exploit all benefits from the Navier-Stokes approach.

Euler Simulations

All results shown in this paper were obtained by the solution of the time-dependent Euler-equations in an explict time-stepping procedure of Runge-Kutta type as described by JAMESON and SCHMIDT [29]. The 3-D version has been further developed [30] and includes a block-structured approach. Using this version and combining it with a new mesh generator the Euler simulated results shown here were calculated [31]. The computations were performed primarily in an H-O-mesh type. Those results for the C-H-mesh are already described in reference 26. The basic H-O-mesh contained approximately 270.000 volumes: in I-direction 132, in J-direction 32 and in K-direction 64 volumes. On the wing lower and upper surface 80 volumes in lengthwise and 32 volumes in spanwise direction were distributed. In normal direction 32 volumes were used. Also a so-called superfine mesh with 540.000 (132*64*64) volumes was tested. Throughout all cases in the numerical scheme a three-step Runge-Kutta type algorithm has been used employing the same filter and enthalpy forcing terms and the same multi-grid parameters for all cases. The results were obtained by the multi-level strategy starting in a coarse mesh which was refined after a set number of computational cycles till the finest mesh was reached.

Figure 1 shows the sharp-edged version of the Vortex Flow Experiment delta wing. For the results shown, the support-fuselage was omitted in the numerical evaluations for reasons of simplicity. The fuselage had only a negligible influence on the vortex flowfield. The C-H-mesh results were dicussed elsewhere [26, 31, 32] and will not be repeated here.

A first insight into the effects of the compressibility on leading-edge vortex flows is provided by the Mach number comparison of figure 2. It depicts the isobars of the 10° angle-of-attack cases. The higher the Mach number the lower the upstream influence of the trailing edge, the more conical the flow looks like. Also the relative suction-peaks of the pressure coefficient Cp become lower, while the physical pressure differences are growing. The vorticity contours (fig. 3) do not show essential differences if the rather arbitrary factor M_∞ is removed by dividing the values by the free stream Mach number. The increasing pressure differences result in lower densities on the upper wing surfaces (fig. 4). The Mach numbers induced on the wing (fig. 2) can be considerably higher than the free-stream Mach number.

Figures 5 and 6 show the Cp-distribution for the Mach numbers 0.85 and 1.2 at 10° angle-of-attack in comparison with the experiments done at the NLR [33]. In all cases the Euler-results in the H-O-mesh show a reasonable simulation of the vortex influence. The suction peaks of Euler-results usually are somewhat higher and the vortices somewhat more outboard compared to the experiment. This is due to the neglect of the viscous secondary separations. At supersonic conditions the experimental results in figure 6 show a steep gradient in the Cp-distribution on the inboard side of the vortex which is not simulated in the Euler-result. This gradient sometimes is supposed to be due to a so-called cross-flow shock but this interpretation appears to be rather doubtful [17, 34]. Figure 7 shows total-pressure losses typical for the calculations. Close by and almost parallel to the wing leading edge high losses can be seen. These losses may indicate a cross-flow shock which, however, may be a numerical substitute replacing some effects of the secondary vortices in reality. A mechanism feeding the total pressure losses into the vortex core cannot be observed. However, within the core their values agree surprisingly well with some measurements [32].

A survey of the flowfield at Mach = 0.85 and $\alpha = 10°$ is given in figure 8. In the vortex the Mach numbers are considerably higher than the free-stream Mach numbers. The patterns of the isobars, iso-Mach contours and total-pressure losses is typical for all other Mach number and angle-of-attack cases below vortex-breakdown conditions. Figure 9 shows reasonable agreement between the Euler-simulations and experiments also for higher incidence. Figure 10 presents the lift versus the angle-of-attack at Mach = 0.85. Here, the tendency of Euler-methods to overpredict the total lift in fine meshes can be seen. A comparison of different mesh refinements in H-O- and C-H-topology results in increasing lift with increasing resolution for both topologies.

At very high angle-of-attack vortex-breakdown occurs above the wing (fig.10). The Euler-simulation follows the same trend, also simulating vortex-breakdown above the wing as shown in the isobar patterns of figure 11. As in the experiments [34], vortex-breakdown crosses the trailing edge at 20° angle-of-attack. At higher angle-of-attack ($\alpha = 24°$) it is situated above the rear part of the delta wing. Figure 12 shows the distinct change of the Cp-distribution due to vortex-breakdown.

A more detailed investigation of the flowfield at the transonic vortex-breakdown conditions (fig. 13) shows the typical axial flow reversal in a surface above the wing. The corresponding iso-Mach contours exhibit a shock close to the centerline, while in surfaces above the wing a shock near the vortex-breakdown position can be seen. Figure 14 shows the same situation in the so-called superfine mesh (increased resolution in the direction normal to the surface). Here, the flow within - but only within - the breakdown flowfield became instationary during the computational cycles. Despite using a local time-stepping procedure and other convergence acceleration features this is a qualitative indication for the unsteadiness observed in vortex-breakdown regions of corresponding experiments. The breakdown position here has moved somewhat forward, which may be due to the flow sensitivity as described also in experiments [34]. Figure 15 reveals a very complex flowfield which contains some features also found in the experimental work. On and above the wing front part the results are very similar to those in figure 9. In the rear part of the wing all variables are considerably different due to the breakdown flowfield. The breakdown bubble as indicated on the starboard side is connected with very low Mach numbers in the rear part. The isobar pattern shows traces of a quite complex shock system connected with the vortex-breakdown flowfield. Reference 31 gives more details also for subsonic vortex-breakdown, while reference 34 gives the experimental findings and their explanation very similar to the theoretical results.

Considerations on the Structure of the Vortex Cores

High pressure losses must not be expected at the vortex cores in Euler solutions since the formation of vortices is also possible in flows which remain homentropic and homenergetic. Then the vorticity vector everywhere must be aligned with the velocity vector (Crocco's law) or the vorticity must be zero. Besides the conical solutions of Brown [18] this fact can easily be studied for the limiting case of cylindrical vortices. Using the cylindrical coordinates x, ϕ, r pointing in the axial, circumferential and radial directions, respectively, and the corresponding velocity components u, v, w the momentum equations reduce to the radial one, if no radial velocity and no gradients in axial and circumferential directions are admitted (conditions for cylindrical vortices),

$$\frac{v^2}{r} = \frac{1}{\rho}\frac{\partial p}{\partial r} \tag{1}$$

with density ρ and pressure p while the continuity equation also is fulfilled. A usual relation of thermodynamics connects enthalpy h and pressure changes at constant entropy s by

$$\left(\frac{\partial h}{\partial p}\right)_s = \frac{1}{\rho}$$

Thus, for homentropic flow we get from eq. (1) the following differential equation

$$\frac{v^2}{r} = \frac{dh}{dr} \tag{2}$$

which easily can be integrated to find h if v is known as a function of r. In case u is known as a function of r we can eliminate v using the energy equation

$$h_t = h + \frac{1}{2}(u^2 + v^2) \tag{3}$$

and thus obtain the general linear differential equation of first order

$$\frac{dh}{dr} + \frac{2}{r}h = \frac{2}{r}h_t - \frac{1}{r}u^2 \tag{4}$$

which generally can be solved with the integrating factor r^2, even if also the total enthalpy h_t is a function of r. If we restrict ourselves to homenergetic flow (h_t = const.) we get from equations (2) and (4) the alternate formulas (using again eq. (3) and assuming $v_a = v(0) = 0$)

$$u^2 = u_a^2 - v^2 - 2\int_0^r \frac{v^2}{r}dr \tag{5a}$$

$$v^2 = -u^2 + \frac{2}{r^2}\int_0^r ru^2 dr - \frac{2C}{r^2} \tag{5b}$$

with the subscript "a" indicating values at the vortex axis. The last term of eq. (5b) with the arbitrary integration constant C shows that in v^2 always a singular term can be added which corresponds to that of a 2-D potential vortex.

Choosing zero for the constant C,the equations (5) represent two alternate formulas for constructing the velocity profiles of those vortices by prescribing one of the velocity components arbitrarily. Surprisingly the formulas (5) do not exhibit a compressibility influence. Hence,they are valid for all speeds including the incompressible case. By fixing the enthalpy at the axis h_a the Mach number profile is determined. A vacuum core is avoided choosing a positive value for h_a, provided the behaviour of v is nonsingular there, i.e. like a positive power of r. If this power is greater than 1/2 also a kink or cusp in u cannot occur. Finally, also for the axial vorticity component

$$\omega_x = \frac{1}{r}\frac{\partial(rv)}{\partial r} = \frac{\partial v}{\partial r} + \frac{v}{r} \tag{6}$$

a naturally smooth behaviour at the axis can be achieved by the condition

$$\frac{\partial \omega_x}{\partial r} = \frac{\partial^2 v}{\partial r^2} + \frac{1}{r}\frac{\partial v}{\partial r} - \frac{v}{r^2} \rightarrow_{r \rightarrow 0} 0 \tag{7}$$

which will be fulfilled by the particular solution $v \sim r$, i.e. rigid body rotation. According to the theory of linear differential equations of second order the second particular solution can be constructed being $v \sim r^{-1}$, the potential 2-D vortex which we will exclude because of its singularity.

Thus, the most natural solution of the inviscid compressible vortex problem seems to be a vortex with rigid body rotation behaviour at its axis in contrast to the conical vortices of Brown [18]. Of course, the gradient of v at the axis must increase with increasing strength of the vortex if the size of the vortex is not sufficiently enlarged at the same time. Since such homenergetic, homentropic vortices could be consistent with Euler solutions in cases where no dissipative mechanisms as e.g. shocks are present, the leading-edge vortices may be supposed to be of similar structure. Then, the only explanation for the high pressure losses observed in the cores is offered by the insufficient resolution of the velocity gradients there by the computational meshes. The increase of the losses with angle of attack is consistent with this supposition because of the growing gradients at the axis while using the same meshes. Nevertheless, the good agreement of total pressure losses calculated with some measured values is still very surprising.

References

[1] Polhamus E.C., "Applying Slender Wing Benefits to Military Aircraft", Journal of Aircraft , Volume 21, No. 8, August 1984

[2] Peckham D.H., Atkinson S.A., "Preliminary Results of Low Speed Wind Tunnel Tests on a Gothic Wing of Aspect Ratio 1.0", ARC CP 508, 1957

[3] Hummel D., "Untersuchungen über das Aufplatzen der Wirbel an schlanken Deltaflügeln", Zeitschrift für Flugwissenschaften , Verlag Friedrich Vieweg & Sohn, Braunschweig, Volume 5, 1965, pp 158-168

[4] Wentz W.H., Kohlman D.L., "Wind-Tunnel Investigations of Vortex-Breakdown on Slender Sharp-Edged Wings", NASA CR 98737, 1968

[5] Hummel D., Redeker G.,"Über den Einfluß des Aufplatzens der Wirbel auf die aerodynamischen Beiwerte von Deltaflügeln mit kleinem Seitenverhältnis beim Schiebeflug", Jahrbuch 1967 der WGLR, 1967, pp 232-240

[6] Hummel D., "On the Vortex Formation Over a Slender Wing at Large Angles of Incidence", AGARD-FDP Symposium, Sandefjord 1978, AGARD-CPP-247, High Angle of Attack Aerodynamics, 1978, pp 15-1 - 15-17

[7] Wilson J.A., Lovell J.C., " Full-Scale Investigation of the Maximum Lift and Flow Characteristics of an Airplane Having Approximately Triangular Plan Form, NACA Research Memorandum No. L6K20, 1947

[8] Hitzel S.M., Schmidt W., "Slender Wings with Leading-Edge Vortex Separation - A Challenge for Panel-Methods and Euler-Solvers", Journal of Aircraft, Volume 21, 1984, pp 751-759, former AIAA 83-262

[9] Johnson F.T., Lu P., Tinoco E.N., Epton M.A., "An Improved Panel Method for the Solution of Three-Dimensional Leading-Edge Vortex Flows", Volume I - Theory Document, NASA CR-3278, 1980

[10] Helmholtz H., "Über Integrale der hydrodynamischen Gleichungen welche den Wirbelbewegungen entsprechen", Journal für die reine und angewandte Mathematik , Band 55, Heft 1, 1858, pp 25-55

[11] von Kàrmàn T., "The Theory of Lift", Aerodynamics, McGraw-Hill Paperback, Cornell University Press, 1954, pp 41

[12] Werle H., "Flow Visualization Techniques for the Study of High Incidence Aerodynamics", AGARD Lecture No. 121, 1982, pp 3-1 - 3-36

[13] Malcolm G.N., Nelson R.C., "Comparison of Water and Wind Tunnel Flow Visualization Results on a Generic Fighter Configuration at High Angles of Attack", AIAA paper 87-2423, Monterey, 1987

[14] Erickson G.E., Peake D.J., Del Frate J., Skow A.M., Malcolm G.N., "Water Facilities in Retrospect and Prospect. An Illuminating Tool For Vehicle Design", AGARD-CP-413, Monterey, 1986, pp 1.1-1.28

[15] Hitzel S.M., "Kompressiblität und kontrollierte Wirbelablösung, zwei Schritte zum modernen Deltaflügler, DM-1 und Nachfolger", unpublished, 1988

[16] Campbell J.F., Osborn R.F., "Leading-Edge Vortex Research: Some Nonplanar Concepts and Current Challenges", NASA Conference Publication 2416, Vortex Flow Aerodynamics, Volume I, 1985, pp 31-64

[17] Schrader K.F., Reynolds G.A., Novak C.J., " Effects of Mach Number and Reynolds Number on Leading-Edge Vortices at High Angle-of-Attack", AIAA paper 88-0122, Reno, 1988

[18] Brown S., "The Compressible Inviscid Leading-Edge Vortex", Journal of Fluid Mechanics , Volume 22, part 1, 1965, pp 17-32

[19] Hall M.G., "A Theory for the Core of a Leading-Edge Vortex", RAE Report No. AEO 2644, 1960

[20] Smith J.H.B, "Numerical Solutions for the Three-D Cases - Delta Wings", AGARD AR-211, Test Cases for Inviscid Flow Field Methods, 1985, pp 8-1 - 8-94

[21] Joint US/European Vortex Flow Symposium in Stockholm, Sweden 1986, Proceedings of the International Vortex Flow Experiment on Euler Code Validation, Stockholm, Sweden, 1986,

[22] Smith J.H.B, "Overview of the Symposium", Proceedings of the International Vortex Flow Experiment on Euler Code Validation, Stockholm, Sweden, 1986, pp 357-361

[23] Hitzel S.M., "Vortex Breakdown of Leading-Edge Vortices above Swept Wings - It's Governing Terms in the Describing Equations", DORNIER Note BF30-2576/84, 1984

[24] Hitzel S.M., Wagner B., "Untersuchungen der dreidimensionalen Vorderkantenablösung an schlanken Flügeln mittels einer Euler-Zeitschrittmethode", 4. DGLR-Symposium on separated flows, Göttingen, 10-12 October 1983

[25] Kumar A., Das A., "Numerical Solutions of Flow Fields Around Delta Wings Using Euler Equation Method", Proceedings of the International Vortex Flow Experiment on Euler Code Validation, Stockholm, Sweden, 1986, pp 175-186

[26] Hitzel S.M., Wagner B., Leicher S., "Euler-Simulation of the Vortex Flow Experiment - A Critical Consideration", Proceedings of the International Vortex Flow Experiment on Euler Code Validation, Stockholm, Sweden, 1986, pp 281-288

[27] Müller B., Rizzi A, "Navier-Stokes Simulation of Laminar Flow Over the 65° Round Leading-Edge Delta Wing at $M = 0.85$ and $\alpha = 10°$", Proceedings of the International Vortex Flow Experiment on Euler Code Validation, Stockholm, Sweden, 1986, pp 269-280

[28] Rieger H., Jameson A., "Solution of Steady Three-Dimensional Compressible Euler and Navier-Stokes Equations by an Implicit LU sheme", AIAA paper 88-0619, Reno, 1988

[29] Jameson A., Schmidt W., "Some Recent Development in Numerical Methods for Transonic Flows", Computational Methods in Applied Mechanics and Engineering 51, 1985, pp 467-493

[30] Leicher S., "Numerical Solution of Internal and External Inviscid and Viscous 3-D Flow Fields", AGARD Application of Computational Fluid Dynamics in Aeronautics 14, Aix-en-Provence France, 1986

[31] Hitzel S.M., "Wing Vortex-Flows Up Into Vortex-Breakdown - A Numerical Simulation", AIAA paper 88-2518, 6th. Applied Aerodynamics Conference, Williamsburg, Virginia, June 1988

[32] Wagner B., Hitzel, S.M., Schmatz M.A., Schwarz W., Hilgenstock A., Scherr S., "Status of CFD Validation on the Vortex Flow Experiment", AGARD Fluid Dynamics Symposium on Validation of Computational Fluid Dynamics, Lisbon, Portugal, May 1988, pp 10.1-10.10

[33] Boerson S.J., Elsenaar A., "Test of the AFWAL 65° Delta Wing at NLR: A Study of Vortex Flow Development between Mach = 0.4 and 4.0", Proceedings of the International Vortex Flow Experiment on Euler Code Validation, Stockholm, Sweden, 1986, pp 23-36

[34] Elsenaar A., Hjelmberg L., Bütefisch K., Bannink W.J., "The International Vortex Flow Experiment", AGARD Fluid Dynamics Symposium on Validation of Computational Fluid Dynamics, Lisbon, Portugal, May 1988

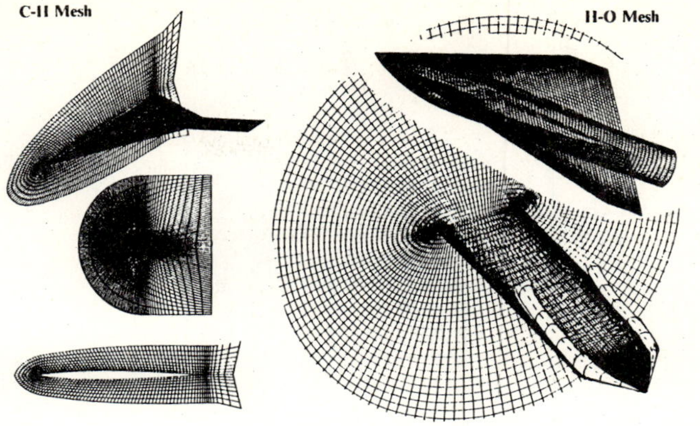

Fig. 1 Euler-meshes of the Vortex Flow Experiment delta wing

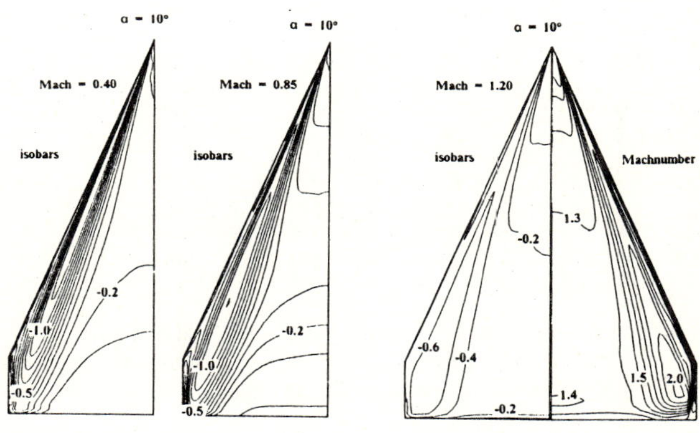

Fig. 2 The delta wing at different Mach numbers at α = 10°

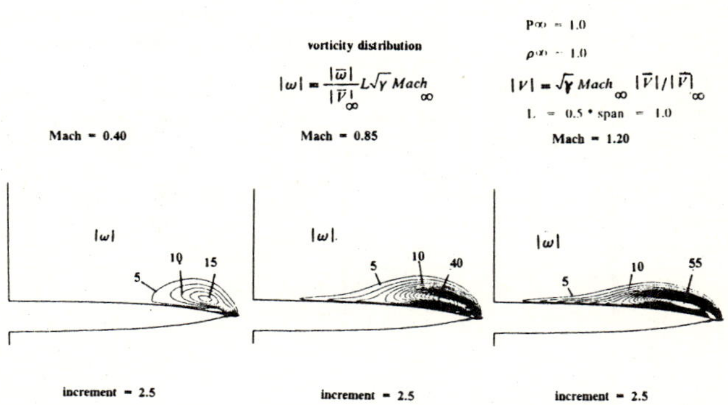

Fig. 3 The vorticity at X/C = 0.6 at α = 10° above the delta wing

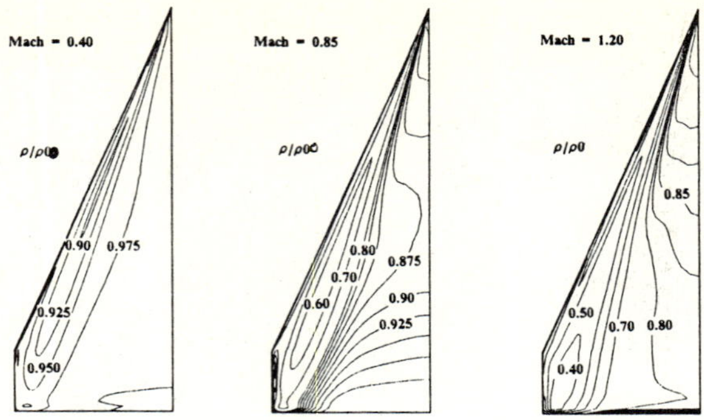

Fig. 4 The density on the delta wing at α = 10°

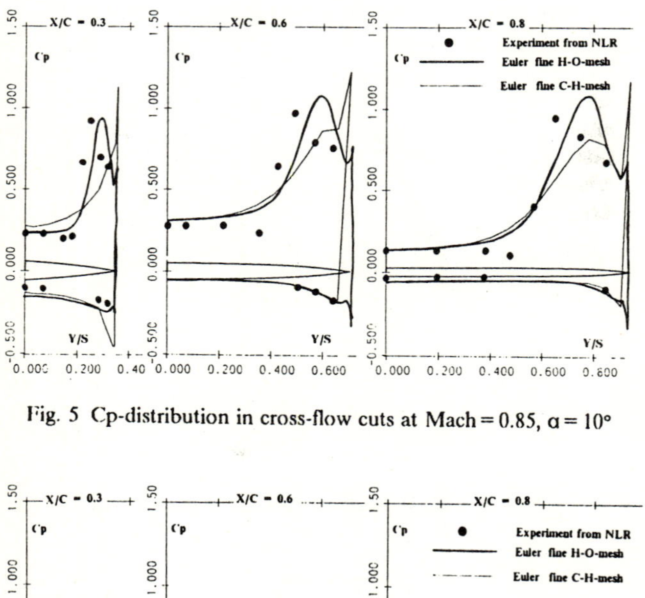

Fig. 5 Cp-distribution in cross-flow cuts at Mach = 0.85, α = 10°

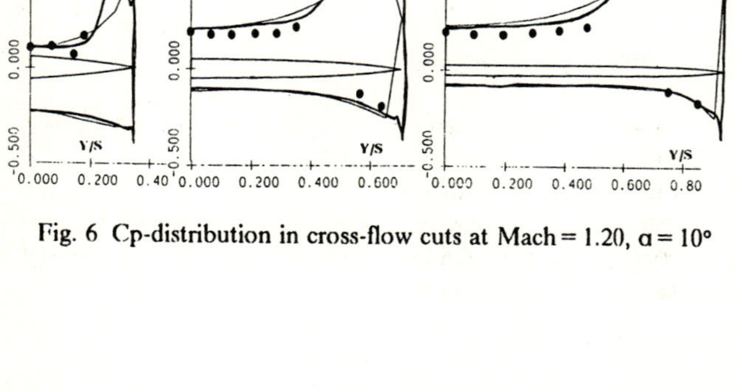

Fig. 6 Cp-distribution in cross-flow cuts at Mach = 1.20, α = 10°

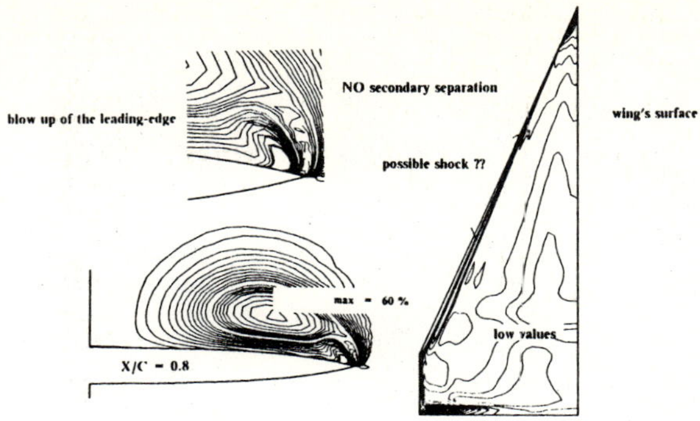

Fig. 7 Total-pressure losses at Mach = 0.85, α = 20°

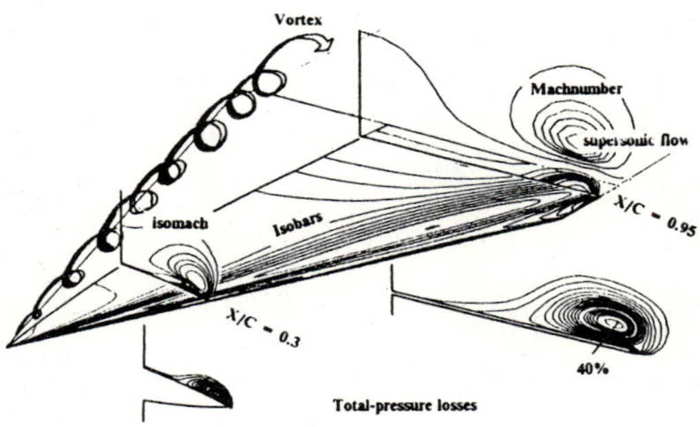

Fig. 8 The flow structure at Mach = 0.85, α = 10°

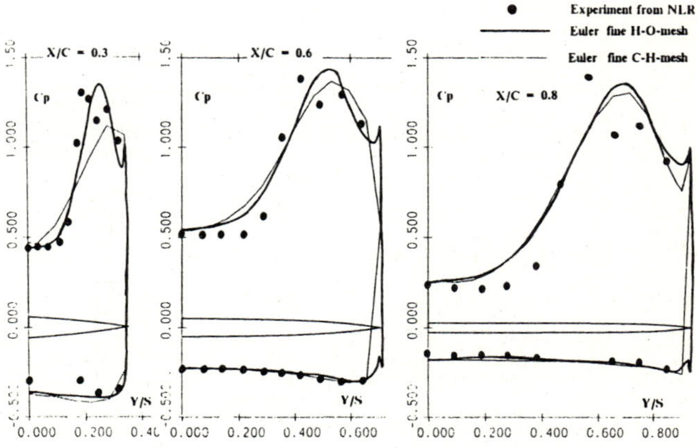

Fig. 9 Cp-distribution in cross-flow cuts at Mach = 0.85, α = 18°

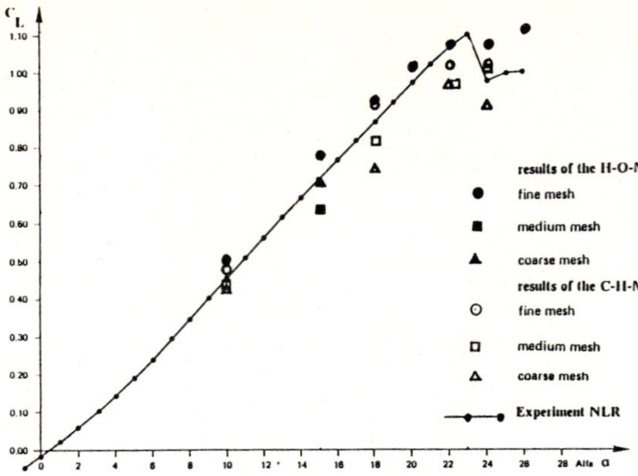

Fig. 10 The lift versus angle-of-attack at Mach = 0.85 of the delta wing

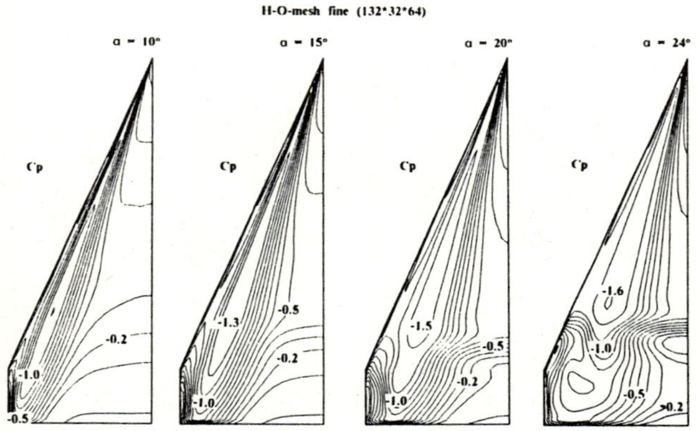

Fig. 11 Isobars on the delta wing at Mach = 0.85 at different α

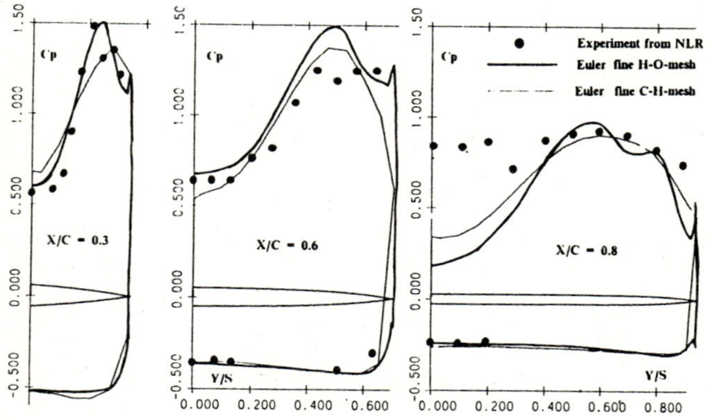

Fig. 12 Cp-distribution in cross-flow cuts at Mach = 0.85, α = 24°

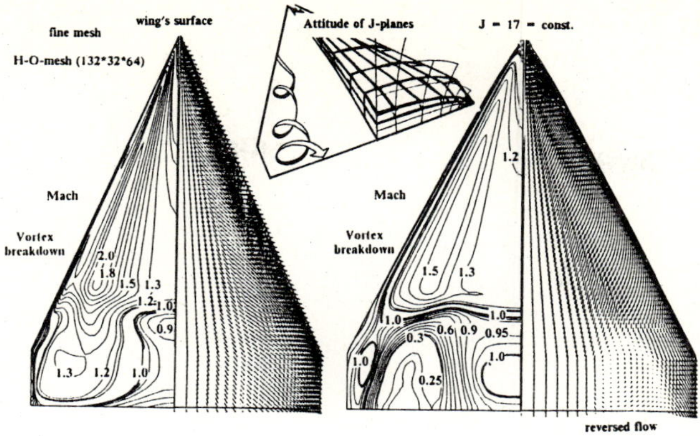

Fig. 13 Vortex-breakdown flowfield at Mach = 0.85, α = 24°
in the fine mesh (132*32*64)

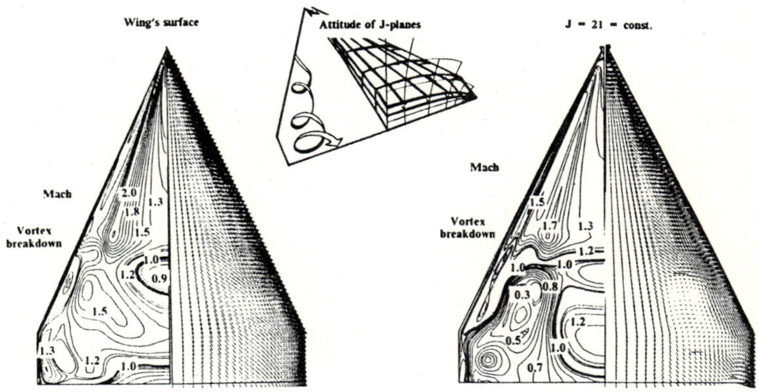

Fig. 14 Vortex-breakdown flowfield at Mach = 0.85, α = 24°
in the superfine mesh (132*64*64)

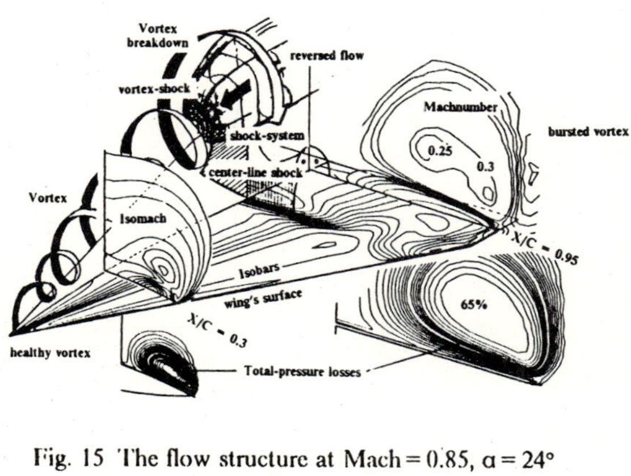

Fig. 15 The flow structure at Mach = 0.85, α = 24°

Navier-Stokes and Euler Solutions for Vortex Flow over a Delta Wing

A. Rizzi*, G. Drougge* and B. Müller**

* FFA, The Aeoronautical Research Institute of Sweden, S-161 11 Bromma, SWEDEN

** DFVLR - Institute for Theoretical Fluid Mechanics D-3400 Göttingen, FRG

Summary

Large-scale Navier-Stokes solutions on a mesh of 128x48x64 cells for transonic flows $M_\infty = 0.85$, $\alpha = 2^o$, 20°, and 24° and $Re_\infty = 2.38 \cdot 10^6$ around a 65° swept delta wing with round leading edge are presented and compared with corresponding large-scale Euler solutions. The viscous results for $\alpha = 2^o$ reveal the presence of a laminar separation bubble along the leading edge. At the higher angle of attack before breakdown, $\alpha = 20^o$, the Euler solution shows the primary vortex splitting into two vortices, apparently due to the strong cross-flow shock under the vortex at $x/c = 0.8$. The Navier-Stokes solution does not indicate a splitting of the vortex. Instead of a strong cross-flow shock, it shows the boundary layer separating. The location of the vortex in each solution is quite different, and comparison with experiment [9] indicates that the Navier-Stokes solution is more realistic. At the highest angle of attack, $\alpha = 24^o$, after breakdown, the Navier-Stokes solution suggests that the secondary separation ceases at about $x/c = 0.8$. This is perhaps one telltale of breakdown.

Introduction

Numerical solution to the Euler equations is currently being proposed as a model for the problem of leading-edge separation from a delta wing and the consequent formation of a vortex over the wing [1-3]. But serious reservations arise with this approach if the edge is round because the action of viscosity in the boundary layer now determines precisely where the flow separates, and not a geometrical singularity as in the case of a sharp leading edge. The comparison with the measured surface pressure, however, is not particularly good because in the experiment the secondary vortex displaces the position of the primary vortex. One concludes that a viscous model is needed for a more realistic simulation.

We present here some of our latest results from large-scale Navier-Stokes simulations of transonic flows $M_\infty = 0.85$, $\alpha = 2^o$, 20°, and 24°, $Re_{\infty, c_R} = 2.38 \cdot 10^6$ over the round-edge delta wing of the International Vortex Flow Experiment [4], and compare them with our Euler solutions and the measured pressure distributions. This is an extension of our earlier results for $\alpha = 10^o$ [5]. Both sets of solutions for the high α cases show the development of a primary vortex sheet from the round leading edge of the wing, but their positions over the wing are different. In the viscous solution the primary vortex lies inboard and above the one seen in the Euler solution. The viscous solution contains the secondary vortex as well as other vortices located between the upper surface of the wing and the shear layer shedding from its leading edge. In the Euler solution the flow compresses under

J. Zierep, H. Oertel (Eds.)
Symposium Transsonicum III
IUTAM Symposium Göttingen, 24.-27. 5. 1988

the primary vortex and a cross-flow shock develops. The comparison with the experiment shows that the position of the primary vortex and the surface pressure under its core agree better with the Navier-Stokes solution than with the Euler solution.

Outline of Numerical Method

The Navier-Stokes equations for an arbitrary stationary cell V with boundary ∂V and outer normal unit vector $\underset{\sim}{n}$ in a Cartesian reference frame read

$$\frac{\partial}{\partial t} \int_V \underset{\sim}{q} \, dV + \int_{\partial V} \underset{\approx}{H}(\underset{\sim}{q}) \cdot \underset{\sim}{n} \, ds = 0 \tag{1}$$

The notation is standard. The column vector q is density, momentum, and total energy. The gas is air and is considered perfect. Newton's law gives the stress tensor, and Fourier's law gives the heat flux.

The viscosity coefficients μ and λ are related by Stoke's hypothesis. The Prandtl number is assumed to be constant, namely $Pr = 0.72$, thus giving a simple relationship between the thermal conductivity coefficient κ and the viscosity coefficient μ. The dependence of μ on the temperature T is obtained from the Sutherland law with the Sutherland constant $S/T_\infty = 0.375$. The Navier-Stokes equations (1) are discretized in hexahedrons using the finite-volume technique [5]. The surface integral in (1) is approximated by averaging the flux tensor in the adjacent cells. We obtain gradients of the velocity components and temperature by the gradient theorem, which yields a surface integral, that is approximated similarly to (1):

$$\text{grad } ø_P = \frac{\int_{\partial V_P} ø \, \underset{\sim}{n} \, dA}{\int_{V_P} dV} \equiv \sum_{k=1}^{6} ø_{Pk} \frac{\int_{\partial V_{Pk}} \underset{\sim}{n} \, dA}{\int_{V_P} dV} \tag{2}$$

where $ø_{Pk}$ is the average $ø_{Pk} = \frac{1}{2}(ø_P + ø_k)$ and the cells adjacent to cell P are denoted by $k = 1$ to 6.

The spatial discretization constitutes the physical difference operator $\underset{\sim}{F}_{PH}$. These central differences do not damp unphysical oscillations caused by flow discontinuities and waves with short wavelengths. The numerical damping terms $\underset{\sim}{F}_N(\underset{\sim}{q})$, which are therefore added to $\underset{\sim}{F}_{PH}(\underset{\sim}{q})$, comprise nonlinear second-order differences sensed by the discretized second order derivative of the pressure, and linear fourth-order differences of the conservative variables [6].

The resulting semi-discrete approximation of the Navier-Stokes equations

$$\frac{d\underset{\sim}{q}}{dt} = \underset{\sim}{F}(\underset{\sim}{q}) \quad \text{where} \quad \underset{\sim}{F} = \underset{\sim}{F}_{PH} + \underset{\sim}{F}_N$$

represents a large system of first-order ordinary differential equations, which is solved for the steady state by a second-order explicit three-stage Runge-Kutta scheme [5]. The no-slip condition holds on the wing surface which is assumed to be adiabatic. The pressure is obtained by $\partial p/\partial n = 0$. The pressure and the stress tensor at the wing interface of the first cell above the wing are approximated by their values in that cell. The boundary conditions at the farfield boundary are based on the theory of characteristics for locally one-dimensional inviscid flow.

Computed Solutions

A hemisphere with a radius of 3 root chord lengths is chosen as the farfield boundary. Then O-O meshes are generated for the Navier-Stokes and Euler solutions. The fine mesh for the former consists of 128, 48, and 64 grid cells in the chordwise, near normal, and spanwise directions, and for the latter, 124x36x64 cells. (Fig. 1)

Low Alpha Case: $\alpha = 2°$

The experiments suggest that a vortex is not shed from the leading edge in this case, and the Navier-Stokes simulation confirms this. The streamlines in Fig. 2 show that a shear layer is shed only from the side edge (cf. velocity vectors), but the skin friction lines do indicate a separation along the leading edge. This feature is studied further below. The comparison of computed and measured surface C_p distribution in four sections $x/c = 0.3$, 0.6, 0.8 and $y/s = 0.55$ (Fig. 3) establishes the overall good agreement of both the Euler and Navier-Stokes simulations with the experiment. The largest differences occur at the leading edge, and we examine this region by velocity vectors drawn in the plane normal to the leading edge at $x/c = 0.7$. They clearly indicate (Fig. 4) a laminar separation bubble at the leading edge which then reattaches inboard. The distribution of C_p versus y/y_{max} in the figure points out the attachment line, the position of minimum pressure, separation, and finally reattachment. The computation is fully laminar whereas the experiment may contain transition to turbulence in the bubble, which may explain the differences in the comparison in Fig. 3.

Similar to Fig. 3, Fig. 5 presents the comparison of C_p for Euler, Navier-Stokes, and measured values for the moderate angle of attack case $\alpha = 10°$ in which a leading edge vortex is shed. Clearly the comparison of measurements with Navier-Stokes values is better than it is with Euler values. This case was thoroughly analyzed before, with the conclusion that the Navier-Stokes solution is the better model for vortex flow [5]

High Alpha: $\alpha = 20°$

The overall aim of this paper is to determine if the same conclusion holds for high angle of attack cases, including vortex breakdown. We consider first $\alpha = 20°$, before breakdown occurs according to the experiments. Figure 6 compares the streamlines and velocity vectors in the $x/c = 0.95$ plane from the Euler and Navier-Stokes solutions. The vortex in the Euler solution is more disordered than the one in the Navier-Stokes solution, but the biggest difference appears in the velocity vectors. They reveal two co-rotating vortices in the Euler solution, but a single vortex in the Navier-Stokes result. The double vortex apparently results from a splitting of the primary vortex slightly ahead of the cropped edge in the Euler results. Lamar et al. [7] have observed two co-rotating vortices in a free flight experiment, but the reason why they appear is still under debate.

As a means to check the overall structure of the flow, we compare (in Fig. 7) the computed skin friction lines with the experimental oil flow pattern of Bannink [4] (model with sharp edge and at a higher Re). The locations of the secondary separation lines agree reasonably well, but surprisingly the skin friction lines turn upstream while converging towards the secondary separation line around $x/c = 0.7$ indicating a nodal point of separation. There also appears to be some structure in the oil pattern at this location which is in broad agreement with this interpretation.

One expects strong shock waves in this flow, and the upper surface isocontours in Fig. 8 present evidence for this. The comparison of C_p shows a strong cross-flow shock under the vortex in the Euler solution, but not in the Navier-Stokes results. Inboard of the vortex the flow recompresses in both solutions, perhaps through a weak shock which in turn may interact with the vortex. The interaction is stronger in the Euler than in the Navier-Stokes results. The loss in total pressure coefficient $1 - p_t/p_{t\infty}$ is greater in the Euler than in the Navier-Stokes results. The comparison of isocontours in the plane $x/c = 0.8$ over the wing (Fig. 9) confirms these overall conclusions. Notice in particular that the cross-flow shock in the Euler isobars in the $x/c = 0.6$ plane appears to produce sufficient vorticity for two vortices to appear at $x/c = 0.8$, i.e. the cross-flow shock is the cause of the second co-rotating vortex. No cross-flow shock occurs in the Navier-Stokes results, instead the boundary layer separates and leads to a secondary vortex.

The comparison of computed and measured surface C_p in Fig. 10 shows that the Navier-Stokes results agree better with the measured values than the Euler results. This suggests that the strong cross-flow shock and extreme total pressure losses may not be realistic. It would appear that the separation of the boundary layer plays an important role in alleviating the shock and the losses. Inboard at $y/s = 0$ and 0.2 the Navier-Stokes chordwise C_p distribution undergoes a supersonic-to-subsonic recompression which perhaps could be called a shock. Surprisingly it re-expands again downstream.

A central question is how well the computational methods predict the location and overall motion of the vortex. We try to answer this by the comparison of measured [8] and computed velocity vectors (Fig. 11) in the spanwise plane $x/c = 0.8$. The laser measurements were not able to determine the velocity in the core, but it is clear that the measurements agree better with the Navier-Stokes results than with the Euler results. The position of the vortex in the Euler solution is more outboard and closer to the wing than in the other two results, which, we conclude, is not realistic.

Velocity vectors can be misleading, but they are less so when they are drawn in a plane normal to the leading edge of the wing. Figure 12 presents the vectors drawn at equi-intervals, not grid points, in the plane normal to the leading edge at $x/c = 0.7$, along with the isocontours of vorticity modulus. They show the primary reattachment point, the primary vortex, and the multiple inflection points in the boundary layer profiles between the primary vortex and the leading edge.

Multiple vortices appear in this region, and the vortical structure is complex. It is similar to what we found for the $\alpha = 10°$ case. The detailed comparison of the profiles of total pressure and vorticity modulus along the dashed line drawn in the isobar contours ($x/c = 0.8$) of the Euler and Navier-Stokes solutions (Fig. 13) points out the large differences in the two results in the vortex core. The Euler profiles indicate two vortices because of the two maxima, while the Navier-Stokes profile indicates only one. The large gain (36%) in total pressure in the Euler solution is also unrealistic. The gradient by which the u profile of the Navier-Stokes solution goes to zero indicates how thin the boundary layer is on the upper and lower surfaces.

Case $\alpha = 24°$ after breakdown

The next question we address is whether we can simulate vortex breakdown. The experiment has found breakdown at $\alpha = 21°$. Here we look at the case $\alpha = 24°$, well beyond breakdown, and try to identify features in the flow that signify breakdown has occurred. Figure 14 presents the streamlines and velocity vectors at $x/c = 0.95$ from the Euler and Navier-Stokes solutions. Only a single primary vortex is apparent now, but as with the case $\alpha = 20°$, the Navier-Stokes vortex is less disordered than the Euler vortex. The skin friction lines show a nodal point of separation and possibly a saddle, and apparently the secondary separation vanishes at $x/c = 0.7$. The velocity vectors in the plane normal to the leading edge at $x/c = 0.7$ confirm these conclusions (Fig. 15). We can only conjecture that this effect might be a product of the breakdown process. The comparison of computed surface C_p with experiment (Fig. 16) also shows that the agreement is better with the Navier-Stokes results than with the Euler values.

Concluding Remarks

We have studied the simulation of flow around a 65° swept delta wing at both low and high angles of attack. At $\alpha = 2°$ the viscous simulation predicts a laminar separation at the leading edge which does not develop into a vortex. At the higher angle of attack $\alpha = 20°$ a vortex is found in both the Euler and the Navier-Stokes simulations. A strong cross-flow shock occurs in the Euler solution, while in the Navier-Stokes solution the boundary layer separates. There also may be an interaction between the primary vortex and a shock that starts at the symmetry plane and extends outboard towards the vortex. The agreement with experiment is better for the Navier-Stokes results. At $\alpha = 24°$ breakdown has taken place and in general the agreement between measurements and the Navier-Stokes solution is worse, although it is still better than the comparison with the Euler results. The secondary separation disappears in the Navier-Stokes solution at about $x/c = 0.7$, and two nodal points of separation are seen in the skin friction lines. We conjecture that this is an effect of breakdown.

Acknowledgements

Thanks are due to a number of people: Charles Purcell, ETA Systems, for producing the Euler solutions. Per Trydell, FFA, for his help in getting the post-processing system to function, and Peter Sacher, MBB, for suggesting the $\alpha = 2°$ case.

References

[1] Smith, J.H.B.: "Vortex Flows in Aerodynamics", Ann. Rev. Fluid Mech., Vol. 18, 1986, pp. 221-242.

[2] Newsome, R.W., and Kandil, O.A.: "Vortical Flow Aerodynamics - Physical Aspects and Numerical Simulation", AIAA Paper 87-0205, Jan. 1987.

[3] Rizzi, A., and Purcell, C.J.: "On the Computation of Transonic Leading-Edge Vortices Using the Euler Equations", J. Fluid Mech., Vol. 181, 1987, pp. 163-195.

[4] Elsenaar, A., and Eriksson, G. (eds), Proc. Intl. Vortex Flow Experiment on Euler Code Validation, FFA Rep., Stockholm, 1986.

[5] Rizzi, A., and Müller, B.: "Comparison of Euler and Navier-Stokes Solutions for Vortex Flow over a Delta Wing", Aero J., April 1988, pp. 145-153

[6] Rizzi, A., and Eriksson, L.-E.: "Computation of Flow around Wings Based on the Euler Equations". J. Fluid Mech., Vol. 148, 1984, pp.45-71

[7] Lamar, J.E., Hallissay, J.B., Frink, N.T., Smith, R.H., Johnson, Jr. T.D. Jr. T.D., Pao, J-L, and Ghaffari, F.: "Review of Vortex Flow Flight Projects on the F-1068", AIAA Paper 87-2346 CP, 1987.

[8] Bütefisch, K.A., Pallek, D., and Sauerland, K.H.: "Three Component LDA Measurements on a 65° Delta Wing", DFVLR Interner Bericht 222-87A34, 1987.

[9] Hartmann, K.: "US / European Transonic Vortex Flow Experiment - Data Lists of Pressure Measurements", DFVLR Interner Bericht 222-86A26, 1986.

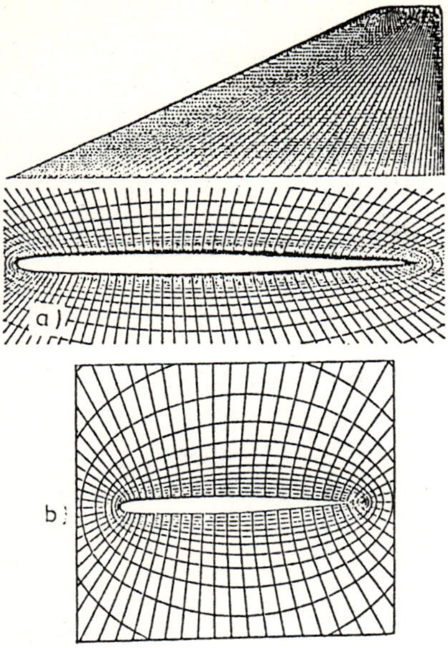

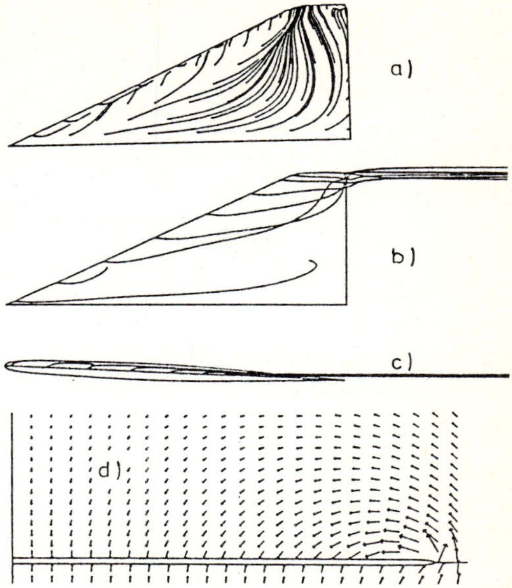

Fig. 1 Surface mesh and chordwise plot of a) Navier-Stokes mesh 129x49x65 and b) Euler mesh 125x37x65

Fig. 2 a) Skin friction lines and b), c), and d) streamlines of the Navier-Stokes solution for $\alpha = 2^o$, $x/C_R = 0.95$

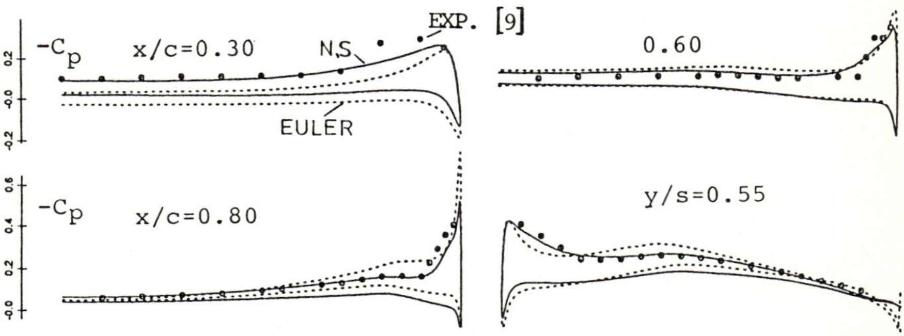

Fig. 3 Comparison of measured and computed C_p distributions of Navier-Stokes and Euler solutions for $\alpha = 2^o$

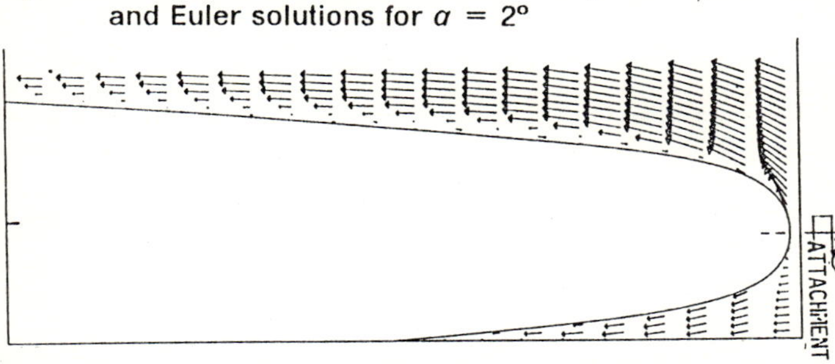

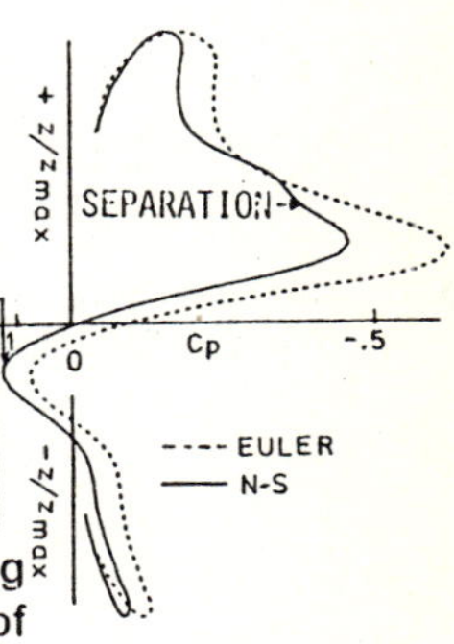

Fig. 4 Velocity vectors in the plane normal to the leading edge at $x/c = 0.7$ and the distributions of C_p versus z/z_{max} of the Navier-Stokes and Euler solutions for $\alpha = 2^o$

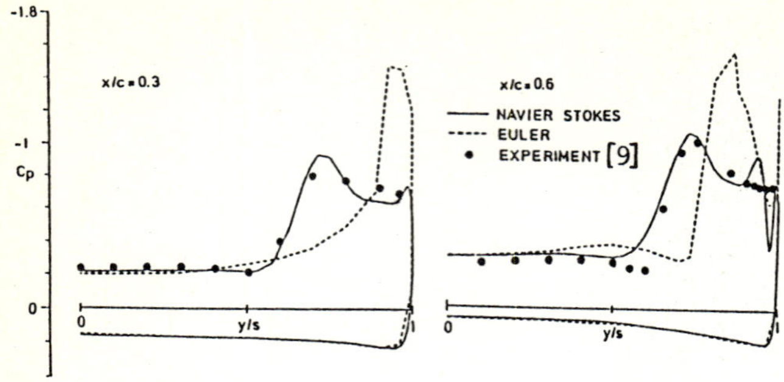

Fig. 5 Comparison of measured and computed C_p distributions of Navier-Stokes and Euler solutions for $\alpha = 10°$

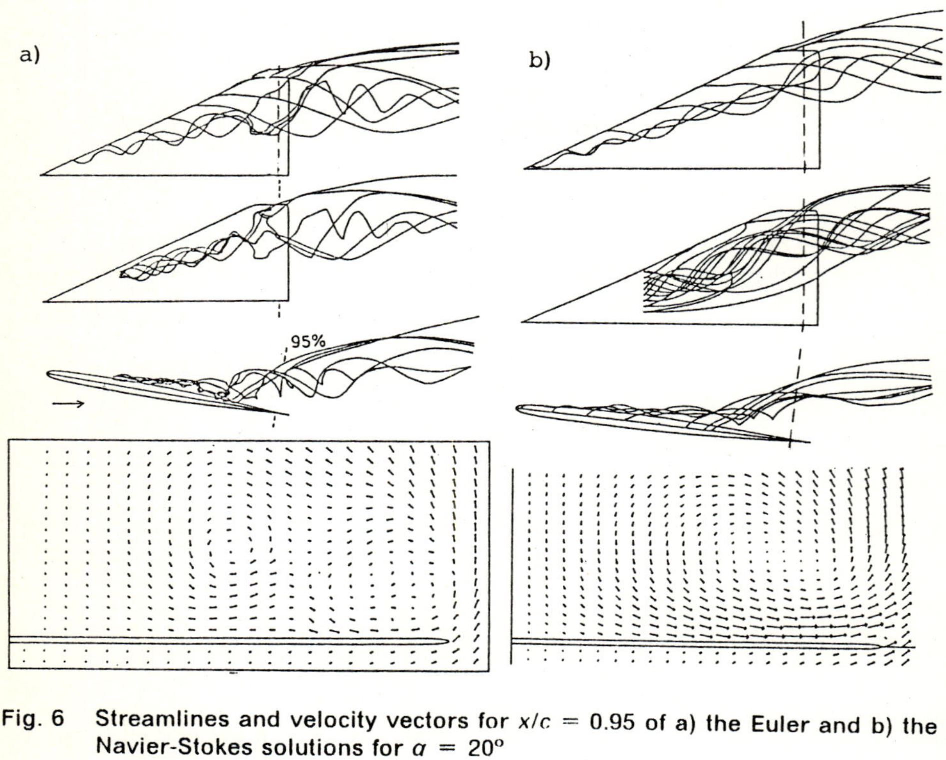

Fig. 6 Streamlines and velocity vectors for $x/c = 0.95$ of a) the Euler and b) the Navier-Stokes solutions for $\alpha = 20°$

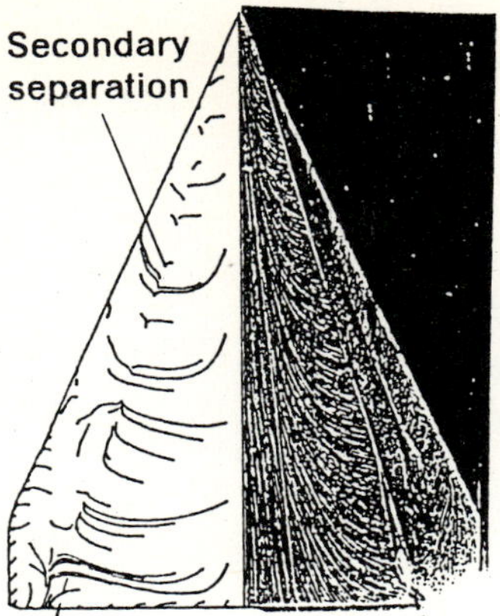

Fig. 7

Skin friction lines of Navier-Stokes solution compared with oil flow pattern for $\alpha = 20°$

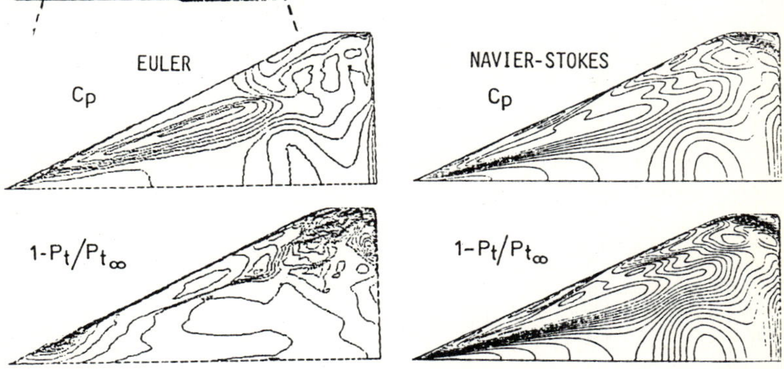

Fig. 8 Contour plots for C_p and for total pressure losses $1 - P_t/P_{t_\infty}$ of the Navier-Stokes solution and the Euler solution for $\alpha = 20°$

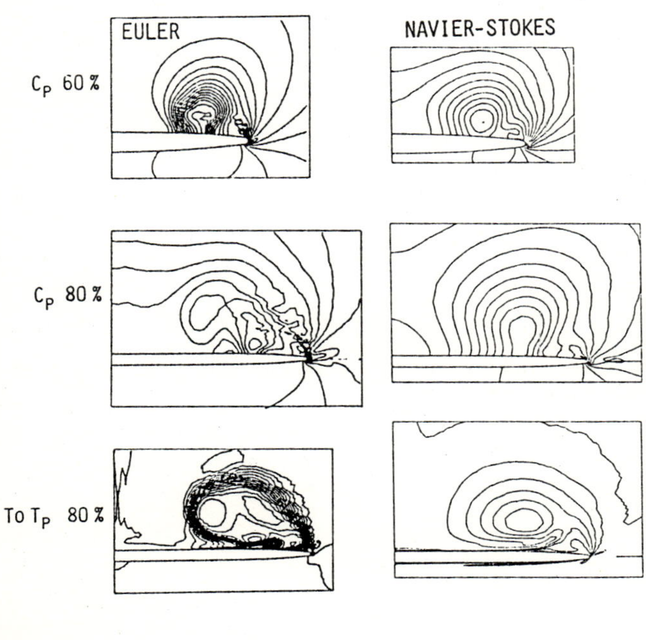

Fig. 9

Contour plots for C_p and for total pressure losses $1 - P_t/P_{t_\infty}$ of the Navier-Stokes solution and the Euler solution for $\alpha = 20°$ in the planes $x/c = 0.6$ and $x/c = 0.8$

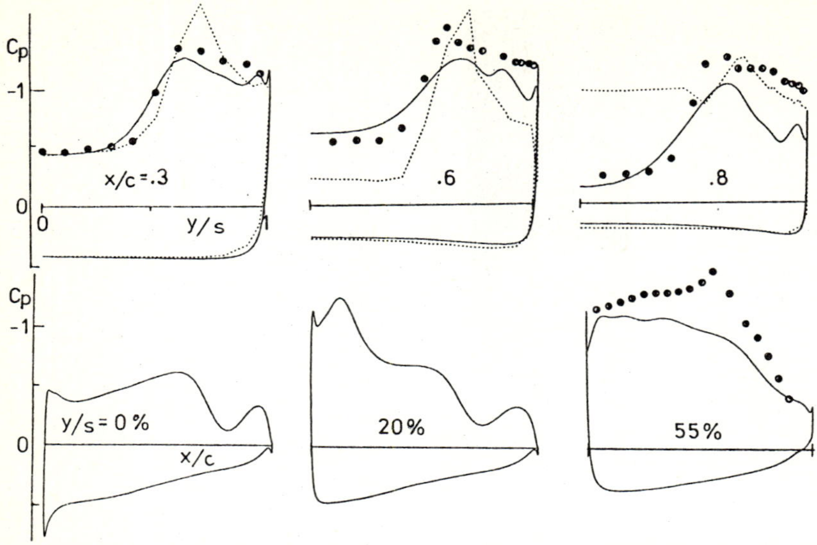

Fig. 10 Spanwise and chordwise plots comparing Navier-Stokes and Euler solutions with experiments for $\alpha = 20°$

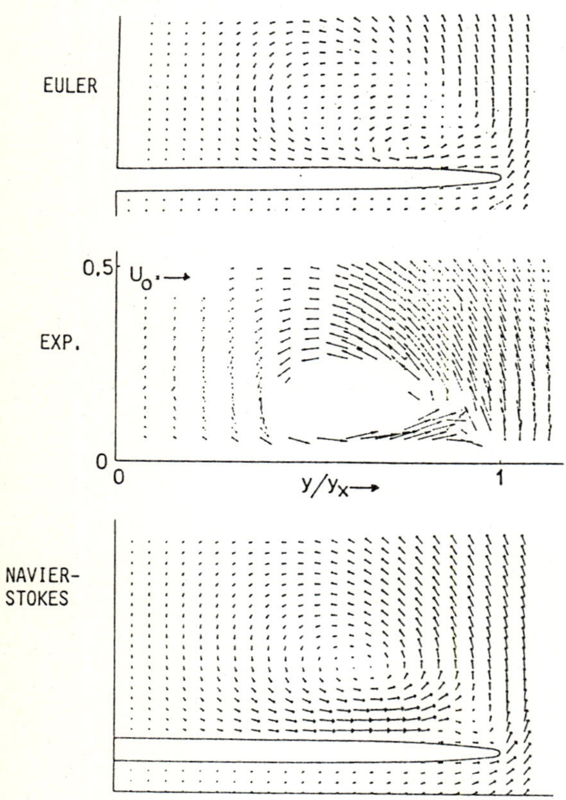

Fig. 11

Velocity vectors in the spanwise plane $x/c = 0.80$ as found from the Navier-Stokes and Euler computations and from laser measurements [8], $\alpha = 20°$

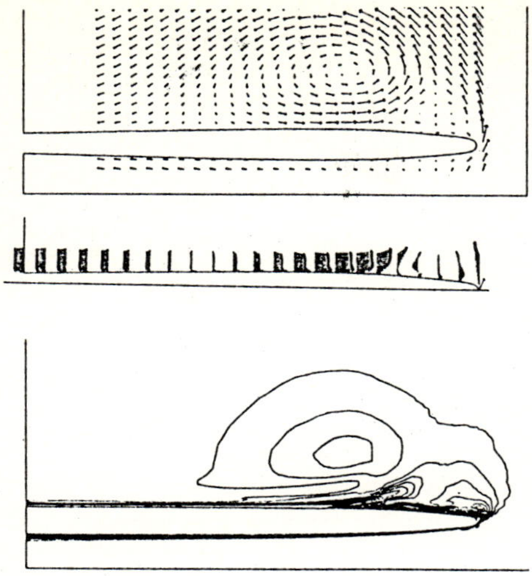

Fig. 12
Velocity vectors in the plane normal to the leading edge at $x/c = 0.7$ and the vorticity contours in the same plane for $\alpha = 20°$

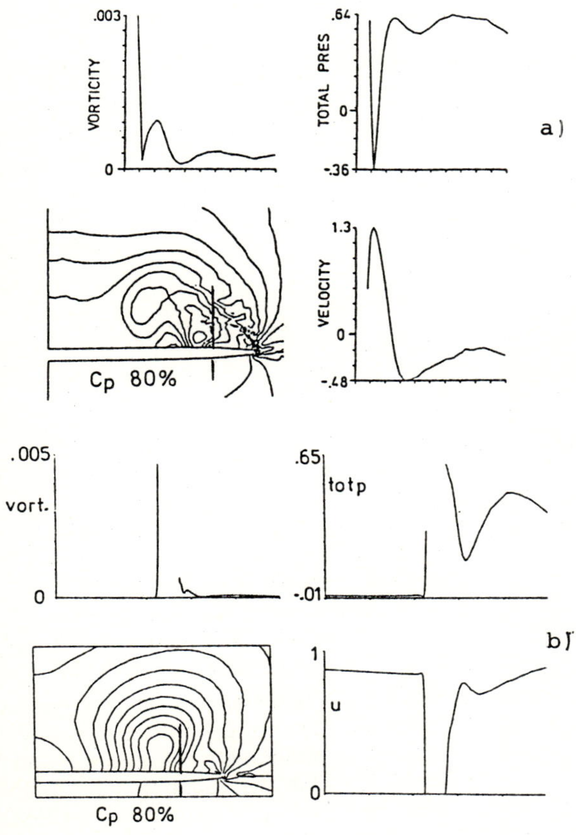

Fig. 13
C_p contours, vorticity, total pressure and u-velocity for a) Euler and b) Navier-Stokes for $x/c = 0.80$ and $\alpha = 20°$

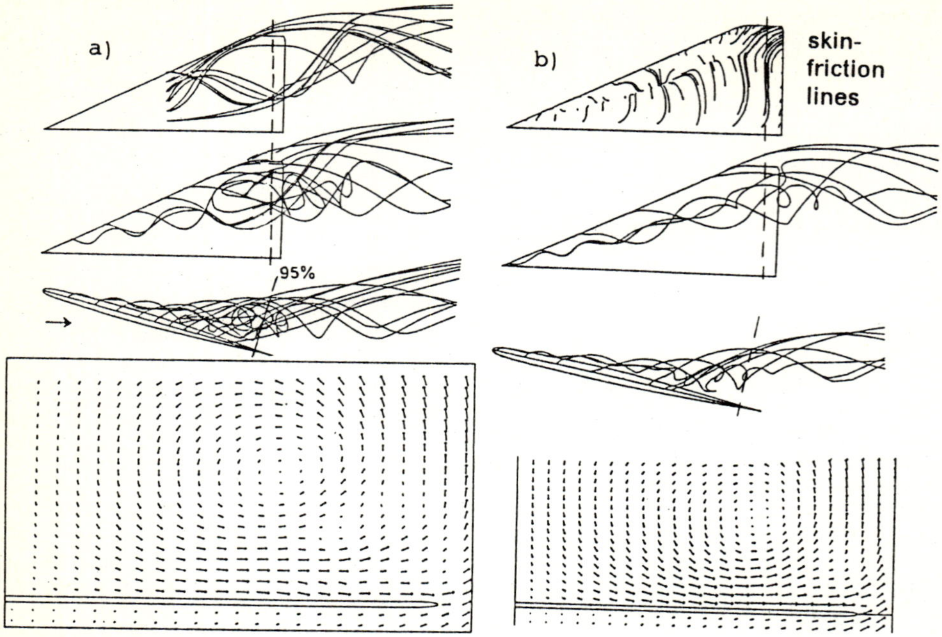

Fig. 14 Streamlines and velocity vectors for x/c = 0.95 of a) the Euler and b) the Navier-Stokes solutions for α = 24°

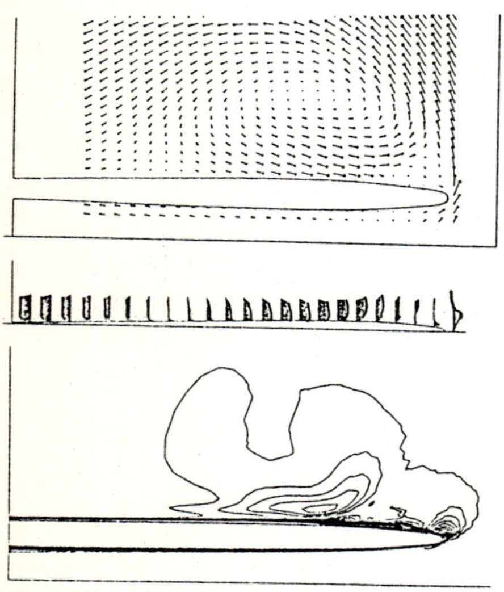

Fig. 15 Velocity vectors in the plane normal to the leading edge at x/c = 0.7 and vorticity contours in the same plane for α = 24°

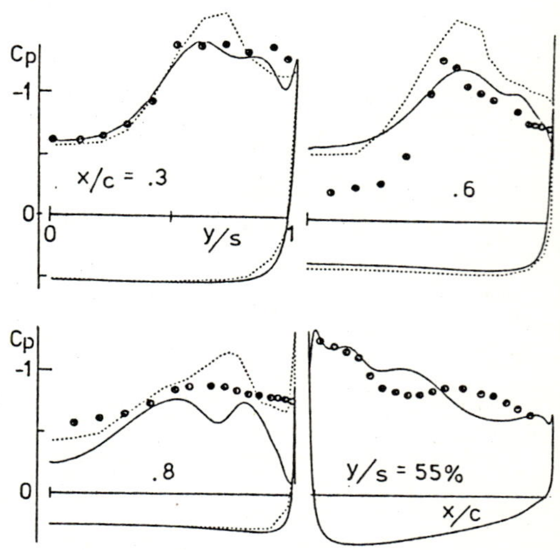

Fig. 16 Comparison of computed surface C_p from the Euler and Navier-Stokes solutions with the experiments for α = 24° in the sections x/c = 0.3, 0.6, and 0.8 and in the section y/s = 0.55

Computation of Vortex Flow on a Delta Wing at Transonic Speed

A. Kumar* and A. Das**

* National Aeronautical Laboratory, Bangalore, INDIA
(Visiting Scientist for 10 mths. 1985/86 in the DFVLR)

** DFVLR - Institute for Design Aerodynamics
D-3300 Braunschweig, FRG

Summary

Computation of the flow field around a 65° cropped delta wing is undertaken by using the Euler equation method. The parameter variation includes two onflow Mach numbers M = 0.4 and 0.85, angles of incidence from $\alpha = 0°$ to $30°$ and two types of leading edges, sharp and round. Field-grids having 0-0 topology have been generated. Influence of grid density and implementation of the surface boundary condition on the numerical solution is discussed. The field solutions depict clearly the nature of vortex formation, the corresponding lift and circulation distribution on the wing, the setting in of the vortex lift and finally the breakdown of the vortices.

Introduction

Flow fields over highly swept slender wings of small aspect ratio at moderate to high angles of incidence are characterized by separation from the leading edges and formation of spiraling vortices on the leeside of the wings. Augmentation of lift due to these vortices is of great aerodynamical use. These wings therefore have been studied extensively since long, both experimentally and theoretically; a recent extensive review of the subject is given by Newsome and Kandil [1].

Numerical computations of vortex flow over slender wings were initially done by using panel methods, which needed a modelling of the vortical flows [2,3]. With the advancements in computational capabilities, it is now practical to investigate such flow fields by using the Euler equations method [6-8]. Being a field method it can yield many details of the flows, including the formation of the spiraling vortices around the wing and their natural changes with incidence. Some further relevant papers are cited in [9 to 14].

J. Zierep, H. Oertel (Eds.)
Symposium Transsonicum III
IUTAM Symposium Göttingen, 24.-27. 5. 1988

In the present paper an Euler equation method is used to investigate the flow over a cropped delta wing having a leading edge sweep of 65° as shown in Fig. 1a. The wing section is symmetric, having maximum thickness of 5%. Wings having round leading edges (RLE) and sharp leading edges (SLE) have been considered. Influence of grid density and implementation of the surface boundary condition on the numerical solution is investigated. Following the numerical solutions, extensive analysis of the flow fields could be taken up illustrating the nature of vortex formation - the local circulation and lift distributions on the wing changing with the angle of incidence, the amount of vortex lift and finally the vortex bursting. The phenomenon which leads to the vortex breakdown has been elucidated in detail, affirming that it arises in inviscid medium.

Euler equations consisting of conservation of mass, momentum and total energy are

$$\int_{\Omega} \frac{\partial \bar{Q}}{\partial t}\, dv + \int_{\partial\Omega} \bar{\bar{H}} \cdot d\bar{S} = 0$$

where

$$\bar{Q} = \begin{vmatrix} \rho \\ \rho u_i \\ \rho E \end{vmatrix}, \quad \bar{\bar{H}} = \begin{vmatrix} \rho\vec{q} \\ \rho u_i \vec{q} + p\vec{i} \\ (\rho E + p)\vec{q} \end{vmatrix}$$

These equations are solved by using the cell centered finite volume spatial discretization with five-stage Runge-Kutta time stepping scheme of Jameson et al. [9]. Careful blending of second and fourth order artificial dissipation values helps to capture schock waves and damp high frequency oscillations, while convergence to steady state is accelerated by using local time steps, enthalpy damping, implicit residual averaging and successive grid refinement. Flux balance at the cells aligned with the body is evaluated by using the no-flow condition through the solid surface and a linear extrapolation of the pressure. An alternate method for determining the surface pressure is to use the normal momentum equation. The merits of the two methods of evaluating surface pressure as reflecting on the solution are also discussed below. The linear extrapolation method is, however, used for the computed results described subsequently. Riemann invariants are employed to implement the far-field boundary condition. No additional condition is used to fix the leading edge separation. Further, the Euler code has a multi-block structure [10]. The multi-block coding is particularly suited to limited-memory facilities since only a single block at a time need reside in a core, and this helps to deal with complex configuration.

Grid Generation

Generation of computational grids is based on the transfinite interpolation method [11,12]. 0-0 topology with parabolic singular lines is chosen for the grids as it has a high mesh efficiency ratio. Grids on the surface of the wing are generated first from the given geometrical data. In the grid topology chosen here a minute rounding of the wing tip is required. Grids on the outer semi-spherical boundary are prescribed analytically. This is followed by the generation of root plane grids using an interpolation in which the function and its first three derivatives on the wing surface and only the function at the outer circular boundary are given. Grids on the 'cut' plane extending from the wing tip to the outer boundary are generated next from an interpolation. Generation of the interior grids is based on the transfinite interpolation in which the function on the lower and upper halves of the root plane and on the cut plane are used for the spanwise interpolation, and the function and its first derivative on the wing surface and the function on the outer surface are used for the interpolation normal to the wing surface. Introduction of the cut plane in the spanwise interpolation is to provide a better control over the interior grids, and therefore only one derivative on the wing surface is used in the interpolation in the direction normal to the wing surface. The grids extend 10 root chords upstream and 15 root chords downstream of the apex. The standard grid used for the computation has 96x24x48 cells;

that is, 96 cells are wrapped around the chord, 24 cells lie between the wing surface and the far-field boundary, and 48 cells are along the semi-span. A fine grid consisting of 128x48x64 cells is also used. Fig. 1b shows a perspective view of the grid system where clustering of cells near the leading edge and towards the tip can be clearly noticed.

Influence of Grid Density on the Solution

The primary feature of the flow under investigation is the vortex sheet separating from the leading edge and coiling to form a vortex core above the wing as is depicted in Fig. 2. The coarseness of the grid can substantially reduce the amount of vorticity that is shed. This is shown in Fig. 3 where total lift coefficient at M = 0.4 and 0.85 is plotted against $N^{-2/3}$, N being the total number of grid cells. N = 96x24x48 is the standard grid, and N = 48x12x24 is the coarse grid on which the solution is first obtained before proceeding to the standard grid. It is seen from the figure that the lift increases with the grid refinement. Flow at α = 20° is also computed on the fine grid of 128x48x64 cells. Fig. 4 shows comparisons of c_p distribution at M = 0.4 and 0.85 in a section of the vortex cut by an internal grid-plane as yielded by the fine grid solution and the standard grid solution. Formation of a cross-flow shock wave is evident in the fine grid solution at M = 0.85. The differences in the total pressure losses $\Delta p_o/p_{o\infty} = (p_{o\infty} - p_o)/p_{o\infty}$ in the vortices as obtained from the standard grid and fine grid solutions are given in the following table:

SLE-Wing at α = 20°	$\Delta p_o/p_{o\infty}$ at M = 0.4		$\Delta p_o/p_{o\infty}$ at M = 0.85	
	Min	Max	Min	Max
96x24x48 Grids	- 0.018	0.161	- 0.037	0.593
129x48x64 Grids	- 0.028	0.194	- 0.055	0.603

The suction and the total pressure loss in the vortex core are seen to be larger in the fine grid solution.

Methods of Implementation of Surface Boundary Condition and the Resulting Influence on the Solution

Numerical implementation of the surface boundary condition needs some especial care as it can influence the solution. In the finite volume method only pressure is required for evaluation of the surface flux due to the no flow condition through the body surface. Two methods of evaluating the surface pressure are considered. Either it can be obtained from a linear extrapolation or by using the normal momentum equation. Comparison of the spanwise lift coefficient distribution from the two methods is shown in Fig. 5. The two methods give nearly identical results for SLE wing at $M = 0.4$, $\alpha = 10°$. The difference in total lift coefficients is within 1%. However, the two solutions differ for RLE wing. This arises essentially from the flow in the neighbourhood of the leading edge. The separation is delayed and a larger suction is obtained when the normal momentum equation is used. At $M = 0.4$ and $\alpha = 15°$ the computed flow is marginally separated thus yielding a difference in the total lift coefficient by about 6%. At $M = 0.4$ and $\alpha = 25°$ the flow undergoes large separation. In this case the normal momentum equation method gives more lift in the outbord portion and a smaller lift in the inboard portion; the difference in the total lift being 0.1%.

Analysis of the Flow Around the Delta Wing Causing Vortex Lift and Vortex Break Down

The flow fields of the delta wing have been calculated both for round and sharp leading edges at $M = 0.4$ and 0.85 and at angles of incidence α ranging from 0° to 30°. The wing with sharp leading edges reveals well defined flow separation from the edge at all angles of incidence, while the wing with the round leading edges indicates flow round the edges - especially in the forward region at angles of incidence less than $\alpha = 15°$. Thus for the wing with the round leading edges the flow resolution near the edge brings in some uncertainties. On this round it was preferred to make the flow field analysis of the wing with sharp leading edges.

The field solutions obtained from the Euler finite volume method yield complete information regarding the flow quantities on and around the wing. This enables to derive all the data needed to calculate the local lift or circulation distribution on the wing and also to follow up the changes occurring with the angle of incidence - both in the vortex formation and the resulting circulations around the wing.

The circulation distribution over the span of the wing and its changes with the angle of incidence are shown in Fig. 6a, which in turn yields the total lift coefficient and the amount of vortex lift as depicted in Fig. 6b. It is remarkable that at higher incidences the circulation distribution deviates significantly from the elliptic distribution and deforms appreciably, specially in the outboard regions. The total lift coefficient $c_L(\alpha)$ indicates clearly a fall in the vortex lift, which sets in at high incidence both for M = 0.4 and M = 0.85.

A close examination of the cross flow fields containing the spiraling vortices shows some significant changes beyond a certain value of the incidence angle — being illustrated in Fig. 7 containing also the pressure contours on the suction side of the wing. While at $\alpha = 20°$ and onflow Mach number 0.85 the spiraling vortices and the pressure contours show regular patterns, the condition is quite different at $\alpha \geq 24°$, indicating high radial velocities in the cross flow plane, thus causing a breakdown of the vortex. At higher angles of incidence when incipient vortex breakdown is approached, the convergence history of the Euler solution shows certain oscillations indicating a pseudo-unsteadiness of the flow condition. The amplitude of the unsteady part remains, however, confined to small deviations about the steady average values.

The Euler finite volume method being based on flow equations in inviscid medium, the vortex breakdown phenomena can essentially be looked for in the structure of the flow field comprising vortex shedding. It is very simple to demonstrate the fact that in the vortex fields of delta wings counter-rotating vortices can come into being - either in the coiled up trailing vortices or in the spiraling leading edge vortices, as depicted in Fig. 8. The region in-between the counter-rotating vortices will be subjected to a high radial velocity as already shown in Fig. 7.

As a result, a significant swelling of the coiled up vortex sheet occurs causing an instability of the vortex core, thus leading to its breakdown. Hence, the vortex lift of the wing starts falling and continues to fall more and more when the region with counter-rotating vortices moves upstream with increasing angle of incidence.

The computations reported here have been carried out on a CRAY 1S computer at the DFVLR. Time taken by the present Euler code is 56 micro-seconds per point per time step. A typical calculation of 600 time steps requires about 1 hour. In some solutions at large incidence up to 1200 time steps are used.

Conclusions

Three-dimensional vortex flow on a cropped delta wing is simulated using Euler equations. No additional condition is used to fix the leading edge separation. While fine grids, especially at the leading edges, are necessary for resolving the details of the flow, such as cross-flow shocks, calculations based on coarse grids can cause underprediction of vortex lift. The way of fulfilling the surface boundary condition has some marked effect on the separation from the round leading edge. The field solution yields the flow quantities required to calculate the circulation and lift distribution of the wing, showing up at the same time the amount of additional lift arising from the vortex roll up and the subsequent breakdown of it at high incidence. The cause of the vortex breakdown is looked for in the structure of the flow field with vorticity shed from the trailing and leading edges of the wing.

References

1. Newsome, R.W. and Kandil, O.A.: Vortical Flow Aerodynamics - Physical Aspects and Numerical Simulation. AIAA paper 87-0205, 1987.

2. Johnson, F.T., Tinoco, E.N., Lu, P. and Epton, M.A.: Three-Dimensional Flow over Wings with Leading Edge Vortex Separation. AIAA J., Vol. 18, pp. 367-380, 1980.

3. Hoeijimakers, H.W.M.: Numerical Computation of Vortical Flows about Wings. VKI Lecture Notes 1984-04, 1984.

4. Rizzi, A.: Euler Solutions for Transonic Flow Around the Dillner Wing - Compared and Analysed. AIAA paper No. 84-2142, 1984.

5. Raj, P. and Sikora, J.S.: Free-Vortex Flows; Recent Encounters with an Euler Code. AIAA paper No. 84-0135, 1984.

6. Thomas, J.L., Taylor, S.L. and Anderson, W.K.: Navier-Stokes Computations of Vortical Flows Over Low Aspect Ratio Wings. AIAA paper 86-0207, 1986.

7. Müller, B. and Rizzi, A.: Navier-Stokes Computation of Transonic Vortices over a Round Leading Edge Delta Wing. AIAA paper 87-1227, 1987.

8. Fujii, K. and Schiff, L.B.: Numerical Simulation of Vortical Flows over a Strake Delta Wing. AIAA paper 87-1229, 1987.

9. Jameson, A., Schmidt, W. and Turkel, E.: Numerical Solutions of the Euler Equations by Finite Volume Methods Using-Runge Kutta Time Stepping Scheme. AIAA paper No. 81-1259, 1981.

10. Radespiel, R. and Kroll, N.: Progress in the Development of an Efficient Finite Volume Code for the Three-Dimensional Euler Equations. DFVLR FB 85-31, 1985.

11. Gordon, W.J. and Hall, C.A.: Construction of Curvilinear Coordinate System and Application to Mesh Generation. Intl. J. Num. Math. Eng'g., Vol. 7, pp. 461-477, 1973.

12. Eriksson, L.E.: Practical Three-Dimensional Mesh Generation Using Transfinite Interpolation. SIAM J. Sci. Stat. Comput., Vol. 6, No. 3, pp. 712-714, 1985.

13. Rossow, C., Kroll, N. Radespiel, R and Scherr, S.: Investigations of the Accuracy of Finite Volume Methods for 2- and 3-Dimensional Flows. Proceedings of AGARD Symposium on Validation of Computational Fluid Dynamics, Lisbon, 1988.

14. Das, A.: On the Kutta Condition for Lifting Aerofoils and Wings. DFVLR FB 87-40, 1987.

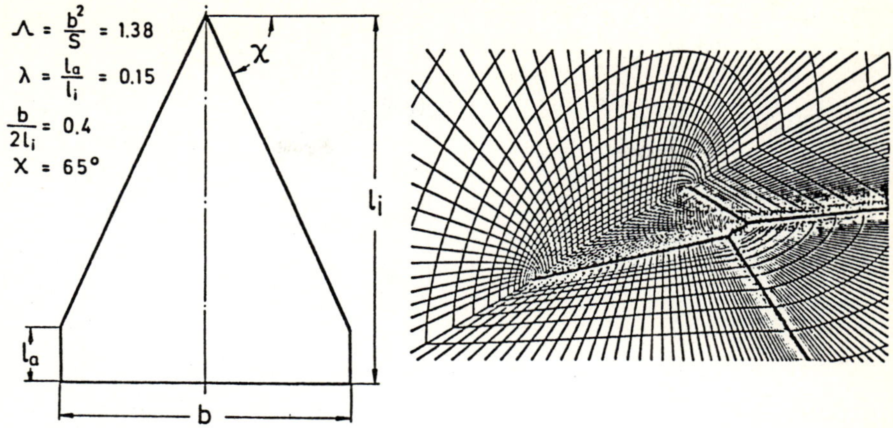

Fig. 1 Geometrical data of the cropped delta wing and a view of the O-O grid generated around it.

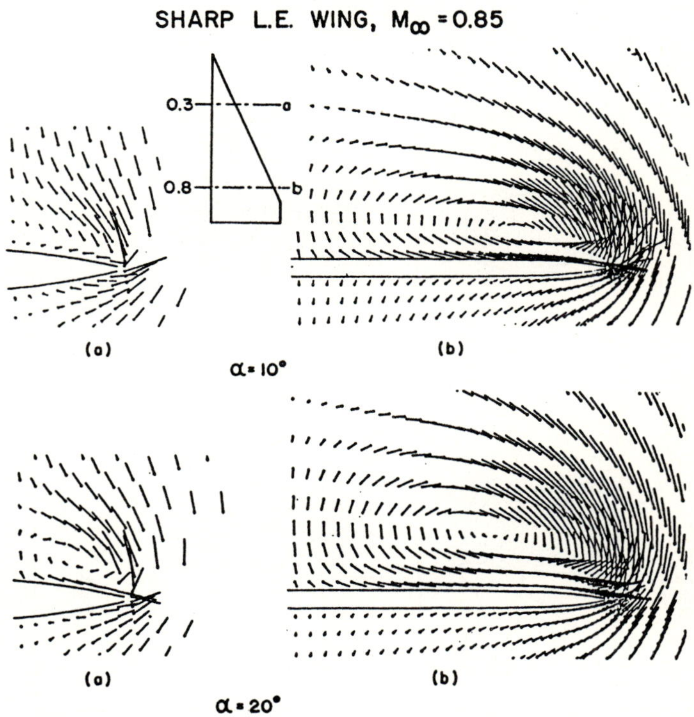

Fig. 2 Cross-flow velocity fields around the sharp leading edge delta wing at M = 0.85 as yielded by the Euler equation method.

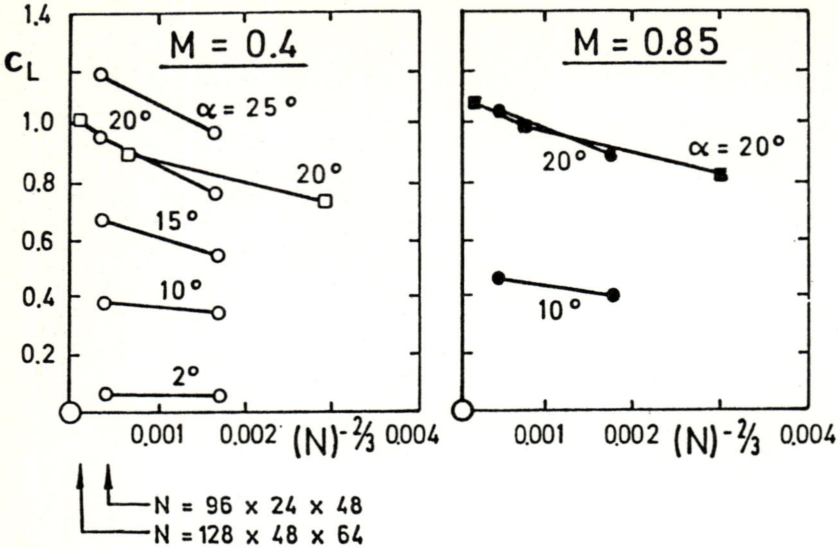

Fig. 3 Influence of the grid size on the total lift coefficient of the sharp leading edge delta wing at onflow Mach numbers of 0.4 and 0.85.

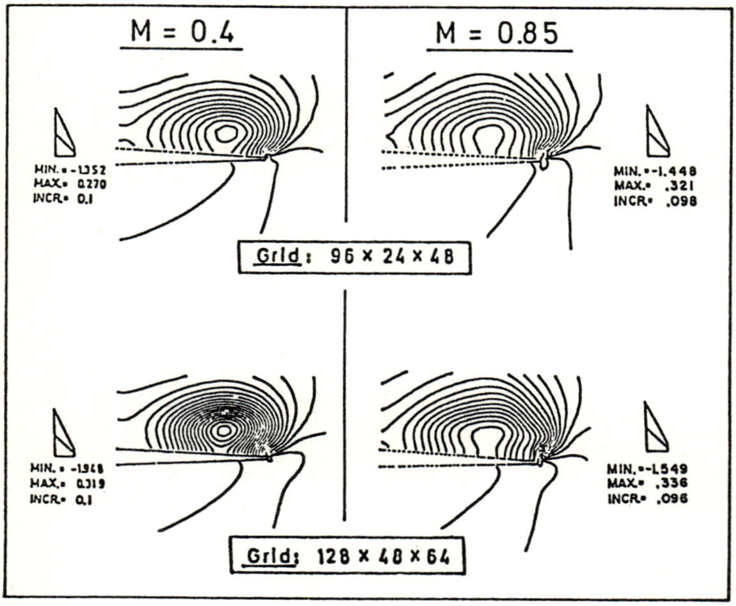

Fig. 4 Pressure contours in the section of a coiled up vortex cut by an internal grid plane as obtained from the fine grid and standard grid solutions.

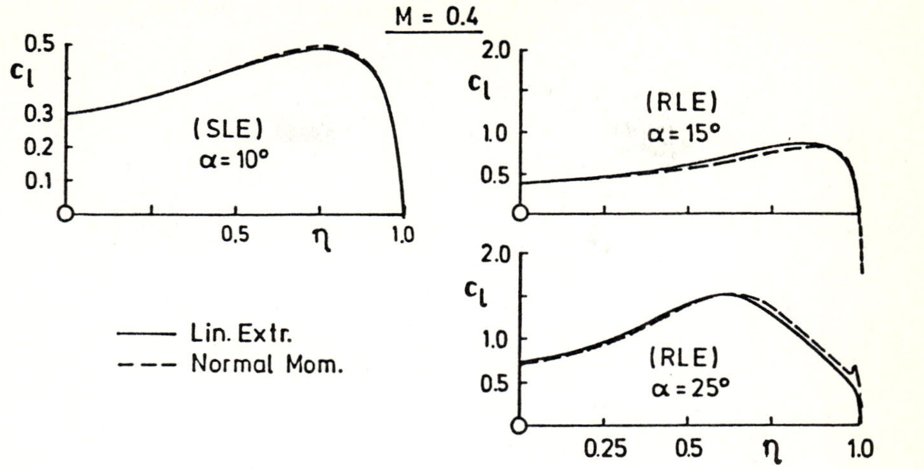

Fig. 5 Comparisons of the spanwise c_l-distributions on the delta wing as yielded by the two different methods of implementing the surface boundary condition.

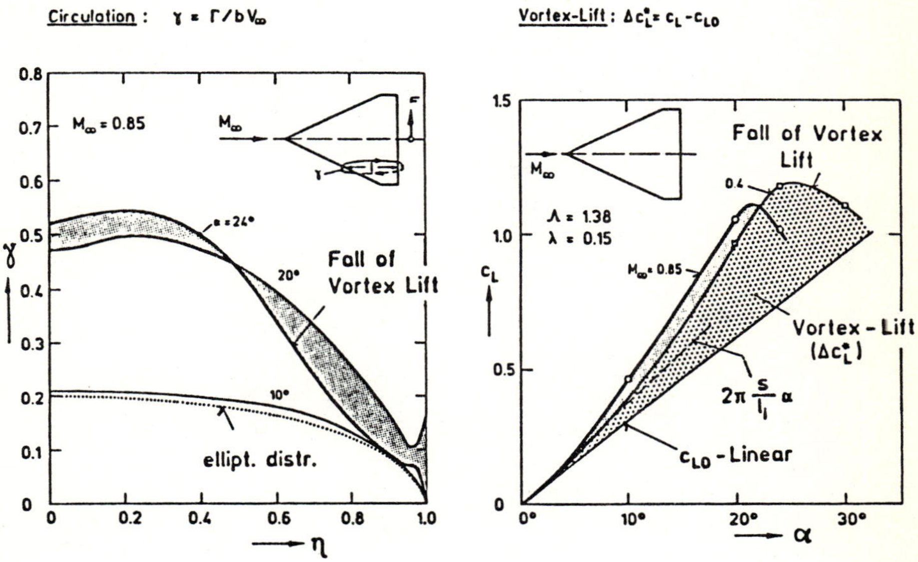

Fig. 6 Circulation distribution and vortex lift of a delta wing based on Euler equation method.

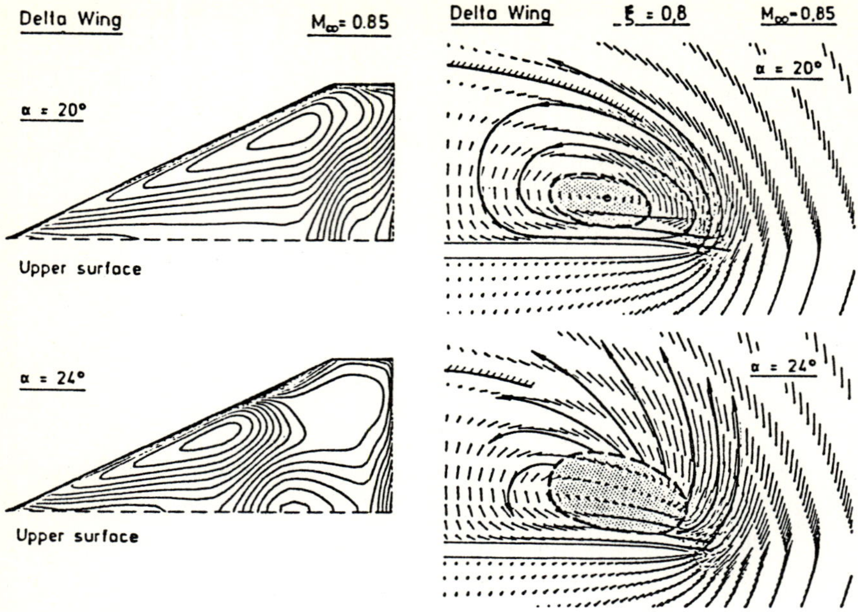

Fig. 7 Changes in the pressure contours and flow field structures of a delta wing with increasing angle of incidence.

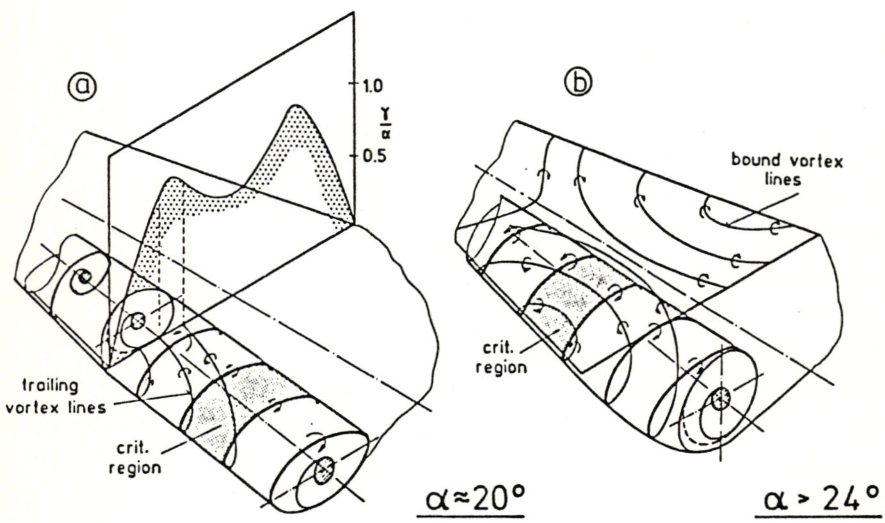

Fig. 8 Origin of counter rotating vortices which lead to high radial velocities and to the breakdown of the coiled up vortices in the flow fields of delta wings.

Vortical Flows on the Lee Surface of Delta Wings

K.Y. Narayan, S.N. Seshadri

Aerodynamics Division
National Aeronautical Laboratory
Bangalore, INDIA

1. INTRODUCTION

In the more than three decades of research on the lee surface flow of delta wings, starting from the identification of a single pair of counter-rotating vortices springing from the leading edges, multiple vortices (secondary, tertiary), "squashed" vortices, vortices embedded in boundary layers, shock waves on the wing, on, underneath and in between the vortices, shock wave boundary layer interaction, shock-induced separation, vortex bursting, asymmetric flow behaviour and so on have been progressively "discovered". In a gross sense, the geometric and free stream conditions under which each of these flow phenomena occurs are reasonably well understood. However, the physical mechanisms underlying some of the flow phenomena remain far from clear. Significant progress has been made in recent times in the computation of the complex three-dimensional flows on delta wings; however, solutions for even the simplest flow type (flow with a pair of leading edge vortices) tax the resources of the largest available computer. Solutions for the more complex flow types including multiple separations, shocks, etc., appear to be realisable only in the distant future.

After more than three decades of research on delta wings, it seems that this is an opportune time to review the progress made in the understanding of complex vortex-dominated flows. In this paper, we have not attempted to critically examine all the literature available on delta wing flow fields, but rather concentrate on some of the most recent work, notably those of Szodruch [1], Miller and Wood [2], Seshadri and Narayan [3], Elsenaar [4] and Narayan and Hartmann [5]. The main objective of this paper is to briefly examine the different flow types that occur on the lee surface of delta wings and to highlight their dependence on the more important geometrical and free stream parameters.

J. Zierep, H. Oertel (Eds.)
Symposium Transsonicum III
IUTAM Symposium Göttingen, 24.-27. 5. 1988

2. TYPES OF FLOW ON LEE SURFACE OF SHARP WINGS

In discussing flow past delta wings, it is often convenient to interpret the results in terms of the flow past sections normal to the leading edge. This is usually done by defining a Mach number M_N and an angle of attack α_N normal to the leading edge. M_N and α_N are given by

$$M_N = M_\infty (1 - \sin^2 \Lambda \ \cos^2 \alpha)^{\frac{1}{2}}$$

$$\alpha_N = \tan^{-1} (\tan \alpha / \cos \Lambda)$$

Here, M_∞ is the free stream number, α is the angle of attack and Λ is the sweepback of the wing.

The most comprehensive sets of measurements on the lee surface flow of delta wings are those of Miller and Wood [2] on thin wings, Szodruch [1] on thick wings and Seshadri and Narayan [3] on both. Based on these results, Seshadri and Narayan [3] identified nine different types of flow on the lee surface. These are discussed in the following paragraphs.

Figure 1 (from [3]) shows surface oil flow patterns, interpreted cross—flow patterns and static pressure distributions corresponding to each of the above types of flow. Figure 2 (also from [3]) shows the types of flow on thick delta wings and the boundaries between them displayed in a plane of M_N and α_N .

Classical leading edge separated flow with its sub-types are shown in Figures 1a-d and 1i. Figure 1a corresponds to a flow with secondary separation, while that shown in Figure 1b corresponds to a flow with secondary and tertiary separation. Flows with shock waves beneath the primary vortices are shown in Figures 1c and d. Figure 1i corresponds to a flow with a shock wave on the wing and in between the primary vortices.

Flows with cross-flow shocks in the outer flow are shown in Figures 1e-h. While a fully attached flow is shown in Figure 1e, Figures 1f and g correspond to flows attached at the leading edge but with inboard shock-induced separation. In addition, the flow shown in Figure 1g corresponds to a mixed type with a "shed" vortex. Figure 1h corresponds to a flow with a leading edge bubble and shock. Details of these flow types are discussed by Seshadri and Narayan [3]. They also showed that the above features do not significantly depend on wing thickness.

3. CROSS-FLOW SHOCKS AND SHOCK-INDUCED SEPARATION

Figure 2 shows that flow attachment at a sharp leading edge occurs for M_N slightly lower than unity. The flow attains supersonic Mach numbers on the lee surface as it expands through a swept Prandtl-Meyer fan at the leading edge. The attached flow, after moving inboard for a certain distance, comes into the influence of a shock wave. This shock wave (called the "embedded" shock wave) is formed as the flow is turned chordwise in response to the flow moving inboard from the other leading edge. The "embedded" shock wave interacts with the boundary layer developing on the lee surface and if strong enough might cause shock-induced separation. Seshadri and Narayan [6] studied this type of flow in detail and proposed that the interaction between the "embedded" shock and boundary layer on the lee surface of a delta wing is not only qualitatively similar to that between a glancing shock wave and boundary layer on a flat plate, but more importantly, exhibits quantitative similarity in certain overall features also like, in particular, conditions for incipient shock-induced separation . Based on this and using the observation that incipient separation on the lee surface of a delta wing occurs if the pressure ratio across the "embedded" shock is about 1.5 (Figure 3), Seshadri and Narayan [6] were able to predict the boundary between fully attached flows and flows with inboard shock-induced separation. Figure 4 shows the good agreement of the predicted boundary with the experimentally determined one.

Flows with leading edge separation under certain conditions, have been known to contain a shock wave beneath the primary vortices. This has been observed by Michael [7], Vorropoulos and Wendt [8] and Seshadri and Narayan [3]. While Michael [7] and Seshadri and Narayan [3] identified this shock from a "kink" in the surface flow visualisation patterns (Figure 1c), Vorropoulos and Wendt [8] did so from velocity field measurements in the cross-flow plane. The static pressure distributions of Seshadri and Narayan [3] also showed a pressure rise (Figure 1c) at the spanwise location of the "kink" in the oil flow patterns. Recent fine-resolution Euler calculations ,Rizzi [9], have also shown the existence of shock waves beneath the primary vortex.

The mechanism of formation of this shock wave is not well understood. But following Vorropoulos and Wendt [8], we might expect that at sufficiently high Mach numbers and angle of attack, local supersonic flow pockets develop on the wing lee surface due to the high velocities induced by the vortex. This supersonic flow passes through a convergent-divergent passage formed between the dividing streamline and the wing surface, and as the flow moves outboard, the pressures have to match with the higher pressures near the leading edge. This pressure adjustment takes place across a shock wave which is formed in between the dividing streamline and the wing surface at a location slightly downstream of the throat of the convergent-divergent passage. Figure 5 shows schematically the possible mechanism of formation of the shock wave.

The above shock wave interacts with the boundary layer developing outboard from the primary re-attachment line, and if strong enough causes shock-induced separation. Classical secondary separation usually occurs close to the leading edge, and as it changes to shock-induced, its location shows a rapid inboard excursion, as shown in Figure 6.

As the angle of attack increases, the two leading edge vortices move closer together in the cross plane. Under these circumstances, a shock wave has been observed to lie above the lee surface in between the vortices (Szodruch and Ganzer [10]). Figure 7 (from [10]) suggests that the shock wave is formed as the induced flow (supersonic) in between the vortices is turned parallel to the lee surface.

4. EFFECT OF LEADING EDGE BLUNTNESS

An important difference between sharp and blunt wings is that in the former, the location of separation is fixed (at the sharp edge) while in the latter, it is not. Thus, in a blunt delta wing, leading edge separation starts first near the tips and progresses upstream with increase in angle of attack. (This, of course, is due to the higher induced angles of attack in the tip region.) Further, since the effect of nose bluntness is to decrease the adverse pressure gradients near the leading edge compared to a sharp wing, the leading edge vortex once formed on the former would be weaker than in the latter.

Since for a rounded leading edge wing the separation line is not fixed, a type of flow where the flow is attached along part of the leading edge and separated over the rest is possible. (We call this type of flow a "mixed" flow and should not be confused with flow type 7 on sharp wings.) This type of flow was first seen in the measurements of Sutton [11] and Hall and Rogers [12]. A similar flow has also been reported by Kulfan [13] based on the measurements of Manro et al. [14]. Recently, results from a systematic series of measurements carried out on a 65 degree swept-back delta wing with rounded leading edge at subsonic, transonic and supersonic speeds have been reported by Elsenaar [4]; Narayan and

Hartmann[5] have analysed these results in some detail. We discuss these results in the following paragraphs.

The effect of Mach number on leading edge vortex development on a blunt wing is shown in Figures 8 and 9 in a plane of $x/c]_S$ and α. $x/c]_S$ is the streamwise distance measured from the wing apex downstream of which a leading edge vortex exists. We see from Figure 8 that at subsonic and transonic Mach numbers, leading edge vortex development starts at a higher angle of attack at lower Mach numbers. Furthermore, the rate of progression with α of leading edge separation from the wing tip to the apex is much slower at the lower Mach number. These trends are due to compressibility; with increase in Mach number, the adverse pressure gradients in the nose region increase, thus resulting in earlier (in terms of α) leading edge separation. At supersonic speeds however, the angle of attack for onset of leading edge vortex development increases with Mach number, as the results shown in Figure 9 for M_∞ = 1.32 and 1.72 show. These results suggest that the type of separation might change as Mach number is increased. Thus, while at subsonic and transonic speeds separation could be of the classical "pressure-induced" type, that at higher Mach numbers could be of the shock-induced type. (This classification of the type of separation was given by Boersen and Elsenaar [15].)

With increase in Mach number (M_∞ > 1.8), the flow is attached all along the leading edge, but with shock-induced separation inboard (similar to flow type 6 for sharp edge wings). Results from Narayan and Hartmann [5] show that shock-induced separation also starts from the tip region and progresses upstream as the angle of attack is increased.

Figure 10 (from [5]) shows in a plane of M_N and α_N the different flow types that occur on the lee surface of a blunt edge delta wing. There are basically three types of flow on the wing; fully attached, mixed flow and fully separated flow. The important difference between sharp and blunt wings, as can be seen from a comparison of Figures 2 and 10, is that a mixed type of flow involving both attached and separated flows exists on the blunt wing over a wide range of parameters, M_N and α_N .

5. EFFECT OF REYNOLDS NUMBER

Figure 11 displays the variation of the spanwise location (η_F) of the "foot" of the pressure rise following suction as a function of angle of attack and Reynolds number. At low angles of attack, η_F is a strong function of Reynolds number; the "foot" of the pressure rise moves outboard with increase in Reynolds number. α_S, the angle of attack at which leading edge vortex development starts also increases with Reynolds number. As the angle of attack increases beyond α_S, η_F becomes less strongly dependent on Re, becoming almost independent of Re, at sufficiently high Re. Based on this and data at other Mach numbers (0.7 and 1.2), Narayan and Hartmann [5] suggest that the flow field of the primary vortex is insensitive to Reynolds number at high angles of attack where a well developed leading edge vortex exists; however, in the attached flow regime and in a regime where the vortex is still developing in some sense, Reynolds number effects are quite significant.

The effect of Reynolds number on the angle of attack for leading edge separation is shown in Figure 12. Here, Reynolds number is based on a mean value of the leading edge radius. The differences in the dependence of α_{NS} (α_N for leading edge separation) with Re at various streamwise stations is perhaps due to: (i) the mean leading edge radius not being the right length scale for the definition of Reynolds number and (ii) effects of the finite tip and the apex.

The effects of Reynolds number on the leading edge vortex development on sharp edged delta wings have not been studied in any detail, particularly at transonic and supersonic speeds. We can expect that since separation is fixed at the leading edge of a sharp edged wing, Reynolds number will have a relatively weaker effect on vortex development compared to blunt wings. However, other parameters like the reattachment line, secondary separation line, etc. may still show significant Reynolds number effect; these aspects have also not been studied in any detail.

6. CONCLUDING REMARKS

Available measurements on the lee surface flow fields of delta wings indicate the existence of an astonishing variety of different flow types involving vortices and shock waves. Although the general features of these flows are fairly well understood, a lot more work, including in particular flow fields measurements in the cross–flow plane, needs to be done to obtain clarity on the physical mechanisms underlying the formation of the cross-flow shocks. Extensive research effort is also required to study the effect of Reynolds number on wings with blunt as well as sharp leading edges.

7. ACKNOWLEDGEMENT

The authors thank Prof.R.Narasimha, Director, National Aeronautical Laboratory,

Bangalore, for many useful discussions on delta wing flow fields.

REFERENCES

1. Szodruch, J.: Leeseiten-Strömung bei schlanken Delta Flügeln endlicher Dicke. ILR-BERICHT 23, Technische Universität Berlin, 1977.

2. Miller, D.S., Wood, M.W.: Lee-Side Flow Over Delta Wings at Supersonic Speeds. NASA TP-2430, 1985.

3. Seshadri, S.N., Narayan, K.Y.: Lee-Surface Flow Over Delta Wings at Supersonic Speeds. NAL TM AE 8610. 1986.

4. Elsenaar, A (Ed).: Proceedings of a Symposium on International Vortex Flow Experiment on Euler Code Validation. Stockholm 1986.

5. Narayan, K.Y., Hartmann, K.: Transonic and Supersonic Flow Past a 65 Degree Delta Wing with Rounded Leading Edges - Analysis of Experimental Data - To be published as DFVLR FB.

6. Seshadri, S.N., Narayan, K.Y.: Shock-Induced Separated Flows on the Lee Surface of Delta Wings. Aero.Jour., Vol.91, No.903 (1987), 128-141.

7. Michael, W.H.: Flow Studies on Flat-Plate Delta Wings at supersonic Speeds. NACA TN 3472, 1955.

8. Vorropoulos, G., Wendt, J.F.: Laser Velocimeter Study of Compressibility Effects on the Flow Field of Delta Wings. AGARD CP-342, 1983.

9. Rizzi, A.: Euler Solutions of Transonic Vortex Flow Around the Dillner Wing. J.Aircraft, Vol.22, No.4 (1985), 325-328.

10. Szodruch, J.,Ganzer, U.: On the Lee-Side Flow Over Delta Wings at High Angle of Attack. AGARD CP-247, 1978.

11. Sutton, E.P.: Some observations of the Flow Over a Delta-Winged Model with 55 Degree Leading-Edge Sweep, at Mach Numbers between 0.4 and 1.8. ARC R&M No. 3190, 1960.

12. Hall, I.M., Rogers, E.W.E.: The Flow Pattern on a Tapered Swept-Back Wing at Mach Numbers between 0.6 and 1.8. ARC R&M No.3271, 1957.

13. Kulfan, R.M.: Wing Airfoil Shape Effects on the Development of Leading-Edge Vortices. AIAA Paper No.79-1675, AIAA Atmospheric Flight Mechanics Conference, 1979.

14. Manro, G.N., Manning, K, Hallstaff, T.H., Rogers, J.T.: Transonic Pressure Measurements and Comparison of Theory to Experiment for an Arrow Wing Configuration. NASA CR-2610, 1976.

15. Boersen, S.J., Elsenaar, A.: Tests on the AFWAL 65 Degree Delta Wing at NLR:A Study of Vortex Flow Development between Mach=0.4 and 4. Proc. Symp. on Intl. Vortex Expt. on Euler Code Validation, Stockholm, 1986.

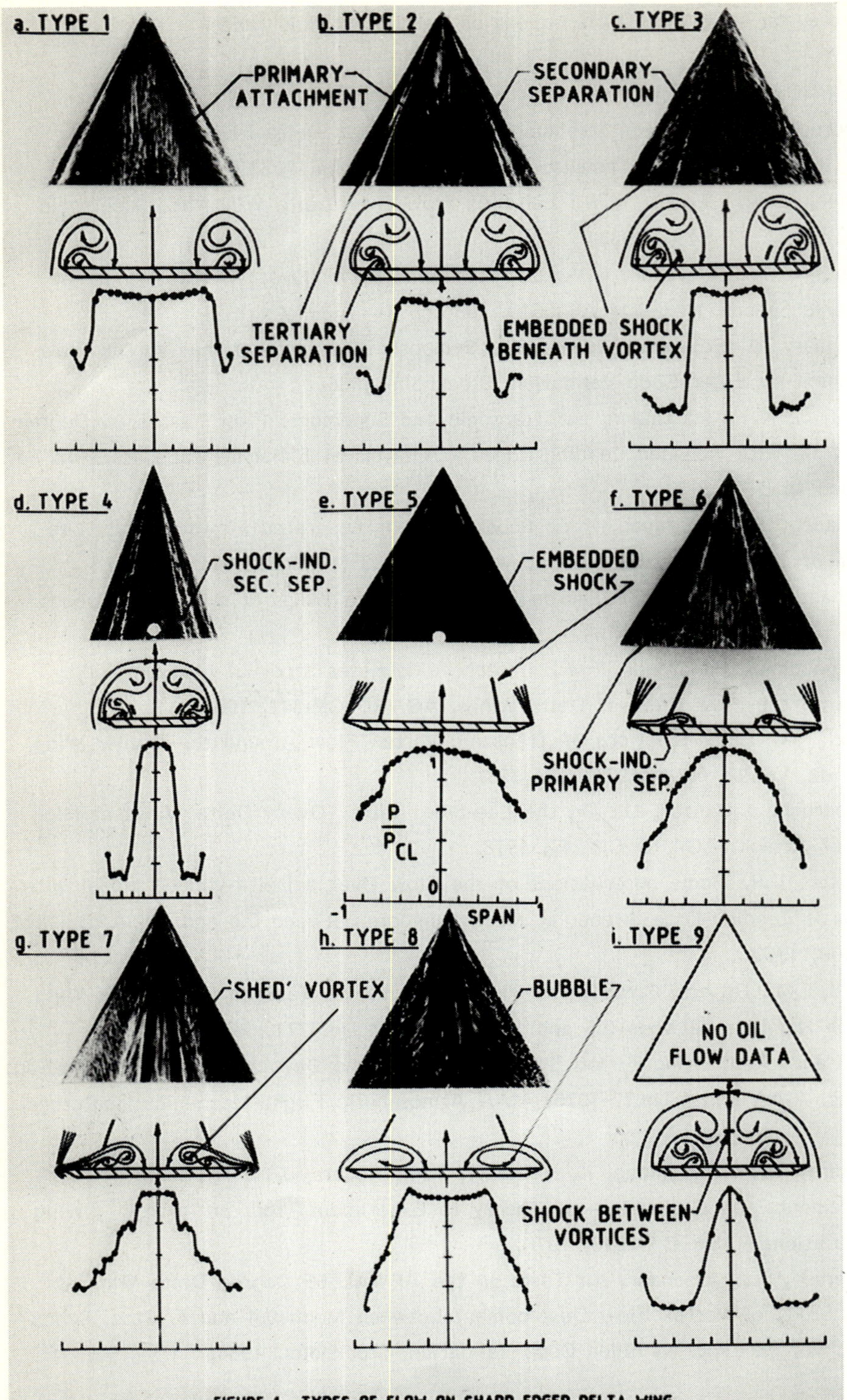

FIGURE 1. TYPES OF FLOW ON SHARP EDGED DELTA WING

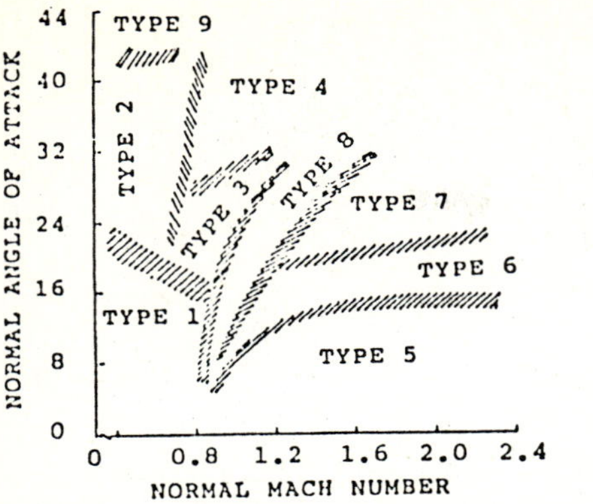

FIGURE 2 FLOW BOUNDARIES FOR SHARP EDGED DELTA WING

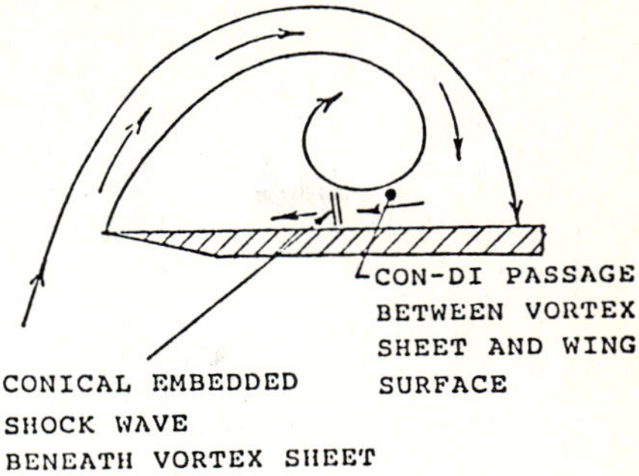

FIGURE 5 POSSIBLE MECHANISM OF FORMATION OF SHOCK BENEATH VORTEX

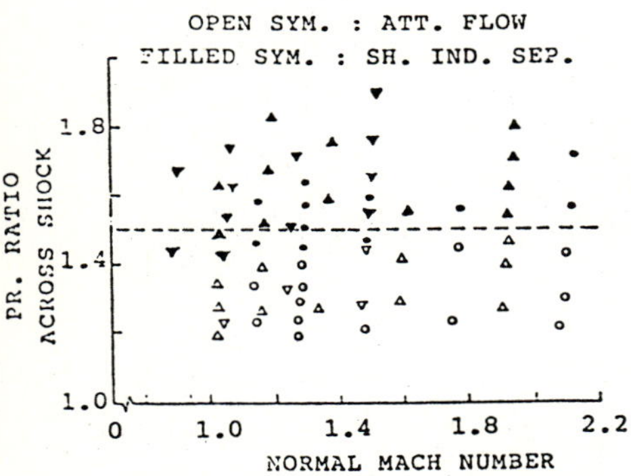

FIGURE 3 ESTIMATED PRESSURE RISE ACROSS EMBEDDED SHOCK

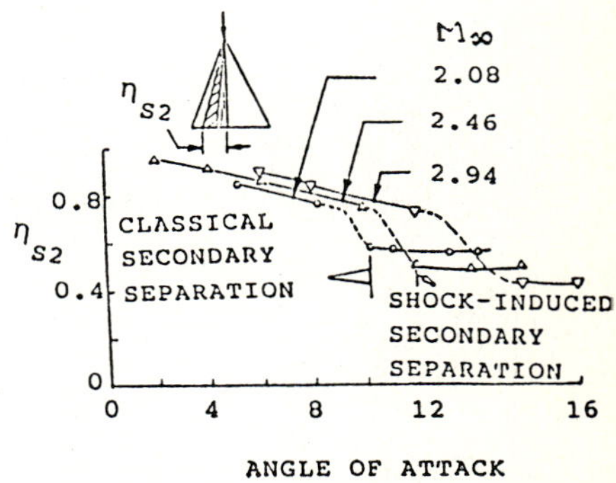

FIGURE 6 VARIATION OF SECONDARY SEPARATION LOCATION

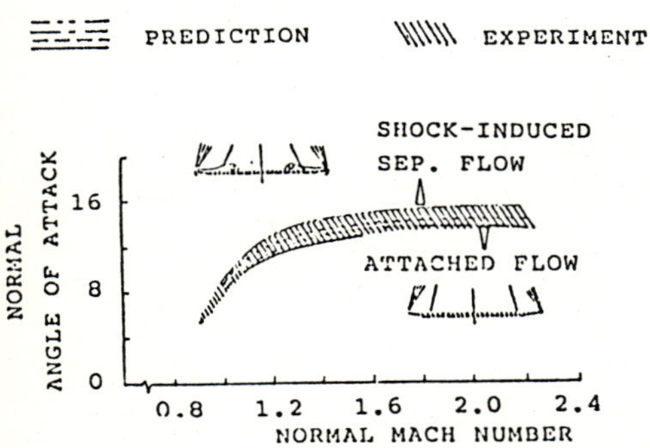

FIGURE 4 COMPARISON OF PREDICTED BOUNDARY WITH EXPERIMENT

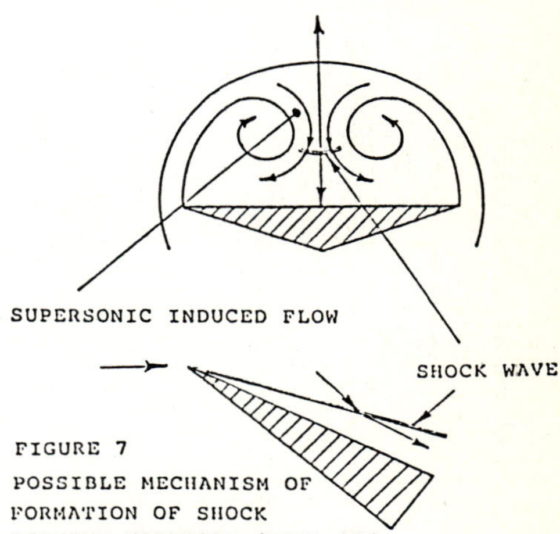

FIGURE 7 POSSIBLE MECHANISM OF FORMATION OF SHOCK BETWEEN VORTICES (REF. 10)

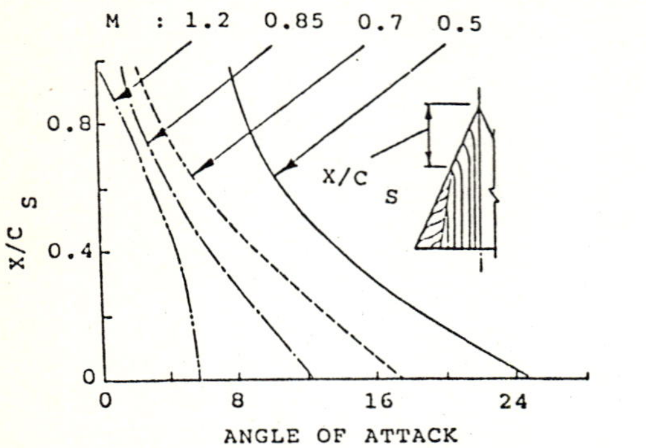

FIGURE 8 EFFECT OF MACH NUMBER ON L.E. VORTEX DEVELOPMENT (SUBSONIC & TRANSONIC SPEEDS)

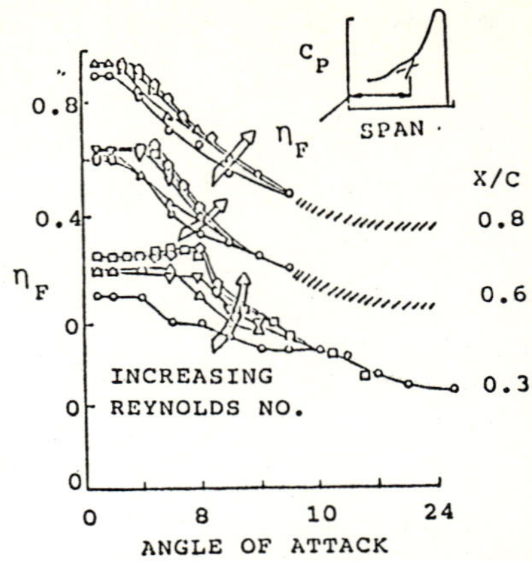

FIGURE 11 EFFECT OF REYNOLDS NUMBER ON LOCATION OF 'FOOT' OF PRESSURE RISE

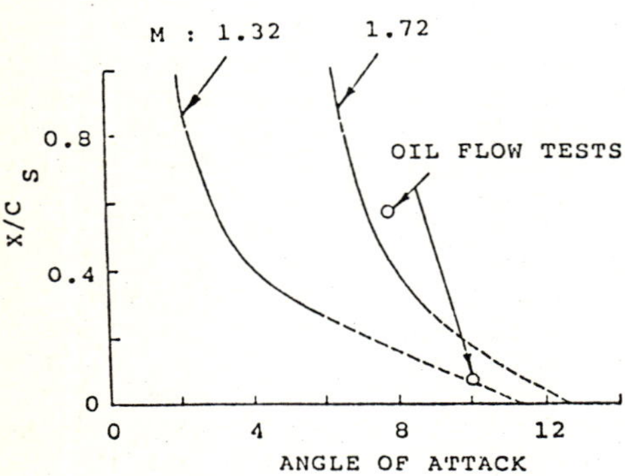

FIGURE 9 EFFECT OF MACH NUMBER ON L.E. VORTEX DEVELOPMENT (SUPERSONIC SPEEDS)

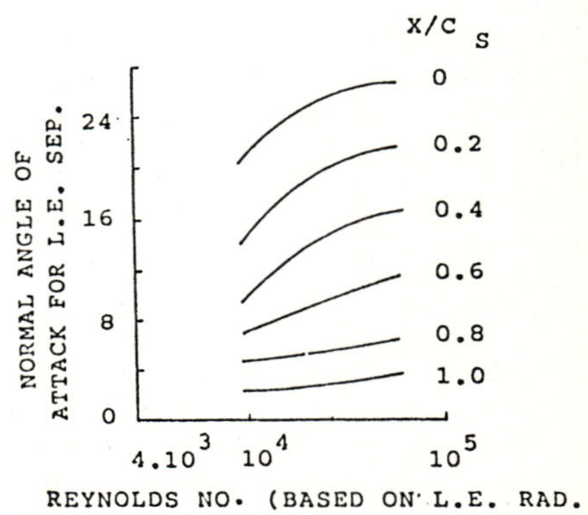

FIGURE 12 EFFECT OF REYNOLDS NUMBER ON ANGLE OF ATTACK FOR L.E. VORTEX DEVELOPMENT

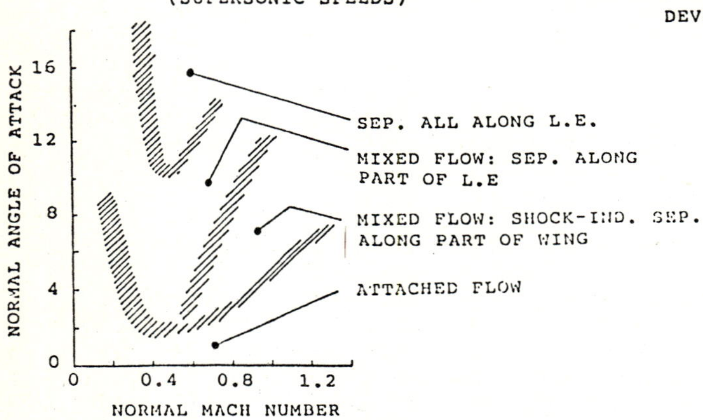

FIGURE 10 TYPES OF FLOW ON BLUNT DELTA WING

Numerical Simulation of Rotational Flows

M. Hafez and J. Ahmad

Department of Mechanical Engineering
University of California
Davis, California 95616, USA

Introduction

In this paper steady, inviscid, compressible two-dimensional and axisymmetric flows (with swirl) are studied. The present calculations are different in many aspects from the majority of the simulations available in literature.

For example, a common practice is to start with time dependent equations and to use explicit schemes, which are fully vectorizable, to reach a steady state. (In other words, the boundary value problem is embedded in an initial boundary value problem.) In the present work, the steady equations are attacked directly. Fully implicit, fully coupled schemes are used, and upon linearization of the discrete equations, a large system of algebraic equations is solved at each (Newton's) iteration. Thus, all admissible stable and unstable solutions can be calculated with quadratic convergence. On the present computers, the storage requirement is affordable, at least for inviscid flow simulation in two dimensions. Although efficiency was not the primary motivation, it turns out that the present technique (banded Gaussian elimination or sparse matrix technology) is indeed competitive to the iterative methods, particularly if unstructured grids are used and highly accurate solutions of the discrete equations are required.

Another common practice, in rotational flow simulation, is to use Euler equations in primitive variables representing conservation of mass, momentum and energy. Explicit or implicit artificial viscosity methods are implemented to capture contact discontinuities and shocks. Extra (numerical) boundary conditions may be needed for the higher order (viscous) equations as for example in the case of central difference schemes. Careful treatment of boundary conditions is essential, otherwise an artificial boundary layer is developed in the neighborhood of a solid surface.

As a consequence of the presence of artificial viscosity, discontinuities are spread over few grids. A wake, for example, will have a finite width with a structure similar to a free shear layer. In sketch (1), the ideal inviscid wake for incompressible flow is compared to a captured wake, where the maximum total pressure loss is about 50%. With mesh refinement, the wake layer becomes thinner (because of less viscosity), but the maximum pressure loss remains the same. For compressible flows, the total pressure is directly related to entropy. A comparison between entropy profiles across ideal and captured shocks is given in sketch (2).

J. Zierep, H. Oertel (Eds.)
Symposium Transsonicum III
IUTAM Symposium Göttingen, 24.-27. 5. 1988

In many cases, the structure of the wake or the shock is not important and the quality of the calculations should be determined by examining the solutions outside these layers. (In some cases, however, the artificial layers cannot be tolerated as in reacting flows, where artificial processes may be triggered due to the presence of the artificial viscosity.)

In the present work, a system equivalent to Euler equations is solved. A stream function is introduced and the stream function equation for compressible flow is solved. The vorticity is convicted along streamlines and the discretization of the momentum equations is completely avoided. In this formulation, a wake must be fitted. On the other hand, shocks may be captured but they must be detected to calculate the entropy jumps and hence the vorticity generated due to the entropy gradient across streamlines.

In the following, the stream function equations are derived for two-dimensional and axisymmetric flows (with swirl). The advantages and disadvantages of this alternate formulation are discussed. A new scheme for the solution of the stream function equation is introduced and some preliminary numerical results are presented.

Stream Function Formulation

To demonstrate some of the features of the present method, a simple case of an incompressible flow is considered first.

The stream function-velocity relation in Cartesian coordinates is given by

$$u = \psi_y, \qquad v = -\psi_x$$

The definition of vorticity yields

$$\psi_{xx} + \psi_{yy} = -\omega \tag{1}$$

Manipulation of the momentum equations results in

$$(u\omega)_x + (v\omega)_y = 0 \tag{2}$$

or, $$\omega = \omega(\psi) \tag{2'}$$

and, $$H = p + \frac{1}{2}(\psi_x^2 + \psi_y^2) = H(\psi) \tag{3}$$

Thus, the pressure is decoupled and the vorticity and the total head are preserved. Combining equations (1) and (2′), yields in general a nonlinear equation in ψ. Central difference approximation results in a nonlinear system of equations which can be solved by Newton's method. At each step, a large linear system of algebraic equations is solved directly by a banded Gaussian elimination procedure. The matrix, in general, is not positive definite, and standard iterative techniques are not applicable.

A classical example, where an exact analytical solution is available, is a cylinder in a constant shear flow. Using complex variables, the corresponding flow around ellipses or Joukowsky airfoils can be easily constructed. Analytical solution, in terms of a series of Bessel functions, is also possible in the case of linear shear flow. Analytical solutions of more general cases can be obtained via perturbation techniques.

For many cases, however, ω is not given as an explicit function of ψ. Instead, both ω and ψ are given as functions of y. A curve fitting procedure or a table can be constructed to obtain the required relation. Alternatively, equation (2) for the vorticity advection may be solved numerically using conservative schemes.

The numerical solutions of the stream function equations are free from the artificial viscosity effects and can be used to assess the quality of the Euler codes for rotational flow simulation,in the same manner the solutions of the potential equation are used for the case of irrotational flows.

In summary, the stream function formulation has many advantages. Mass is automatically conserved. The equation is second order with Dirichlet boundary condition for the analysis problem. (The formulation is also attractive for the inverse [design] problem since a body is a streamline.) The stream function is continuous across wakes (and shocks). The vorticity and the total head are functions of ψ, and can be easily conserved provided wakes (and shocks) are fitted.

The above formulation is limited to two-dimensional and axisymmetric flows (and in case of variable density, the flow must be steady). For three-dimensional flows, more than one stream function must be used; such extensions will not be discussed here.

An important issue is the treatment of recirculation regions and closed streamlines where levels of ω and H cannot be determined from the inlet conditions. This issue will be discussed with connection of axisymmetric flows with swirl.

In the case of compressible flow, there is another important issue, namely, the nonunique relationship between the density and the flux.

<u>Two-Dimensional Transonic Flows</u>

The stream function-velocity relation is

$$\rho u = \psi_y, \qquad \rho v = -\psi_x$$

The stream function equation becomes

$$(\psi_{x/\rho})_x + (\psi_{y/\rho})_y = -\omega \tag{4}$$

The vorticity is related to the entropy and the total enthalpy gradients via Crocco's relation

$$\omega = p\,\frac{d\frac{\Delta S}{R}}{d\psi} - \rho\,\frac{dH}{d\psi} \tag{5}$$

where $H = H(\psi) = \dfrac{\gamma p}{(\gamma-1)\rho} + \dfrac{1}{2}\dfrac{(\psi_x^2 + \psi_y^2)}{\rho^2}$ (6)

and $p = \dfrac{\rho^\gamma}{\gamma M_\infty^2} e^{\Delta S/cv}$ (7)

H is continuous across shocks, but ΔS jumps. The entropy rise is a function of the local Mach number and the shock inclination. Three special cases of equation (5) can be identified: for isentropic flows, ω/ρ is constant along streamlines while for isoenergetic flows, ω/p is a function of ψ. If the flow is both isentropic and isoenergetic it is also irrotational. Even with such simplifications, there is a serious problem. Combining equations (6) and (7) gives

$$\frac{\rho^{\gamma-1}}{(\gamma-1)M_\infty^2} + \frac{1}{2}(\psi_x{}^2 + \psi_y{}^2)/\rho^2 - H_\infty = 0 \tag{8}$$

For a given flux below a critical value, there are two solutions; one corresponds to a subsonic branch and the other belongs to a supersonic one. The two branches join at the sonic condition (critical value) with a square root singularity. The flux cannot attain values greater than the critical one; the density becomes imaginary. (In other words, the slope of the curve changes direction in the transonic regime and the Jacobian is singular at the sonic condition.) This was the stumbling block in the development of numerical techniques for the solution of the stream function equation for many years.

One of the first attempts (before computers) was reported by Emmons [1] in 1948. Mitchell [2] in 1951 obtained solutions using Southwell relaxation and a trial and error procedure near the sonic line. Sells [3] (1968) obtained solutions only for subcritical flows. Steger [4] (1969) tried to reproduce Emmons results with man-machine interaction. Recently, Kopriva [5] (1982), Morice [6] (1984), Amara [7] (1983) and Dubois [8] (1988) used the artificial density method to calculate transonic flows. To avoid the use of excessive artificial viscosity, the first author proposed in 1981 [9] an iterative procedure where the density is calculated as a function of the speed which is obtained as the solution of a first order partial differential equation consistent with the definition of vorticity.

The vorticity equation is

$$u_y - v_x = -\omega \tag{9}$$

Equation (9) can be rewritten in the form

$$(q \cos \alpha)_y - (q \sin \alpha)_x = -\omega \tag{9'}$$

where q is the speed and α is given by

$$\tan \alpha = -\frac{\psi_x}{\psi_y}$$

Equation (9′) is solved for q, hence the density is uniquely determined via Bernoulli's law. The stream function equation, with a retarded density is then used to update the stream function using a line relaxation procedure and the whole process is repeated until convergence.

With the present strategy (Newton's method and direct solver), there is no need to use q as one of the variables and the problem can be solved in terms of ψ and ρ only. In the following, a new scheme is introduced such that the artificial viscosity does not produce any artificial vorticity.

The continuity equation is modified by adding a term proportional to the Laplacian of the density. In Cartesian coordinates, it reads

$$(\rho u)_x + (\rho v)_y = \nu(\rho_{xx} + \rho_{yy}) \tag{10}$$

A stream function is introduced to satisfy equation (10), namely

$$\psi_y = \rho u - \nu\rho_x, \qquad -\psi_x = \rho v - \nu\rho_y \tag{10′}$$

$$\text{or,}\qquad u = \frac{\psi_y + \nu\rho_x}{\rho}, \qquad v = \frac{-\psi_x + \nu\rho_y}{\rho} \tag{10″}$$

The stream function equation is obtained by substituting u and v from equation (10") in the definition of vorticity

$$\left(\frac{\psi_y + \nu\rho_x}{\rho}\right)_y + \left(\frac{\psi_x - \nu\rho_y}{\rho}\right)_x = -\omega \tag{11}$$

The contribution of the artificial viscosity in equation (11) cancels identically and indeed equation (11) is the same as equation (4).

To construct a finite difference scheme with the same property, consider the stencil in sketch (3). The density is defined at the center of the cell, while the stream functions are located at the corners.

A discrete form of equation (10") is given by

$$u_{i,j+1/2} = \frac{(\psi_{i,j+1} - \psi_{i,j})/\Delta y}{(\rho_{i+1/2,j+1/2} + \rho_{i-1/2,j+1/2})/2}$$

$$+ \frac{\nu(\log \rho_{i+1/2,j+1/2} - \log \rho_{i-1/2,j+1/2})}{\Delta x}$$

and

$$v_{i+1/2,j} = \frac{-(\psi_{i+1,j} - \psi_{i,j})/\Delta x}{(\rho_{i+1/2,j+1/2} + \rho_{i+1/2,j-1/2})/2}$$

$$+ \frac{\nu(\log \rho_{i+1/2,j+1/2} - \log \rho_{i+1/2,j-1/2})}{\Delta y} \tag{12}$$

Using the above definitions of $u_{i,j+1/2}$ and $v_{i+1/2,j}$ in the discrete form of the vorticity definition, leads to a discrete form of equation (11), where the contribution of the artificial viscosity terms cancel again identically.

To avoid taking logarithms, the following approximation may be used

$$\log \rho_a - \log \rho_b = \log\frac{\rho_a}{\rho_b} = \log\left(1 + \frac{\rho_a - \rho_b}{\rho_b}\right)$$

$$\approx \frac{\rho_a - \rho_b}{(\rho_a+\rho_b)/2} + O(\rho_a - \rho_b)^3$$

Alternatively, equation (12) is simply modified to read

$$u_{i,j+1/2} = \frac{(\psi_{i,j+1} - \psi_{i,j})/\Delta y + \nu(\rho_{i+1/2,j+1/2} - \rho_{i-1/2,j+1/2})}{(\rho_{i+1/2,j+1/2} + \rho_{i-1/2,j+1/2})/2}$$

$$v_{i+1/2,j} = \frac{-(\psi_{i+1,j} - \psi_{i,j})/\Delta x + \nu(\rho_{i+1/2,j+1/2} - \rho_{i+1/2,j-1/2})}{(\rho_{i+1/2,j+1/2} - \rho_{i+1/2,j-1/2})/2} \tag{12'}$$

Equation (12) (or [12']) is used to evaluate the density in terms of the speed, coupled with the discrete form of the stream function equation (4). The resulting large system of nonlinear equations, including the discrete form of the boundary conditions, is solved by Newton's method and a banded Gaussian elimination procedure (Linpack).

Two examples are calculated by the present scheme. The first is irrotational isentropic flow over NACA 0012. The results are in agreement with the potential calculations based on the same artificial viscosity formulation (see Hafez et al., 1988 [10]). For convenience Cartesian coordinates and linearized boundary conditions are used in both cases. The surface pressure distribution and the convergence history for different free stream Mach numbers are plotted in fig. (1).

The second example is an airfoil (NACA 0012) in a shear flow of boundary layer type (see sketch 4) where at the inlet

$$u = \sin\alpha y, \qquad \psi = -(\sin\alpha y)/\alpha$$

hence,

$$\omega(\psi) = \alpha^2\psi.$$

At the inlet, the flow is assumed to be isentropic with uniform pressure (and density). Hence,

$$H = \frac{1}{(\gamma-1)M_\infty^2} + \frac{1}{2} u^2$$

and therefore, $H(\psi) = \dfrac{1}{(\gamma-1)M^2} + \dfrac{1}{2} (1-\alpha^2\psi^2)$,

(M_∞ is the Mach number of the flow outside the boundary layer.)

Neumann boundary conditions are used on the top and the bottom and at the exit of the computational domain.

If strong shocks are developed, the entropy rise must be accounted for. Some preliminary results, for only weak shocks, are shown in figure (2). Again Cartesian coordinates are used, but the exact Dirichlet boundary condition, ψ = constant on the surface of the airfoil, is implemented in these calculations. In general, Kutta condition is imposed such that the flow leaves the airfoil in a direction which bisects the trailing edge angle. This requirement determines the value of the stream function on the surface of the airfoils. In this particular example, the airfoil is at zero angle of attack; the lift generated depends on ω and M_∞.

Axisymmetric Flow with Swirl

In cylindrical coordinates (x,r,θ), the velocity components (w,u,v) are given in terms of Stokes stream function.

$$w = \psi_r/(\rho r), \qquad u = -\psi_x/(\rho r)$$

The stream function equation becomes

$$r(\psi_r/\rho r)_r + (\psi_x/\rho)_x = -r\omega \tag{13}$$

where

$$-r\omega = \rho\left(r^2 \frac{dH}{d\psi} - \frac{1}{2}\frac{d\Gamma^2}{d\psi}\right) - pr^2 \frac{d\frac{\Delta S}{R}}{d\psi} \tag{14}$$

$$H = \frac{\gamma p}{(\gamma-1)\rho} + \frac{1}{2}(u^2 + v^2 + w^2) \tag{15}$$

$$\Gamma = rv \tag{16}$$

Γ is the circulation (due to the swirl).

From the inlet conditions, ΔS, H and Γ can be obtained as functions of ψ. These equations provide an inviscid model for vortex breakdown simulation. For incompressible flows, the authors (1987) [14] showed that beside the trivial solution, there are many admissible solutions with multiple bubbles. In these calculations, it is assumed that H and Γ have the same functional dependence on ψ, outside and inside the recirculation regions.

The problem of the closed streamlines was discussed first by Prandtl [11] in 1922 and Batchelor [12] in 1956 for two-dimensional flows. In their model, ω is constant in the recirculation zone. Guderley [13] (1977) extended their analysis to the case of compressible axisymmetric flows. It is argued that, in flows with closed streamlines, entropy, total enthalpy and circulation are determined by the requirement that the cumulative effects of viscosity and heat conduction vanish. The necessary conditions can be expressed as the balance equations for energy, entropy and angular momentum which must be satisfied for each volume bounded by an axisymmetric surface ψ = constant. (If the Navier-Stokes equations were exactly satisfied, then corresponding conditions would hold for <u>any</u> toroidal surface in the flow field. According to Guderley's analysis, this requirement is weakened; only surface ψ = constant need be considered.)

The implementation of Guderley's conditions requires the solution of a system of three integral equations and will be discussed somewhere else.

In the present work, it is assumed that ΔS, H and Γ have the same functional dependence on ψ, outside and inside the recirculation regions. The compressibility effects are manifested through the variation of the density; equation (13) is no longer elliptic, its type depends on the Meridian Mach number. This can be shown by rewriting equation (13) in a quasilinear characteristic form

$$(1 - \frac{w^2}{a^2})\psi_{xx} - \frac{2uw}{a^2}\psi_{xr} + (1 - \frac{u^2}{a^2})\psi_{rr} - \psi_{r/r} = -r\omega \qquad (13')$$

The type of equation (13') (Or [13]) does not depend on the swirl component v.

Central difference schemes can be used everywhere as long as the Meridian Mach number is less than one. For mixed type flows, artificial viscosity is used as before. The continuity equation is modified with a term proportional to the Laplacian of the density.

$$(r\rho w)_x + (r\rho u)_r = \nu((r\rho_r)_r + (r\rho_x)_x) \qquad (17)$$

A stream function is introduced such that equation (14) is automatically satisfied

$$\psi_r = r\rho w - \nu r\rho_x, \qquad -\psi_x = r\rho u - \nu r\rho_r \qquad (17')$$

or,

$$w = \frac{\psi_r + \nu r\rho_x}{r\rho}, \qquad u = \frac{-\psi_x + \nu r\rho_r}{r\rho} \qquad (17'')$$

Substituting equation (17") in the definition of vorticity, $w_r - u_x = -\omega$, one obtains

$$\left(\frac{\psi_r + \nu r\rho_x}{r\rho}\right)_r + \left(\frac{\psi_x - \nu r\rho_r}{r\rho}\right)_x = -\omega \qquad (18)$$

The artificial viscosity terms cancel identically and equation (18) is the same as equation (13). A finite difference scheme is introduced, where the same property is valid on the discrete level, as in the two-dimensional case. The following case is simulated. At the inlet, the velocity components are w=1, u=0, and v is given by a blending of a rigid body rotation and a potential vortex.

$$v = Vr(2-r^2) \qquad \text{for } r<1$$
$$v = V/r \qquad \text{for } r\geq 1 \tag{19}$$

These are the same inlet conditions the authors (1987) used for the incompressible flow simulation where Γ and H were easily obtained as functions of ψ. The trivial solution corresponds to $u_x = 0$, $\omega = 0$, and $\psi = r^2/2$. In terms of a perturbative stream function, $\psi = r^2/2-\phi$, the ψ-equation becomes

$$r(\phi_r/r)_r + \phi_{xx} = -V^2\alpha^2\phi \tag{20}$$

where α^2 is a function of ϕ and r. Non-trivial solutions of Eq. (20) can be obtained where ϕ (or its derivatives) vanish on the boundaries. For compressible flow simulation, it is assumed that ρ=1 at the inlet. Hence, the same trivial incompressible solution u_x=0, ω=0, and $\psi=r^2/2$, is admissible. Moreover, at the inlet, the pressure is obtained from the momentum equation, $\partial p/\partial r = v^2/r$ with $P = 1/\gamma M_\infty^2$ in the far field. The entropy profile is calculated from the relation $P = \rho^\gamma e^{\Delta S/cv}/\gamma M_\infty^2$. Hence, H and $\Delta S/R$ can be calculated as functions of ψ. Γ has the same form as in the incompressible flow case. Thus, the right hand side of Equation (13) can be calculated in terms of r, ψ and ρ. The discrete forms of the stream function equation and the energy (Bernoulli's) equation are solved by Newton's method and a direct solver and the results for different free stream Mach numbers are plotted in Figure (3). To avoid the trivial solution, Hill's vortex or the incompressible solution is used as an initial guess, and a continuation process is adopted for higher Mach number cases. Again only flows with weak shocks where the entropy jump is negligible are considered, otherwise a shock fitting procedure is needed.

Conclusion

Two-dimensional and axisymmetric inviscid rotational flows are simulated, based on an alternate formulation using stream function equations. Unlike Euler equations, central differences are used everywhere (without any artificial viscosity) for subsonic flows. For transonic flows, a new scheme is introduced, where the artificial viscosity does not introduce any artificial vorticity (provided proper numerical boundary conditions are implemented). The velocity vector is modified by a gradient of a scalar function of the density; the vorticity field, however, is unaltered since the curl of the gradient vanishes identically. The main advantage of the present approach is that the discretization of the momentum equations is completely avoided. Extension of the stream function formulation to three-dimensional flows is possible, but it is not straightforward because of the boundary conditions.

References

1. Emmons, H.: Flow of a compressible fluid past a symmetrical airfoil in a wind tunnel and in free air. NACA Technical Note 1746 (1948).

2. Mitchell, A.: Application of relaxation to the rotational field of flow behind a bow shock wave. Quart. J. Mech. and Applied Math., Vol. IV, Pt. 3, (1951), pp. 371-383.

3. Sells, C.: Plane subcritical flow past a lifting airfoil. Proc. Royal Soc. A. 308 (1968) 377-401.

4. Steger, J.: Ph.D. Thesis, Iowa State University, 1969.

5. Kapriva, D.: Ph.D. Thesis, University of Arizona. 1982.

6. Morice, Ph.; Cambier, L.; Veuillot, J. Numerical computation of transonic flow past an axisymmetric nacelle. ICAS Paper 84-1.10.2.

7. Amara, M.: Ph.D. Thesis, Uiversity of Paris VI. 1983.

8. Dubois, F.: Ph.D. Thesis, University of Paris VI. 1988.

9. Hafez, M. and Lovell, D.: Numerical solution of transonic stream function equation. AIAA Computational Fluid Dynamics Conference. Palo Alto, CA. (July 1981). See also:

9'. Hafez, M. and Lovell, D.: Entropy and vorticity corrections for transonic flows. AIAA Computational Fluid Dynamics Conference, Danvers, Mass. (July 1983).

10. Hafez, M.; Palaniswamy, S.; and Mariani, P.: Calculations of transonic flows with shocks using Newton's method and direct solver, Part II. AIAA Paper 88-0226.

11. Prandtl, L.: Motion of fluids with very little viscosity. NACA Technical Memorandum No. 452 (1928).

12. Batchelor, G.: On steady laminar flow with closed streamlines at large Reynold's number. JFM, Vol. 1 (1956) 177-190.

13. Guderley, K.: Steady compressible swirl flows with closed streamlines at high Reynold's numbers. JFM, Vol. 82 (1977) 321-347.

14. Hafez, M.; Ahmad, J.: Kurvila, G.; and Salas, M.: Vortex breakdown simulation, Part I. AIAA Paper 87-1343.

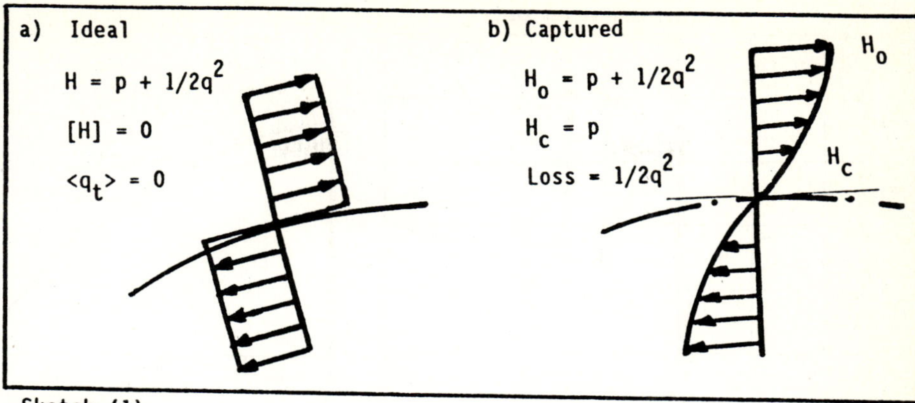

Sketch (1)
Comparison between Ideal and Captured Wakes for Incompressible Flows.

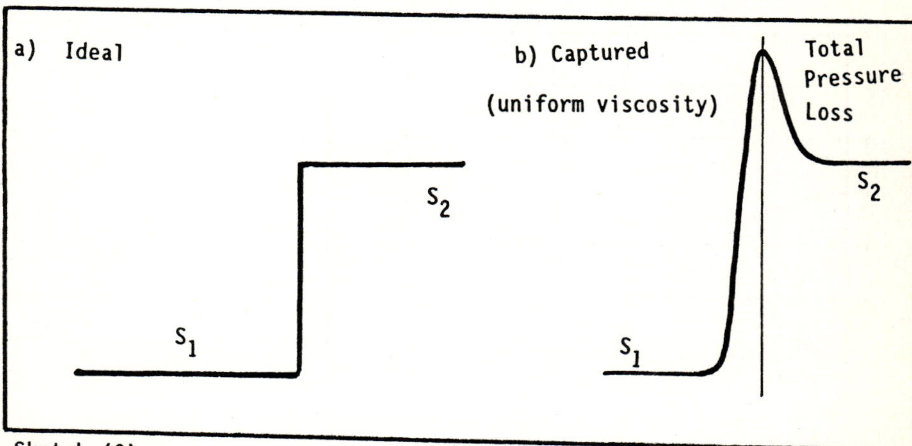

Sketch (2)
Comparison between Ideal and Captured Shocks.

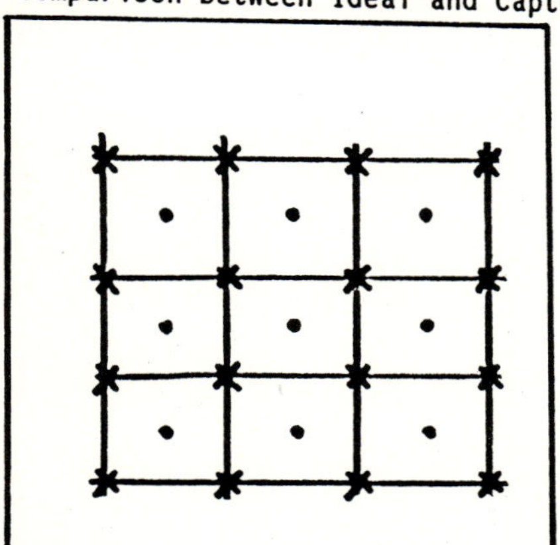

Sketch (3)
The grid for ψ and ρ.

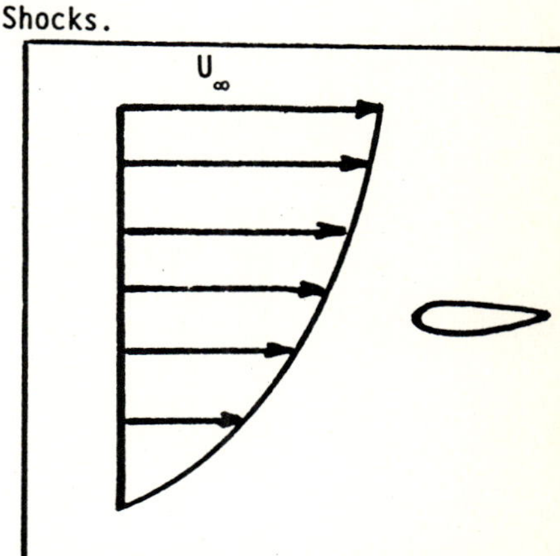

Sketch (4)
Airfoil in Inviscid Shear Flow.

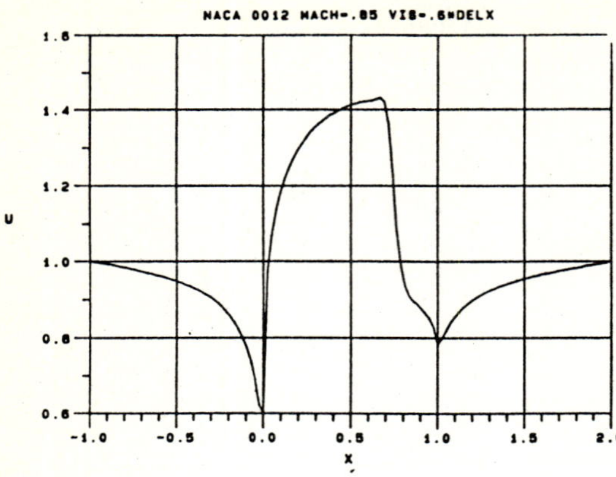
NACA 0012 MACH=.85 VIS=.6*DELX
U
X

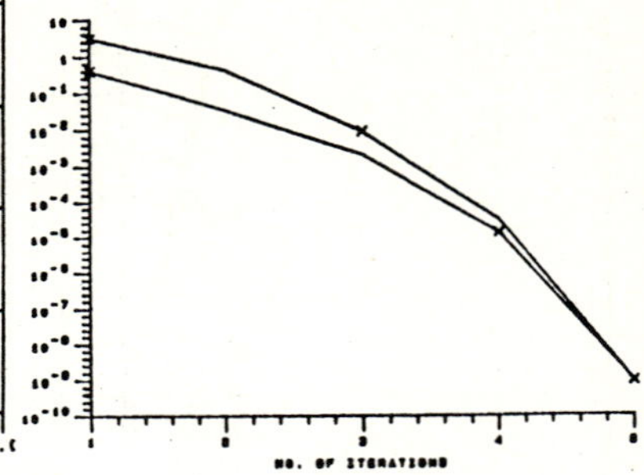
NO. OF ITERATIONS

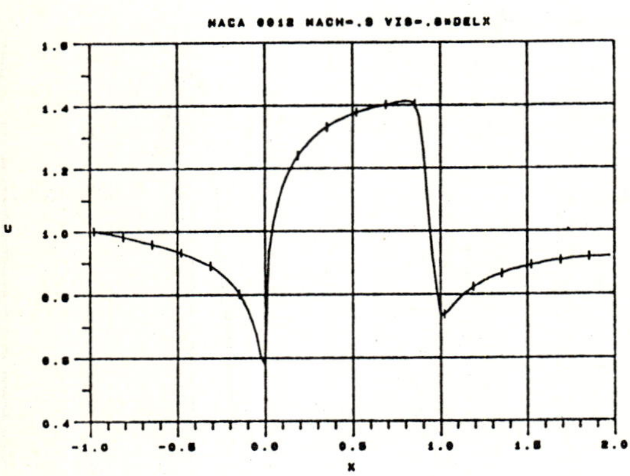
NACA 0012 MACH=.9 VIS=.6*DELX
U
X

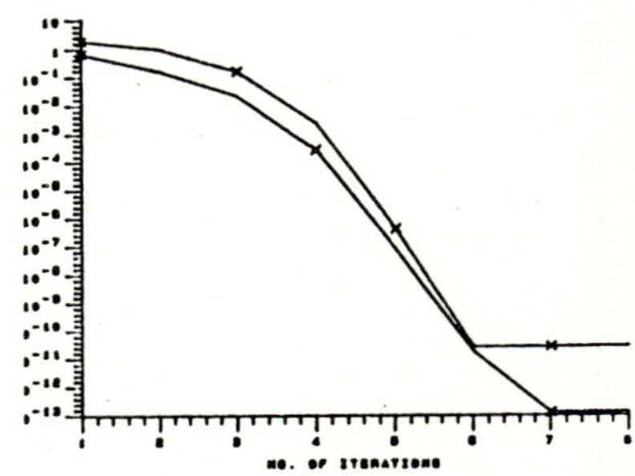
NO. OF ITERATIONS

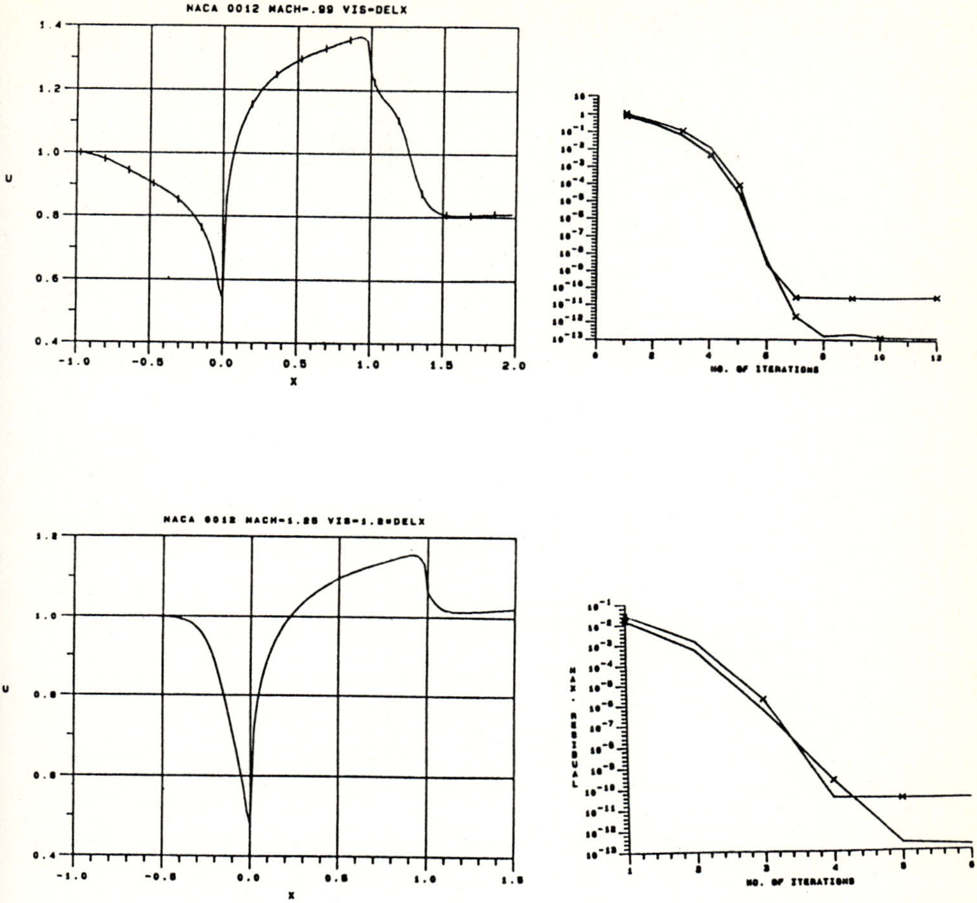

Figure (1)
Calculations of Transonic Flows over NACA0012 Airfoil
(Domain: 3 x 2;   Grid: 125 x 30).

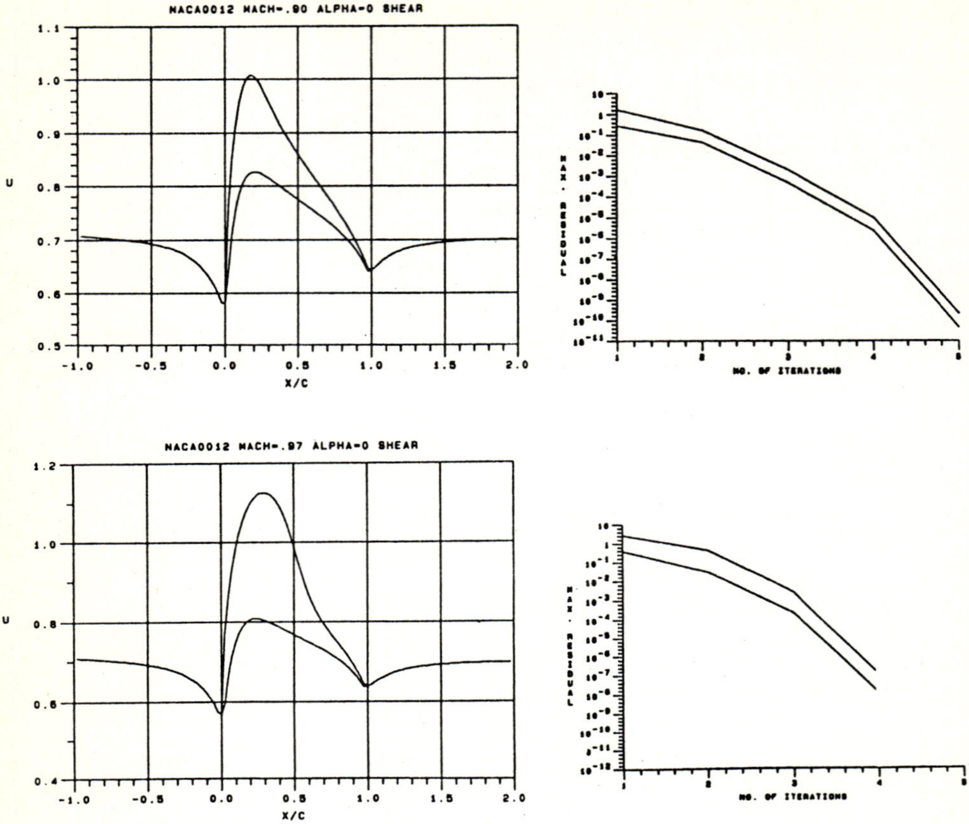

Figure (2)
Calculations of Rotational Flow over NACA0012 Airfoil
(Domain: 3 x 6; Grid: 125 x 30).

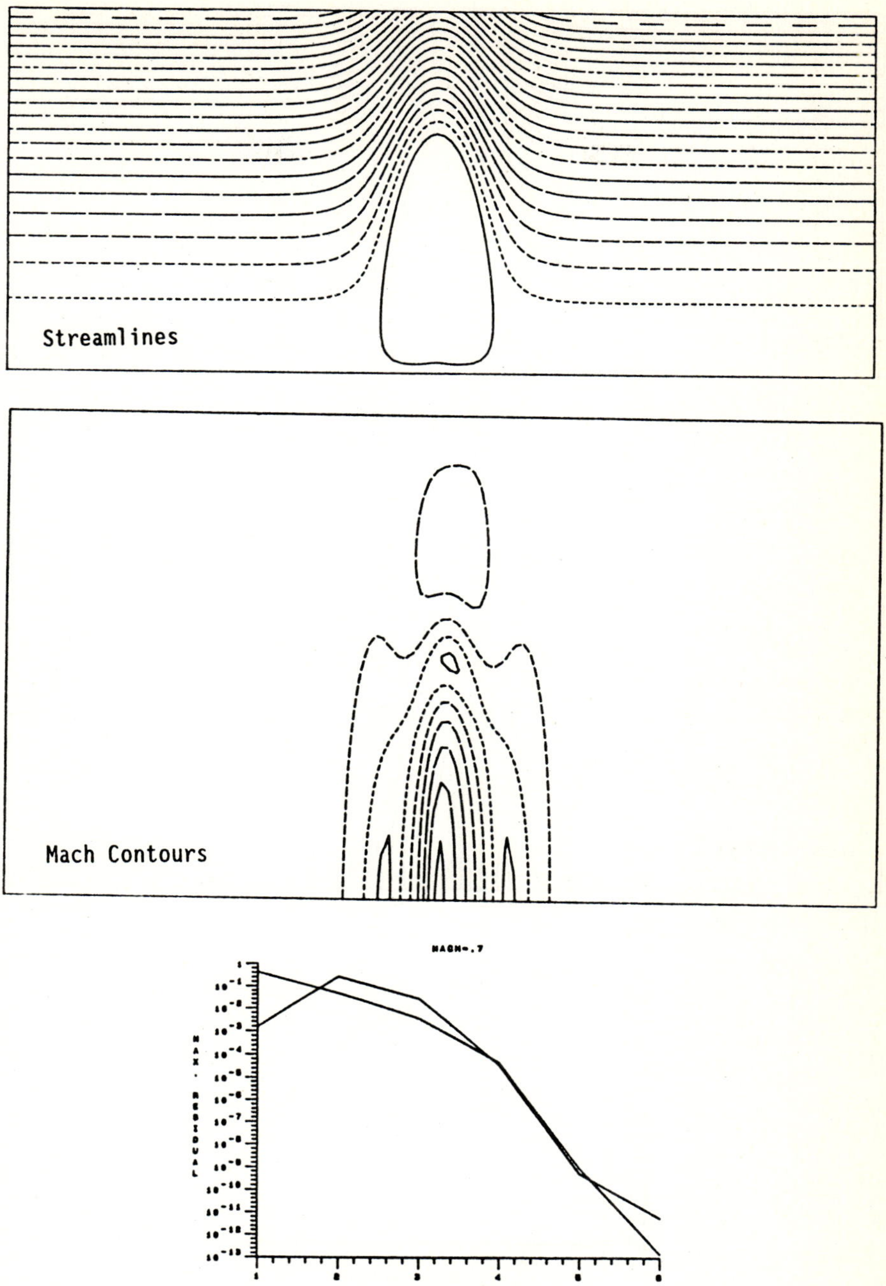

Figure (3-a)
Vortex Breakdown Simulation
(Domain: 20 x 2; Grid: 120 x 32)

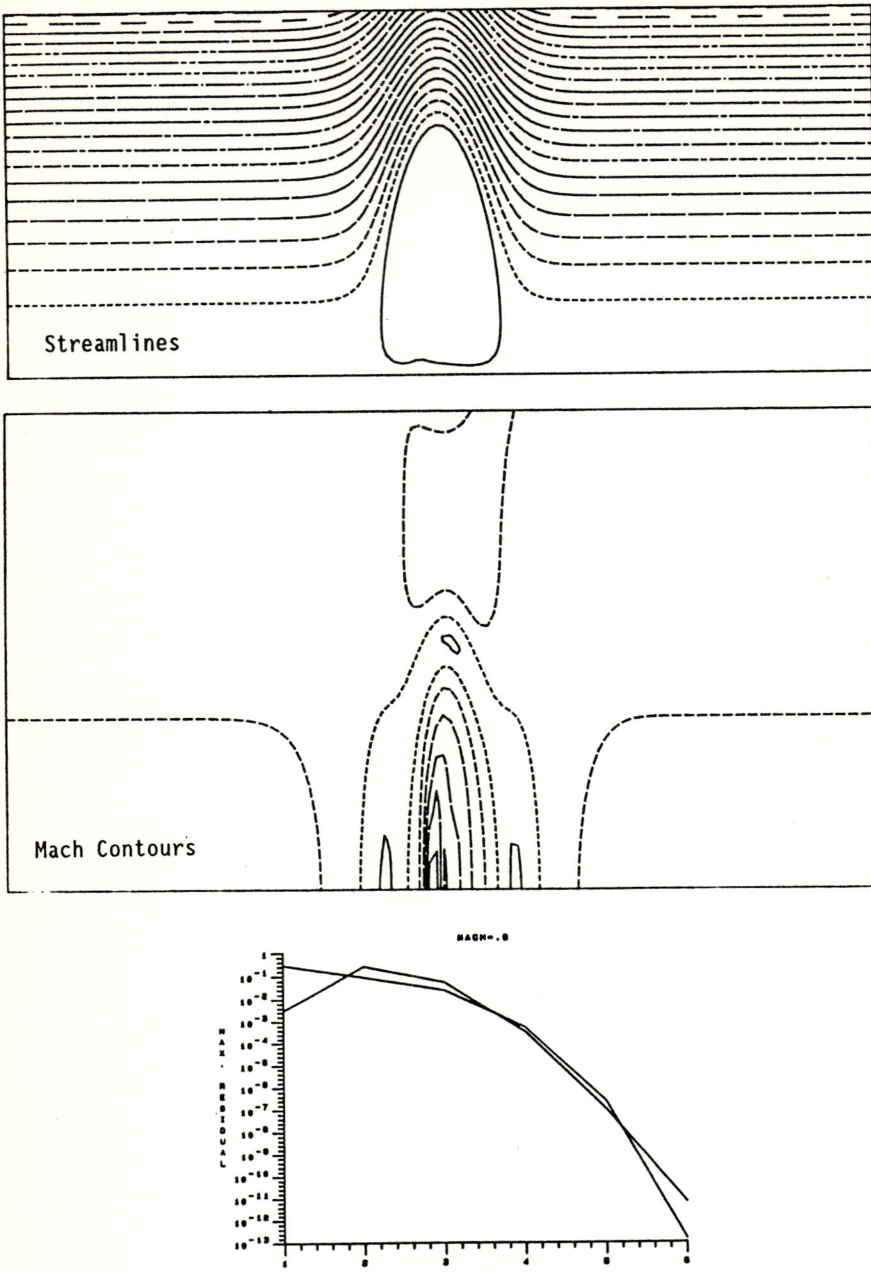

Figure (3-b)
Vortex Breakdown Simulation
(Domain: 20 x 2; Grid: 120 x 32)

Turbulent Shock –
Boundary Layer Interaction

Recent Basic Studies on Transonic Shock-Wave/ Turbulent Boundary-Layer Interactions

J. Délery

ONERA
Aerodynamics Department
92320 Chatillon, FRANCE

Summary

Basic and well-documented experiments are still needed to help in the physical understanding of transonic interacting flows so as to provide test cases for validating theoretical predictions. Three basic studies are presented. The first one deals with turbulence modelling in 2-D shock-induced interactions, the second is a detailed investigation of the structure of a 3-D channel flow, the third study is relative to passive control of a shock-induced interaction occurring in a 2-D channel flow.

1 - Introduction

Since the last Symposium Transsonicum, already held at Göttingen, in 1975, considerable progress has been made both in the physical understanding and in the theoretical prediction of turbulent shock-wave/boundary-layer interactions. On the experimental side, the advent of Laser Doppler Velocimetry (LDV) has operated a break-through in our capacity to investigate complex turbulent flows. With LDV, it has become possible to routinely perform reliable and accurate measurements in high speed flows containing shock waves and large separated regions. Not only the mean flow structure can be described in detail but also the complete Reynolds stress tensor can be determined, even in 3-D configurations [1].

It is remarkable that the spectacular advances in experimental techniques were accompanied by the considerable development of the theoretical tools which now permit a detailed description of complex shock-induced interactions. Inviscid-viscous interactive methods are now able to accurately predict a large variety of strongly interacting flows [2]. Analytical methods, derived from the triple-deck theory, have been extended to turbulent interactions and have largely contributed to our understanding of the fine structure of this kind of flows [3]. The

J. Zierep, H. Oertel (Eds.)
Symposium Transsonicum III
IUTAM Symposium Göttingen, 24.-27. 5. 1988

considerable increase in computer performance and the parallel progress in numerical methods now render possible the solution of the full (time averaged) Navier-Stokes equations for predicting 2-D as well as 3-D transonic viscous flows [4]. Mesh refinements presently achieved allow an accurate resolution of the smallest mean features of a turbulent flow. However, all the above theoretical approaches are hampered by the difficulty of correctly representing the turbulent terms in the time averaged equations. In the domain of turbulence modelling, progress is much slower and important work has still to be done [5].

The aim of this paper is to present three basic studies undertaken with a view to helping in the improvement of predictive methods applied to transonic shock-induced interactions.

2 - Turbulence Modelling in 2-D Transonic Interactions

The experimental part of this study has been carried out in the transonic channel shown in Fig. 1. The test facility consists of a 2-D converging-diverging channel whose top and bottom walls support interchangeable bumps allowing the flow to accelerate to slightly supersonic speeds. A second throat, with an adjustable cross-section located downstream of the test section, produces a shock wave whose interaction with the boundary layer developing on the lower wall of the channel is studied.

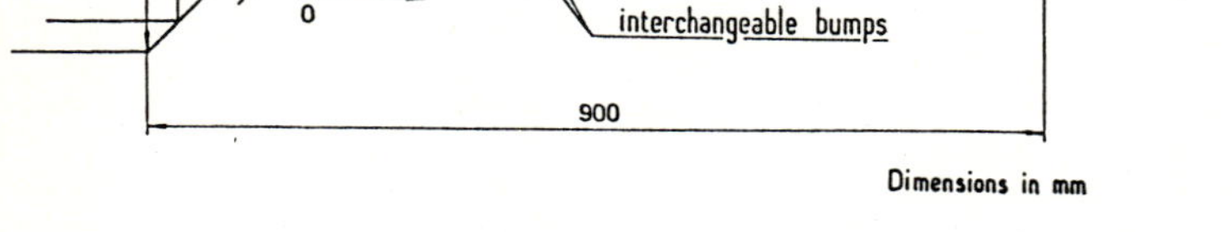

Fig. 1 – Definition of the two-dimensional transonic channel.

The turbulence model analysis focused on three interactions which were carefully probed by using a two-component LDV system. The experimental results having been published in several circumstances [6,4], we shall concentrate here on the conclusions of a theoretical study in which these results were used to test six turbulence models chosen among the most commonly used in Navier-Stokes calculations [7].

As solution of the Navier-Stokes equations is still costly in terms of computing time, the present study has used a boundary-layer approach. Since separation occurs in some of the computed interactions, an inverse mode of solution has been adopted, the measured boundary-layer displacement thickness being prescribed as boundary condition. Careful comparisons with Navier-Stokes calculations have shown that the boundary-layer approach was legitimate in the present transonic applications.

The turbulence models considered are the equilibrium algebraic models of Michel, Quémard and Durant, of Alber and of Baldwin and Lomax, the non-equilibrium algebraic model of Johnson and King, the $[k, \epsilon]$ transport equation model of Jones and Launder and the Algebraic Stress Model (ASM). Details about these models and corresponding references can be found in [7]. The ASM can be considered as an intermediate step before the complete second order closure in which transport equations are used to compute the Reynolds stresses [8].

Here, we will only examine results relative to the most severe case in which an extended separated zone forms. Figure 2 shows an interferogram of the flowfield and a tracing of iso-Mach lines revealing the existence of a strong interaction leading to the formation of a large separated region with the accompanying lambda shock system. The computed wall pressure distributions are compared to experiment in Fig. 3a (equilibrium models) and Fig. 3b (non-equilibrium models). In the present case, where a large separation bubble forms, the rapid pressure rise at separation is followed by a well–defined plateau. Equilibrium models all indicate far too high a compression at the beginning of interaction and the plateau level, when it is predicted, is highly overpredicted. As can be seen in Fig. 3b, the models incorporating a history effect lead to a much better agreement with experiment. The result obtained with the ASM is even spectacular.

The mean streamwise velocity distributions at several X-wise locations in the interaction domain are plotted in Figs. 4a to d. As can be seen, the shape of the velocity distribution is predicted poorly by the equilibrium models, both in the separated region and downstream of reattachment where the profiles are not sufficiently filled. In spite of a clear improvement, the models with history effect also exhibit defects such as an underprediction of the reversed flow amplitude, to slow a filling near the wall downstream of reattachment and an underpredicted thickening of the boundary-layer.

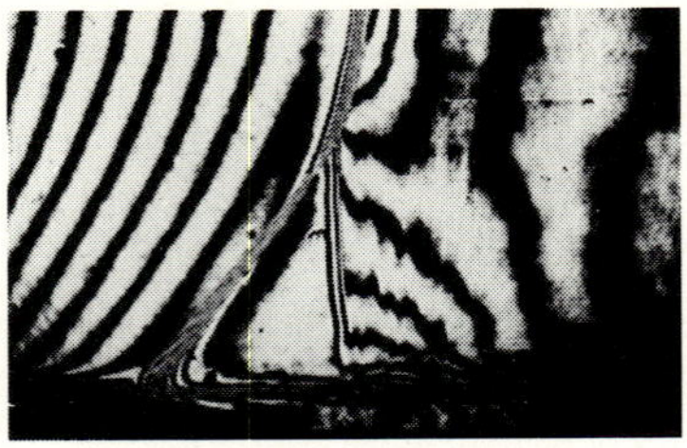

a) Interferogram of flowfield

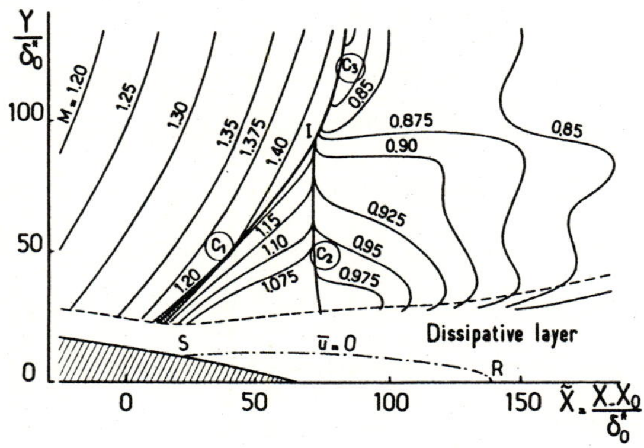

b) Iso-Mach lines

Fig. 2 - 2-D transonic interaction. Inviscid flowfield structure.

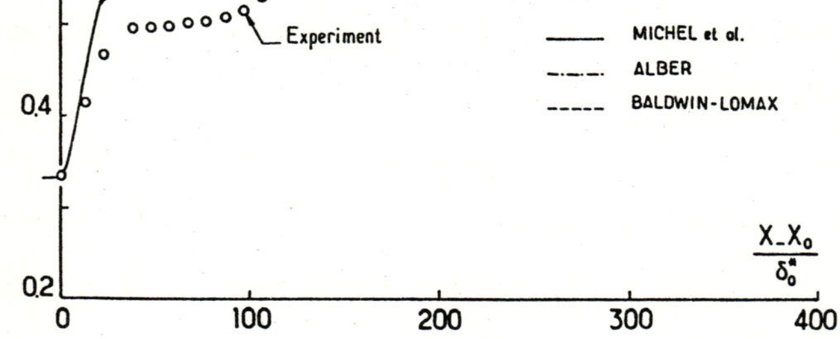

a) Equilibrium models

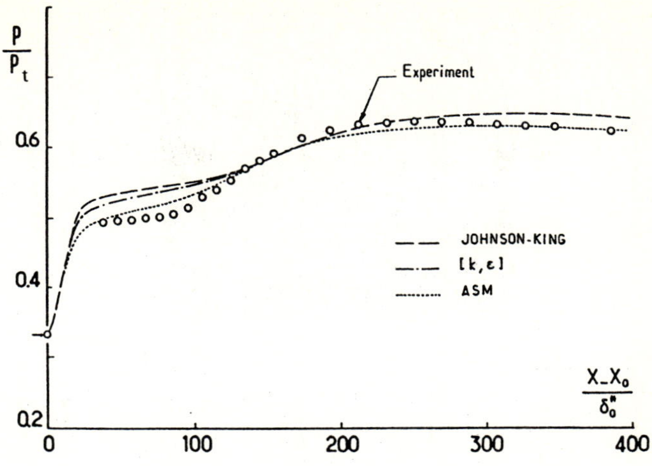

b) Non-equilibrium models

Fig. 3 – Computed wall pressure distributions.

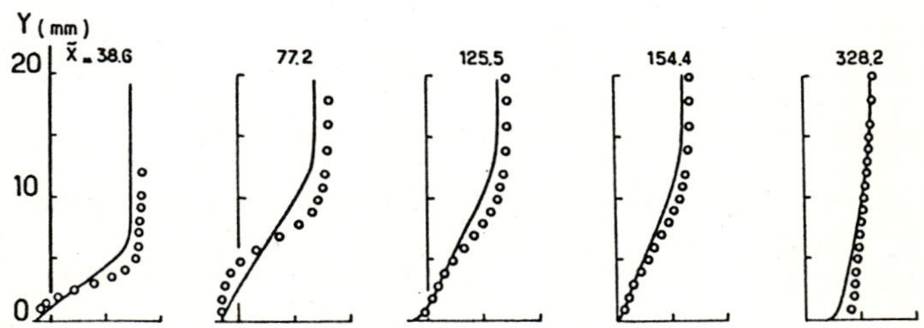

a) Michel et al.

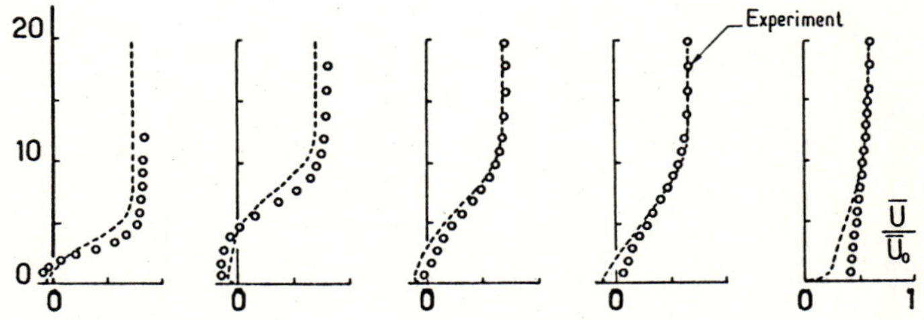

b) Baldwin-Lomax

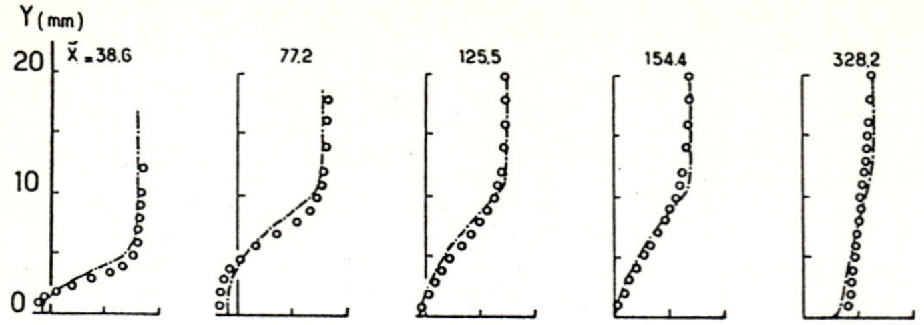

c) [k, ϵ] model

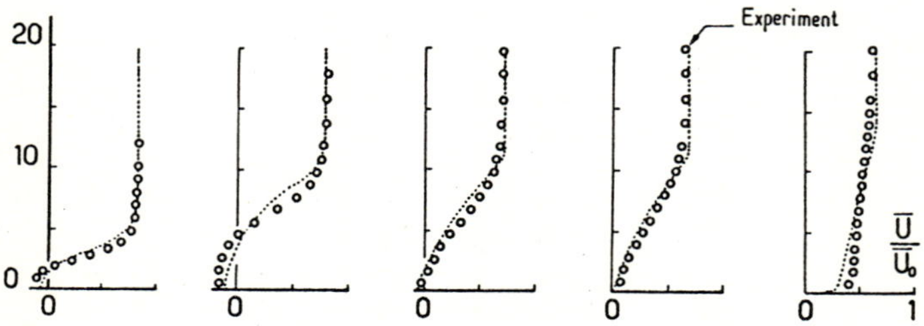

d) ASM

Fig. 4 – Computed mean streamwise velocity profiles.

As far as the turbulent shear-stress is concerned (see Figs. 5a to d), no model gives a really satisfactory result. As for the [k, ϵ] model and the ASM, the prediction, although acceptable in the upstream part of the interaction, strongly underpredicts the maximum levels reached during the interaction. The computed turbulence relaxation is much too fast. However, a good agreement between the computed and measured Reynolds shear-stress profiles is observed for the most downstream stations, but the lateral diffusion is poorly represented.

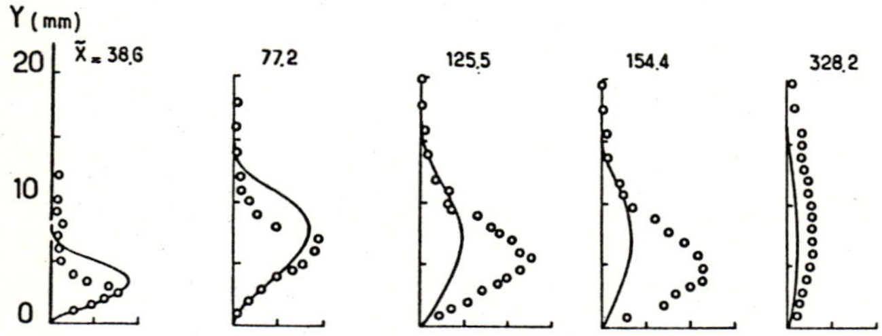

a) Michel et al.

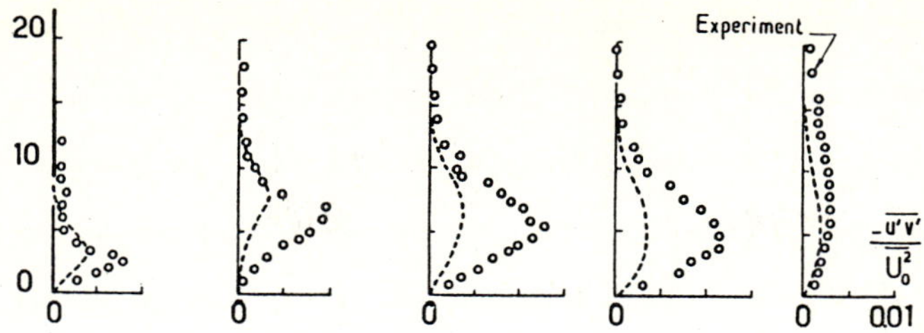

b) Baldwin-Lomax

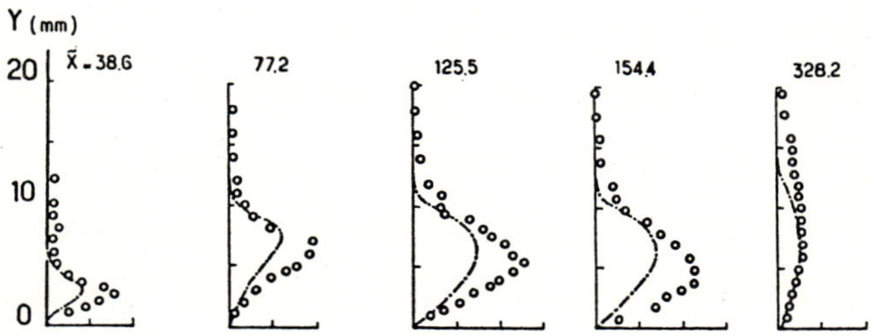

c) [k, ε] model

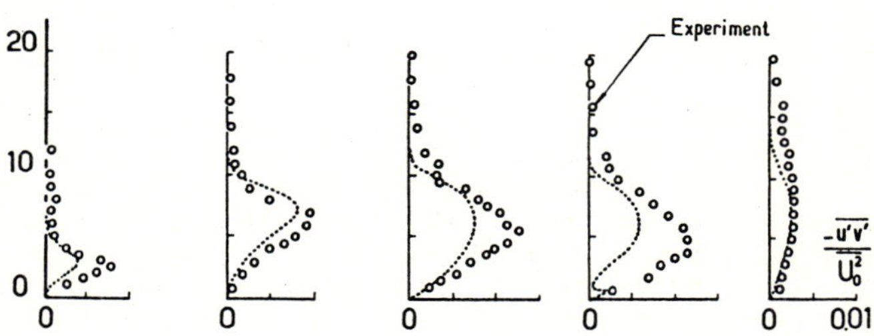

d) ASM

Fig. 5 – Computed turbulent shear stress profiles - Flow C.

The above results lead to the following conclusions. It is clear that a history effect must be included in any turbulence model aiming at the prediction of shock-induced interactions. The best results are obtained with the [k, ε] model and the ASM which use transport equations for turbulence quantities. In particular, the wall pressure distribution and the mean flow properties are predicted with honorable accuracy. However, the turbulence field is still far from being correcly predicted. Thus, no model is capable of accounting for the slow relaxation of turbulence downstream of the interaction region. In fact, in such a flow, a coupling between the boundary-layer turbulence and the shock wave is established at the shock foot causing an oscillation of the shock wave. This phenomenon acts as a turbulence amplifier which results in the formation of large structures which persist far downstream of the interaction region. Under these circumstances, it does not seem probable that models which do not take this mechanism into account can give a truly realistic prediction. The analysis of this shock-wave/turbulence interaction, on which research is only beginning, deserves a thorough investigation [9].

3 - Transonic Interaction in a Three-Dimensional Channel

Analysis of transonic shock-wave/boundary-layer interactions is a far more complicated task in 3-D flows than in 2-D flows. Both conceptual and practical difficulties are encountered by experimentalists and theoreticians in the investigation of 3-D configurations [10]. As a matter of fact, accurate calculations of 3-D interacting flows not only require realistic turbulence models but also necessitate large computer capacities enabling the solution of the governing equations on a grid containing a very large number of points (10^5 to 10^6). On the other hand, arriving at a clear physical grasp of the structure of a 3-D separated flow is difficult, specially in transonic situations.

The present experimental investigation has been carried out on the same transonic facility as that used for the above 2-D study. A bump, whose crest was swept at an angle of 30° relative to the incoming flow, is mounted on the test section lower wall, as shown in Fig. 6. The three other faces of the channel are flat. The upstream subsonic flow accelerates in the converging part of the channel to reach a sonic state in the vicinity of the bump crest. The flow becomes supersonic downstream and then, due to the choking effect set up by a second throat (not represented in Fig. 6), it slows down through shock waves interacting with the boundary layers of the channel walls.

The mean flow structure was carefully investigated by means of surface pressure measurements, surface flow visualizations and field measurements performed with a three-component LDV system. The volume explored with the LDV system is indicated in Fig. 6. Explorations were made along vertical lines contained in 5 vertical planes parallel to the test section sidewalls [11].

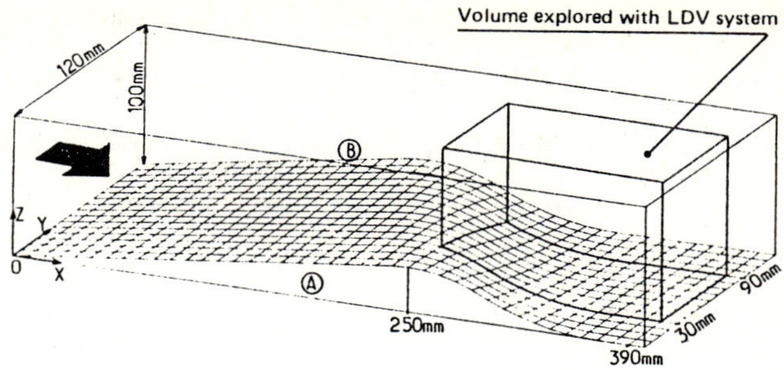

Fig. 6 – 3-D interaction - Definition of transonic channel.

Surface patterns obtained by visualization techniques provide a precious and nearly indispensable help in understanding the physics of separated 3-D flows. In effect, such patterns can be assimilated with a good level of confidence to the skin-friction line patterns. It is by inspecting the organization of the skin-friction lines and observing the singular points in their pattern that the concepts of separation and reattachment can be defined in 3-D flows where these notions are trickier to grasp than in 2-D flows. Interpretation of surface patterns is essentially based on the analysis of Legendre [12] inspired by the work of Poincaré [13] on the singularities of differential systems. Let us simply mention here the possible existence of isolated singular points where the friction and vorticity vectors both cancel out. Of these points, we should distinguish: nodes, foci and saddle points.

We are then brought to introduce the fundamental concept of separation line (S) which is defined as a skin–friction line passing through a saddle point. The separation line, in the terminology Legendre introduced, can be associated just as well with an attachment (or reattachment) or with what is really separation. In the separation case, the flow in the vicinity of (S) "leaves" the wall according to the process described by Lighthill [14], the skin-friction lines having the tendency to converge towards (S). On the other hand, for an attachment line, a little above (S), the outer flow "dives" in the direction of the wall and the skin-friction lines have a tendency to diverge from (S). A true separation line is the trace on the body surface of a separation sheet in the field which rolls up to constitute what is commonly called a vortex. The trace of a vortex on the wall being a focus, a separation line, origin of a separation sheet, must end up in this focus. There is no place here to comment any further on these concepts which are essential for a rational interpretation of separated 3-D flowfield.

The diagrams in Figs. 7a to d have been established from photographs and careful examination during the test of the surface flow visualizations made on the four faces of the channel. Although these diagrams are meant to be qualitative, they do keep closely with the proportions of reality. Here, we will only examine in some detail the flow over the bump (Fig. 7a). Its skin-friction line pattern contains two foci F_1 and F_2 clearly visible in the left hand half of the channel. The separation line (S_1) passing through saddle-point C_1 and spiralling around F_1 and F_2 constitutes a "barrier" approximately at the foot of the oblique shock wave of the lambda system attached to the interaction (see below). The skin-friction lines coming from upstream that are contained in the domain delimited by the separation line (S_2) passing through saddle-point C_2 are going to wind around either F_1 or F_2 depending on whether they are located to one side or the other of separation line (S_3) which carries both C_1 and C_2. The structure made up of (S_1), (S_2) and the two foci F_1 and F_2 is classically observed in shock-wave/boundary-layer interactions in a 2-D channel with strong 3-D effects [15]. In such a situation, (S_1) can be associated to a well-characterized separation and (S_2) to an attachment. This makes C_1 a separation saddle-point and C_2 an attachment saddle-point.

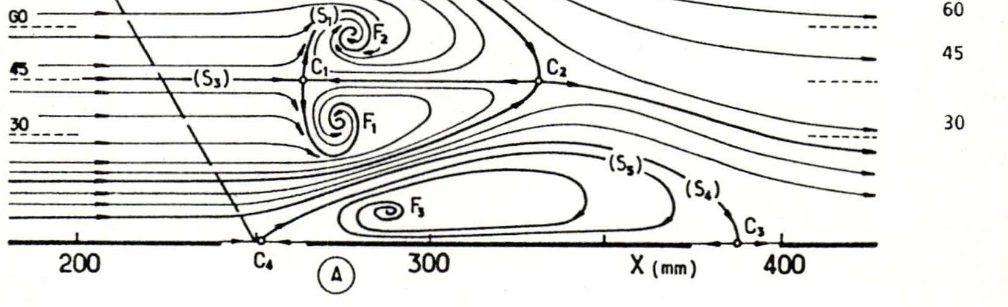

a) Lower wall

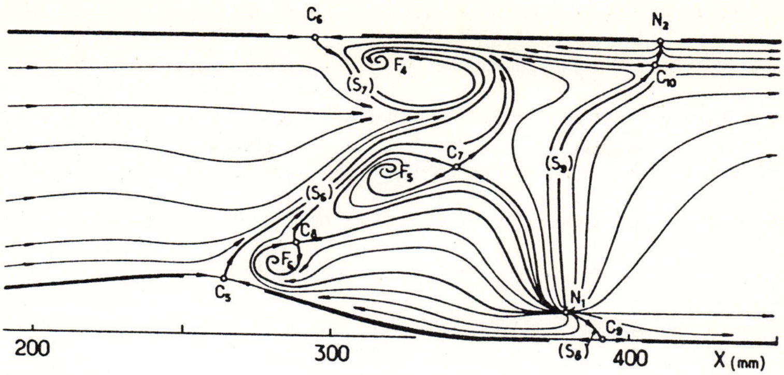

b) Side wall A

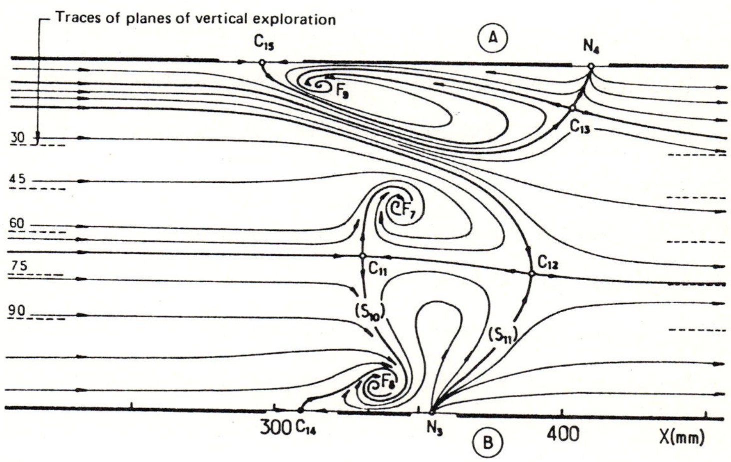

c) Upper wall

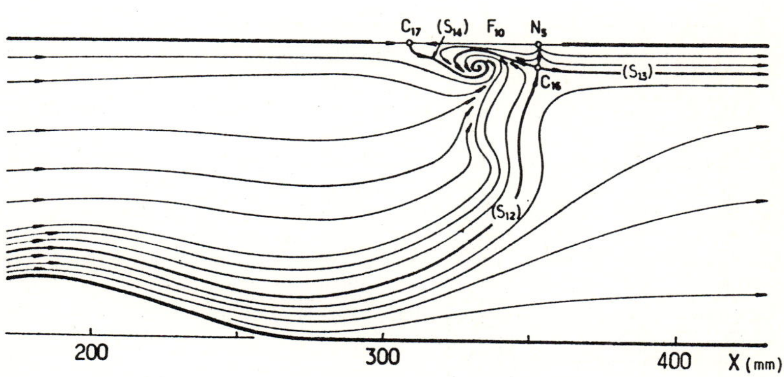

d) Side wall B

Fig. 7 – Skin-friction line pattern.

The spirals winding around foci F_1 and F_2 are the traces on the wall of vortices escaping into the flow with separation line (S_1) being the origin of a separation sheet that winds around the axes of the vortex structure. Returning to the diagram in Fig. 7a, we observe that the skin-friction lines from upstream divide into two families. The first, near face B, all flow continuously downstream. The other, near wall A, separate in turn into two subfamilies:

- those lines included between (S_2) and the separation line (S_4), leading to half-saddle C_3, that continue their path downstream;
- those lines between wall A and (S_4) that wind around a third focus F_3 to form a vortex delimited by the separation line (S_5) coming from half-saddle C_4.

A similar description of the wall patterns in terms of singular points and separation lines can be made for the other faces of the channel, care being taken to insure continuity of patterns at the crossing of the corner lines.

Now, let us consider the structure of the outer flow by considering the iso-Mach lines in the longitudinal planes (X,Z) which have been probed with the LDV system (see Figs. 8a to e). The spanwise locations of these planes are indicated in Fig. 7a. In each of these planes, a lambda shock pattern is built up that includes a leading oblique shock (Γ_1), a quasi-normal shock (Γ_2), occupying about half the channel, the leg of the lambda (Γ_3) being very weak and hardly discernible. The oblique shock (Γ_1) is clearly pronounced in the Y = 30, 45 and 60mm planes. But at Y = 90mm, it is considerably weakened and replaced by a continuous compression. It should be noted that the feet of shock-waves (Γ_1) and (Γ_2) vary little spanwise, which shows that these shocks are not very sharply inclined with respect to the velocity at the channel inlet. The iso-Mach lines also reveal the large region of separated fluid starting from the foot of shock (Γ_1). This separated zone is well visible in planes Y = 30 and 45mm It disappears completely when one tends towards face B. A separated zone is also clearly visible on the upper wall, induced by shock (Γ_2). These outer field features are of course correlated with the surface flow patterns.

The above flow is a typical example of a 3-D shock-induced interaction occurring in a transonic channel. Its thorough experimental investigation, which has to be completed by turbulence measurements, can provide instructive information on the fine structure of such flows and furnish a particularly well documented test case for validating 3-D Navier-Stokes calculations.

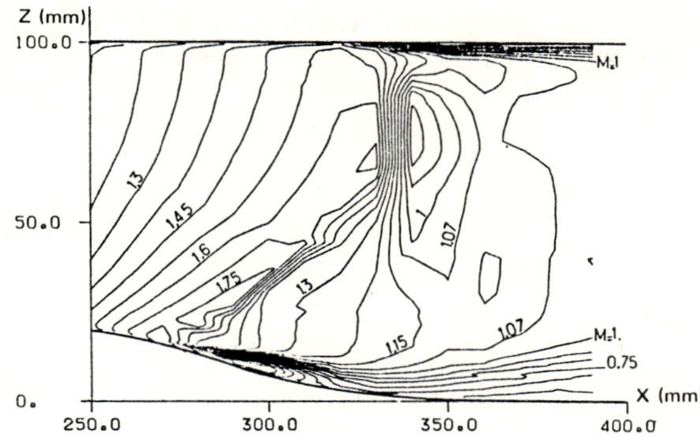

a) Y = 30 mm

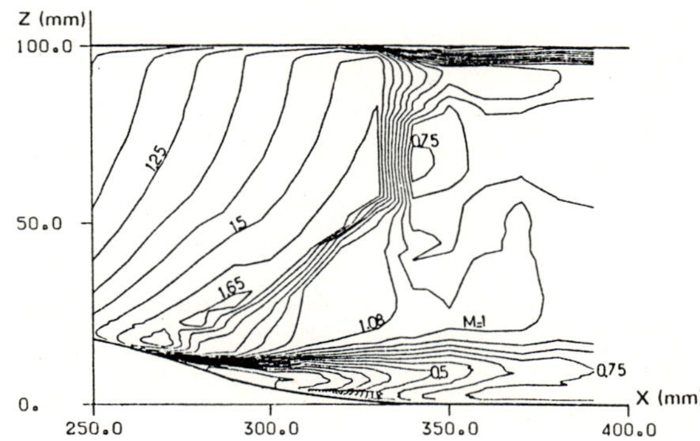

b) Y = 45 mm

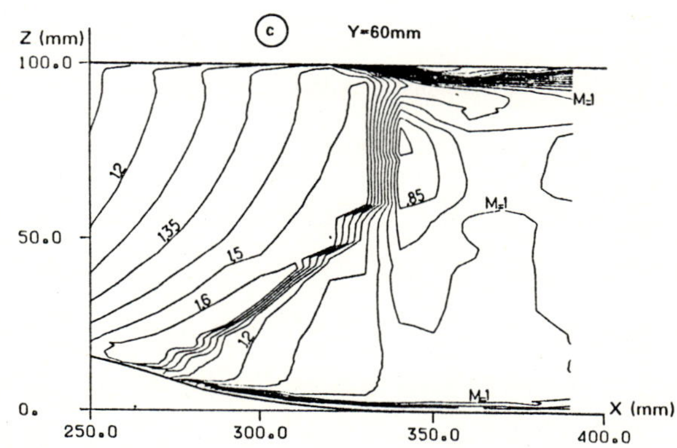

c) Y = 60 mm

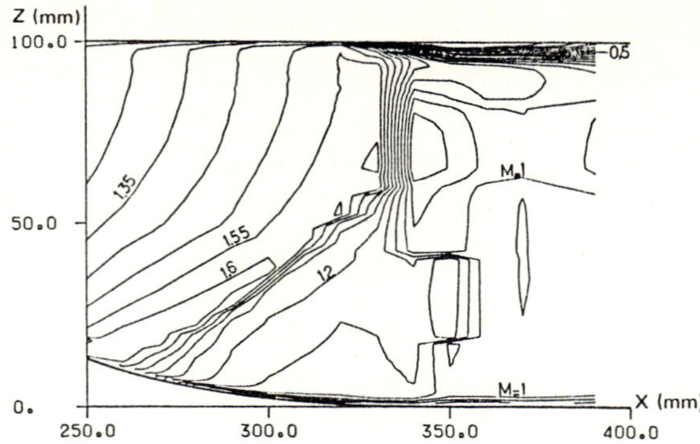

d) Y = 75 mm

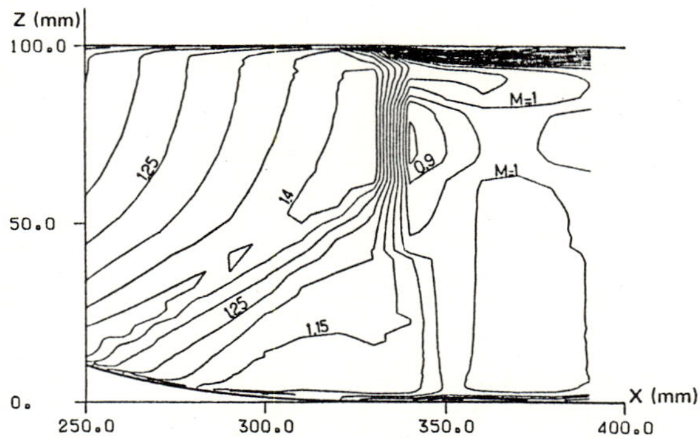

e) Y = 90 mm

Fig. 8 – Iso-Mach lines in longitudinal (X, Z) planes.

4 - A Shock-Induced Interaction with Passive Control

The third basic research considered here is an experimental investigation of passive control applied to a transonic shock-induced interaction. This technique has been suggested rather recently as an economical means to improve airfoil performance at transonic speeds [16]. Basically, passive control is applied by replacing a part of the airfoil upper surface by a porous plate in communication with a closed chamber installed inside the airfoil. The porous plate being implemented in the shock region, there occurs a natural flow circulation, via the cavity, from the high pressure downstream part to the low pressure upstream part of the interaction. Systematic and careful experiments have shown that passive control can be very effective in improving supercritical airfoil performance at off-design conditions [17].

The experiments presented here were executed in the same facility as the one used for the preceding studies. As shown in Fig. 9, the test set up consists of a 2-D channel with a flat upper wall and a contoured lower wall designed so as to produce a supersonic uniform flow with a nominal Mach number equal to 1.4. The passive control region was implemented at the downstream extremity of the contoured lower wall at a place where the local Mach number is close to 1.3. The streamwise extent of the control zone was equal to 70mm, which approximately represents 14 times the incoming boundary-layer thickness. Control was performed through a perforated plate having a porosity equal to 0.08. Three different cavity depths have been considered: 10, 30 and 50mm. The results obtained being nearly the same for the three cases, here we will only examine those relative to the deepest cavity. The interacting flows realized without (solid wall) and with control have been carefully investigated by means of schlieren visualizations, surface pressure measurements and LDV surveys [18].

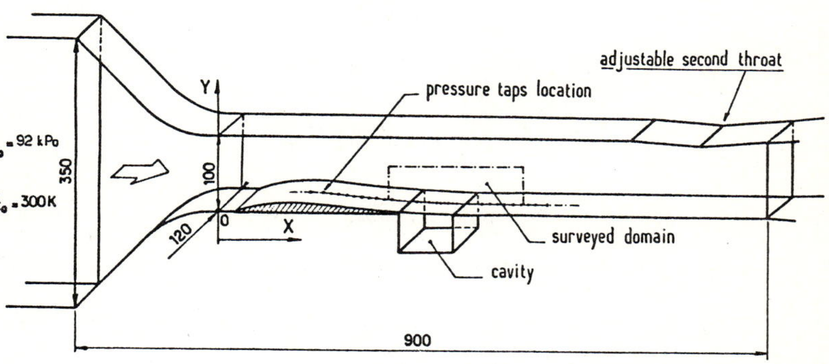

Fig. 9 – Passive control - Experimental arrangement.

The effect of passive control on the flowfield overall structure is shown by the iso-Mach lines traced in Figs. 10a and b. In the reference case (see Fig. 10a), the iso-Mach lines tracing reveals the existence of a nearly normal shock which occupies practically the entire section of the transonic channel. At the boundary-layer level, an interaction takes place, typical of a shock wave at a Mach number of 1.3. When passive control is applied (see Fig. 10), the interaction domain is strongly affected. Now, an oblique shock (C_1) emanates from a station located slightly upstream of the cavity origin. The still supersonic zone following (C_1)

is terminated by a weak quasi-normal shock wave (C_2). Shocks (C_1) and (C_2) meet at triple point I from which shock (C_3) starts. The present shock pattern is similar to the lambda shock system observed in transonic interactions with extended separation (see Fig. 2).

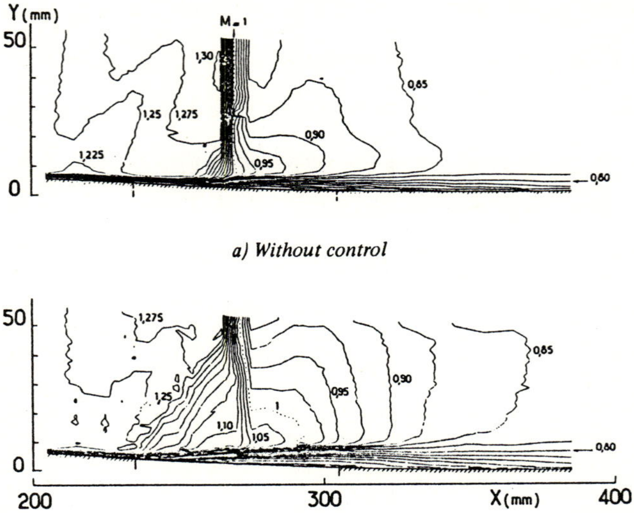

a) Without control

b) With passive control - large cavity

Fig. 10 – Iso-Mach lines

The mean streamwise velocity profiles across the boundary layer are plotted in Figs. 11a and b.These results show the destabilizing effect of the shock wave and particularly the strong retardation of the flow in the near wall region. When control is applied (Fig. 11b), the retardation effect is substantially magnified and a region exists very close to the wall where the streamwise velocity component is negative. Also, the thickening of the boundary-layer is now much more important. The behavior of the boundary-layer mean flow can be best illustrated by the tracing of the lines of constant value of the quantity ψ defined in Fig. 12 (ρ_{t_0} and a_{t_0} are, respectively, the density and sound velocity corresponding to freestream stagnation conditions, H = 100mm is the channel height). When there is no control, ψ is the streamfunction and the iso-ψ lines are mean flow streamlines. On the other hand, when passive control is applied, since the velocity component normal to the wall is non–zero over the porous portion, ψ has no longer a well–defined physical meaning. In this case, the iso ψ -lines tracing has to be considered as a way to visualize the mean flow structure.

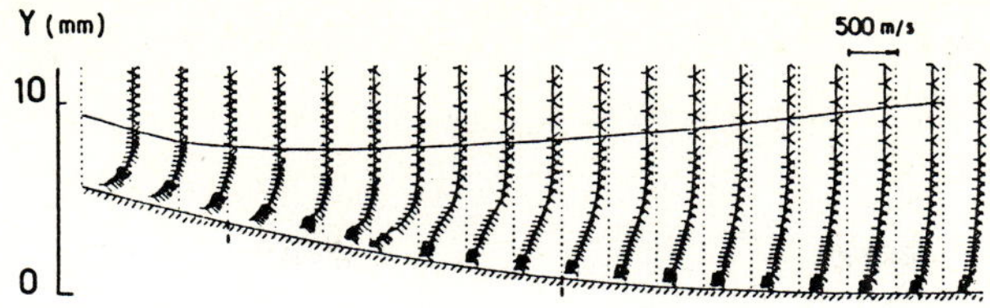

a) Without control

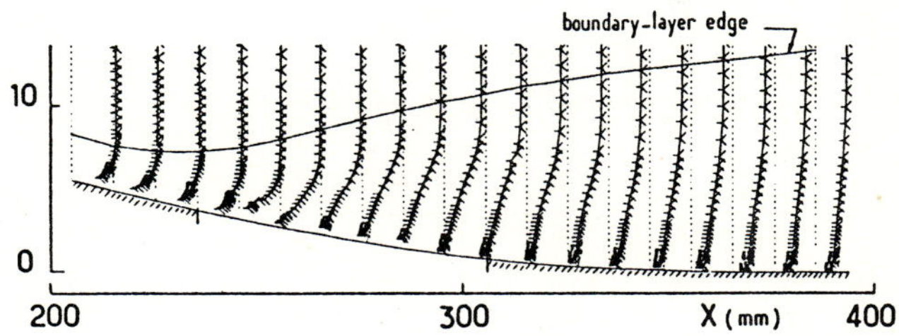

b) With passive control - large cavity

Fig. 11 – Distribution of streamwise velocity component.

The streamlines traced in Fig. 12a for the reference case (solid wall) reveal the dilatation of the inner part of the boundary layer under the action of the shock. The large number of outer flow streamlines entering the boundary layer through its outer edge show the intensification of entrainment occurring during the interaction. When passive control is applied (see Fig. 12b), intensification of entrainment is still more pronounced. Also a region close to the wall forms where the iso-ψ lines are closed curves. This feature indicates the existence of a reversed flow streaming in a direction opposite to that of the main stream.

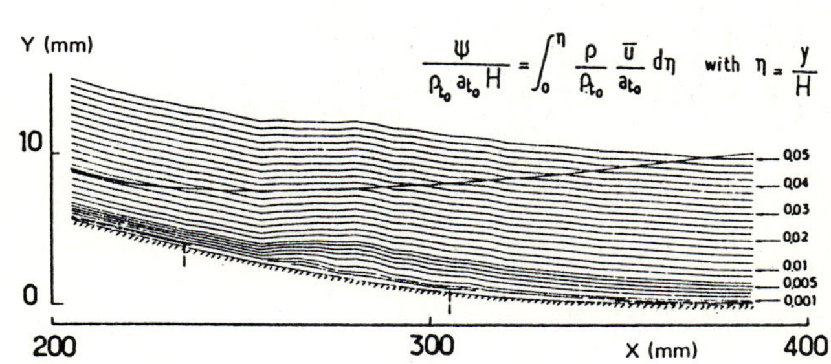

a) Without control (enlarged Y-scale)

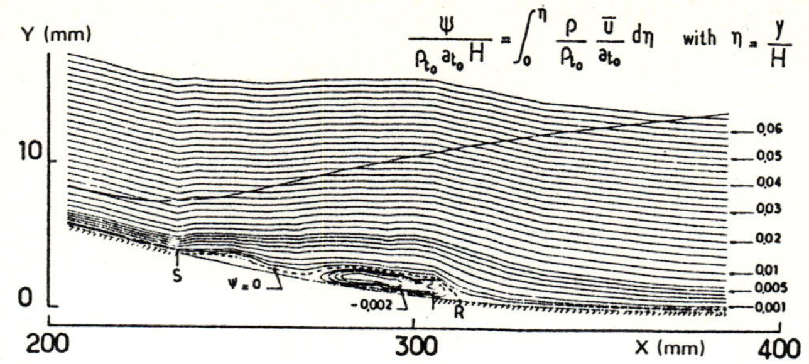

b) With passive control - Large cavity (enlarged Y-scale)

Fig. 12 – Iso-ψ lines

Figure 13 shows the mean velocity field measured inside the cavity in association with the outer flowfield (velocity scales are different in the two flow domains). Three regions can be distinguished inside the cavity:

- In its downstream part, large inflow velocities are observed near the perforated plate surface. They denote the strong suction effect existing in the high pressure region of the interaction.
- In its upstream part, the velocity vector orientation denotes the transpiration effect occurring in the low pressure part of the interaction.
- The overall cavity flow is organized into a single vortex whose center is displaced towards the right upper corner of the cavity.

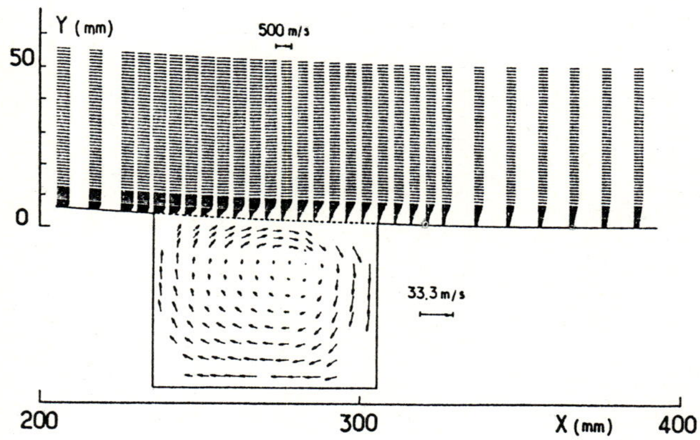

Fig. 13 – Flow inside the large cavity.

The spatial distributions of the normalized Reynolds shear stress are represented in Figs. 14a and b. For the reference configuration, there is an important rise in the turbulence level during the first part of the interaction process. This well-known tendency is largely amplified when passive control is applied. Then, the maximum turbulence levels are twice those measured for solid wall conditions. However, this amplification of turbulence is less important than it would have been for a separated flow. Thus, turbulence in a control zone tends to develop properties intermediate between an attached flow situation and a separated flow case.

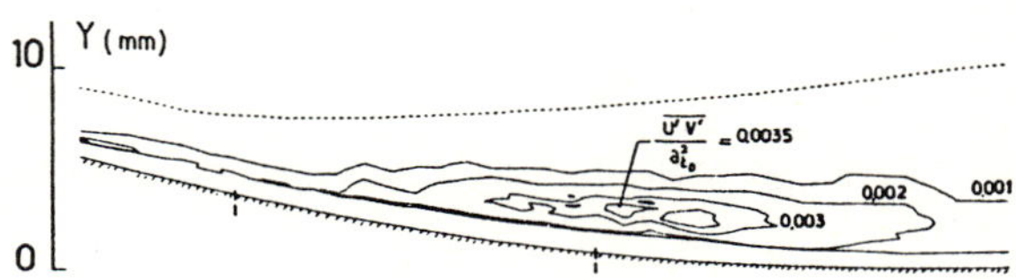

a) Without control

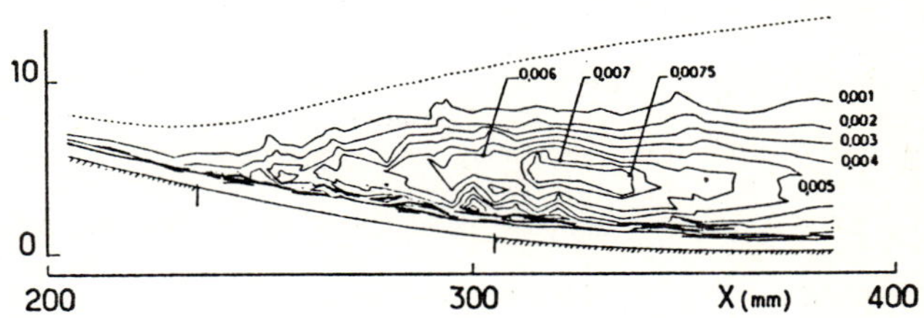

b) With passive control - large cavity

Fig. 14 – Spatial distribution of shear stress

In conclusion, the following remarks can be made. The injection effect taking place in the upstream part of the control zone induces an abrupt rise in the rate of displacement thickness increase. The resulting "viscous ramp" effect leads to the formation of an oblique shock wave originating from the front edge of the porous plate. Thus, the strong single normal shock that would form on a solid wall is replaced by a multiple shock system producing a smaller entropy rise; hence, a substantial reduction in wave drag. However, the intensification of dissipative effects caused by passive control may entail an increase in friction drag outbalancing the wave drag reduction.

5 - Concluding Remarks

In spite of the spectacular advances in computational fluid dynamics, basic experiments are still needed to help in the physical understanding of transonic shock-wave/boundary-layer interaction and to provide well documented test cases to validate theoretical models.

Experiment can help greatly in the definition of turbulence models to be introduced in the time averaged Navier-Stokes equations, the problem of turbulence modeling in such complex flows being still far from being completely solved. In spite of honorable results given by transport equation models in the prediction of some mean flow properties, the evolution of turbulence during the interaction process is poorly predicted, which reveals some basic defficiency of these models.

The structure of the flow resulting from shock-induced interactions in a 3-D transonic channel is far more complicated than that of a nominally 2-D interaction. Examination of skin-friction line patterns reveals a complex topology containing a large number of singular points and separation lines. Very detailed and well–interpreted experiments will be necessary to validate 3-D Navier-Stokes codes applied to such configurations.

Passive control entails a strong and unusual alteration of the boudary-layer mean and turbulent properties. In this case also, only fine experiments will allow the grasping of the physics involved in this process and the development of realistic theoretical models essential for optimization purposes.

6 - References

1. Boutier, A.; d'Humières, Ch. and Soulevant, D.: Three-Dimensional Laser Velocimetry: a Review. 2nd Int. Symp. on Applications of Laser Anemometry to Fluid Mechanics, Lisbon, 2-4 July, 1984

2. Le Balleur, J.-C.: Viscous Inviscid Interaction Solvers and Computation of Highly Separated Flows. Studies of Vortex Dominated Flows, Chapter 3, pp. 158-192, Springer-Verlag, 1987

3. Kluwick, A.: Interacting Boundary-Layers. ZAMM 67 (1987) 4, T 3-13

4. Délery, J. and Marvin, J.G.: Shock-Wave/Boundary-Layer Interactions. AGARDograph N° 280 (1986)

5. IUTAM Symposium on: Turbulent Shear-Layer/Shock-Wave Interactions, J. Délery Ed., Springer-Verlag (1986)

6. Délery, J.: Experimental Investigation of Turbulence Properties in Transonic Shock/Boundary-Layer Interactions.AIAA Journal, Vol. 21, N° 2, pp. 180-185 (Feb. 1983)

7. Benay, R.; Coët, M.-C. and Délery, J.: Validation of Turbulence Models Applied to transonic Shock-Wave/Boundary-Layer Interaction. La Recherche Aérospatiale, N° 1987-3, pp. 2-16 (english edition)

8. Vandromme, D. and Ha Minh, H.: Physical Analysis of Turbulent Boundary-Layer/Shock-Wave Interactions Using Second Order Closure Predictions. IUTAM Symposium on Turbulent Shear-Layer/Shock-Wave Interactions, Springer-Verlag, 1986

9. Debiève, J.-F. and Lacharme, J.-P.: A Shock-Wave/Free Turbulence Interaction. IUTAM Symposium on Turbulent Shear-Layer/Shock-Wave Interactions, Springer-Verlag, 1986

10. Dallmann, U.: Topological Structures of Three-Dimensional Vortex Flow Separation. AIAA Paper N° 83-1735

11. Benay, R.; Délery, J. and Pot, T.: Experimental Analysis of the Flow Through a Three-Dimensional Transonic Channel. La Recherche Aérospatiale, N° 1986-6, pp. 11-26 (English edition)

12. Legendre, R.: Lignes de courant d'un écoulement permanent: décollement et séparation. La Recherche Aérospatiale, N° 1977-6

13. Poincaré, H.: Les points singuliers des équations différentielles. C.R. Académie des Sciences, t. 94, pp. 416-418 (Feb. 13, 1882)

14. Lighthill, M. J.: Attachment and Separation in Three-Dimensional Flows. Section II 2-6 of Laminar Boundary-Layers, Oxford University Press, pp. 72-82 (1963)

15. Doerffer, P. and Dallmann, U.: Separation Structures Produced by Normal Shock-Wave/Turbulent Boundary-Layer Interaction in a Narrow Wind Tunnel. AIAA Paper N° 87-1370

16. Bahi, L.; Ross, J.M. and Nagamatsu, H.T.: Passive Shock-Wave/Boundary-Layer Interaction Control for Transonic Airfoil Drag Reduction. AIAA Paper N° 83-0137

17. Thiede, P.; Krogmann, P. and Stanewsky, E.: Active and Passive Shock/Boundary-Layer Interaction Control on Supercritical Airfoils. AGARD CP N° 365 (1984)

18. Chanetz, B. and Pot, T.: Expériences fondamentales sur le contrôle passif de l'interaction onde de choc-couche limite en transsonique. 24ème Colloque d'Aérodynamique Appliquée de l'AAAF, 26-28 Oct. 1987, Poitiers (France)

Passive Control of Transonic Shock/Boundary Layer Interaction

P. Thiede* and P. Krogmann**

* Messerschmitt-Bölkow-Blohm GmbH
Unternehmensgruppe Transport- und Verkehrsflugzeuge, D-2800 Bremen

** DFVLR - Institut für Experimentelle Strömungsmechanik
D-3400 Göttingen, FRG

Summary

Experimental investigations are presented, aiming at the improvement of transonic wing performance through passive control of shock/boundary layer interaction (SBLI) by boundary layer ventilation in the shock region.

In the first part of the paper, results of 2-D experiments are summarized, carried out in the DFVLR Transonic Windtunnel Göttingen (TWG) using the supercritical MBB airfoil VA-2. To allow measurements with and without control, the airfoil model was equipped with an exchangeable insert in the shock region. It is shown that ventilation in the shock region mainly weakens the shock wave and delays shock-induced separation, resulting in substantial improvements of the off-design performance and boundaries.

In the second part of the paper, 3-D experiments on the MBB half wing ZKP-6V model are described, performed in the NLR High Speed Tunnel (HST) Amsterdam. The wing model was - as in the 2-D case - equipped with an exchangeable insert for measurement with and without control. The results show that the effectiveness of passive SBLI control depends largely on the pressure type of the wing sections. At high-loaded sections, the passive effect can be of the same order as in the 2-D case.

Introduction

The aerodynamic performance of supercritical wings at transonic speeds is strongly influenced by shock/boundary layer interaction (SBLI). Not only the off-design characteristics suffer from this phenomenon but also buffet and drag rise boundaries are a consequence of shock-induced separation. Therefore, the development of practicable SBLI control methods with the aim to weaken the shock wave and to delay shock-induced separation is an important objective of transonic wing aerodynamics.

J. Zierep, H. Oertel (Eds.)
Symposium Transsonicum III
IUTAM Symposium Göttingen, 24.-27. 5. 1988

In the past, different methods have been investigated to control actively SBLI on supercritical airfoils by tangential blowing (Ref. 1) or suction in the shock region (Refs. 2, 3). But the easiest way SBLI control can be accomplished is passively by boundary layer ventilation in the shock region through a porous surface with a cavity located underneath (passive SBLI control, Refs. 4 to 7), Fig. 1.

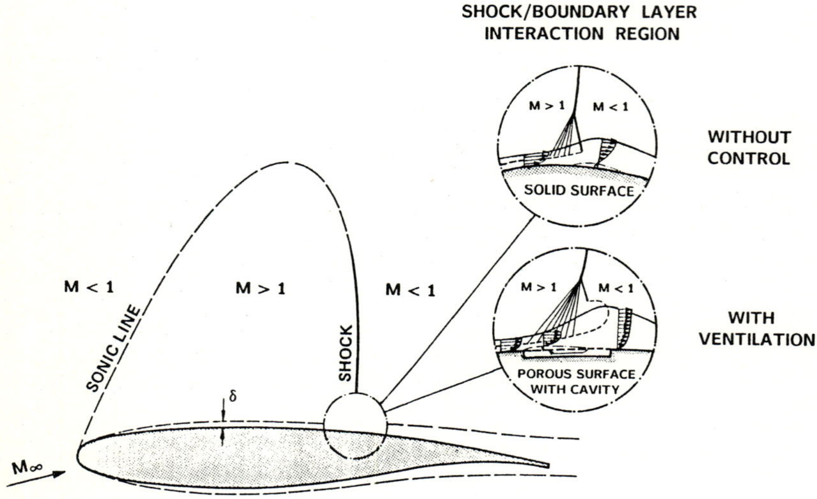

Fig. 1 Passive control of shock/boundary layer interaction at transonic airfoil flow

The self-acting ventilation flow (blowing upstream and sucking downstream of the shock) is induced by the pressure rise in the shock region. With LDA-measurements from ONERA (Ref. 8), Fig. 2 shows the ventilation effect on the flow structure in the SBLI region. As a consequence of shock-induced ventilation, the wall streamline curvature in the shock region is increased, leading to a weakened shock near the wall with an extended isentropic recompression zone and an additional rear-expansion region.

Passive control strongly intensifies the capability of the boundary layer to weaken the shock wave and to delay the onset of shock-induced separation, resulting in substantial wave and shock-induced viscous drag reductions, see also Refs. 9 and 10.

LDA - MEASUREMENTS OF DELERY (ONERA)

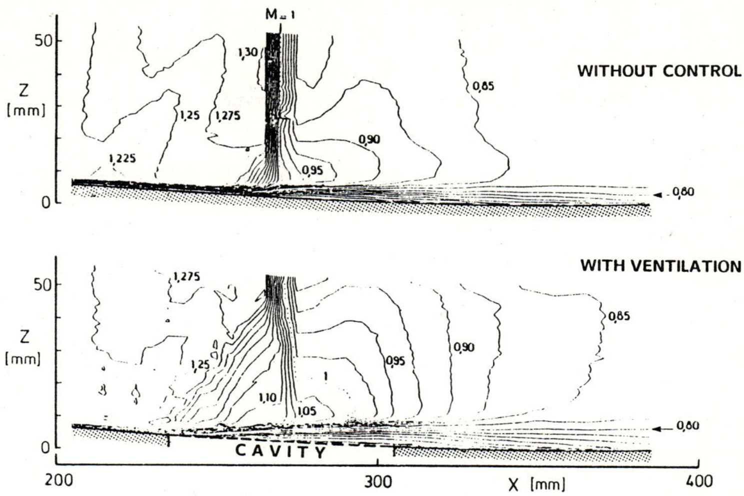

Fig. 2 Shock/boundary layer interaction with/without passive control

2-D experimental investigations

In the first part of the paper, results of 2-D experimental investigations are summarized aiming at the improvement of transonic airfoil performance through passive SBLI control by boundary layer ventilation in the shock region (already published partially in Ref. 11).

These experiments were carried out in the DFVLR 1m x 1m Transonic Windtunnel Göttingen (TWG), using the supercritical MBB airfoil VA-2. This airfoil was chosen because of its relatively small shock shifting with Mach number at off-design conditions. The airfoil model was equipped with an exchangeable insert in the shock region to allow measurements with ventilation through slotted or perforated strips and measurements without control. Fig. 3 shows the model configurations tested with passive SBLI control and the model arrangement in the test section. A double slot insert (slot width 0.6 mm, slot distance 7 % chord) and two perforated strip inserts with the same porosity (8.16%) but different width (7.5 % and 15 % chord) were investigated. The experimental investigations include surface pressure distribution, boundary layer and wake measurements as well as Schlieren observations.

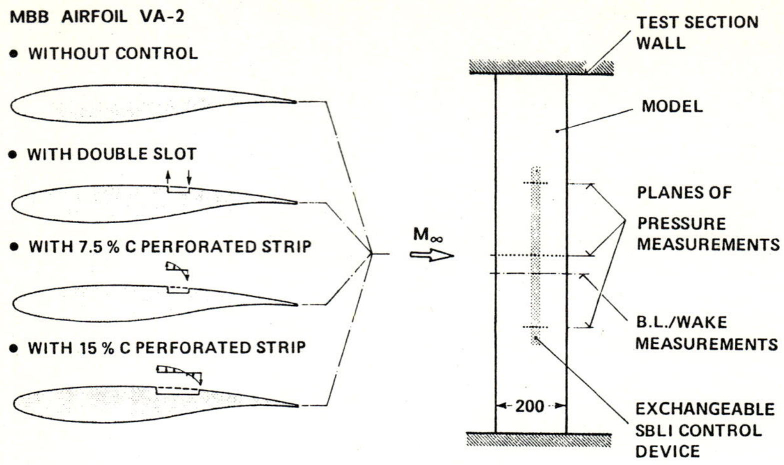

Fig. 3 2-D model configurations tested at DFVLR-TWG with passive SBLI control

The influence of boundary layer ventilation in the shock region on the complex transonic airfoil flow development can be analysed from typical surface pressure distribution and wake pressure measurements, concentrating here on the 15 % chord perforated insert results, Fig. 4. In the case of attached transonic flow ($\alpha_g = 3°$), the passive effect causes a decreased pressure rise in the shock region and consequently a reduced pressure loss in the wake. In the second case with shock-induced separation ($\alpha_g = 4°$), the shock-induced pressure rise is flattened by ventilation, resulting in a delay of separation with a substantially reduced pressure loss in the wake. In the case of higher transonic loading finally ($\alpha_g = 5°$), where the buffet boundary is clearly exceeded (shock located far upstream and shock-induced separation reaching downstream to the trailing edge), buffeting is completely suppressed by passive control, showing a pressure distribution with the shock stabilized immediately upstream of the control device and the flow attached downstream as well as a wake profile with a drastically reduced pressure loss. The described ventilation effects were also indicated by Schlieren observations not presented here.

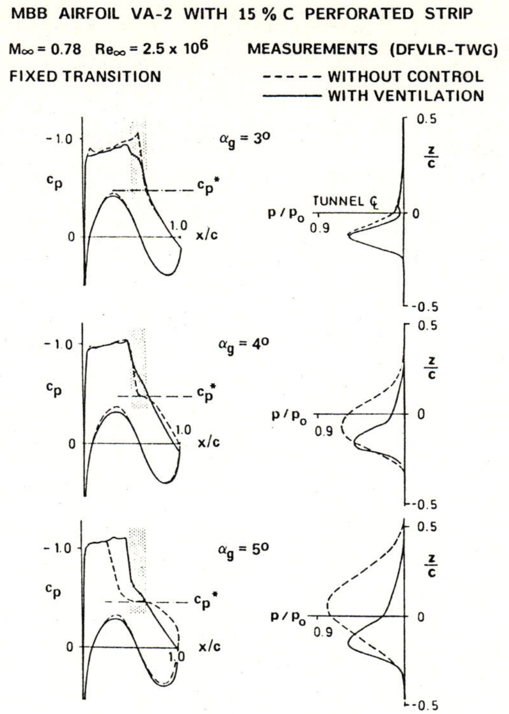

Fig. 4 Wall/wake pressure distribution with/without passive SBLI control

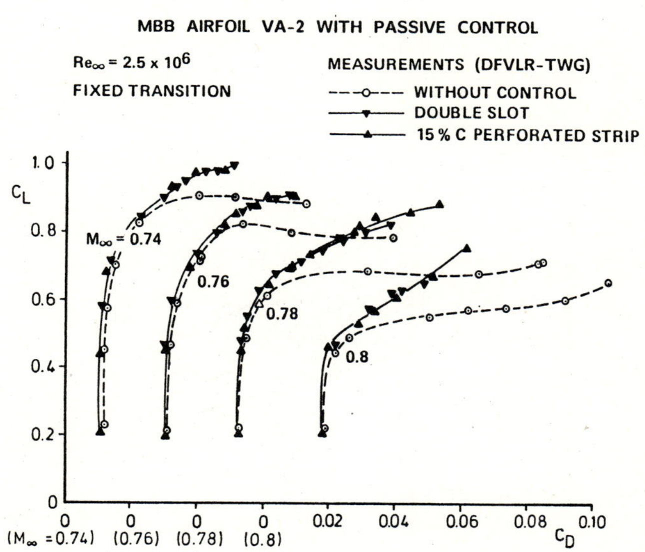

Fig. 5 Drag polar improvements by passive SBLI control

The positive effects of passive SBLI control on surface pressure distribution and wake pressure profiles result in substantial improvements of the airfoil characteristics at off-design conditions. In Fig. 5 measured drag polars of the airfoil model equipped with the double slot and the perforated inserts are compared with the basic model results without control. For both ventilated configurations a favourable passive effect was obtained at all Mach numbers tested. Especially at off-design conditions with strong shocks drastic drag reductions were measured.

Besides the drag polar improvements, a favourable passive effect on the buffet and drag rise boundaries was obtained by ventilation in the shock region, see Ref. 11. In some hard off-design cases, buffeting could be completely ceased although the flow was separated downstream of the shock. It seems that the strong coupling between shock and trailing edge flow behaviour is largely suppressed by boundary layer ventilation.

3-D experimental investigations

Because of encouraging 2-D results, 3-D experiments were performed to investigate the effectiveness of passive SBLI control on a modern supercritical wing (Ref. 12).

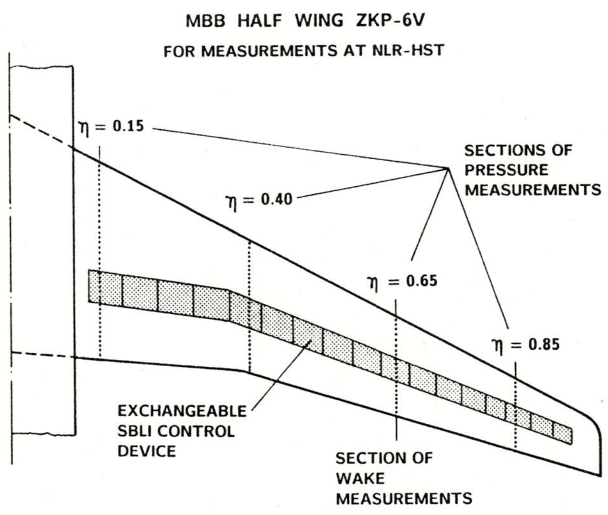

Fig. 6 Half wing model tested with passive SBLI control

For these experiments the MBB half wing ZKP-6V model was equipped with an exchangeable insert in the shock region to allow measurements with and without SBLI control, Fig. 6. The width of the insert had to be chosen rather large (25 % chord at the outer wing, between 15 and 25 % chord at the inner wing) because of the relatively strong shock shifting at off-design conditions of the used wing not designed to meet the special requirements of SBLI control. A slotted as well as a perforated insert with the same porosity (8.16 %), each consisting of 3 individual panels, were investigated. The plenum underneath the porous surface was divided into separate cavities to avoid spanwise ventilation. The experiments were carried out in the High Speed Tunnel (HST) of the NLR Amsterdam. The experimental investigations include surface pressure distribution (4 sections), wake pressure (1 section) as well as 3-component balance measurements.

Fig. 7 shows measured wing pressure distributions and in addition at the section η = 0.65 wake profiles with and without passive SBLI control. As in the 2-D case, the shock wave is weakened by ventilation at all wing sections investigated. But, the effectiveness of passive control is quite different at various wing sections depending on its type of pressure distribution. At the high-loaded outer wing section η = 0.65 with a strong shock, the passive effect is similar to the 2-D case: the weakened shock results in an improved flow recovery downstream of the shock indicated by an increased trailing edge pressure, and leads consequently to a reduced pressure loss in the wake. Near the kink (η = 0.4) and further outside (η = 0.85) of the wing, however, the stress on the flow is higher in the trailing edge region than in the shock region. This means that in these wing sections the weakening of the shock wave by ventilation is largely compensated by a worsened recovery process behind the shock, indicated by an unchanged trailing edge pressure. At the inner wing (η = 0.15) with a double shock system finally, the rear-loading dominates over the shock loading so that ventilation cannot be effective.

In Fig. 8 measured drag polars of the wing section η = 0.65 equipped with the slotted insert are compared with solid model results without control. A favourable passive effect was only obtained at high Mach numbers with strong shocks. At M_∞ = 0.84 the drag reductions by passive

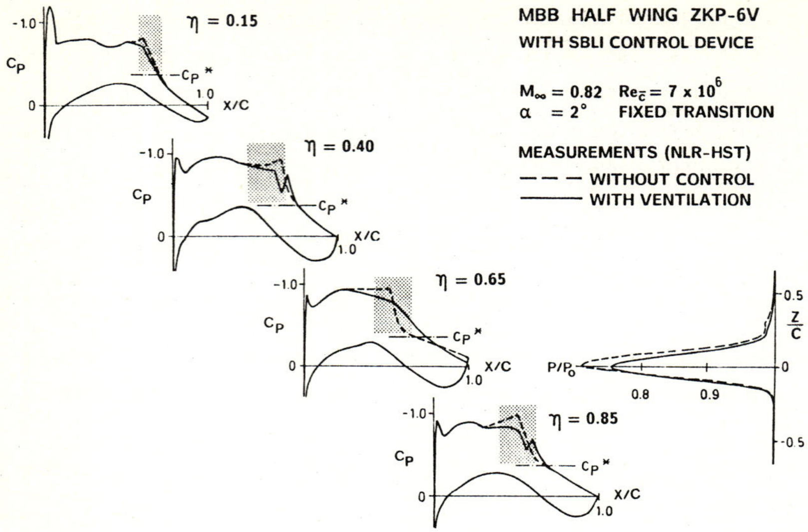

Fig. 7 Wall/wake pressure distribution
with/without passive SBLI control

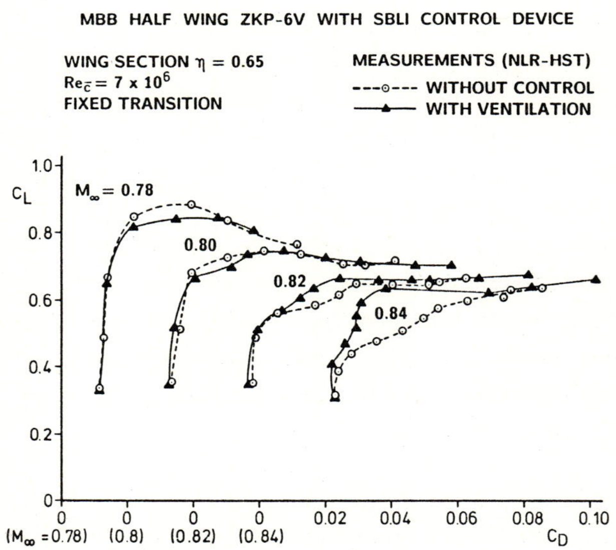

Fig. 8 Wing section drag polars
with/without passive SBLI control

control are of the same order as in the 2-D case. At lower Mach numbers (M_∞ < 0.82), the shock-induced stress on the flow is smaller than that of rear-loading so that in the ventilation case wave drag reductions are largely compensated by viscous drag rises due to a worsened recovery process in the trailing edge region.

Because of the penalties of boundary layer ventilation especially at the inner wing, only minimal drag polar improvements of the ZKP-6V half wing through passive SBLI control could be obtained from the balance measurements.

Conclusions

The present 2-D investigations have in principle confirmed that the transonic airfoil performance can be considerably improved through passive SBLI control. The results show that passive control by boundary layer ventilation in the shock region mainly weakens the shock wave and delays shock-induced separation, leading to substantial drag polar improvements at off-design conditions as well as to extensions of the buffet and drag rise boundaries.

3-D investigations of passive SBLI control on a transonic wing have shown that the effectiveness of passive control depends largely on the pressure type of the wing sections. At high-loaded outer wing sections with strong shocks, the passive effect can be of the same order as in the 2-D case. But at inner wing sections with a double shock, where rear-loading dominates over shock loading, in the ventilation case wave drag reductions are largely compensated by viscous drag rises produced by a worsened recovery process in the trailing edge region.

An effective application of passive SBLI control necessitates a wing design that meets the special requirements of SBLI control: dominant shock loading combined with small shock shifting at off-design conditions.

References

1. K. Finke: Stoßschwingungen in schallnahen Strömungen. VDI-Forschungsheft 580 (1977).

2. P. Thiede: Supercritical airfoil flow control by slot suction in the shock region. Proc. of the 5th USAF/FRG DEA-Meeting on "Viscous and Interacting Flow Field Effects", Annapolis, Md., AFFDL-TR-80-3088 (1980).

3. P. Krogmann, E. Stanewsky, P. Thiede: Effects of suction on shock/boundary-layer interaction and shock-induced separation. Journal of Aircraft, Vol. 22 (1985).

4. L.Bahi, J.M. Ross, H.T. Nagamatsu: Passive shock wave/boundary layer control for transonic airfoil drag reduction. AIAA Paper 83-0137 (1983).

5. P. Thiede, P. Krogmann, E. Stanewsky: Active and passive shock/boundary layer interaction control on supercritical airfoils. AGARD-CP-365, Brussels (1984).

6. P. Krogmann, E. Stanewsky, P. Thiede: Transonic shock/boundary layer interaction control. ICAS-Paper 84-2.3.2, Toulouse (1984).

7. G. Savu, O. Trifu: Porous airfoils in transonic flow. AIAA-Journal, Vol. 22 (1984).

8. B. Chanetz, T. Pot: Experiences fondamentales sur le controle passif de l'interaction onde de choc-couche limite en transsonique. 24ème Colloque d'Aerodynamique Appliquèe de l'AAAF, Poitiers (1987).

9. A. Bertelrud: Passive shock modification in flight. EUROMECH 181, Saltsjöbaden, Sweden (1984).

10. S. Raghunathan , D.G. Mabey: Passive shock wave boundary layer control experiments on a circular arc model. AIAA Paper 86-0285 (1986).

11. P. Thiede, P. Krogmann: Improvement of transonic airfoil performance through passive shock/boundary layer interaction control. IUTAM-Symposium on "Turbulent Shear Layer/Shock Wave Interactions", Palaiseau, France (1985).

12. R. Henke: Ergebnisse der Windkanalmessungen am Halbmodell mit ZKP-6V Flügel mit glatten und perforierten Einsätzen (Phase 1). MBB-Bericht TE 2-1516 (1986), nicht veröffentlicht.

13. R. Bohning, J. Zierep: Calculation of 2-D turbulent shock/boundary layer interaction at curved surfaces with suction and blowing. IUTAM-Symposium on "Turbulent Shear Layer/Shock Wave Interactions", Palaiseau, France (1985).

14. C.L. Chen, C.Y. Chow, W.R. Van Dalsem, T.L. Holst: Computation of viscous transonic flow over porous airfoils. AIAA Paper 87-0359 (1987).

Turbulent Shock - Boundary-Layer Interaction with Control

Theory and Experiment

R. Bohning and H. Jungbluth

Institut für Strömungslehre und Strömungsmaschinen
Universität Karlsruhe, Kaiserstraße 12
D-7500 Karlsruhe, FRG

Summary.

Experimental and theoretical investigations are presented concerning the two-dimensional turbulent shock-boundary-layer interaction at a curved wall with passive flow control through ventilation in the shock region. The theoretical flow simulation is based on an iteration method, coupling a boundary layer integral solution with an embedded interference model for the viscid flow and the non-linear potential solution for the inviscid flow.

1. Introduction

Figure 1 shows the transonic flow past a profile with a high subsonic free stream Mach number. The local supersonic region is closed by a normal shock wave. Because the surface of the profile is curved, the Mach number behind the shock near the surface may be greater than one due to the post-expansion. The boundary layer thickens near the shock and may even separate totally or only locally. Our institute has been dealing with this shock-wave-boundary-layer interaction for a long time [1,2,3]. We are now trying to influence the flow field, the shock and the boundary layer through a cavity which is placed in the wall underneath the shock and which is covered by a perforated wall. Through such a setup we are

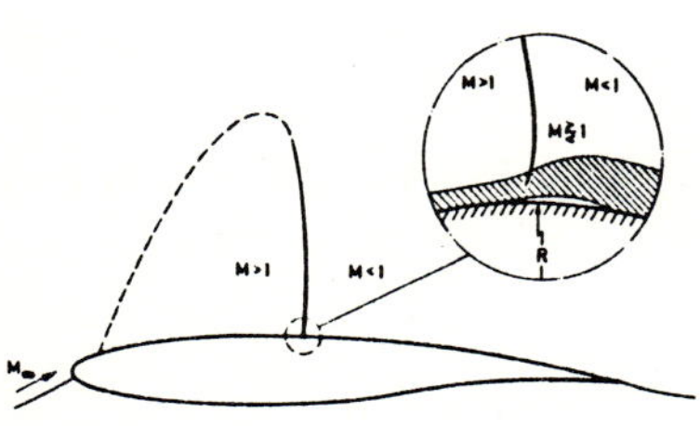

Fig. 1: Transonic flow around a profile

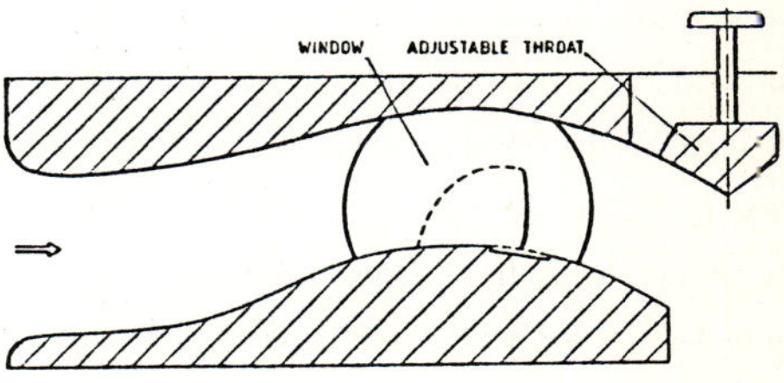

Fig. 2: Test section (schematic)

J. Zierep, H. Oertel (Eds.)
Symposium Transsonicum III
IUTAM Symposium Göttingen, 24.-27. 5. 1988

trying a) to weaken the shock, b) to fix the shock in its position and c) to reduce the drag of the profile. In years past initial steps concerning this experimental and theoretical research topic were made by a number of other collegues around the world [4-10].

2. Experimental setup

Figure 2 shows the setup used in the supersonic wind tunnels of our institute. Instead of a profile we use a curved wall to simulate the upper surface of the profile. The advantage of this is the greater scale of the whole setup and the thicker boundary layer near the shock. The adjustable throat is used to change the Mach number. With a Mach-Zehnder interferometer, adjusted to "infinity string", we obtain the iso-Machlines in the whole flow field. The cavity is covered with a perforated metal sheet with a perforation of 2.8%, 8.2% or 22.7%. In the following, experiments with 8.2% and 2.8% perforation are presented.

3. Experimental results

Figure 3 shows a first Mach-Zehnder interferogram with closed cavity. The Mach number in front of the shock is 1.22. The dark and bright lines are lines of equal velocity. The rising lines in front of the shock show the acceleration. Behind the shock the rising and descending of the iso-Machlines show the post-expansion caused by the wall curvature and the displacement effect of the boundary layer. In figure 4 the cavity is open. From the beginning of the cavity a weak oblique shock starts and is reflected at the lower end of a normal shock which is placed inside the flow field. The oblique shock is further reflected from the wall and the sonic line and ends in a weak normal shock located a little bit behind the end of the cavity. The flatness of the iso-Machlines shows that the velocity is nearly constant along the cavity. Only small jumps of the lines, when the shocks are being crossed, show that those shocks are weak. The maximum Mach number in the flow field is now only 1.16. The distribution of the Mach number along the wall is shown in figure 5. When the cavity is open, there is only a soft decrease of the velocity at the end of the cavity, but a strong decrease in a shock when the cavity is closed. To demonstrate the flow through the cavity in figure 6, the difference in pressure within the flow field less the pressure in the cavity is plotted. At the end of the cavity the pressure in the flow field is higher; in this part medium will enter the cavity. In front of the cavity it is vice versa.

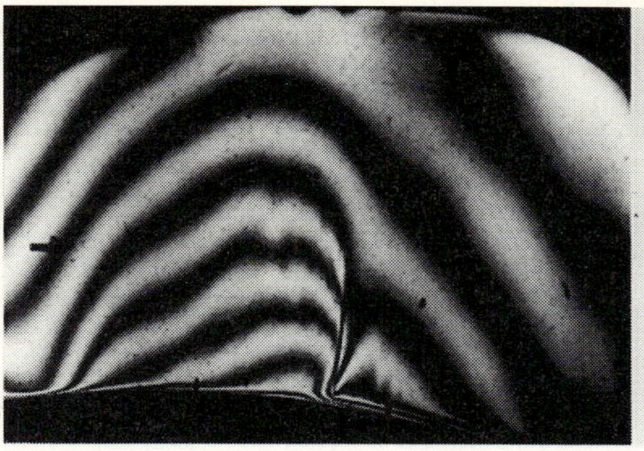

Fig. 3: Mach-Zehnder Interferogram
M = 1.22, cavity closed

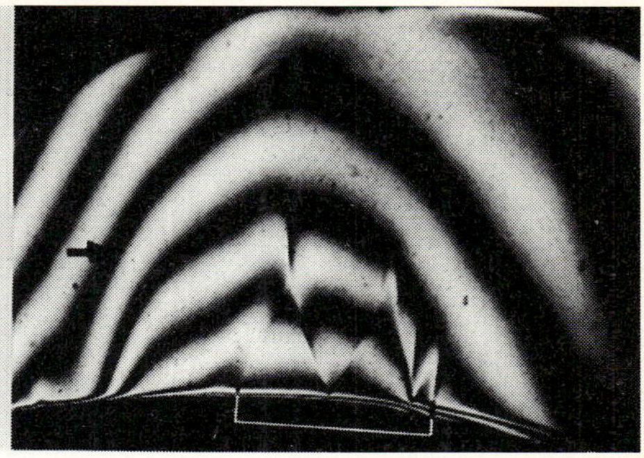

Fig. 4: Mach-Zehnder Interferogram
M = 1.16, cavity open

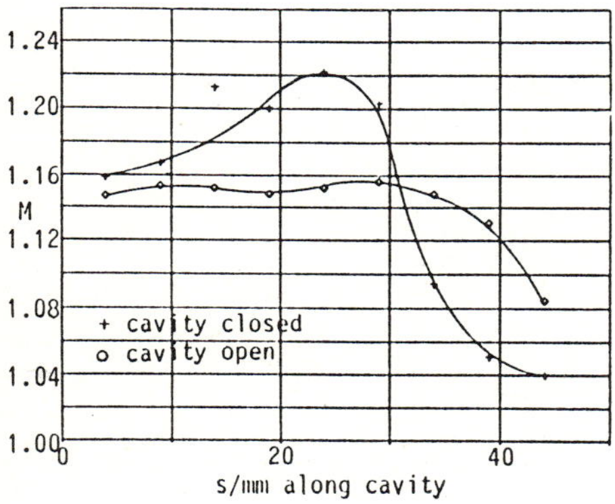

Fig. 5: Mach number on wall along cavity

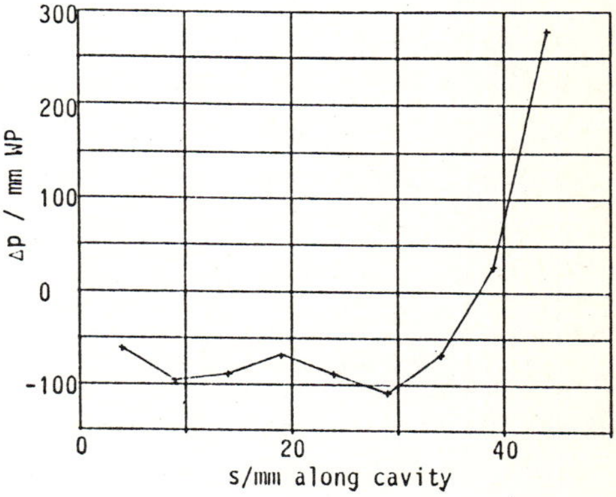

Fig. 6: Pressure difference between flow field and cavity

Fig. 7: Mach-Zehnder Interferogram
M = 1.21, cavity open
Perforation 2.8%

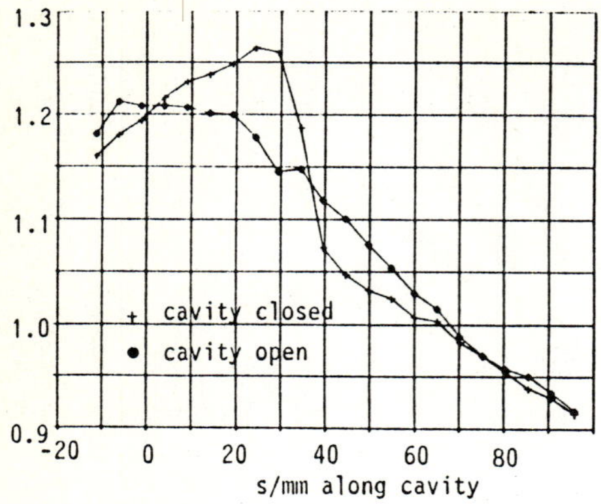

Fig. 8: Mach number on wall along cavity

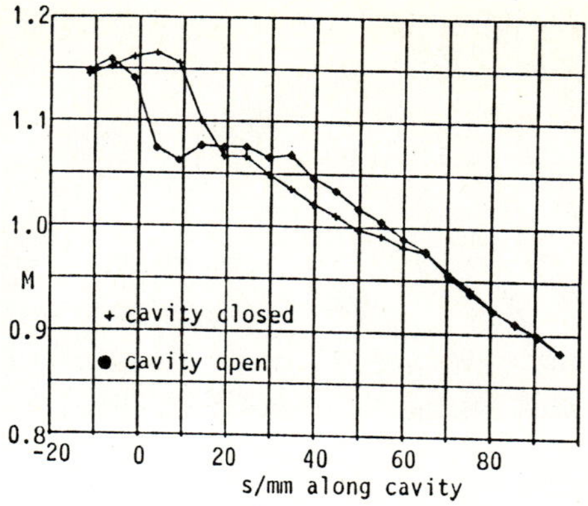

Fig. 9: Mach number on wall along cavity

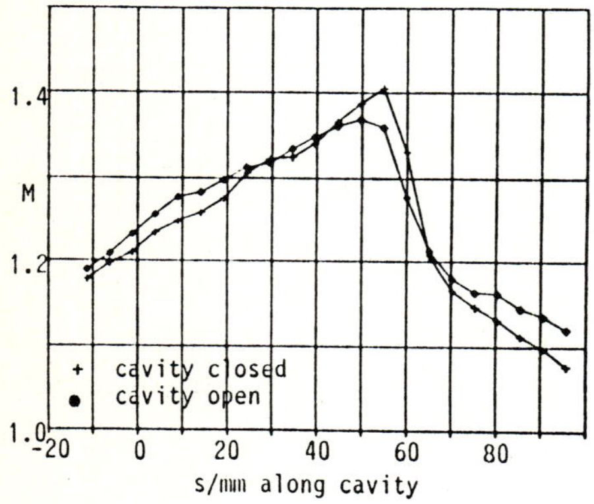

Fig. 10: Mach number on wall along cavity

All pressure differences are really small. In figure 5 and 6 the cavity begins at s = 0 mm and ends at s = 50 mm. The perforation of the metal sheet in this experiment was 8.2%. If the perforation is only 2.8%, we get a flow field which is presented by the interferogram in figure 7. The cavity is now longer but the structure of the flow field remains the same: an oblique shock starts at the beginning of the cavity and is reflected from a short normal shock in the flow field. Two weak normal shocks are located on the wall behind the λ-shaped shock system. Figure 8 shows the Mach number distribution along the wall. The cavity begins at s = 0 mm and ends at s = 72 mm. Without a cavity, a strong normal shock with a pre-shock Mach number of 1.26 closes the supersonic region. When the cavity is opened, the Mach number distribution is smooth. If the pre-shock Mach number is reduced, in the case of a closed cavity, the shock moves upstream and becomes weaker (figure 9). Now, when the cavity is open, this shock moves upstream to the beginning of the cavity (s = 0 mm), but the intensity remains the same. On the other hand, if the pre-shock Mach number is increased (figure 10), the shock moves downstream and is only a little bit weaker when the cavity is opened, because the perforated area on the wall behind the shock is now too small.

To conclude these experiments, it can be said that the shock which closes the local supersonic region is weakened by a cavity if it is located near the middle of the cavity. If the shock is located in the front part of the cavity it moves upstream to the beginning of the cavity, if it is located near the end, the influence of the cavity is only small.

4. Theory

When theoretically describing this problem, the transonic flow through a channel analogous to the experiment is being examined. The flow field is divided into several areas according to its physical character, a laminar/turbulent boundary layer, a shock-boundary-layer interference area and an inviscid flow outside of the boundary layer. We reach the final global solution through iterative coupling.

Boundary layer calculation. From the channel inlet to the interference area the boundary layer is calculated by means of the traditional integral methods. Downstream from the interference area we use an inverse boundary layer procedure which also permits to calculate locally-separated boundary layers. The interference area itself, in which the usual boundary layer assumptions are not valid, is being described through the analytical three-layer model of Bohning/Zierep [1-3]. In this model we simulate the passive control of the interference region with a new boundary condition at the wall, namely, a normal velocity distribution at the wall which is different from zero. A calculated typical result is shown in figure 11. It shows lines of constant Mach numbers which represent the structure of the flow field within the boundary layer without ventilation (figure 11) and with ventilation (figure 12). One can see that the upstream influence of the shock is intensified when there is ventilation which leads to a reduction of speed in front of the shock and a reduction of shock strength. The post-shock expansion is intensified too. Figure 13 shows the normal velocity v/c^* at the wall and at the edge of the boundary layer. At the wall we chose a simplified v/c^*-distribution because of the improved representation. At the edge of the boundary layer (as seen above in fig.

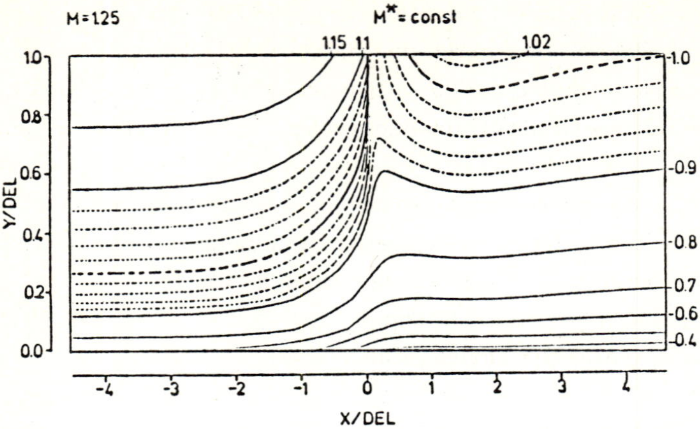

Fig. 11: Lines of constant Mach number without ventilation (M_δ = 1.25, R/δ = 175, Re_δ = $2 \cdot 10^4$)

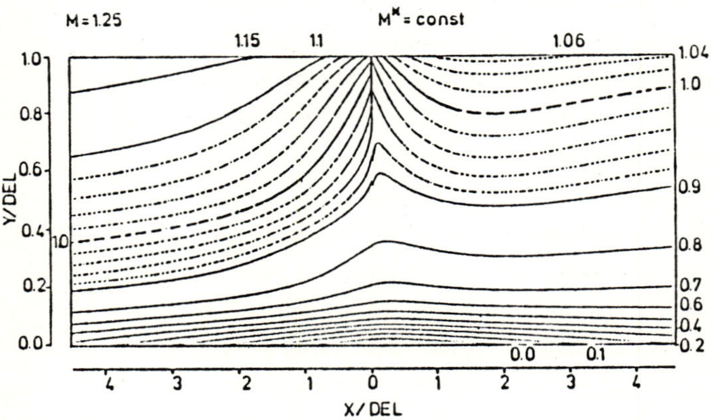

Fig. 12: Lines of constant Mach number with ventilation

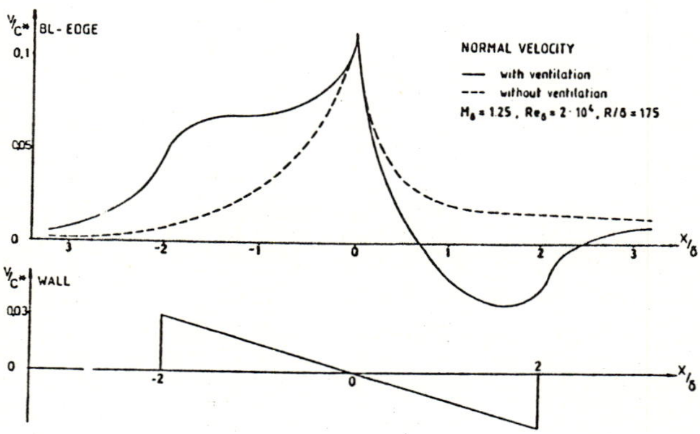

Fig. 13: Normal velocity at the wall and boundary layer edge

13) the v/c*-distribution is raised in front of the shock and lowered behind it because of the blow- and suction effect coming from the wall. The streamlines at the edge of the boundary layer with ventilation and without ventilation react accordingly (figure 14). The stronger upstream influence and the enforced convex streamline curvature in the case of ventilation explain the development of the experimentally observed oblique pre-shock at the beginning of the cavity and the intensification of the post-shock expansion.

Inviscid outer flow. The inviscid flow is calculated with a difference method. In order to do this, an orthogonal grid is generated initially (figure 15), whose lines are the equipotential- and the streamlines of the incompressible flow. The density of the lines can be varied.
The non-linear potential equation is now being solved on this grid. As boundary condition we demand tangent flow at the walls and a constant potential at the beginning and the end of the channel. According to the flow behaviour in the sub- and supersonic area different differential approximations are needed. We use central differential approximations in the elliptic part, backward approximations streamwise and central ones crosswise in the hyperbolic part of flow. The solution takes place iteratively and begins with an incompressible initial solution. To improve

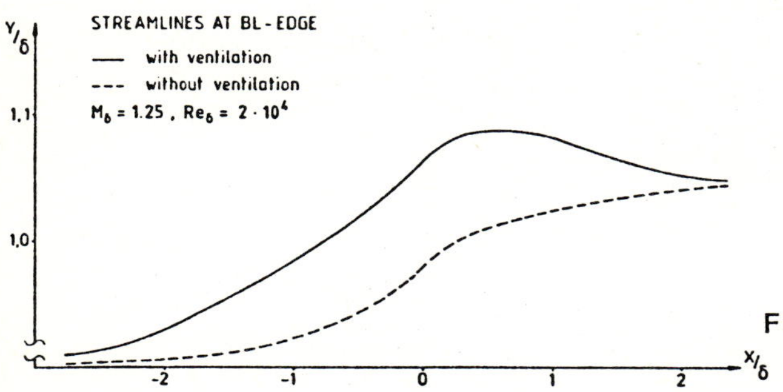

Fig. 14: Streamlines at the edge of the boundary layer

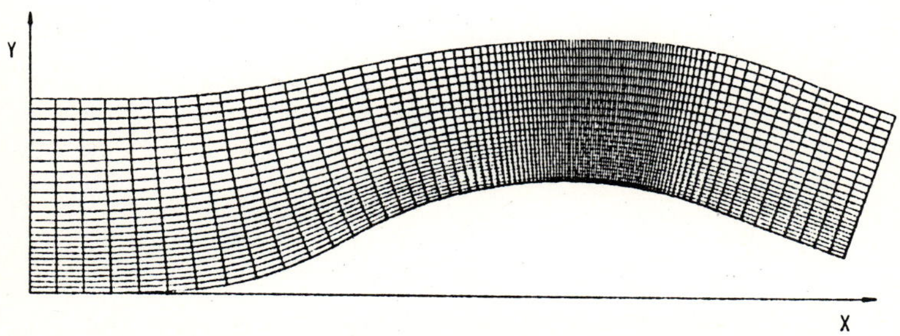

Fig. 15: Orthogonal grid with variation of line density

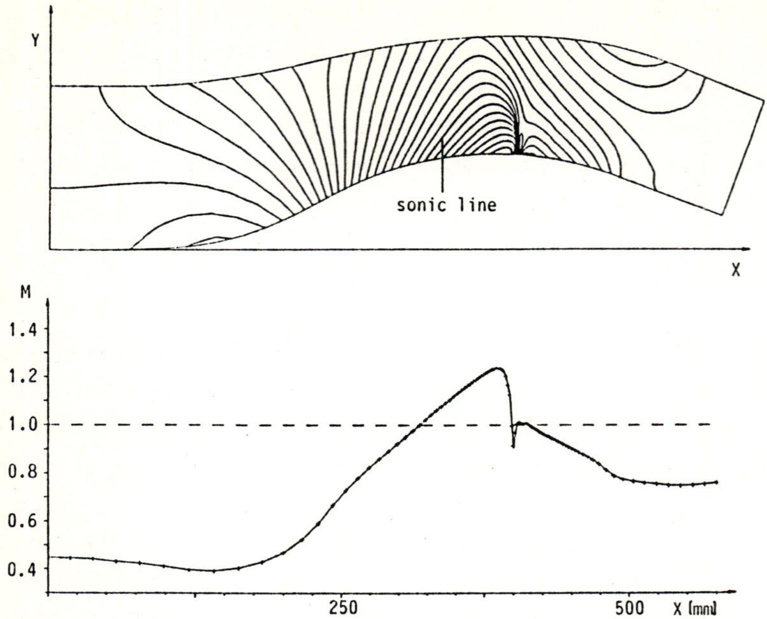

Fig. 16: Mach number contours and Mach number distribution near the surface without ventilation

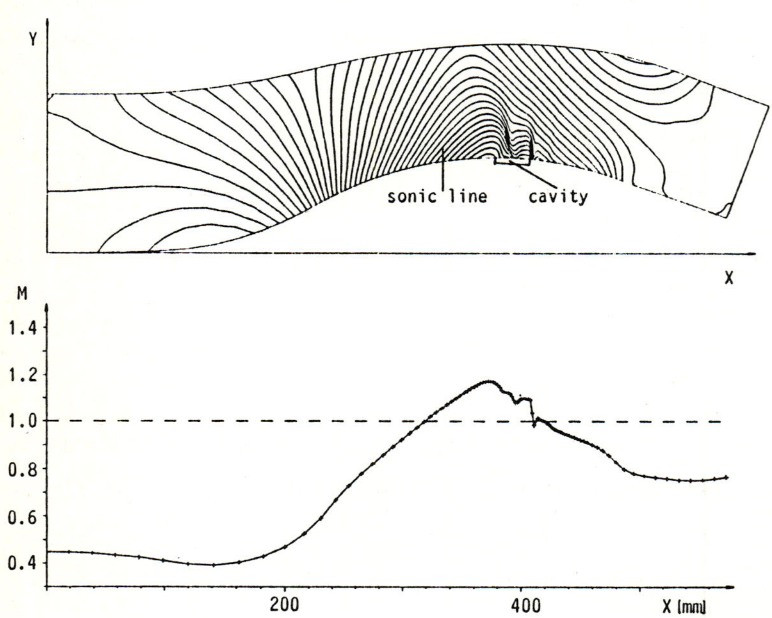

Fig. 17: Mach number contours and Mach number distribution near the surface with ventilation

convergence and subsequently a shortening of the computing time, we introduce a suitable line relaxation. A special shock operator sees to it that, based on controlled mass increase at the shock, an adequate description of the shock strength and shock position is obtained. The boundary-layer and the inviscid solution are iteratively coupled. With each iteration step the channel outline is being corrected to the degree of the calculated displacement thickness and the grid is adjusted.

5. Theoretical results

Figures 16 and 17 show the calculated iso-Machline distribution in the curved channel and the accompanying Mach number distribution at the bottom edge of the calculation area. The difference of the Mach numbers in the isoline fields amounts to 0.03 in figure 16 and 0.02 in figure 17.
With the example chosen in figure 16 without ventilation we achieve a pre-shock Mach number of 1.25. We can clearly see the distinct shock with the post-expansion following.

In the case we use ventilation (figure 17), the flow structure changes drastically; it corresponds in all details to the experimental observations. The porosity of the wall is the same as used in the first experiment, 8.2%. The normal velocity distribution at the wall which simulates the ventilation is derived from the calculated wall pressure distribution. The maximum normal velocity amounts to 3.3 m/s in the front part of the cavity and 6.6 m/s in the back part. Figure 17 (above) shows that the original shock has been split into two weaker shocks, one visible in a short pre-shock inside the flow field with a compression area in front of it caused through the blow from the wall and a weak after-shock at the back part of the cavity.

6. Conclusion

Experimentally as well as theoretically we can see that if the cavity is positioned properly, the shock strength on the one hand and the pressure gradient accompanying it are weakened and the shock position is fixed.

The experimental and theoretical investigations were made with assistance of our coworkers W. Braun and Th. Breitling.

References

[1] Bohning, R.: Zierep, J.: Der senkrechte Verdichtungsstoß an der gekrümmten Wand unter Einfluß der Reibung.
ZAMP 27, 225-240 (1976)

[2] Bohning, R.; Zierep, J.: Normal Shock-Turbulent Boundary-Layer Interaction at a Curved Wall.
AGARD CP No. 291, Computation of Viscous-Inviscid Interactions, pp 17-1 - 17-8, 1980

[3] Bohning, R.; Zierep, J.: Calculation of 2D Turbulent Shock/Boundary-Layer Interaction at Curved Surfaces with Suction and Blowing.
Turbulent Shear Layer/Shock Wave Interactions IUTAM Symposium Palaiseau 1985, pp 105-112, Ed: J. Délery, Springer-Verlag

[4] Thiede, P.; Krogmann, P.: Stanewsky, E.: Active and Passive Shock/Boundary Layer Interaction Control on Supercritical Airfoils.
AGARD-FDP-Symposium, Brussels, May 1984, pp 24-1 - 24-13

[5] Thiede, P.; Krogmann, P.: Improvement of Transonic Airfoil Performance through Passive Shock-Boundary-Layer Interaction Control.
Turbulent Shear Layer/Shock Wave Interactions IUTAM Symposium Palaiseau 1985, pp 113-123, Ed: J. Délery, Springer-Verlag

[6] Bahi, L.; Ross, J.M.; Nagamatsu, H.T.: Passive Shock Wave/Boundary Layer Control for Transonic Airfoil Drag Reduction.
AIAA Paper 83-0137, Jan. 1983

[7] Chung-Lung Chen: Computation of Transonic Flow over Porous Airfoils.
Dissertation, University of Colorado, 1986

[8] Délery, J.; Chanetz, B.; Pot, T.: Fundamental Experiments on Passive Control Applied to Transonic Shock-Wave/Boundary-Layer Interaction.
3. STAB-Workshop, 10.-11.Nov.1987, DFVLR Göttingen

[9] Raghunathan, S.; Mabey, D.G.: Passive Shock-Wave/Boundary-Layer Control on a Wall-Mounted Model.
AIAA-Journal 25, 275 ff (1987)

[10] Liu, X.; Squire, L.C.: An Investigation of Shock/Boundary-Layer Interactions on Curved Surfaces at Transonic Speeds.
J. Fluid Mech. 187, 467-468 (1988)

Passive Control of Shock/Boundary Layer Interactions: Numerical and Experimental Studies

N. Chokani (Graduate Student) and L.C. Squire (Reader)

Department of Engineering
University of Cambridge
Cambridge, UK

1. Summary

The passive control of transonic shockwave/turbulent boundary layer interaction on a porous surface has been experimentally studied on wall-mounted models in a wind tunnel. The transonic flowfield has been measured using laser holographic interferometry. The numerical predictions of a two-dimensional time-marching Navier-Stokes method are compared with the quantitative details of the interferogram, and found to be in good agreement with the experimental data.

2. Introduction

The passive control of shock/boundary layer interactions, whereby part of the solid surface of the aerofoil is replaced by a porous surface over a shallow cavity, Figure 1, has been shown to be a means of improving the aerodynamic characteristics of supercritical aerofoils, Thiede et al. [1]. These experimental results showed considerable drag reduction while calculations using a viscous-inviscid coupling procedure suggested that shock-induced separation on a porous aerofoil may be eliminated, Olling et al. [2]. In order to gain some understanding of the mechanisms involved in the interaction between a shockwave and a boundary layer on a porous surface, an experimental and numerical study of the flow over a simple porous model has been undertaken. The experimental flow was generated over a circular-arc model mounted on the floor of a rectangular section wind tunnel, Figure 2, which was non-porous apart from the control region. The tunnel flow was choked by using an aerofoil at incidence in the tunnel diffuser to fix the back-pressure, thus for a given tunnel settling chamber stagnation pressure the flow conditions were well-defined. The results presented here were obtained by a non-intrusive technique based on holographic interferometry.

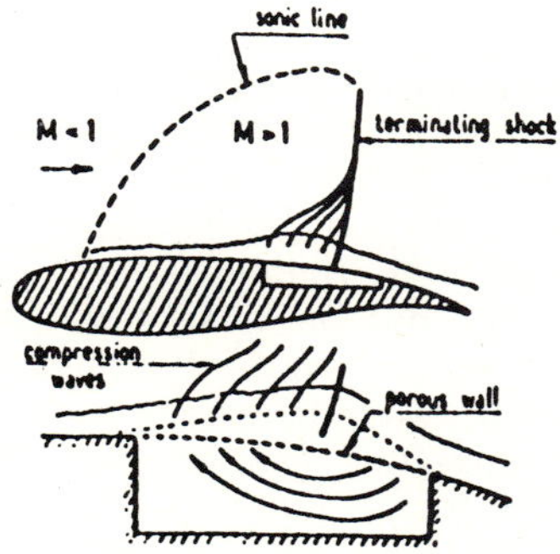

Figure 1 Passive control on supercritical aerofoil

J. Zierep, H. Oertel (Eds.)
Symposium Transsonicum III
IUTAM Symposium Göttingen, 24.-27.5.1988

The shockwave/boundary layer interaction was computed with a two-dimensional Navier-Stokes solver. The original program was written for the blade-to-blade regions in turbine cascades, and is suitable for computing the choked flow; it was reprogrammed to incorporate the injection and suction effects of passive control in the interaction region. The results are shown to be in good agreement with the experiment and enable a detailed study of the interaction to be made.

3. Experimental Details

The experiments were conducted in a supersonic tunnel of the Aerodynamics Laboratory, Cambridge University Engineering Department. The tunnel was of the intermittent blowdown type, and exhausted directly to atmosphere. The tunnel working-section was 114*mm* wide and 63*mm* high. Interchangeable circular-arc models, of chord 80*mm*, were mounted on the tunnel floor, 668*mm*downstream of the tunnel entrance. The tunnel roof was flat. The working-section was closed by solid doors equipped with 203*mm* diameter optical windows for the flow visualisation. The tunnel floor was gently tapered to account for the boundary layer growth on all four sidewalls. The boundary layer developed from the tunnel entrance, and was fully turbulent - the Reynolds number based on momentum thickness ahead of the circular-arc model was greater than 10^4.

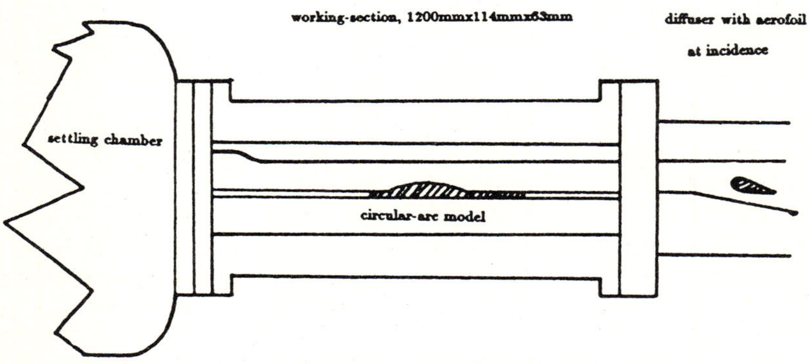

Figure 2 The wind tunnel

A porous surface, 104*mm* wide and 20*mm* long, with holes of diameter 0.813*mm*, drilled normal to the surface was placed in the shockwave/boundary layer interaction region. The holes were spaced 1.5*mm* apart in the spanwise direction, and 2.5*mm* apart in the chordwise direction. The porous open area, ratio of open area to "control surface area", was 13.6%. The mean cavity depth was 4mm. Two cases were investigated: Case A - the porous surface was mounted on a circular-arc model, radius of curvature 401*mm*; Case B - the porous surface was mounted on the flat tunnel roof, and the shockwave/boundary layer interaction generated by a circular-arc model, radius of curvature 163*mm*, mounted on the tunnel floor. The test details are shown in Table 1.

Table 1 Operating conditions

	Case A	Case B
Circular-arc model radius, *mm*	401	163
Stagnation pressure, *kPa* (gauge)	96.5	48.3
Stagnation temperature, °*K*	292	292
Unit Reynolds number	$2.8 \cdot 10^7$	$2.1 \cdot 10^7$
Freestream Mach number, M_∞	0.853	0.763

The use of the laser holographic interferometry technique to measure the density flowfield is described in detail by Vest [3]. A pulsed ruby laser, with a maximum output of 1.2*J* in 30*ns*, was used for holographic formation. The laser wavelength was 0.69μm. The double exposure technique was used. The first exposure was made prior to the start of the tunnel flow, in which the density was uniform normal to the beam passing through the working section. A second exposure on the same holographic plate was made during the run, and in this case the density varied normal to the beam, but was assumed to be constant along it. The fringes of the resulting interference pattern are of equal density change, $\Delta \rho = 0.027 kg/m^3$, and shown in Figures 3-6. The measured wall static pressure, tunnel settling chamber stagnation pressure and temperature, and the isentropic flow relationship between density and Mach number are used to calculate the density of the reference fringe labelled 0. By counting fringes, the whole flowfield is quantitatively detailed.

4. Experimental Results

Case A

The shockwave/boundary layer interaction occurs near the end of the porous surface on the tunnel floor. Figures 3 and 4 show the interaction without and with the passive control, respectively. Upstream of the shock, the flow is isentropic and the fringes correspond to Mach number contours as well as density contours. The Mach number at the shock foot is calculated to be 1.25 without the passive control, and 1.22 with the passive control. This weakening of the shock strength would suggest a reduction in the wave drag. In the mainstream flow the shockwave is thin. As the wall is approached, the shockwave/boundary layer interaction smears out the shock; with the passive control the smearing is greater than without the control. The increased smearing of the shock foot indicates that the pressure rise at the wall is less steep, and the boundary layer less likely to separate. Measurements of the surface static pressure confirmed the reduction in the wall pressure rise. Immediately upstream of the shock, kinks are observed in the fringes and are a result of Mach waves emanating from the holes in the porous surface.

Case B

The shockwave/boundary layer interaction occurs near the mid-section of the porous surface on the tunnel roof. Figures 5 and 6 show the interaction without and with the passive control, respectively. The Mach number at the shock foot with the passive control is calculated to be 1.24, and the shockwave is nearly normal to the surface. With the passive control a strong weakening of the normal shock is observed - an oblique shock, with Mach number 1.18 ahead of it emanates from near the leading edge of the porous surface; assuming an isentropic relationship across the weak shock, the Mach number immediately downstream of it is calculated to be 1.12. The substantial reduction in shock strength would suggest a very significant reduction in the wave drag component of total drag. Bahi et al. [4] reported observing a similar shockwave structure in their Schlieren observations. This shockwave structure is not observed in Case A where the shockwave is located near the rear of the porous region.

5. Numerical Method

The blade-to-blade two-dimensional compressible Navier-Stokes solver of Dawes [5], was used to calculate the flow. The flow equations used are the two-dimensional, isenthalpic, thin-layer Navier-Stokes equations; a fully implicit numerical algorithm is used to time-march the flow equations. The two-layer algebraic eddy-viscosity model of Baldwin and Lomax [6] was used to close the set of equations. The van Driest damping term, D, in the inner Prandtl eddy-viscosity formulation and in the outer Baldwin-Lomax function, $F(y) = y|\omega|D$, was modified for the injection and suction effects using an approach of Cebeci and Smith [7]:

$$D = 1 - \exp\left(-y^+/A^+\right)$$

where

$$A^+ = A_0 \exp\left(-5.9 v_w^+\right);\ A_0 = 26$$

$$y^+ = y u_r/v;\ v_w^+ = \frac{v_w}{u_r}$$

$$|\omega| = \left|\frac{\partial u}{\partial y} - \frac{\partial v}{\partial x}\right|$$

and the other parameters their usual meaning.

The wall shear stress was evaluated using wall functions, modified for the mass transfer effects over the porous surface. The normal wall velocity over the porous region was evaluated using the linear form of the Darcy law suggested by Savu et al. [8]:

$$v_w = \frac{\overline{\sigma}}{\rho_\infty U_\infty} \Delta p$$

where $\Delta p = p_c - p_w$; p_c, p_w are the cavity and wall pressures, respectively. As observed experimentally, the cavity pressure was held spatially constant. The constant, $\overline{\sigma}$, for the 13.6% porous area was evaluated as 0.62 in a separate experiment, (cf. 0.6 for 10% geometrical porosity of Savu et al. [8]).

Grids of 95x40 and 97x40 mesh points were used for Case A and Case B, respectively. A solution was obtained in about 320 timesteps using about 11 minutes of CRAY-1S CPU time. The computed flowfields are shown in Figures 7-10. The density contours have been chosen to have the same interval as the interferogram.

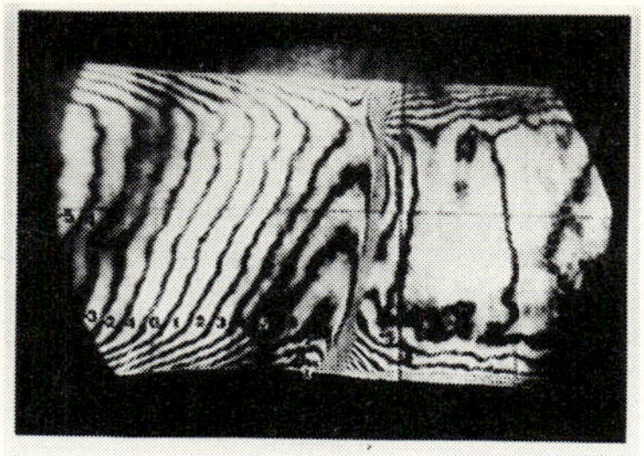

Figure 3 Case A. Interferogram - no passive control.

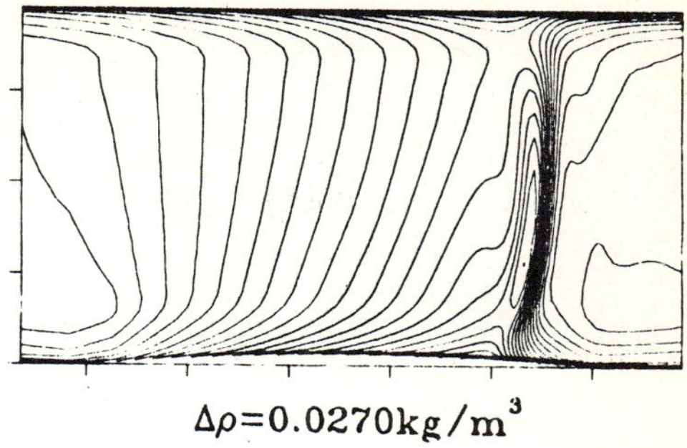

Figure 7 Case A. Computed density contours - no passive control

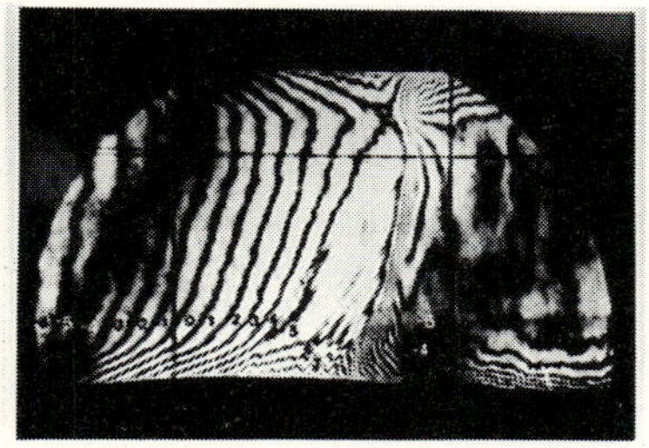

Figure 4 Case A. Interferogram - passive control.

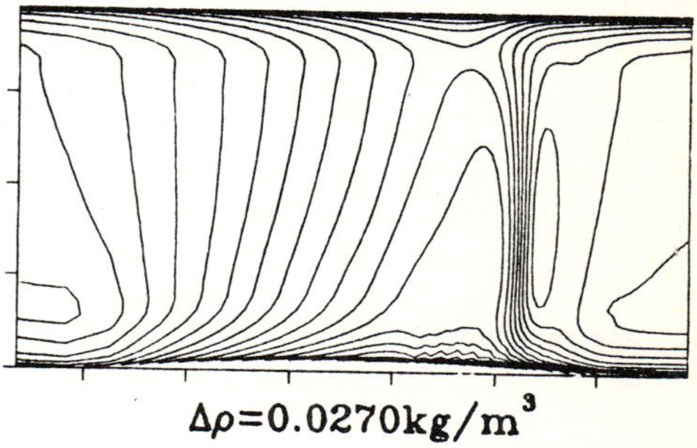

Figure 8 Case A. Computed density contours - passive control

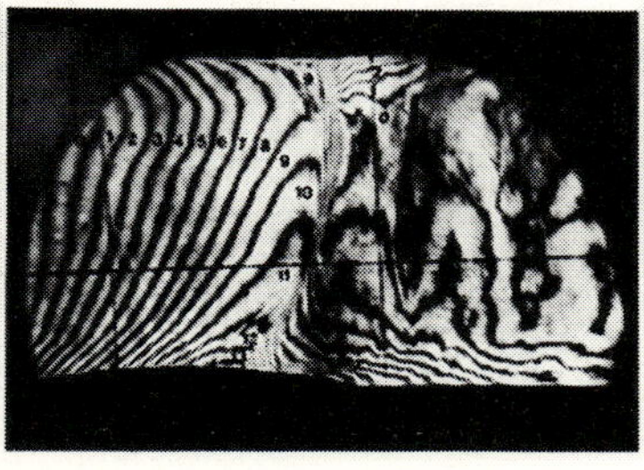

Figure 5 Case B. Interferogram - no passive control

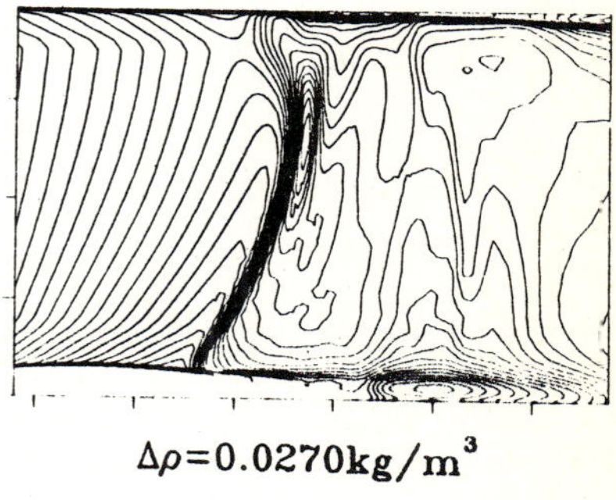

Figure 9 Case B. Computed density contours - no passive control

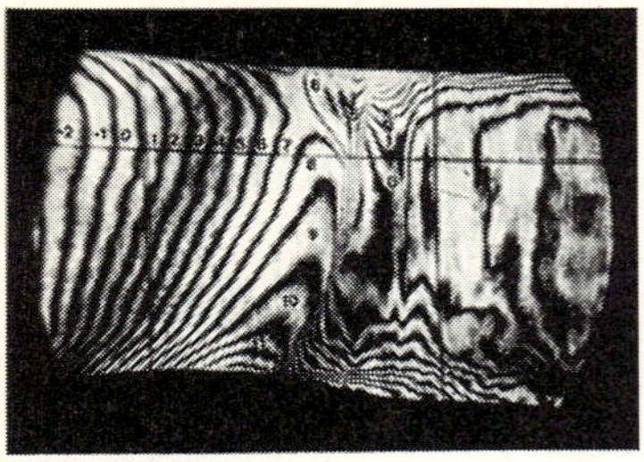

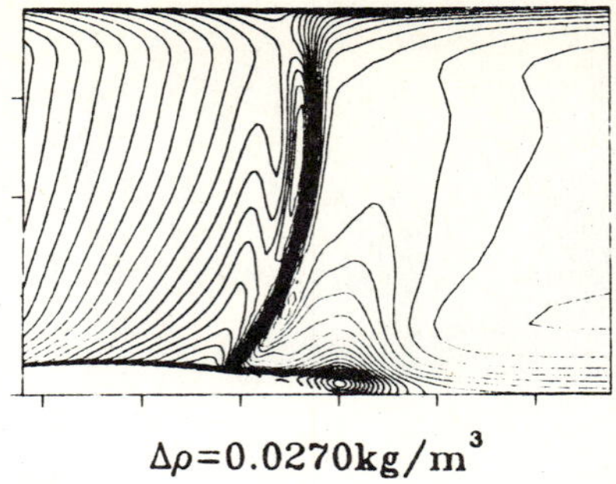

Figure 6 Case B. Interferogram - passive control

Figure 10 Case B. Computed density contours - passive control

The computed density flowfields of Case A are observed to be in good agreement with the experiment. Without the passive control, the shockwave location is in very good agreement with the experiment. The density contours upstream of the shockwave have the same slope as the experiment. With the passive control the agreement is very good, and the further smearing out of the shock foot well represented. On the tunnel roof the shockwave/boundary layer interaction is predicted slightly upstream of the experiment. From the computed results the normal wall velocity may be determined; this is shown in Figure 11, normalised as a mass flux transfer parameter. The results indicate that the injected velocities are very small compared with the freestream velocity, 1-3%.

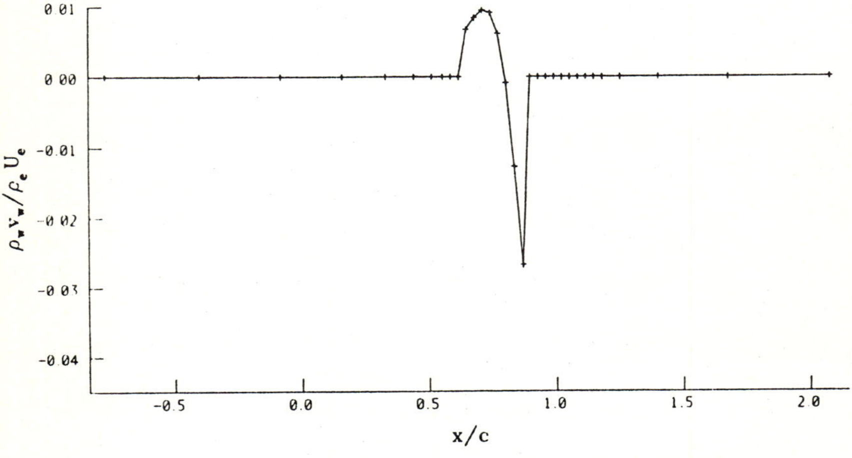

Figure 11 Case A. Computed wall velocity distribution

The computed density flowfields of Case B are shown in Figures 9 and 10. On the opposite side of the passive control surface, the tunnel floor, the flow separation has been overpredicted. The shockwave on the tunnel roof occurs slightly downstream of the experiment. The weakening of the shockwave with the passive control is evident, but the weak oblique shock is not observed. The algebraic turbulence model is inadequate in dealing with the separated flow, and produced an interaction on the porous surface similar to Case A.

6. Conclusions

A flow under well-defined conditions has been used to study the passive control on transonic shockwave/boundary layer interactions. For both the porous and the non-porous surfaces, the tunnel back-pressure was held fixed and the incoming choked flow ahead of the porous surface was identical. These flow conditions enabled the passive effect to be reliably observed. Laser holographic interferometry has been used to study the flow.

In the experiment, the injection is observed to lead to a dramatic weakening of the shockwave when it is the mid-porous section; the passive effect is less significant when the normal shockwave is at the rear of the porous section. The dramatic reduction in shock-strength is accompanied by a bifurcation of the shock.

The interaction has been computed with a two-dimensional Navier-Stokes code. The general features of the interaction are well-predicted - in particular, the weakening of the shockwave and relaxation of flow gradient ahead of it. The computed results are able to provide details of the interaction, such as the magnitude of the normal velocitiy, which is observed to be 1-3% of the freestream velocity. Because of the limitations of the turbulence model, it was impossible to predict well the separated flow on the tunnel floor.

So far the results of the experimental and numerical studies have shown to be promising, and further work is continuing into this research.

Acknowledgements

The authors would like to thank Dr. Dawes, Whittle Laboratory, Cambridge University, for providing a copy of his Navier-Stokes program, and for discussions regarding its use.

References

[1] Thiede, P., Krogmann, P., Stanewsky, E.: Active and passive shock/boundary layer interaction on supercritical airfoils. AGARD CP-365 (1984), Paper no. 25.

[2] Olling, C.R., Dulikravich, G.S.: Porous airfoil analysis using viscous-inviscid coupling at transonic speeds. Int. J. Num. Flds. 7 (1987), 103-129.

[3] Vest, C.M.: Holographic interferometry. London New York. J. Wiley & Sons. (1979).

[4] Bahi, L., Ross, J.M., Nagamatsu, H.T.: Passive shockwave/boundary layer control for transonic airfoil drag reduction. AIAA Paper 83-013 (1983).

[5] Dawes, W.N.: Efficient implicit algorithm for the equations of 2-D viscous compressible flows: application to shock-boundary layer interaction. Int. J. Heat Fluid Flow (1983), 17-26.

[6] Baldwin, B.S., Lomax, H.: Thin layer approximation and algebraic model for separated turbulent flows. (1974), AIAA Paper no. 78-257.

[7] Cebeci, T., Smith, A.M.O.: Analysis of turbulent boundary layers. New York. London: Academic Press. (1974).

[8] Savu, G., Trifu, O., Dumitrescu, L.Z.: Suppression of shocks on transonic airfoils. Proc. 14th Int. Symp. Shock Tubes & Shock Waves. (1983), 92-101.

Transonic Inviscid-Viscous Interactions over Porous Surfaces

C.-Y. Chow*, C.-L. Chen**
T.L. Holst*** and W.R. Van Dalsem***

* Department of Aerospace Engineering Sciences
University of Colorado, Boulder, Colorado

** Sterling Software, Palo Alto, California

*** Applied Computational Fluids Branch
NASA Ames Research Center, Moffet Field, California, USA

Summary

The viscous effects on transonic flow past an airfoil which contains a shallow cavity beneath a porous surface are studied numerically. The porous region occupies a small portion of the airfoil surface, and is near the shock. A thin-layer Navier-Stokes algorithm in combination with a modified Baldwin-Lomax turbulence model is used in computing the outer flow, whereas a stream-function formulation is used to model the inner flow in the cavity. The two flow regions are coupled at the porous surface through Darcy's law. Shock structure and flow patterns have been computed to show the influence of shock/boundary-layer interactions. In agreement with experiments, numerical results show that the total drag of a porous airfoil is reduced at higher transonic speeds but is increased at lower Mach numbers.

Introduction

The transonic flow past an airfoil in flight at a high-subsonic Mach number is characterized by a shock wave standing on the upper surface. The adverse pressure gradient of a strong shock can induce boundary-layer separation, which may increase form-drag and reduce lift.

A passive control device (Fig. 1), which provides blowing and suction without externally supplied power, is an attempt to improve the performance of transonic airfoils by controlling boundary layer separation and reducing shock-wave strength. Indeed such an expectation has been proven in the laboratory under certain conditions [1-6]. In addition, inviscid-flow computations [7,8] also indicate that a porous surface can reduce the wave-drag. However, inviscid simulations cannot predict any effects of porosity on the viscous drag. In particular, when the shock and boundary-layer interaction becomes strong, the inviscid-flow analysis is not sufficient to describe the flow about an airfoil. It is the purpose of this paper to study the effects of shock-wave/boundary-layer interaction in transonic flow past porous airfoils.

J. Zierep, H. Oertel (Eds.)
Symposium Transsonicum III
IUTAM Symposium Göttingen, 24.-27. 5. 1988

A numerical simulation was done by Olling [9] by using an interactive boundary-layer procedure with the assumption of constant pressure in the cavity. In the present work, the cavity flow is modeled and the outer flow is computed using a thin-layer Navier-Stokes procedure which is less restrictive than the boundary layer formulation.

Outer Flow Field

The flow field outside the airfoil is obtained by solving the thin-layer Navier-Stokes (TLNS) equations. A main feature of these equations is the presence of a nonzero normal-pressure gradient, which is necessary to couple the viscous and inviscid regions. By comparison with the Navier-Stokes equations, these composite equations require less computational effort because they contain fewer terms.

The C-type grid used for numerical solution of the outer flow is obtained by a hyperbolic grid generation procedure [10]. The governing equations are transformed from Cartesian coordinates in the physical domain to the computational domain by transformations $\zeta = \zeta(x,y)$ and $\eta = \eta(x,y)$, with the $\eta = 0$ curve coinciding with the airfoil surface. As shown in Fig. 2, the grid spacing is clustered near the airfoil surface. For the test cases presented here, the total number of grid points is 251 x 65 and the outer boundary is set at 16 chord lengths away from the airfoil.

The TLNS procedure (ARC2D) developed at NASA Ames Research center is adopted for this study. This implicit code is based on the Beam-Warming approximate-factorization algorithm. Euler implicit time-differencing and second-order central-spatial differencing are utilized. A nonlinear artificial dissipation model is added to capture non-oscillatory shock profiles and maintain stability. The diagonal version developed by Pulliam and Chaussee [11] is applied in the present study. Also, a space-varying time-step and a coarse- to fine-grid sequencing are used to accelerate the convergence rate.

On the airfoil surface, the normal wall velocity v_n is specified and the tangential velocity is set to zero. The pressure boundary condition at the airfoil surface is taken to be $\partial p/\partial\eta = 0$, since the gradient η lines are nearly orthogonal to the airfoil surface and the grids are very fine in that direction near the airfoil surface. The adiabatic wall condition is used to obtain the density at the surface, and total energy is obtained from the equation of state. The far-field boundary conditions are obtained by imposing a compressible potential vortex solution on the free-stream quantities [12]. All the boundary conditions are updated explicitly in the TLNS procedure.

The Baldwin-Lomax turbulence model [13] used in the TLNS procedure is patterned after that of Cebeci-Smith, with modifications that make it unnecessary to find the outer edge of the boundary layer. However, the Baldwin-Lomax turbulence model did not consider blowing (or suction) effects on A^+, and the damping factor A^+ is redefined by Bushnell et al. [14] as $A^+ = 26/(\tau^+)^n$. The value $n = 0.7$ is used in the present analysis, which is determined by comparing with data measured in boundary layers with blowing or suction.

Cavity Flow

The compressible flow in the cavity is assumed to be inviscid and isentropic, which has a different total pressure from that of the outer flow. The total temperature is assumed to be the same as that of the free stream.

The cavity flow is described by a stream-function formulation. On an algebraically generated grid, the governing equations with appropriate boundary conditions are solved using an approximate-factorization scheme (AF1), which is a reformulation of an ADI scheme. Detailed descriptions of all numerical schemes and computational procedures are referred to Ref. 15.

Porous Interface of Outer and Inner Flows

The outer flow and the flow in the cavity are matched at the porous surface shown in Fig. 3. By Darcy's law, the transpiration velocity through the porous surface is proportional to the pressure difference across it,

$$v_n = -\frac{\bar{\sigma}}{\rho_\infty U_\infty}(p_{outer} - p_{inner}) \tag{1}$$

where $\bar{\sigma}$ is the porosity distribution function, which is determined by viscosity as well as by the size and spacing of the holes in the porous surface.

Two cavity models have been used in this work. The first model assumes constant pressure in the cavity, which has been applied in Refs. 7-9. For the second model considered, the total pressure in the cavity is assumed to be constant.

For passive flow through the cavity, the net mass flow through the porous surface of length S must be zero, which leads to the relation

$$(p_t)_{inner} = \frac{\int_S \bar{\sigma}(\rho p)_{outer} ds}{\int_S \bar{\sigma}\rho_{outer}(\frac{p}{p_t})_{inner} ds} \qquad (2)$$

The pressure variation in the cavity $(p/p_t)_{inner}$ can be determined by computing the cavity flow.

The stream-function is set to zero at both the vertical sidewalls and at the bottom wall of the cavity. Along the porous surface, ψ is determined by the integration of mass flow rate from the outer flow by

$$\psi(x) = \int_{x_1}^{x} (\rho v_n)_{outer} ds \qquad (3)$$

where x_1 is the coordinate of the leading edge of the cavity (Fig. 1).

The $(p/p_t)_{inner}$, $(p_t)_{inner}$, and v_n can be determined iteratively. The process uses the first cavity model to obtain an initial v_n, then integrates the mass flow rate by Eq. (3) to obtain the boundary conditions for the cavity and solves the cavity flow to update $(p/p_t)_{inner}$. $(p_t)_{inner}$ is then updated by Eq. (2), and v_n is updated by using Eq. (1) to finish the first iteration. The iterative procedure is repeated to update the boundary stream-function until the converged values of $(p_t)_{inner}$, $(p/p_t)_{inner}$, and v_n are obtained and V_n is available for the next update of the outer flow solver.

Numerical Results

Effects of various porous surfaces on lifting and nonlifting airfoils have been examined [8,15]. Presented here are the results for a symmetrical airfoil that is generated by reflecting the upper surface of an RAE 2822 airfoil to the lower surface. The porous surfaces are from x = 0.615 to 0.805 with a constant porosity $\bar{\sigma} = 0.4$, and the second cavity model is used on both upper and lower surfaces. The bottom of the cavity is at y = 0, the angle of attack α is 0^0, Reynolds number Re is 6.5×10^6, and the transition is fixed at x = 0.03.

Case (I) for $M_\infty = 0.82$

The comparison of pressure coefficient plots (Fig. 3) for the solid and porous airfoils shows that relative to the solid airfoil, the shock is weakened near the airfoil surface, and the pressure on the porous airfoil is lower at the trailing edge. The C_p plots also indicate that the original, one strong pressure jump becomes two consecutive weaker jumps, which represent the leading and rear legs of a lambda shock. The rear leg is slightly downstream from the shock position on the solid airfoil. The lowered pressure at the trailing edge has also been observed by Raghunathan and Mabey [4] in the flow about a 6%-thick half-circular-arc model mounted on the upper wall of a wind tunnel.

The comparison of the Mach-contour plots (Fig. 4) indicates that a lambda-shock structure does occur on the porous airfoil. The leading leg of the lambda shock slants at the beginning of the porous surface. This shock structure has been observed by Nagamatsu et al. [3] in the laboratory.

A comparison of streamline patterns near the porous region is shown in Fig. 5. The passive flow through the porous surface generates a bump-like bubble (not a separation bubble) where the lambda shock is situated.

Blowing at the leading part of the porous region reduces C_f, and suction at the rear part of porous region makes the local C_f larger in the suction region, as shown in Fig. 6. But downstream of the suction region, the flow has difficulty overcoming the adverse-pressure gradient and becomes separated; as a result, the values of C_f are smaller than they are for the solid airfoil.

After combining the effects of pressure distribution and skin friction, the total drag coefficient C_D is found to increase from 0.0258 for the solid airfoil to 0.0273 for the porous airfoil.

Case (II) for $M_\infty = 0.85$

Both the C_p plot (Fig. 7) and Mach-contour plot (Fig. 8) indicate that the shock is weakened and is moved slightly upstream on the porous airfoil. At this Mach number, the computed result shows a 9% reduction in drag (C_D decreases from 0.055 for the solid airfoil to 0.050 for the porous airfoil).

The separated region above the porous surface is enlarged and the boundary layer becomes thicker, in agreement with experiments by Krogmann et al. [5,6]. Figure 9b indicates that part of the fluid that is blown out at the upstream end of the porous surface encloses a dead air region before going back into the cavity.

As shown in Fig. 10, the suction at the rear part of the porous region makes C_f more negative over the suction region, which is opposite to the effect shown in Fig. 6 of case (I).

Case (III) for Extended Porous Surface

In this case the same modified RAE 2822 airfoil considered in cases (I) and (II) is used for numerical study, except that the downstream end of the porous surface has been extended from x = 0.805 to 0.880. The C_D variation of this airfoil having a 39% longer porous surface with increasing flight Mach number is plotted in Fig. 11. Drag reduction by the porous surface, which can be as much as 20%, occurs at Mach numbers higher than 0.84; below that Mach number the porous surface has an adverse effect of increasing the total drag. Similar phenomena have been observed and drag reductions of the same order of magnitude have been found by Nagamatsu et al. [3] in the study of a supercritical airfoil mounted on the bottom wall of a wind tunnel.

Discussion and Conclusion

Using the present turbulence and porosity modeling, the structure of shock and boundary-layer interactions, flow pattern near the porous surface, as well as the pressure distribution and skin friction on the airfoil surface can be computed for transonic airfoils at small angles of attack. The numerical results so obtained improve the inviscid-flow solutions and qualitatively agree with many of the available experimental data.

In all cases based on the second cavity model of constant total pressure in the cavity, computed results show that the cavity flow is slow and pressure variation in there is negligibly small. Thus, whether the first model of constant cavity pressure or the second model is used does not affect the outer flow solution.

At low transonic Mach numbers, the passive blowing and suction through the porous surface weaken the shock strength (Fig. 3) but the accelerated flow caused by suction increases the skin friction (Fig. 6), resulting in a total drag that is slightly higher than that of a solid airfoil. On the other hand, at higher Mach numbers the porous surface causes extensive boundary layer separation (Fig. 9) to form a region of negative skin friction near the trailing edge (Fig. 10), and to move the shock upstream so that the size of the supersonic bubble becomes smaller and the shock strength becomes much weaker (Fig. 7). The net result is a reduction in total drag.

References

1. Bahi, L.; Ross, J.M.; Nagamatsu, H.T. Passive shock wave/boundary layer control for transonic airfoil drag reduction. AIAA Paper 83-0137.

2. Nagamatsu, H.T.; Orozco, R.D.; Ling, D.C. Porosity effect on supercritical airfoil drag reduction by shock wave/boundary layer control. AIAA Paper 84-1682.

3. Nagamatsu, H.T.; Dyer, R.; Ficarra, R.V. Supercritical airfoil drag reduction by passive shock wave/boundary layer control in the Mach number range .75 to .90. AIAA Paper 85-0207.

4. Raghunathan, S.; Mabey, D.G. Passive shock wave boundary layer control experiments on a circular arc model. AIAA Paper 86-0285.

5. Krogmann, P.; Stanewsky, E.,; Thiede, P.: Effects of suction on shock/boundary-layer interaction and shock-induced separation. J. Aircraft 22 (1985) 37-42.

6. Thiede, P.; Krogmann, P.; Stanewsky, E.: Active and passive shock/boundary-layer interaction control on supercritical airfoils. AGARD CP-365 (1984) 24.1 - 24.13.

7. Savu, G.; Trifu, O.: Porous airfoils in transonic flow. AIAA Journal 22 (1984) 989-991.

8. Chen, C.-L.; Chow, C.-Y.; Holst, T.L.; Van Dalsem, W.R. Numerical simulation of transonic flow over porous airfoils. AIAA Paper 85-5022.

9. Olling, C.R. Viscous-inviscid interaction in transonic separated flow over solid and porous airfoils and cascades. Ph.D. dissertation, University of Texas at Austin, Dec. 1985.

10. Barth, T.J.; Pulliam, T.H.; Buning, P.G. Navier-Stokes computations for exotic airfoils. AIAA Paper 85-0109.

11. Pulliam, T.H.; Chaussee, D.S.: A diagonal form of an implicit approximate-factorization algorithm. J. Comp. Phys. 39 (1981) 347-363.

12. Salas, M.; Jameson, A.; Melnik, R. A comparison study of the nonuniqueness problem of the potential equation. AIAA Paper 83-1888.

13. Baldwin, B.S.; Lomax, H. Thin-layer approximation and algebraic model for separated turbulent flows. AIAA Paper 78-257.

14. Bushnell, D.M.; Cary, A.M., Jr.; Harris, J.E. Calculation methods for compressible turbulent boundary layers - 1976. NASA SP-422 (1977).

15. Chen, C.-L. Computation of transonic flow over porous airfoils. Ph.D. dissertation, University of Colorado at Boulder, Aug. 1986.

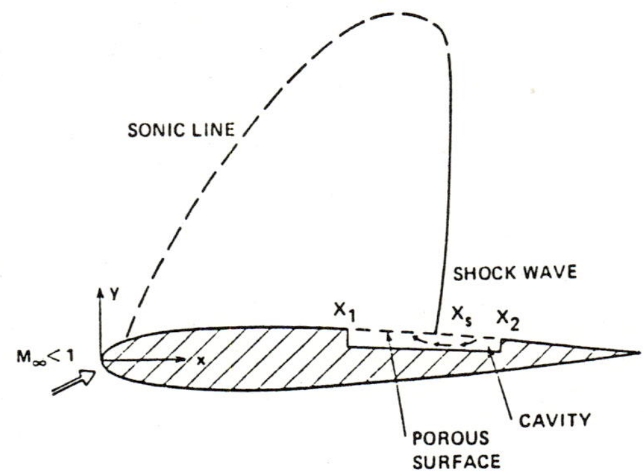

Fig. 1 Porous airfoil in transonic flow

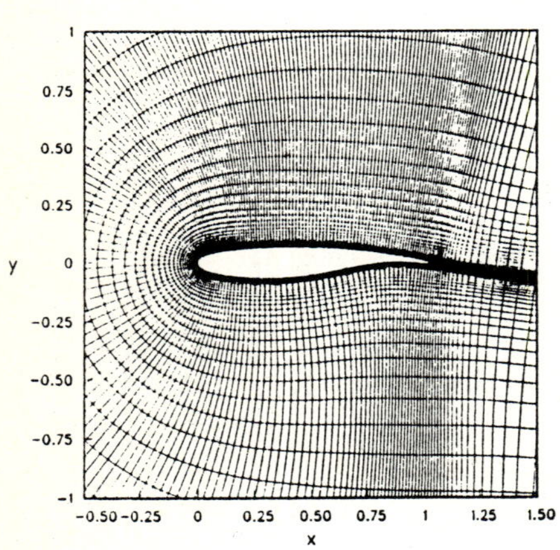

Fig. 2 Grid for TLNS calculation

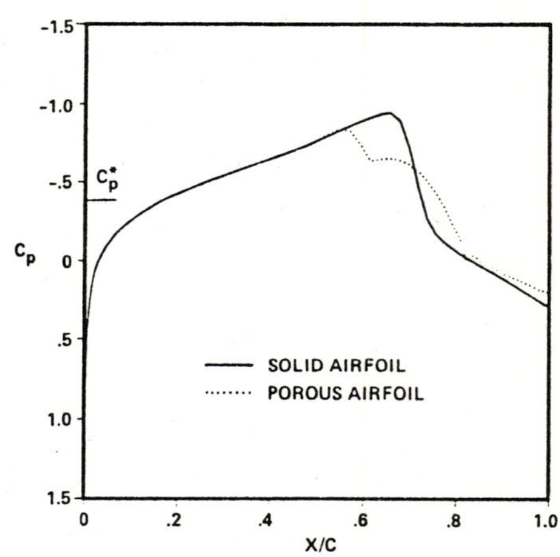

Fig. 3 Surface pressure, $M_\infty = 0.82$

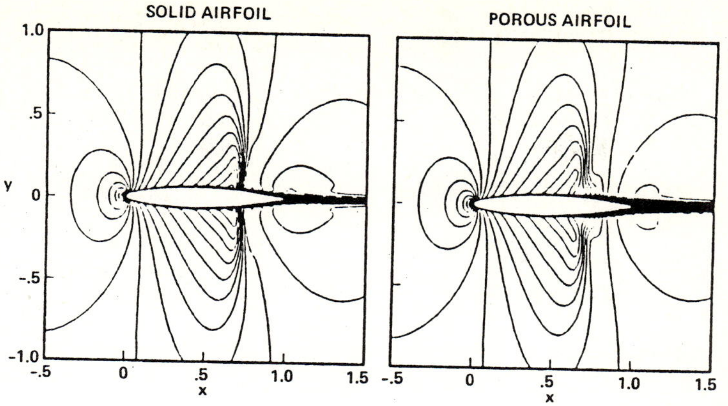

Fig. 4 Constant Mach-number contours, $M_\infty = 0.82$

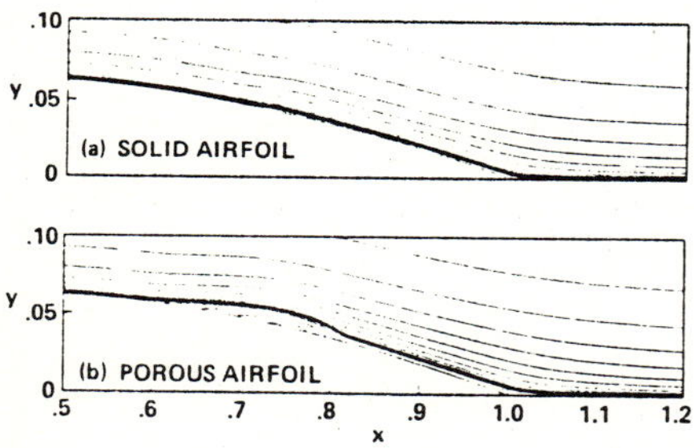

Fig. 5 Streamline patterns, $M_\infty = 0.82$

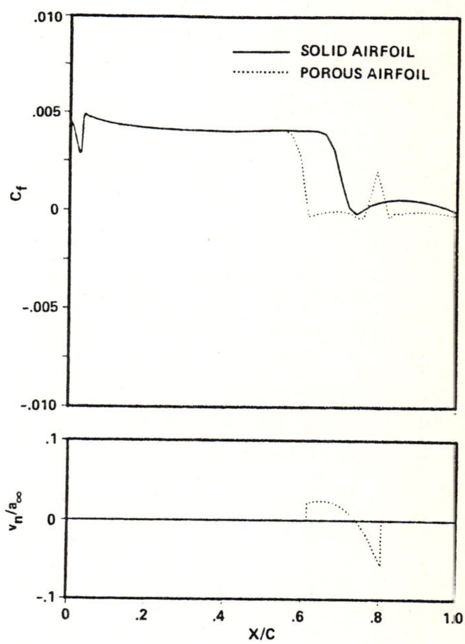

Fig. 6 Skin friction, $M_\infty = 0.82$

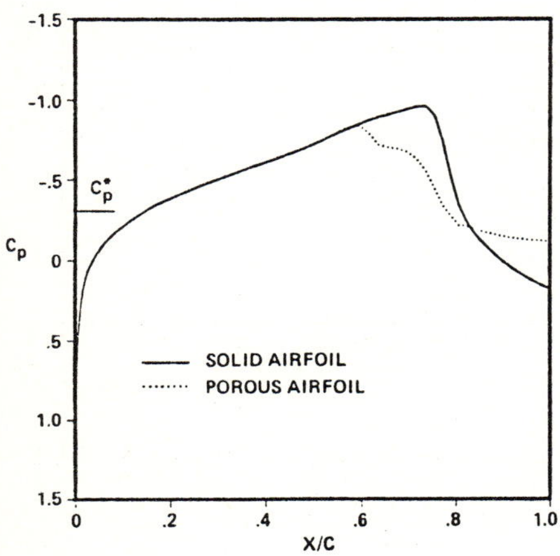

Fig. 7 Surface pressure, $M_\infty = 0.85$

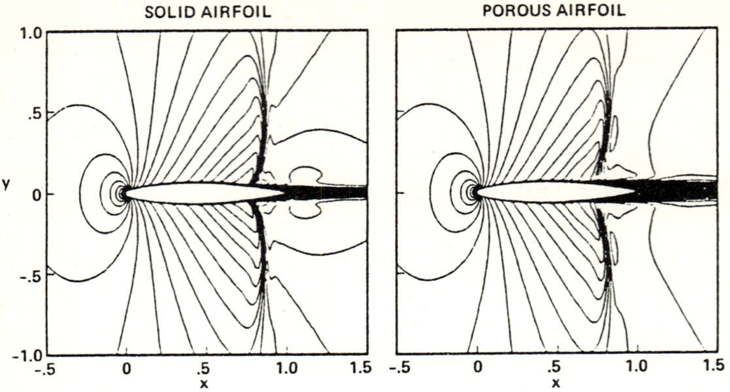

Fig. 8 Constant Mach-number contours, $M_\infty = 0.85$

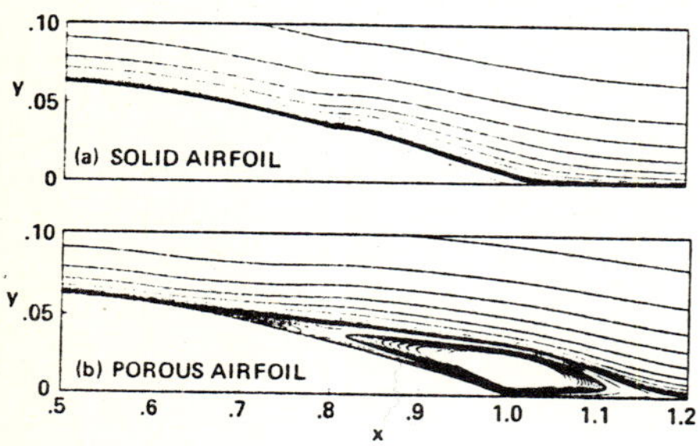

Fig. 9 Streamline patterns, $M_\infty = 0.85$

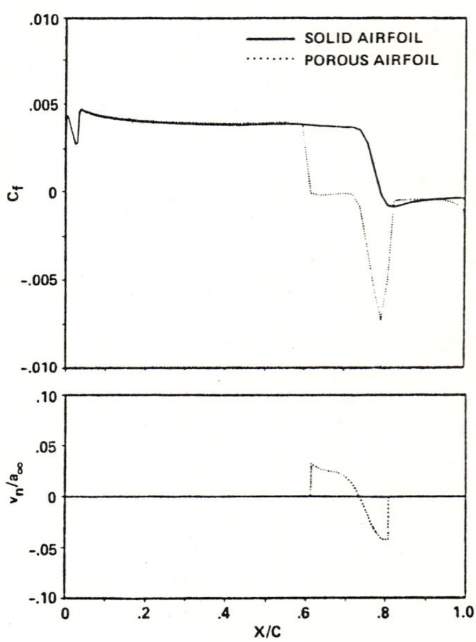

Fig. 10 Skin friction, $M_\infty = 0.85$

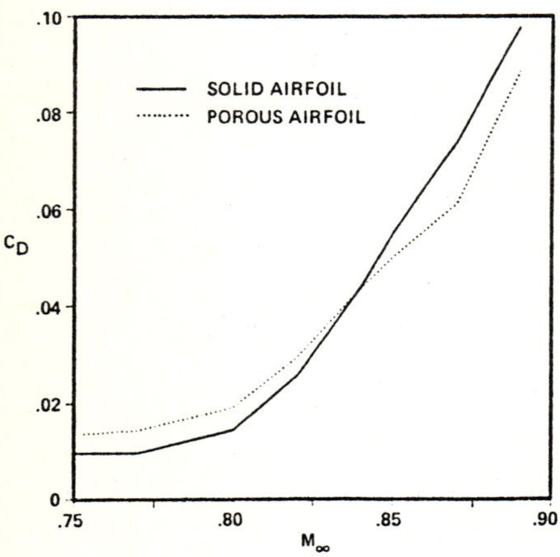

Fig. 11 Variation of drag with M_∞

Wind Tunnel Studies of Natural Shock Wave - Separation Instabilities for Transonic Airfoil Tests

J.B. Dor, A. Mignosi, A. Seraudie, B. Benoit

ONERA/CERT - Aerothermodynamics Department
2, Avenue Edouard Belin, 31055 Toulouse

ONERA OA 29, Avenue de la Division Leclerc, 92320 Chatillon, FRANCE

Summary

Transonic wind tunnel tests on 2D airfoils with tripped transition and unsteady wall pressure or laser velocimetry measurements show that buffeting appears, at high lift coefficient, like a well-established oscillation of the upperside shock wave and the separated flow which is born at its foot. During cryogenic tests at high Reynolds numbers, the buffeting onset conditions and the nature and the intensity of the phenomenon are not clearly modified, but a specific temperature effect on the oscillation frequency, through sound speed variation, is observed. For free transition tests at low Reynolds number, the boundary layer remains laminar up to the shock and the oscillation does not appear in buffeting conditions, there is only a strong increase of the pressure fluctuation level on the airfoil, without any particular frequency. The characteristics (extent, sensitivity) of the separated flow which is generated by the interaction between shock wave and boundary layer seem to govern these instabilities.

1 - Introduction

This paper summarizes aerodynamic wind tunnel studies on stiff 2D airfoils, conducted by ONERA from 1984 to 1987, as a first step to understand the mechanisms of the buffeting phenomenon in high subsonic flight (mechanical vibrations of the plane, due to aerodynamic instabilities at high lift coefficient). These studies were started by tests in the S3MA transonic wind tunnel at Modane, and were largely carried on in the T2 wind tunnel of CERT at Toulouse. §2 gives some indications about the tunnels, the models and the instrumentation.

For the tested airfoils, when transition is tripped near the leading edge, buffeting appears at high Mach numbers or angles of attack as a periodic instability of the upperside shock wave and the downstream separated flow region. §3 presents the first tests in T2 or S3, which were based on unsteady pressure measurements and allowed us to identify this phenomenon. §4 is devoted to laser anemometry measurements in the unsteady upperside flow. §5 gives the main results which were obtained about buffeting during cryogenic airfoil tests at high Reynolds numbers.

It must be noticed that such transonic unsteady flows have already been detected by other scientists on particular airfoils (Ref.4).

2 - General test conditions

The T2 transonic wind tunnel (M=0.6 to M=0.9) / Ref. 7/ is driven by induction and operates by short runs (1 or 2 mn). Its test section (0.39m wide ; 0.37m high) is equipped with 2D adaptive walls which are adjusted by an iterative process during the first part of the run. In order to obtain Reynolds number variations, the tunnel is cryogenic and pressurized and can operate on the ranges : Tt from 300K to 100K, Pt from 1.5bars to 4bars.

S3MA is a transonic and pressurized wind tunnel which operates at ambient temperature. Its test section is larger (0.56m wide ; 0.78m high) and is equipped with perforated top and bottom walls.

J. Zierep, H. Oertel (Eds.)
Symposium Transsonicum III
IUTAM Symposium Göttingen, 24.-27. 5. 1988

The supercritical airfoil RA16SC1 (thickness 16% ; chord 180mm) is a model which is not designed for cryogenic use, and is only tested at ambient flow temperature. Along the middle section, the unsteady wall pressures are measured on upper- and lowerside, until frequencies greater than 1kHz, by 30 transducers which are located in the model (holes $\emptyset$ 0.8mm ; small cavities). An accelerometer detects the mechanical vibrations of the model.

The ONERA/Aérospatiale airfoil (thickness 12% ; chord 150mm) is designed for cryogenic flows and is equipped with one similar transducer to measure the unsteady wall pressure on the downstream part of the upperside (X/C=81%). The transducer operates at variable temperature.

On both models the pressure distributions are also measured in a more classical way, with long response time, by scanivalve systems and transducers which are located outside of the test section. The transition is fixed at X/C=7% by lines of carborundum grains on upper- and lowerside. The unsteady pressures are measured on the range (0Hz ; 800Hz) with the rate 2000 points per second.

Laser anemometry measurements are made at the T2 tunnel, for ambient flow temperature, in the upperside flow of the RA16 model (Ref. 5 and 6). The two component velocimeter bench uses a two colours DISA system (total laser power 1W), in forward scattering mode, with and without Bragg cells. To measure a separated transonic boundary layer, two explorations are necessary : with Bragg cells to get the low and negative speeds near the wall, without them to get the high speed parts. Tracking oil particles of 1µm diameter are injected in the flow. Mean measurement rates of about 5000 points/s are obtained, so the fixed rate 2000 points/s can be used for unsteady velocity measurements and buffeting study.

3 - Natural shock wave - separation instability in tripped transition conditions

3.1 - First results on RA16 C=180mm

The first tests on RA16 are made at low chord Reynolds number, Rec=4. 10^6 , corresponding to ambient temperature and to the total pressure 1.7 bars. Many (M, α) test points (figure 1) are explored through unsteady pressure measurements in both wind tunnels, S3 and T2.

Buffeting first appears like a strong and continuous increase of the measured pressure fluctuation level, all over the airfoil, when the angle of attack or the test Mach number are increased beyond a well defined limit. Distributions of RMS values on upperside and lowerside are shown on figure 3 for α =2° and increasing Mach numbers. The highest fluctuation levels are always measured at two locations on the upperside : around the shock wave and on the downstream part of the model, near the trailing edge, where the transducer X/C=95% is chosen to represent the general buffeting level. Variation curves of this RMS value with Mach number for each angle of attack (α=0° to α=4°; figure 4) lead to the straight line buffeting onset limit in the (M, α) test diagram (figure 1) which corresponds to the beginning of the fluctuation increase.

When the Mach number is increased for a fixed angle of attack (for example α=2°, figure 6) the unsteady pressure signals keep a very low turbulence level until the onset limit is crossed, then their turbulence increases and they rapidly become quite periodic, with the same frequency all over the model, and different forms (figure 7). A significant form is always observed around the upperside shock wave location (X/C=40% for α =2°), a roughly rectangular signal between two levels, the static pressure upstream and downstream of the shock. A characteristic frequency corresponding to the signal's period is also clearly detected on the energy spectra : just beyond the buffeting onset, the first aerodynamic frequency which is detected is about 80Hz for all angles of attack ; in buffeting conditions, this frequency increases with Mach number or angle of attack, to 110-120Hz (figure 8). This frequency is also measured by the accelerometer, but it is independent of the mechanical vibration frequencies of the RA16 model, which are 155Hz and 475Hz. Its variations with test conditions seem to prove its aerodynamic origin. This frequency is also independent of any aerodynamic wind tunnel instability because the same values are measured in both tunnels S3 and T2, which are rather different.

The Mach number distributions on the upperside (figure 2 ; α=2°) which are obtained by long response time pressure measurements show that below the buffeting limit the shock wave moves downstream and becomes stronger when the Mach number increases, and beyond the limit the rapid speed decrease through the shock progressively spreads out chordwise.

The first conclusion of these results is that, on this airfoil, buffeting corresponds to a periodic unsteady flow whose most evident part is an oscillation in position of the shock wave. This shock movement is directly shown by a high speed Schlieren film (400 pictures/s) which was produced at S3.

Oil visualizations on the upperside for M=0.725 and various angles of attack around the onset limit α=1.5° show that this limit corresponds to the appearance of complete separation between the shock wave and the trailing edge : for α=1° and α=1.375° separation is detected downstream of the shock foot, with reattachment at X/C=61% and X/C=67% , and without any separation downstream of this location ; for α=1.5° there is no reattachment downstream of the shock.

So, the second conclusion is that the motor of airfoil buffeting is the growth of a large unsteady separated flow downstream of the shock wave : the induced aerodynamic deformation of the airfoil is sufficient to move the shock upstream and to weaken it ; after that, the separated region is strongly reduced because it is no more provided by the shock foot, and the shock moves downstream again.

For all angles of attack, when the Mach number grows far beyond the buffeting onset limit, the pressure fluctuation increase stops (figure 4). For these test conditions, particularly at high angles of attack, an important separated flow exists, but it probably can no more be strongly reduced and its unsteadiness decreases. So, another important characteristic of buffeting conditions is the high sensitivity of the separated flow to the intensity and location of the shock wave.

A last remark concerns these first test results. The pressure fluctuation level on the downstream upperside gives a good indication of the general buffeting intensity, but it appears impossible to qualify this unsteady flow with one parameter : frequency, location and intensity of the shock wave must be taken in account. Also, the phenomenon depends on the (M, α) test point and particularly on the angle of attack. For high angles, the shock corresponding to buffeting onset is stronger and located more upstream (figure 5), higher fluctuation levels are obtained when the Mach number is increased, corresponding to a large separated region downstream of the shock, with strong unsteady effects on the flow around the airfoil.

3.2 - Laser velocimetry measurements on RA16 C=180mm

Detailed 2D anemometry measurements are made for the test point (M=0.725 ; α=2°) corresponding to a well established buffeting oscillation at the frequency 83Hz (Ref. 6). The boundary layer and the wake are explored at various locations downstream of the shock wave to describe the separated region. Chordwise explorations at different altitudes above the upperside are made to define the shock movement. At each exploration point an unsteady velocity measurement is performed during a sufficient time (100 oscillation periods). During the whole exploration, unsteady pressure on the airfoil is continuously measured by the transducer X/C=40%, which is covered by the shock movement and which is chosen to follow the buffeting oscillation in time. For all measurements the rate is fixed to 2000 points/ second. Of course it can happen that no laser measurement is available (§2).

The average unsteady signal of the velocity components is calculated at each exploration point by a particular method, which is allowed by the periodic nature of the flow (figure 9). All measured cycles (the rectangular pressure signal at X/C=40% is used as a trigger) are divided in 20 intervals and the average value is calculated on each interval over 100 periods. So the average unsteady velocity component is obtained by 20 phase points covering one mean cycle. The real laser measurement rate is also available for each phase point. At all the explored points, these average signals are consistent in time or phase. Phase point 1 always corresponds to the shock moving upstream because the increasing pressure front is used to detect the period beginning. One advantage of this method is to give in a clear way the velocity distributions in space at fixed phase point during the oscillation cycle. The chordwise distributions at different altitudes above the upperside are used to define the instantaneous form of the shock wave and its movement (figure 11). The shock location versus time is given by the velocity distribution at 5mm above the upperside on figure 10, where the phase points are also represented. The instantaneous distributions of the velocity vector in the upperside boundary layer and the wake are presented on figure 12 for characteristic phase points. These measurements are performed with Bragg cells. It must be noticed that the laser measurements are difficult in the lower or separated parts of thin boundary layers, for different reasons : dimension of the measuring volume, reflections on the wall, miss of tracking particles. However, the boundary layer exploration results are rather consistent, and some information about the behaviour of the unsteady separated flow could be obtained.

The shock wave oscillates in position between approximately X/C=34% and X/C=51%, at the frequency 83Hz. The upstream and downstream movements are regular and quite symmetrical (figure 10) but the shock form is not the same (figure 11): the shock is right during the downstream movement and incurvated during the upstream movement, the shock foot moving first. Schematically, the form changes during the upstream and downstream stations.

The important separated flow region is detected during the upstream movement of the shock wave (figure 12). Its extent is maximum, when the upstream position is reached (phase point 4) and decreases rapidly at the upstream station. During the downstream movement (phase points 8 and 12) no important separated flow is measured, and when the downstream shock station is reached (phase points 12 or 16), the separation extent is minimum. Beyond this station separated regions are measured in the downstream boundary layers (phase points 20 and 4).

So the separation extent grows during the upstream movement of the shock wave and it can be supposed that during this part of the oscillation, the shock foot generates separation. Also, the shock moves forward, as the induced airfoil deformation becomes more important. At the upstream station, it can be supposed that the separation production in the weakened shock foot stops, and the shock moves downstream as the remaining separated region evacuates, or is very strongly reduced. So, the downstream station seems to roughly correspond to the shock position without separation, for the (M, α) test point. But in buffeting conditions, the shock is too strong, this position is only reached in an unsteady way, and the production of a new important separated flow by the shock foot begins.

The significant conclusion about the buffeting oscillation mechanism is that the physical parameter which determines the frequency is the growing and decreasing rate of the separated flow region which is generated in the shock foot. Acoustic travel of aerodynamic information coming from the downstream upperside seems to be rapid, compared to the oscillation time.

3.3 - Cryogenic tests at high Reynolds numbers on the ONERA/Aérospatiale airfoil C=150mm

The Reynolds number based on the chord covers the range (3.10^6; 11.10^6) which corresponds to total temperature variation from ambient to 145K, and to total pressure variation from 1.6 to 3bars.

This airfoil is equipped with one transducer located at X/C=81% on the upperside to get comparison with the unsteady pressure measurements on RA16. The first result is that the previous measurements are confirmed : buffeting occurs in the same way, by strong increase of the fluctuation level, with appearance of an aerodynamic frequency which is of about 105Hz near the onset conditions and at ambient flow temperature. Buffeting frequency increases when the Mach number or the angle of attack are increased (figure 14). Of course, the buffeting onset limit (M, α) at high lift coefficient depends on the airfoil shape and is different.

In addition to that, long response time wall pressure measurements are generally used on airfoils, and the apparent spreading out of the shock wave at high lift coefficient, which occurs on the RA16 and ONERA/Aérospatiale airfoils, has also been observed on other airfoils (CAST7, CAST10). So, in tripped transition conditions, buffeting on airfoils seems to generally correspond to a shock-separation oscillation.

When the Reynolds number is increased, no significant variation of the buffeting onset limit is observed and there is generally no important effect on the pressure fluctuation level when the unsteady flow is well established : figure 13 gives the variation curves of the RMS/Ptotal values with the test Mach number for the angles of attack 3° and 4° and the different Reynolds numbers. Only for $\alpha=3°$ and $Rec=3.10^6$ the buffeting fluctuation level is lower than for higher Reynolds numbers. No explanation could be found for this particular effect. It can be noticed that, on this airfoil and below the buffeting limit, the measured effect of the Reynolds number increase is rather classical and corresponds to the thinning down of the airfoil boundary layer : the upperside shock wave progressively moves downstream ($\Delta X/C$ is of about 4% from 3.10^6 to 11.10^6) and becomes a little stronger, and the lift coefficient is increased. Of course, this effect modifies slightly the buffeting conditions.

However, for all cryogenic tests in buffeting conditions, a specific temperature effect is detected (figure 14). For each test, the aerodynamic frequency is well defined but significant variations are systematically observed between the tests for a same (M, α) point at different flow temperatures : the frequency appears to be proportional to $\sqrt{T}$, which represents sound celerity or speed for a fixed Mach number. Indeed, the physical parameter through which frequency is modified seems to be speed and not sound celerity, because speed is one of the parameters which govern the growth of the separated region. The oscillation cycle is determined by this growing rate, but not by acoustic travel of information (§3.2).

It can also be noticed that a similarity problem for such unsteady tests is brought to light by this temperature effect. The reduced frequency or Strouhal number FxC/U must be considered for analysis. So the frequency F depends on a characteristic flow speed U and, in an inverse way, on a characteristic dimension C, which is the model chord.

The buffeting frequency increase with Mach number at fixed angle of attack (figure 8 and 14) perhaps includes a proportional speed effect, but this variation is too important to be explained by such an effect. It seems to be related to a more important production of separated flow in the foot of the shock, which becomes stronger. For this same reason, frequency also increases with the angle of attack, at fixed Mach number.

A last piece of information is given by this airfoil through natural transition tests, for which the upperside boundary layer remains laminar up to the shock wave, for the low Reynolds numbers 3.10^6 and 6.10^6. In these conditions, buffeting appears through strong increase of the pressure fluctuation, but without appearance of any characteristic frequency (figure 15). The RMS levels are generally lower but of the same order as those observed with tripped transition (figure 13) and the onset conditions are near the tripped transition limit. At higher Reynolds numbers, natural transition occurs near the airfoil leading edge and the oscillating flow is observed in buffeting conditions, without any difference to the phenomenon which is measured in tripped transition. So, the shock - separation oscillation only appears when the boundary layer interacting with the shock is turbulent. The separated flows which are produced by a laminar or a turbulent interaction are different, and the laminar separation probably cannot extend so largely on the downstream upperside or is not so sensitive to the shock wave characteristics.

4 - Conclusion

On 2D stiff airfoil the aerodynamic mechanism of buffeting is rather clear and some interesting information about it could be obtained through these wind tunnel studies. But the well established oscillation of the upperside shock wave and the downstream separated flow is a schematic buffeting case which does not occur on 3D plane models. Even on cut airfoils, when one end of the model is in the free stream, the phenomenon is not so simple. However, the local mechanisms which generate the buffeting in flight are probably similar to this oscillation, but their development is limited by the general 3D flow conditions. At the present time, buffeting studies are continued at ONERA by development of calculation models, the 2D oscillation being the first step before extension to 3D flows (Ref. 1).

References

1 - GIRODROUX-LAVIGNE P. and LE BALLEUR J.C.
Unsteady viscous-inviscid interaction method and computation of buffeting over airfoils
Proceeding Springer Verlag Joint IMA/SMAI Conference "Computational Methods in Aeronautical Fluid Dynamics", University of Reading, April 1987
ONERA TP N° 1987-58

2 - BENOIT B. and LEGRAIN I.
Buffeting prediction for transport aircraft applications based on unsteady pressure measurements
AIAA 5th Applied Aerodynamics Conference, Monterey (California), August 1987
ONERA TP N° 1987-122

3 - DESTUYNDER R., SCHMITT V., BERGER J. and BARREAU R.
Des difficultés de prédire en soufflerie le tremblement d'un avion moderne. Remèdes proposés
Meeting FDP/AGARD "Symposium on Aerodynamic Data Accuracy and Quality : Requirements and Capabilities in Wind Tunnel Testing". AGARD-CP 429 (1987)
ONERA TP N° 1987-159

4 - B.H.K. LEE, L.H. OHMAN
Unsteady pressure and force measurements associated with transonic buffeting of a two-dimensional supercritical airfoil
Aeronautical Note NAE-AN-14 NRC N° 22448, June 1983

5 - MIGNOSI A., SERAUDIE A., BONNET J.L. and ARCHAMBAUD J.P.
La vélocimétrie laser. Critique des mesures
Conférence "Progrès et problèmes des moyens d'essais industriels". 8ème Congrès Français de Mécanique. Nantes (France), September 1987

6 - SERAUDIE A., MIGNOSI A., DOR J.B. and PAYRY M.J.
Etude d'un écoulement instationnaire avec un système de vélocimétrie laser bi-composante dans le cas d'un phénomène de tremblement sur un profil RA16 de 180mm de corde dans la veine de la soufflerie T2
Rapport technique OA N° 31/3423 AYD (DERAT N° 12/5017.12), February 1988

7 - DOR J.B.
The cryogenic induction tunnel T2 at Toulouse
AGARD/FDP Special Course on Cryogenic Technology for Wind Tunnel Testing. Report N° 722. VKI, Rhode St Genèse (Belgium), April 1985

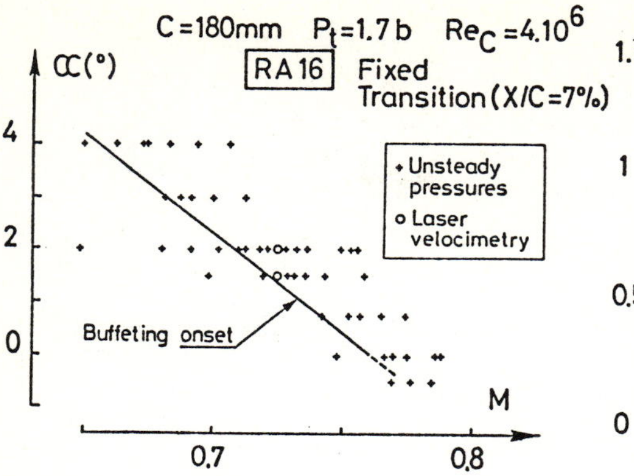

Figure 1 : Test diagram and buffeting onset limit on RA16.

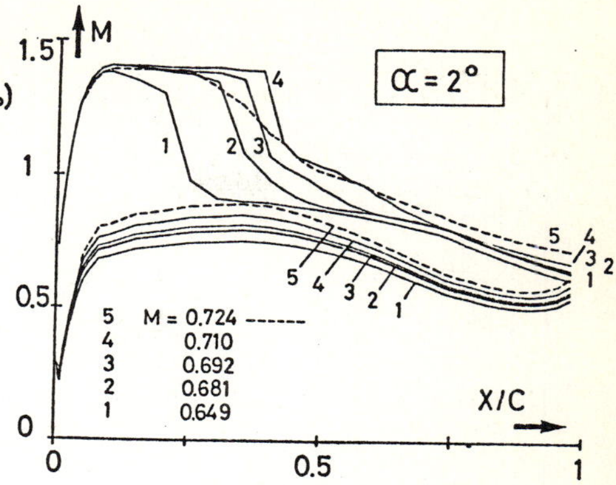

Figure 2 : Mach number distributions on RA16, for α=2°.

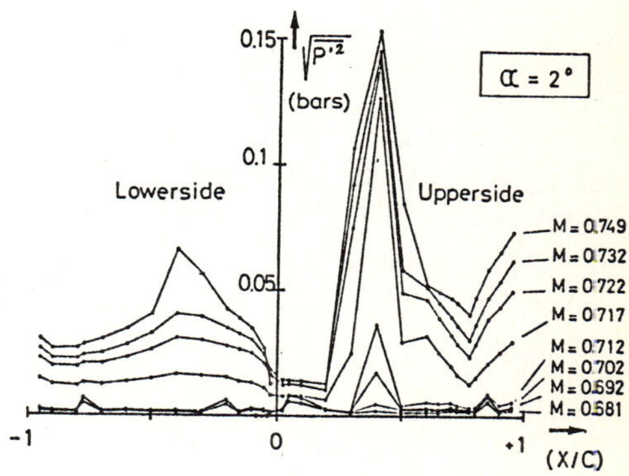

Figure 3 : Distributions of the pressure fluctuation level on RA16, for α=2°.

Angle of attack	Mach number	Upperside shock wave location (X/C)	Local Mach number, upstream of the shock
4°	0.654	25%	1.45
3°	0.680	30%	1.43
2°	0.710	39.5%	1.39
1.5°	0.722	42%	1.37
0°	0.760	52%	1.33

Figure 5 : Upperside shock wave characteristics, for the buffeting onset conditions on RA16.

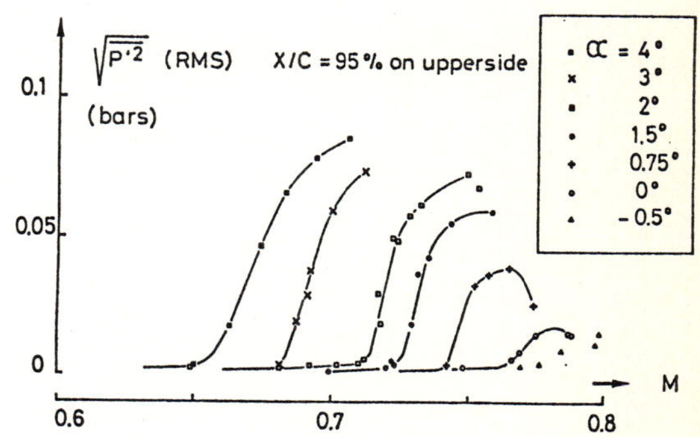

Figure 4 : Pressure fluctuation increase on the downstream upperside, on RA16.

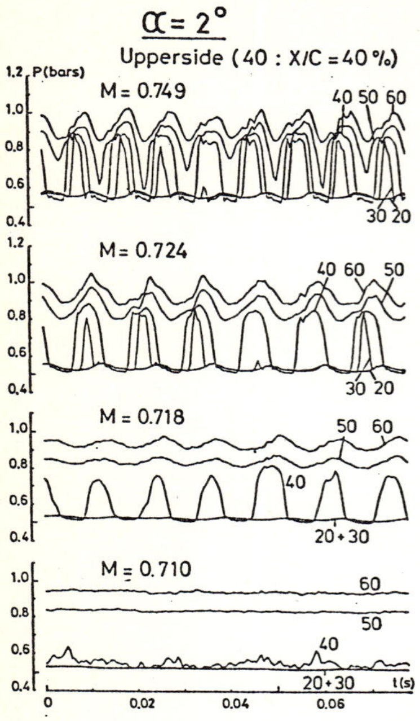

Figure 6 : Upperside pressure signals on RA16, for increasing Mach numbers and for α=2°.

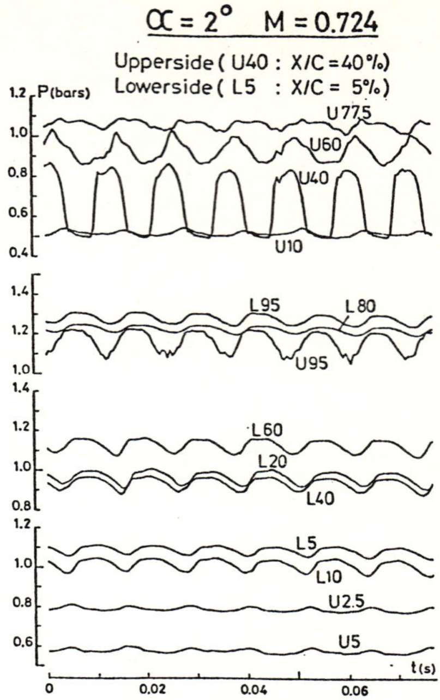

Figure 7 : Pressure signals around the RA16 airfoil, for established buffeting conditions.

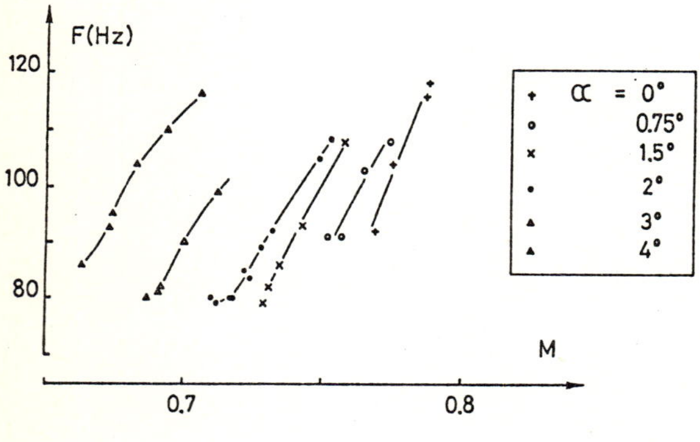

Figure 8 : Variation of the aerodynamic buffeting frequency on RA16, with Mach number and angle of attack.

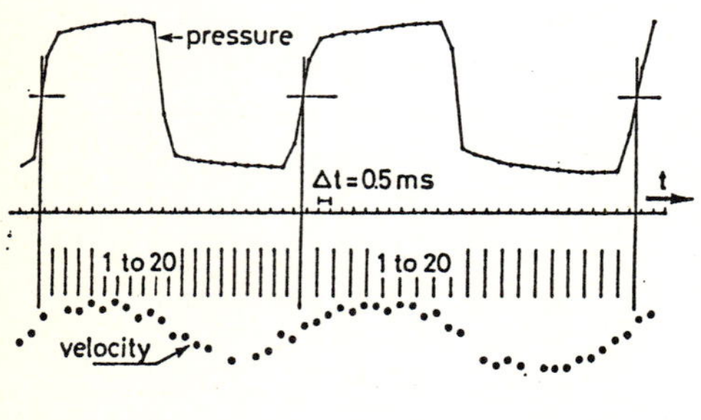

Figure 9 : Calculation of the average unsteady velocity signals.

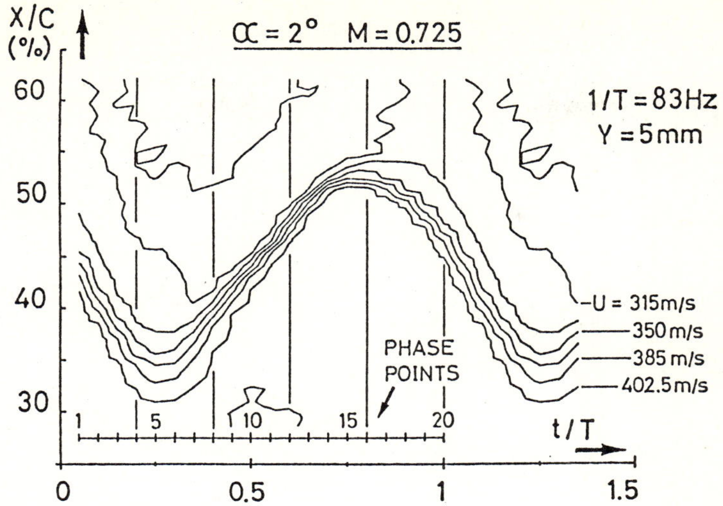

Figure 10 : Chordwise velocity distributions versus time, in the oscillating shock region, 5mm above the upperside, on RA16.

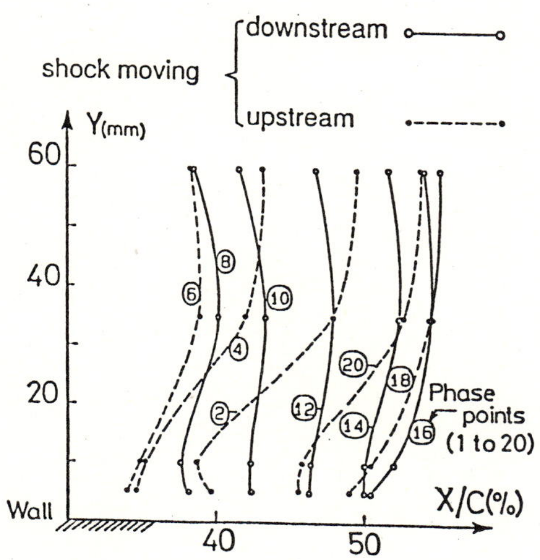

Figure 11 : Form and movement of the upperside shock wave on RA16, for M=0.725 and α=2°.

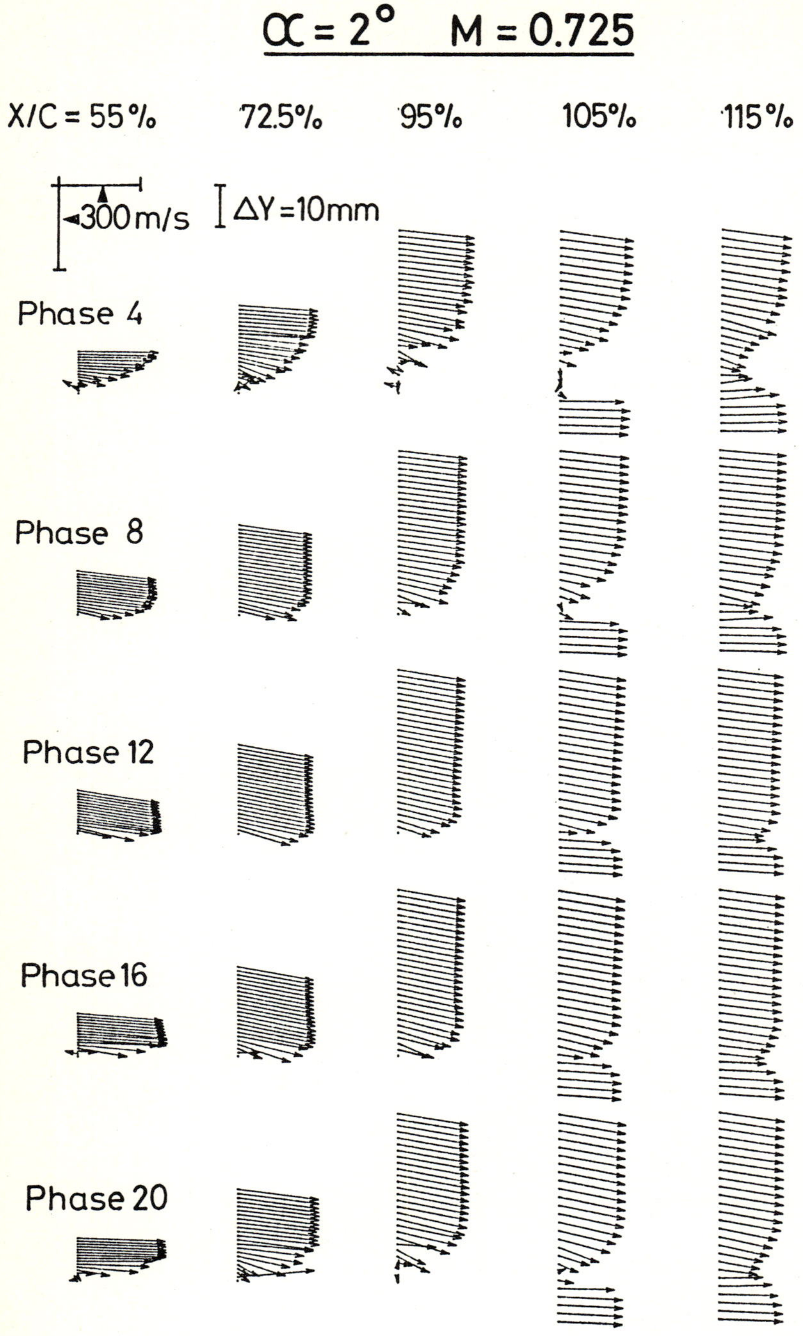

Figure 12 : Average instantaneous velocities in the upperside boundary layer and in the wake, on RA16, for M=0.725 and α=2°.

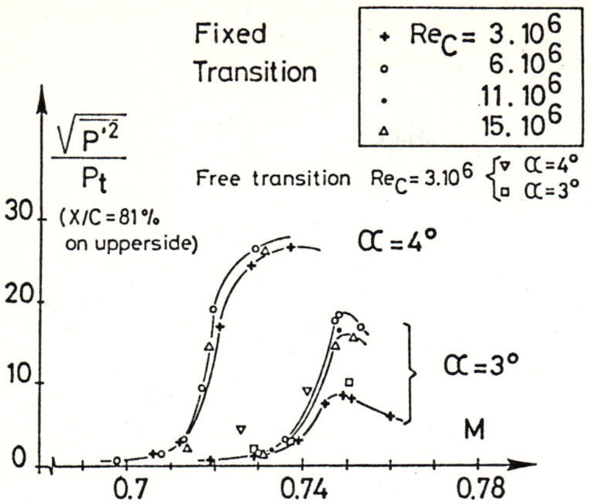

Figure 13 : Pressure fluctuation increase on the downstream upperside, on ONERA/Aérospatiale. Reynolds number effect.

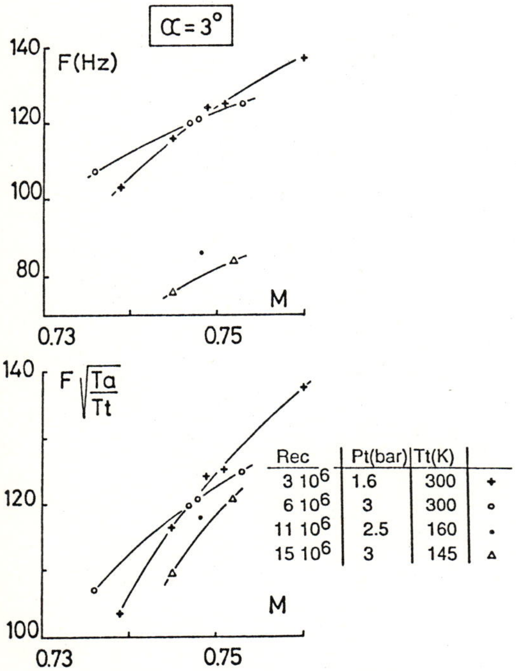

Figure 14 : Temperature effect on the aerodynamic buffeting frequency, on ONERA/Aérospatiale. Frequency variation curves with Mach number, for α=3°.

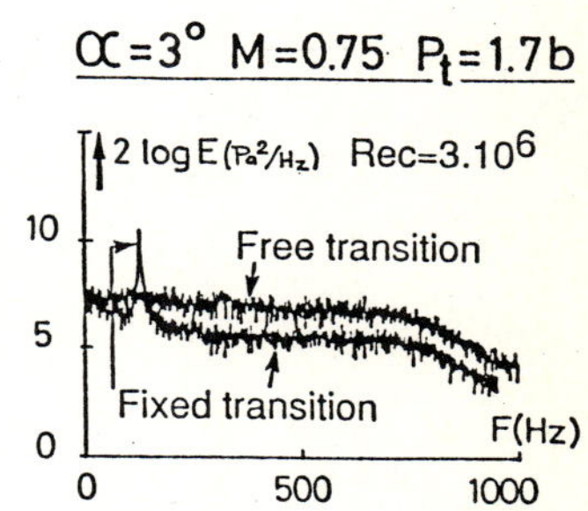

Figure 15 : Comparison of the pressure fluctuation spectra in buffeting conditions, for a laminar and a turbulent boundary layer interacting with the shock wave, on ONERA/Aérospatiale.

Three-Dimensional Flow Separation caused by Normal Shock-Wave / Turbulent Boundary-Layer Interaction

U. Dallmann*, P. Doerffer**

* DFVLR, Institute for Theoretical Fluid Mechanics
D-3400 Göttingen, FR Germany

** Polish Academy of Science
Institute of Fluid Flow Machinery
PL-80-952 Gdansk, Poland

Summary

Experimental results concerned with the determination of shock-induced separation are presented. The effects of Reynolds number and Mach number are analyzed in terms of topological changes of the separated flow structures. Principles of topological flow analysis are explained. These are used to provide better insight into the complicated three-dimensional separated flow structures. A preliminary discussion of the detected incipient separation is provided. The behaviour of the skin-friction coefficient, displaying a plateau just upstream of the location of separation, is revealed. Boundary layer measurements confirmed by oil-flow visualizations prove that the skin-friction plateau is associated with the three-dimensionality of separation.

1. Introduction

The ever increasing interest in normal shock-wave / turbulent boundary-layer interaction (NSW/TBLI) stems from the applicational importance of this phenomenon and from the difficulties experienced in the experimental as well as in the theoretical approach. One can distinguish three aims of experimental investigations found in the literature:

- the first is to analyze certain details of the interaction flow field,
- the second is the determination of the Mach number (M) and Reynolds number (Re) effects on the interaction,
- the third is to verify some theoretically obtained results.

The information obtained from literature concerns mostly arbitrarily chosen cases of flow parameters such as M and Re. Due to the variety of the used test section arrangements, the results are sometimes inconsistent and the influence of flow parameters is studied for so few cases that the conclusions made can only provide trends. Our investigations are based on the measurements carried out at the Institut für Strömungslehre und Strömungsmaschinen in Karlsruhe, FR Germany, and published by Doerffer and Zierep [1] and Doerffer [2]. In our analysis we concentrate on the influence of Reynolds number and Mach number on NSW/TBLI observed by structural changes of the separated flow field. Various aspects of such changes due to Reynolds number variations have already been published and for details and further references see Doerffer and Dallmann [3]. The analysis of the Mach number effect will follow, and a more detailed consideration of the formation of separated flow structures due to NSW/TBLI will be published elsewhere.

J. Zierep, H. Oertel (Eds.)
Symposium Transsonicum III
IUTAM Symposium Göttingen, 24.-27. 5. 1988

2. Interaction on curved surfaces

Undertaking an experimental investigation of the NSW/TBLI one has to decide how the interaction region is to be produced. It is quite obvious that in technical applications the interaction takes place on convex surfaces. However, due to convenient simplifications most of the NSW/TBLI experimental work has been carried out on flat walls. It is generally recognized, although not understood in every respect, that the test section configuration may have a very significant or even qualitative effect upon separating flows.

It seems therefore, that we should try, by all means, to produce an interaction field as a termination of a local supersonic region which forms due to flow acceleration along a convex wall.

3. Test section configuration

Fig. 1 presents the chosen test section configuration. The flow in the curved duct forms a local supersonic area that is terminated by a normal shock which together with the boundary layer on the convex wall produces the interaction field – the objective of our research. The supersonic region does not reach the upper wall and therefore the mass flow rate, hence the supersonic area size and the Mach number in front of the interaction can be controlled by means of the throat adjustment device. The characteristic Mach number M_w is based on the static wall pressure minimum measured just upstream of the interaction zone.

As our facility is a suck down impuls wind tunnel, the stagnation parameters are constant and cannot be used as a means of Reynolds number variation. Therefore, the only possibility to change the Reynolds number is a variation of boundary-layer thickness δ_u upstream of the interaction zone which is the characteristic length scale for the interaction. Several methods of δ_u-change have been tested and the one selected is presented on fig. 1. It is a step-slot arrangement that allows to affect the boundary layer in two ways. One is by a downstream facing step of variable height and the other is by an additional stream bleeding along the convex wall through the slot opened by the step. The here presented test section configuration allows to control M_w and Re_{δ_u} independently within a certain range.

4. Flow two-dimensionality and tunnel narrowness

The theoretical analysis carried out by Bohning and Zierep [4] provides information about *Re*- and *M*-effects upon the flow field and also predicts so-called incipient separation. This and other theoretical work treat purely two-dimensional cases and therefore it is common practice to give two-dimensional interpretations of wind-tunnel measurements. However, there are significant difficulties in this respect. It is generally believed that the width of a tunnel plays a dominating role. Our test section is relatively narrow, the tunnel width is a third of the test section height. It should be mentioned that the increase of tunnel width allows to improve two-dimensionality of the flow only at separation in some small middle part of the test wall, while the separated flow itself usually displays highly three-dimensional effects.

A Schlieren visualization, as the one shown in fig. 2, provides useful information about the flow behaviour. It shows the shock structure and also displays the qualitative behaviour of the boundary layer. Fig. 2a shows that for strong interaction the shear layer lifts off the test wall under the action of the shock wave. One also recognizes large scale turbulent structures along the outer edge of the shear layer downstream of the shock waves. These structures are likely to be generated by the shock induced flow separation. Fig. 2b shows a lower Mach number case where the shock effect upon the boundary layer is not so strong and where it is very difficult to judge about the existence of flow separation. The flow visualization by means of Schlieren pictures or interferograms unfortunately integrates across the whole channel width and therefore does not provide any information about flow two-dimensionality.

5. Scope of the experiment

The aim of our investigation was to perform experiments of NSW/TBLI for a set of flow parameters shown in fig. 3, carried out in the same facility, measurements done with the same technique and data reduced in the same manner. Such a set of measured cases allows a detailed comparison of flows in every respect and gives a chance for quantitative investigations of Mach and Reynolds number effects. The Mach number range has been chosen so to include the formation or disappearance of flow separation.

The measurements which were carried out in the central plane consisted of:

- static wall pressure measurements and
- static and pitot pressure surveys above the wall.

This allowed the determination of velocity profiles, contour plots of flow parameters and the boundary-layer characteristics.

6. Measurement results

In this paper we concentrate on one of the most interesting results obtained. The measurements, carried out in a central plane, revealed a quite interesting behaviour of the skin friction coefficient. To illustrate this, we show in fig. 4 the assembly of data for the case $M_W = 1.47$, $Re_{\delta_u} = 5.5 \times 10^4$. The curves of static wall pressure p_S, displacement thickness δ_1, boundary-layer profile shape factor H_{12} and skin friction coefficient c_f are presented in relation to the shock location.

Following the information from literature flow separation should occur where:

- the static wall pressure gradient starts to decrease after its first nearly linear rapid increase, and where
- the streamwise gradient of the displacement thickness reaches its maximum and remains constant for some distance, and also where
- the shape factor H_{12} is above 2.6 .

It is seen in fig. 4 that all this happens shortly downstream of the front compression zone. It is worth pointing out that an extrapolation of the c_f-drop would reach zero in the same place where all the above mentioned parameters infer flow separation. However, in this region the skin friction curve levels off and reaches zero further downstream. At this place δ_1 is nearly five times larger than its undisturbed value and H_{12} has reached about 3.5 for this particular M_W-case. The behaviour shown in fig. 4 turned out to be qualitatively preserved under the variations of Re_{δ_u} and M_W within the range of our measurements, provided that, within the central plane, flow separation took place. In every case the c_f-plateau shown in fig. 4 was obtained and c_f reached zero in most cases even far downstream of the shock system.

All this seems rather strange and puts up several questions: Is there a failure in c_f-determination at this low value range? Is this behaviour due to some complexity of the separation structure? Does separation happen only on the test wall or also on the tunnel side walls? As a matter of fact, flow field measurements would be needed but the only available means to investigate those questions was a surface oil-flow visualization. The oil-flow visualization applied to the side walls has proven that flow separation is limited only to the test wall.

7. Surface-flow visualization

An example of oil-flow visualization applied to the convex test wall is presented in fig. 5a. The flow is from left to right and the double line gives an approximate location of the front branch of the shock system at the wall. Since the short blow down time did not allow for complete oil evaporation all photographs had to be taken during the blow downs. The photographs therefore show simultaneously wave fronts where oil was accumulated and striations which indicate trajectories of the wall shear stress in other places. Therefore also a movie film was taken in order to better follow the surface flow direction.

8. Topological analysis of surface flow patterns

One of the first tasks in describing the structure of a separating flow is to reduce the information contained in the striation patterns seen in fig. 5 and 7 and to give a description of the surface flow topology. We have to locate the so-called critical points. Critical points are defined by all those locations where the direction of wall shear stress or the velocity direction above the walls are non-uniquely defined and their magnitudes drop to zero. In two-dimensional flows the topological structure can exhibit only saddle points and centers as critical points in sectional planes above the bounding walls; on the walls only so-called degenerate lines of separation or attachment can form. Three-dimensional flows can form additional nodal points. Since most flows do not have local flow symmetries around nodal points it is most likely that the advective transport of vorticity towards these nodal points gives rise to spiralling, helical, swirling, vortical motions near those nodal points. Such foci will then indicate footprints of vortices. According to the clearly visible nodal points in fig. 5 we recognize immediately that the separated flow induced by the interaction must be highly three-dimensional with vortices leaving and others reaching the bottom wall or the side walls along the tunnel corners.

A complete picture of the surface flow topology is provided by the connecting trajectories between all of the critical points. This gives the pattern shown in fig. 5b. Topological and kinematic constraints force us to find extra critical points located either on the walls or even within the flow field which have not been identified in the flow visualization at a first sight. Sometimes a non-unique interconnection of the critical points has to be taken into account due to lack of resolution.

9. On the definition of three-dimensional separation

If one now analyzes the kinematically possible local flow behaviour in the vicinity of every critical point one can deduce whether the flow can leave (separate) or reach (attach) the wall there. Only certain sequences of critical points may be connected following kinematics and topology. In fig. 5 we identify sequences of saddles and nodal points where the flow leaves the tunnel wall. The connections of these critical points define lines of flow separation. Similarly lines of attachment can be identified. However, one may also define flow separation solely on the kinematic fact that due to a local convergence of trajectories of the wall shear stress the flow has to leave the vicinity of the wall there. No critical points of flow separation are required to exist by that definition. However, the associated gradual formation of such so-called open separation lines and open (re-)attachment lines is nothin but a footprint of a vortical flow above the wall. We therefore emphasize that separation as well as reattachment are details of what we may call "formation of vortices and vorticity structures in boundary layers". The topological structure of the velocity field based on critical point analysis therefore has to be accompanied by a similar description of the vorticity field, pressure gradient field etc.

Vortices with vorticity are not necessarily created on the walls. Even within the boundary layers swirling, helical, vortical flow regions, i.e., vortices may form simply due to Biot-Savart induction of the vorticity field which is created either along the rigid bounding surfaces or within curved parts of the shock waves which dominate our separated flow field. For basic features of vortex structures and vorticity topology we refer the reader to Dallmann [5,6] and the references given there.

Dallmann [5,6] has also shown that certain elementary topological structures consisting of certain ensembles of critical points exist, which are connected by well-defined trajectories and which form so-called singular streamsurfaces. These elements also show that there is a non-unique relationship between the surface-flow information contained in oil-flow visualizations and the three-dimensional flow field. Those elementary structures provide a building block system to the construction of the kinematically possible flow structures which fit to an observed surface-flow pattern. A complete topological description of the flow structure (in terms of instantaneous or time-averaged streamlines) requires only the description of those singular streamsurfaces (one of which is always the bounding wall itself) which are given by the set of all of those streamlines which exactly run out or run into the critical points.

The principles which we summarized in the last two sections allow to describe the three-dimensional time-averaged flow structures caused by the NSW/TBLI.

10. Topological structure of the three-dimensional flow

In fig. 5a we identify a "separation line" across the whole tunnel. It is sketched in fig. 5b as the thick line which starts in the corners where the first disturbances enter the flow field at S and it spirals into the large foci F_{S1} and F_{S2}. The location of S coincides with the minimum of the static wall pressure measured along the center line of the test wall. The separation line connects these two foci through the saddle point of separation S_2. Along the central part of tunnel we recognize flow reversal between S_2 and S_3 and one might be led to conjecture that S_3 is a point where the flow reattaches. But we point out that this would only be correct if the flow were two-dimensional. We emphasize that a three-dimensional, time-averaged flow doesn't necessarily "reattach" at all although the surface flow changes again into the main flow direction downstream of S_3. Although the actual time-averaged or instantaneous flow structure may be extremely complicated the underlying principle that allows to explain this feature of three-dimensional flow separation has been explained by Dallmann [5, 6].

As we mentioned above, there is no unique relationship between the surface flow and the flow structure above the wall. However, there is only a finite number of kinematically correct three-dimensional structures associated with the wall pattern of fig. 5. In fig. 6a,b we show a view into the test section from above, onto the separated flow, for two such possiblities. The surface flow pattern suggests that the same flow topology exists on both sides of the central plane. We have used the topological principle that allows to stretch the connections of critical points, i.e., the singular streamsurfaces to an arbitrary degree (as if it were made of rubber) in order to give an impression of the complexity of the flow due to the separation structures that are formed along the bottom wall - side wall corners. It follows from the topology of the ensemble of critical points identified in fig. 5 that there is local flow attachment at F_{A1} and F_{A3} as indicated in fig. 6. The separating singular surface that leaves the front-separation line is necessarily deformed such that it shows a focal winding around a mid-air focus at F_M which one has to conjecture in order that the oncoming flow has access to the attachment foci F_{A1} and F_{A3}.

In fig. 5 one can see strong convergence of wall shear stress trajectories towards the center line downstream of S_3. Therefore the flow leaves the vicinity of the bottom wall within the central plane downstream of S_3. However, we do not have any information about the flow behaviour above S_3. Kinematics allow that the flow in the middle plane of the tunnel may equally well separate at S_2 and reattach at S_3 (fig. 6b) or separate at S_2 and also at S_3 without reattachment taking place as indicated in fig. 6a. This non-uniqueness indicates that flow-field measurements or visualizations would be required for further flow analysis. However, our topological analysis tells us where and what kind of information is missing for a unique flow description.

In high Reynolds number flows, like those investigated here, shear layers exist along the separating singular streamsurfaces shown in fig. 6. Such shear layers are known to be hydrodynamically unstable. Hence, they form the time-averaged skeleton along which shear instability waves propagate. It is likely that these are the origin of those large scale turbulence structures seen in most three-dimensional separating turbulent flows. Hence, in agreement with the statement made in section 4, our description of the time-averaged, separating streamsurface sketched in fig. 6 supports the idea that also those turbulent structures which are seen in the Schlieren picture shown in fig. 2a indeed indicate flow separation.

11. Structural flow changes due to (Re, M)-variation

Since one aim in studying NSW/TBLI is to better understand the influence of Reynolds number and Mach number on the induced flow separation, it is important to further reduce the information contained in the variety of flow fields which form for various Reynolds numbers and Mach numbers. The interest in a topological description of a flow structure is due to the fact that structural changes in real physical space accompany flow bifurcations which take place in the (M, Re)-parameter space. The structural changes can be identified if we vary the Reynolds number and the Mach number (and, of course, also if we change the tunnel configuration). Structural changes occur only at certain critical sets $(M, Re)_{crit}$. This allows us to concentrate on those cases where different topologies can be observed, at least in the surface flow pattern and we may disregard all the other (M, Re)-cases for the purpose of such a study.

In fig. 5 and fig. 7 we show examples of topologically different surface flow patterns. We recognize that within the range $1.35 < M_w < 1.39$ so-called incipient separation occurs along the center line. This feature will be further analyzed in another paper. Here we only mention that the structural change associated with incipient separation is a highly three-dimensional phenomenon and that the incipiently formed separated flow in the central region of the tunnel is one of those elementary flow structures derived by Dallmann [5]. The three-dimensionality is dominated by flow separation along the tunnel corners which is present in the whole (M, Re)-range under investigation.

The Re-variation of the flow structures has already been analyzed by Doerffer and Dallmann [3]. Fig. 7a,b gives two examples where the asymmetry of the separated flow (which is enhanced by an increase of Reynolds number) is clearly visible and only a local structural change is observed near S_3. It seems, that the separated flow induced by NSW/TBLI in a narrow tunnel, forms an asymmetric streamwise oriented vortex configuration which is more stable at high Reynolds numbers than a symmetric one.

12. Conclusions

1. The skin-friction plateau is caused by the three-dimensional structure of the separated flow.
2. The increase of Reynolds number does not cause structural changes at separation but increases flow asymmetry.
3. The decrease of Mach number causes a downstream shift of flow separation until separation disappears far downstream of the shock.
4. At incipient separation the flow is strongly three-dimensional and forms one of the elementary topological structures.

References

1. Doerffer, P., Zierep, J.: An Experimental Investigation of the Reynolds Number Effect on a Normal Shock Wave - Turbulent Boundary Layer Interaction on a Curved Wall. Acta Mechanica 73, 77-93 (1988)

2. Doerffer, P.: An Experimental Investigation of the Mach Number Effect on a Normal Shock Wave - Turbulent Boundary Layer Interaction on a Curved Wall. submitted to Acta Mechanica.

3. Doerffer, P., Dallmann, U.: Separation Structures Produced by Normal Shock-Wave / Turbulent Boundary-Layer Interaction in a Narrow Wind Tunnel. AIAA-87-1370, 1987.

4. Bohning, R., Zierep, J.: Stoß-Grenzschichtinterferenz bei turbulenter Strömung an gekrümmten Wänden mit Ablösung. Z. Flugwiss. Weltraumforsch. 6 (1982), Heft 2, pp.68-74.

5. Dallmann, U.: Three-Dimensional Vortex Structures and Vorticity Topology. Proc. IUTAM Symposium on Fundamental Aspects of Vortex Motion, Tokyo, 31.8.-4.9.1987, Fluid Dynamics Research, The Japan Society of Fluid Mechanics, North Holland Physics Publ.

6. Dallmann, U.: Topological Structures of Three-Dimensional Vortex Flow Separation. DFVLR-IB 221-82 A 07, 1982; see also AIAA-83-1735.

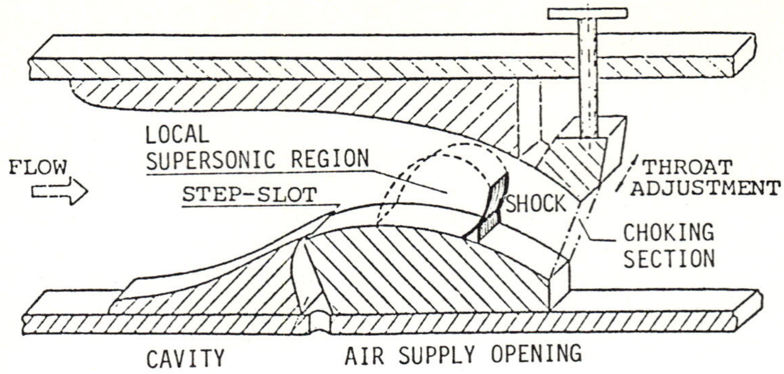

Fig. 1 Test section arrangement

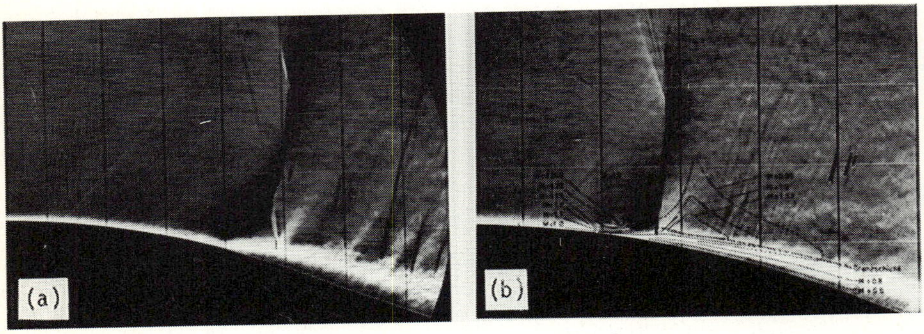

Fig.2 Schlieren pictures for strong interaction $M_w = 1.47$ (a) and weaker interaction $M_w = 1.35$ (b).

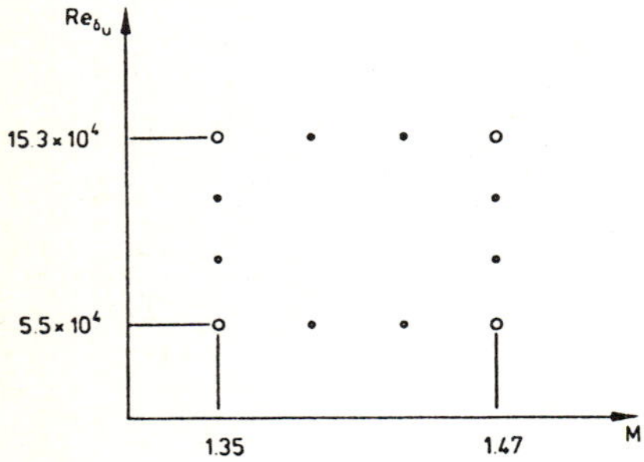

Fig.3 Flow parameters (Re_{δ_u}, M_w) for the investigated cases.

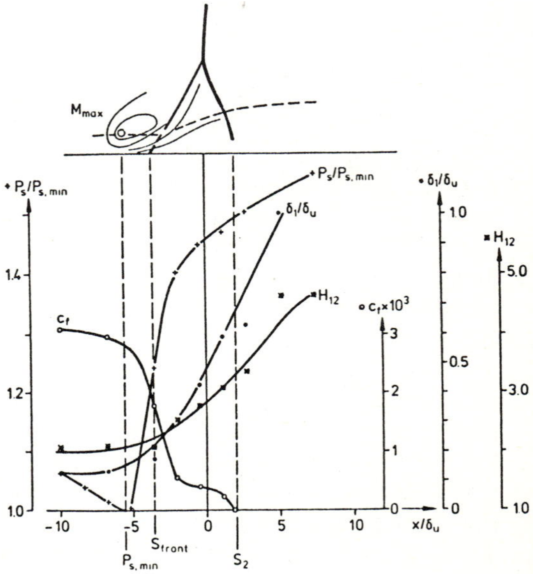

Fig.4 Measurement results for $M_w = 1.47$, $Re_{\delta_u} = 5.5 \times 10^4$ in relation to the shock location.

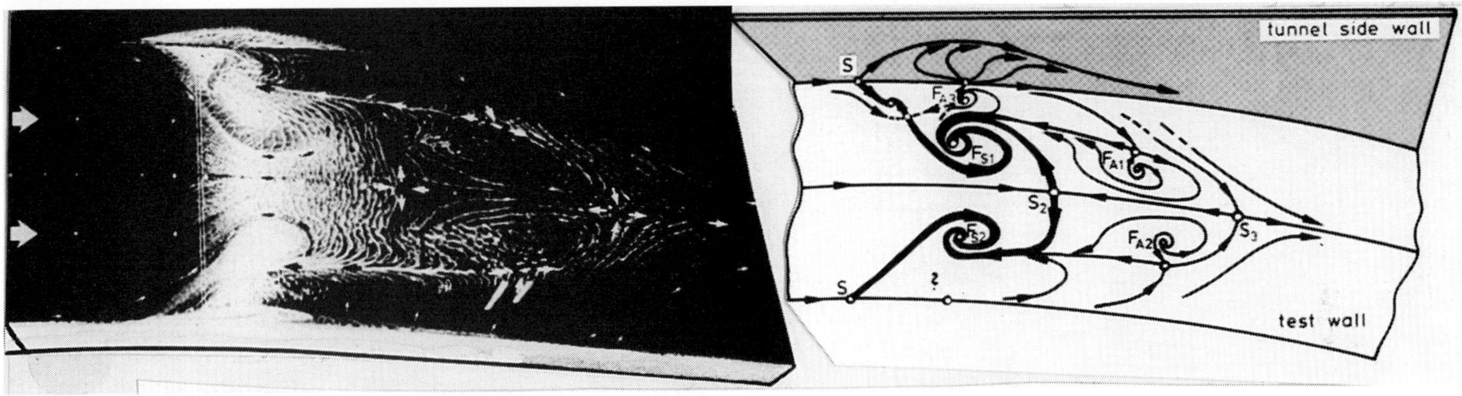

Fig.5 Surface oil-flow visualization for $M_w = 1.39$, $Re_{\delta_u} = 5.2 \times 10^4$ (a) and its topological structure (b).

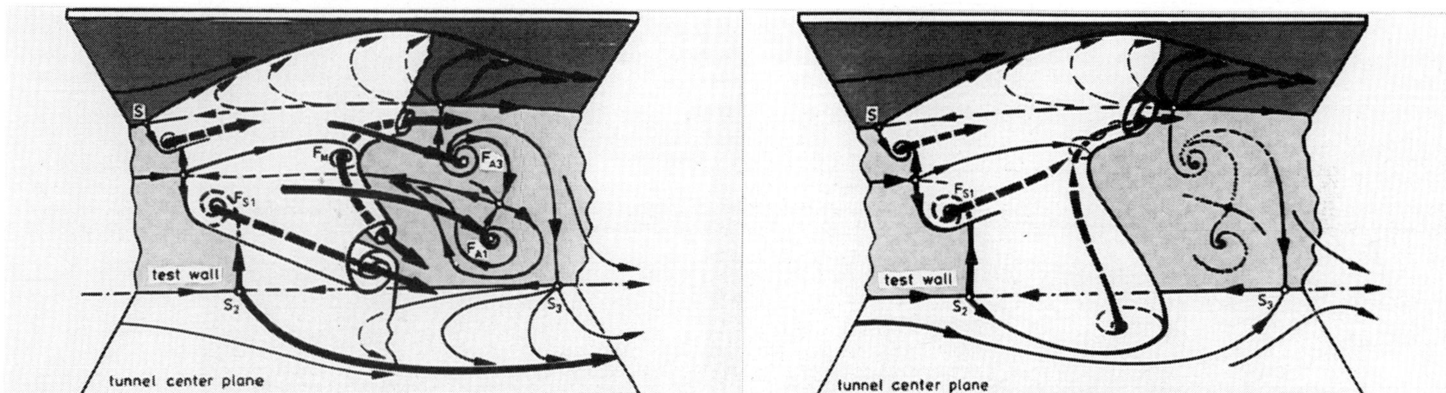

Fig.6 Time-averaged, three-dimensional flow structures (associated with the surface flow pattern of fig.5) which allows in the central plane either total flow separation (a) or flow-reattachment (b).

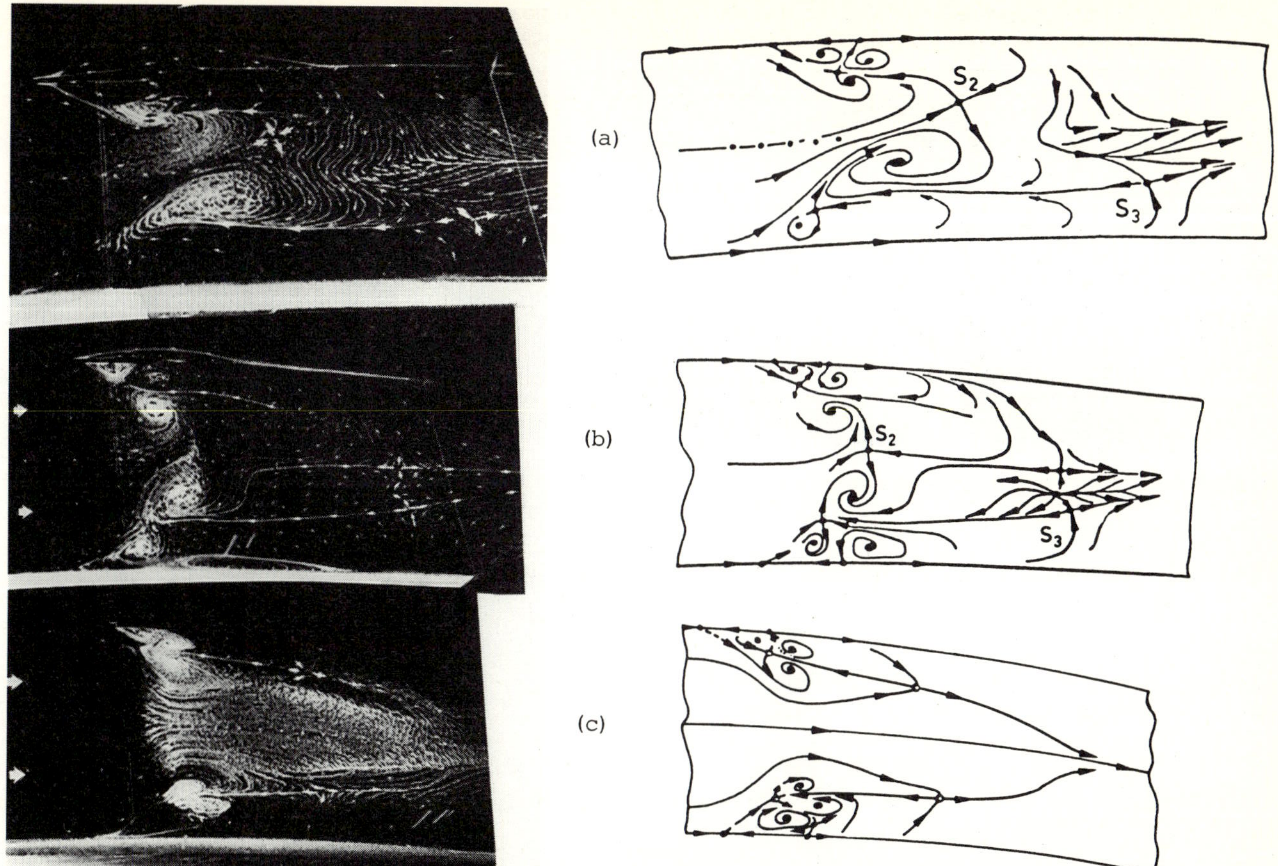

Fig.7 Different topological structures identified in the surface flow patterns for:
(a) $M_w = 1.47$, $Re = 15.3 \times 10^4$, (b) $M_w = 1.47$, $Re = 5.5 \times 10^4$, (c) $M_w = 1.35$, $Re = 5.6 \times 10^4$.

Unsteady Flows

Unsteady Phenomena

G.E.A. Meier

Max-Planck-Institut für Strömungsforschung,
D-3400 Göttingen, FRG

Summary

The large variety of unsteady phenomena in transonic flow is divided in four main groups depending on if there is a crucial influence of viscosity or not on the mechanism of unsteadiness and if it is a forced or self-excited non-stationary behaviour. The main forced, inviscid interactions are due to gusts, shock waves and changes of gas properties. Forced viscid interactions are unsteady separations by shocks and vortices. The inviscid self-excited instabilities can be seen in flutter, cavity oscillations and condensation oscillations. Self-excited instabilities of the viscid type occur both for shock boundary-layer and shock shear-layer interaction.

Introduction

Unsteady transonic flow is, in the usual definition, a non-stationary, mixed subsonic-supersonic flow field with high subsonic flow Mach numbers in the surrounding space. Unsteady means more exactly that the flow is non-stationary either by time-dependent boundary and initial conditions, respectively ("forced"), or by flow instabilities at constant boundary or initial conditions ("self-excited").

For the mathematical and physical treatment it makes a great difference if the interactions or phenomena are of the viscid or inviscid type. The inviscid problems can be solved nowadays with sufficient accuracy by numerical solution of the Euler-equations together with an energy and continuity equation and the appropriate boundary and initial conditions. In contrast to that, the numerical solution of a set with the full Navier-Stokes equations including compressibility is not yet possible with sufficient space and time resolution.

J. Zierep, H. Oertel (Eds.)
Symposium Transsonicum III
IUTAM Symposium Göttingen, 24.-27. 5. 1988

Therefore, it is justified to divide the whole problem into two branches which consist, on the one hand, of the problems with negligible viscosity influence on the unsteady behaviour and, on the other hand, of those which have to be treated by physical modelling because of significant influence of viscous effects.

The physical elements or flow phenomena which play an important role in the unsteady flows are gusts, vortices, turbulence, oscillations, pressure- and sound-waves, shock waves, boundary layers, shear layers, entropy distortions and condensation. External heating, chemical reactions (like combustion), ionisation or electromagnetic effects which can also be the reason for unsteady behaviour are not considered in this paper.

Table 1 gives a collection of the most important non-stationary phenomena divided in four groups which can be defined by the existence of forced or self-excited flow unsteadiness and the presence and importance of viscid or inviscid interactions.

Table 1: Unsteady Behaviour, Instabilities and Oscillations

I. Forced Inviscid Interactions	III. Self-Excited Inviscid Interactions
1. Gusts, Unsteady Boundaries	1. Transonic Flutter
2. Shockgenerated Flow	2. Cavity Oscillations
3. Vortices	3. Condensation Oscillations
4. Quantities (p,s.T)	4. Vortex Vortex Interaction
5. Parameters (k,c,x)	5. Shock Turbulence Interaction
II. Forced Viscid Interactions	**IV. Self-Excited Viscid Interactions**
1. External Turbulence	1. Turbulence
2 Slip Streams	2. Vortex Shedding
3. Entropy Distortions	3. Reattachment Oscillations
4. Shock Generation	4. Shock Shear-Layer Interaction
5. Shock-Induced Separation	5. Shock Boundary-Layer Interaction
6. Vortex Generation	

Since it is not possible to treat all the problems named in Table 1 in this paper, most of them will only be mentioned in the following text with reference to the corresponding literature. In the following four chapters some new examples from each section will be treated more extensively to give an impression of the related physical effects.

The interaction of the above-mentioned flow structures with each other and the boundaries leading to unsteady behaviour is not typical of transonic flow exclusively. But the increasing influence of compressibility results in especially strong pressure and density changes in transonic flow when these interactions occur, and consequently, also result in severe loads on the boundaries and strong pressure waves and soundfields in the flow. The first appearance of supersonic flow regimes additionally results in the existence of terminating shock waves which are an essential part of several mechanisms of instability. So far the transonic flow regime has some special types of unsteady behaviour. Consequently the emphasis of this paper is directed towards these transonic specialities.

I. Forced excitation of unsteady inviscid phenomena

In the field of forced unsteady behaviour, which can be treated successfully in an inviscid approach, the transonic flows on helicopter blades or induced by gusts on airfoils are very important. Also the influence of unsteady boundaries like flaps, oscillating wings or moving walls in a Laval nozzle have been described in the literature [1-16]. In the case of airfoil flow, small upward or downward wind velocities can generate large changes in lift and momentum. Unsteady boundaries are present especially in helicopter rotor blade flows in the tip area where the Mach number can change dramatically during blade rotation. The great influence of a change of the boundary on the shock position in transonic flow is shown in Fig. 1 for a Laval nozzle flow with oscillating nozzle walls.

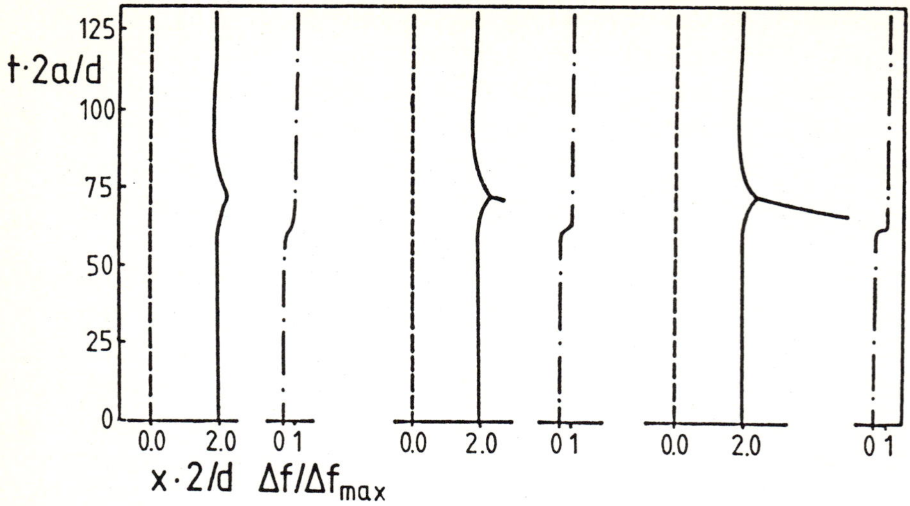

Fig. 1: Shock motions in time t and position x induced by a change of the downstream contour (change in cross-section + 7 %), of a Laval nozzle [12].
(---- sonic line; ——— shock; —·— cross-section)

These numerical results achieved with the method of characteristics show that the shocks have a strong dependence on the steepness of the contour change of the nozzle.

Shock-generated transonic flows occur in shock tubes, detonations and some wind tunnels, where the flow is established by a strong pressure wave. As an example we show in Fig. 2 two frames from a shock tube experiment where an airfoil flow is established by a passing shock wave.

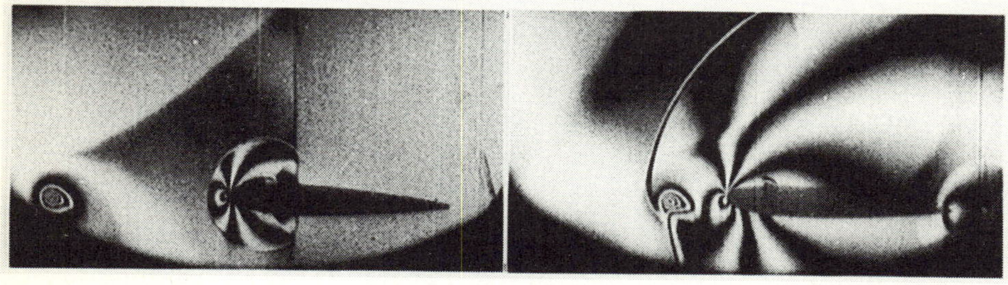

Fig. 2: a. Flow field around an airfoil generated by a passing shock wave. b. Interaction of the scattered shock wave with an incoming vortex (fringes are lines of constant density).

In Fig. 2 the scattered wave from the leading edge area and the very quick generation of the airfoil flow is visible. In the second frame the unsteady interaction of a shock and a vortex is shown.

Vortices can generate unsteady transonic flow fields passing a transonic airfoil at a short distance, too. If the Strouhal number St = u r/L, defined by the circulation L of the vortex, the flow velocity u and the nose radius of the airfoil r is sufficiently large (compared to one), then the interaction can be usually treated as an inviscid one.

Strong vortices are able to generate extended transonic flow areas on airfoils, blades and duct boundaries. Especially in the case of the helicopter where the tip vortex of a passing blade interacts with the transonic flow of the following blade, the generation of supersonic time-dependent flow areas is possible. Fig. 3 shows an example where a vortex rotating clockwise, passing underneath an airfoil generates a supersonic flow regime in the leading edge area, which later comes to a breakdown with associated radiation of a shock wave upstream. This phenomenon we have called generation of transonic waves [11].

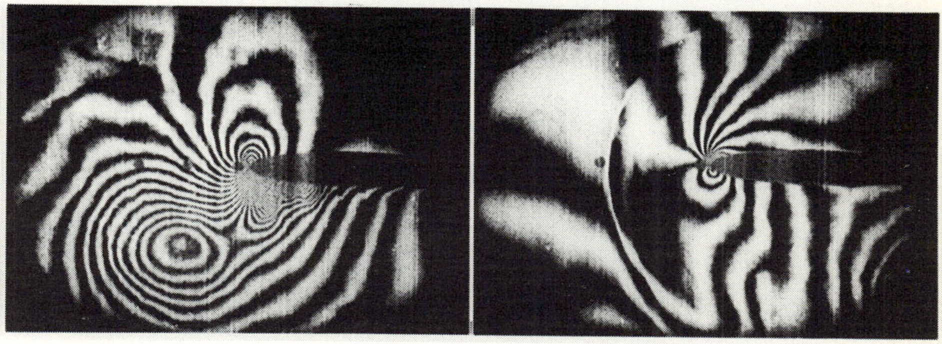

Fig. 3: Generation of a radiated shock wave by transonic interaction of a vortex and an airfoil flow (M = 0.6; airfoil: OLS, chord length l = 12 cm; circulation of the vortex L = 41 m^2/s; interframe time: t = 0.6 ms.).

There are also special transonic stagnation effects in the case of the interaction of a vortex and an airfoil flow, where a compression wave is radiated because of a choking effect when the velocity field of an incoming vortex interacts with a transonic airfoil flow. The waves radiated by this mechanism are called compressibility waves [11].

A similar kind of phenomenon occurs in transonic flows if, for instance, flow quantities (pressure, entropy, temperature) or some characteristic parameters of the flowing gas (ratio of the specific heats k, speed of sound c, or the void fraction x in the case of condensation) are changed in the incoming flow. We have studied some of these effects for a Laval nozzle flow, and it is obvious that especially in vicinity of Mach number 1 in the smallest cross-section the above-mentioned changes in the incoming medium result in strong changes in the flow field, and also in a scatter of waves and distortions. Fig. 4 shows the incoming, the radiated and the permanent distortions in the case of a change in the specific heat of the gas [28].

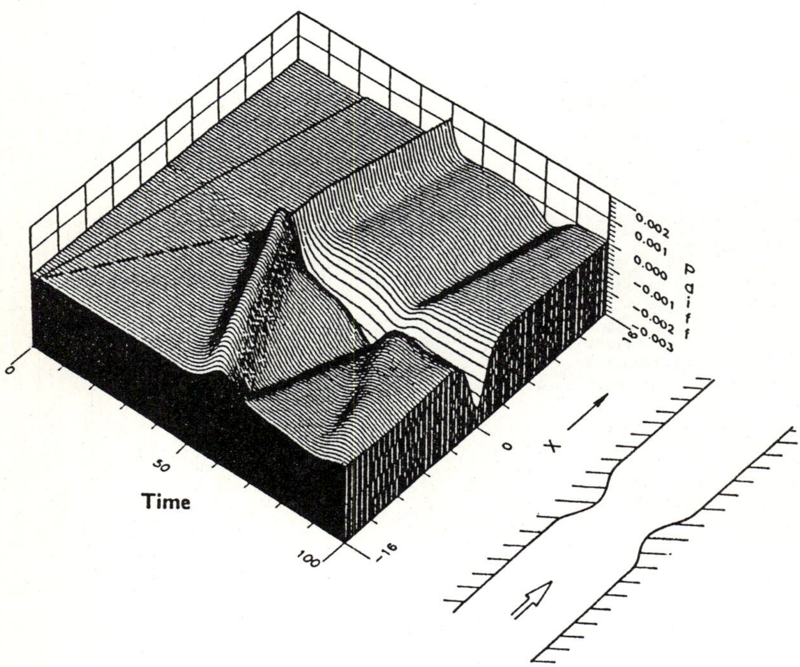

Fig. 4: Radiation of pressure disturbances from the smallest cross-section of a Laval nozzle flow (M = 0.6), induced by an incoming front (---) with a change in the ratio of specific heats of the gas from k = 1.4 to k = 1.33.

A lot of the unsteady loads in technical machinery and aircraft elements which introduce vibration and finally material fatigue are related to the above-mentioned problem. Also the noise generation of flow machinery is often based on these reactions of the flow field to changes or fluctuations in quantities and parameters.

II. Forced unsteady behaviour with viscid interactions

Externally generated turbulence and gusts can also interact in viscid reactions with a transonic flow field. For instance, the transition to turbulence is clearly influenced by the turbulence level of the duct or incoming flow. Shock vibration and vortex shedding can also be influenced by these externally generated fluctuations. Fig. 5 shows some interferograms where a small airfoil is exposed to the vortex street of a rectangular cylinder. In this case the unsteady fluctuations even generate supersonic flow fields with vibrating terminating shock waves [29].

Fig. 5: Generation of oscillating transonic fields on an airfoil by incoming distortions from a turbulent wake of a cylinder (M = 0.65; airfoil: NACA 0018, chord length l = 2 cm; square cylinder: d = 20 mm; interframe time: t = 0.15 ms).

These interactions very often cause flow separation or secondary vortex shedding which cannot be treated in an inviscid approach. Therefore, the theoretical treatment of these flow fields is especially difficult because of the undefined structure of the incoming flow and the viscous effects in the interaction area.

As a consequence of separation and triple points of shock waves (Mach-reflexion) in transonic flows, very often unsteady slip streams are induced which are not only instationary because of external excitation but also because of their own instability (Fig. 5).

In case of forced vortex shedding in the transonic flow regime especially strong entropy distortions occur in the vortex street. Fig. 6 shows an example where between the clearly distinguishable vortices an entropy (low density) distortion is floating together with the shedded vortices. This sheet of lower density material stems from the rear side of the vortex shedding device, where a large loss of total pressure is observed.

Fig. 6: A sheet of increased entropy observed between the vortices of an acoustically driven vortex separation with transonic flow velocity in the wake. (M = 0.27; rectangular cylinder: 70 mm x 15 mm).

At the end of supersonic flow regimes embedded in a subsonic field often the terminating shock waves are followed by a series of less strong shock waves which terminate small supersonic flow regimes generated by the thickened boundary layer downstream of the main shock. So, in the transonic flow regime often a sequence of shock waves is observed in forced unsteady and also steady flow fields. This configuration is highly unstable to small distortions and so these shock systems are always in rapid motion. Fig. 7 shows two

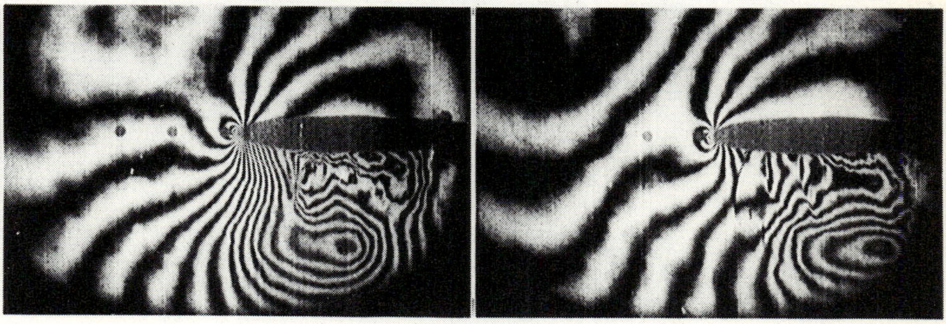

Fig. 7: Unsteady shock sequence on a thickened boundary layer downstream of a supersonic flow regime (M = 0.6; airfoil: OLS, chord length l = 12 cm; circulation of the vortex L = 41 m^2/s; interframe time: 0.1 ms).

interferograms where such a situation is shown. The sequence of shock waves is not steady at all and during the decay of the supersonic flow regimes all the shocks move upstream and merge eventually.

By the forced motion of shock waves due to outer distortions, also the boundary layer separation at the shock's root or the pressure downstream of the shock wave can be influenced. This results in a severe unsteady change of the flow field downstream of the shock boundary interaction. For the case of a Laval nozzle flow, we have used the model of forced shock oscillations to perform stability considerations for this system against downstream pressure fluctuations. This is very important for the understanding of the self-excited oscillations of unstable systems which will be reviewed in the last chapter of this paper. Fig. 8 shows the shock motion and the velocity for a periodic pressure distortion at the nozzle end. This example, calculated by the method of characteristics, shows the high non-linearity of this problem.

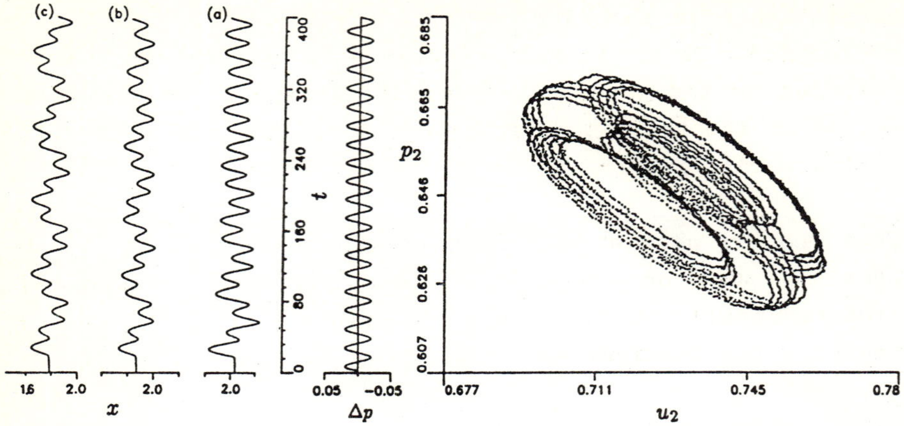

Fig. 8: Shock oscillations in position x and velocity u for a sinusoidal pressure distortion at the end of a Laval nozzle. (a-c) different shock positions [28].

In unsteady flow fields like gusts or passing vortices, the high speed flow regimes at boundaries with a small radius of curvature often tend to be separated in the area of pressure increase downstream of the point of highest velocity. In a highly unsteady flow with suitable time constants the high velocity regime is then transformed into a vortex of adverse sense of rotation compared to the incoming vortex. Fig. 9 shows some interferograms which visualize such a

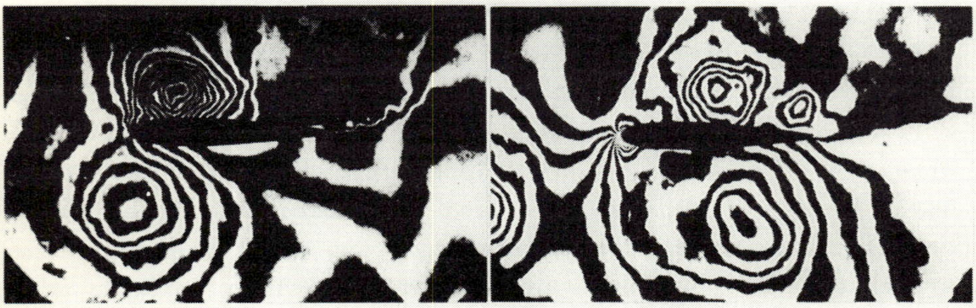

Fig. 9: Generation of a secondary vortex by separation under a high speed flow regime (a). Generation of an additional vortex from the boundary layer (b). (M = 0.5; airfoil: NACA 0012, chord length l = 6 cm; circulation of the vortex L = 5 m^2/s, interframe time: t = 0.3 ms).

generation of a secondary vortex in the leading edge area of an airfoil. This secondary vortex with an adverse rotation compared to the incoming vortex is later able to generate a vortex of the third generation from the boundary layer of the airfoil. This is visible in the second frame.

This Fig. 9a also shows that the shear layer at the trailing edge of such an unsteady airfoil flow will occasionally produce small vortices as a consequence of the unsteady velocity fluctuations caused by outer distortions.

III. Self-excited inviscid oscillations

There are some kinds of self-excited flow oscillations in transonic flows where the viscosity plays no essential role in the mechanism. These oscillations can be treated in models with which the gasdynamic equations, without introducing viscosity terms, are able to predict the unsteady behaviour with sufficient accuracy. If shock waves occur in these flow fields, they have to be treated with the suitable shock relations but all other viscous effects can often be neglected.

A typical example of this kind is the transonic flutter where flexible boundaries of the flow interact with the transonic flow field. These oscillations are especially dangerous and stressing for aircraft and turbo-machinery because of the large pressure fluctuations in unsteady transonic flow [30-40].

Another case where the viscosity plays only a minor role is given by the transonic cavity oscillations. Here, the supersonic flow or jet interacts with the mainly subsonic flow inside a cavity in such a way that the unsteady behaviour is mainly dominated by the acoustic resonances of the cavity (e.g. Helmholtz-Resonator). A typical case is the Hartmann-Oscillator, where a supersonic jet is fed into a cavity. The heavy oscillations of such a system can be explained and modelled by the shock motions and related entropy losses caused by the varying shock strength during the periodic change of the flow field in the cycle of oscillation. Models are described in some papers (see [35]). Fig. 10 shows also the typical fluctuation of the pressure and velocity signals measured in the experiment and calculated by the inviscid model.

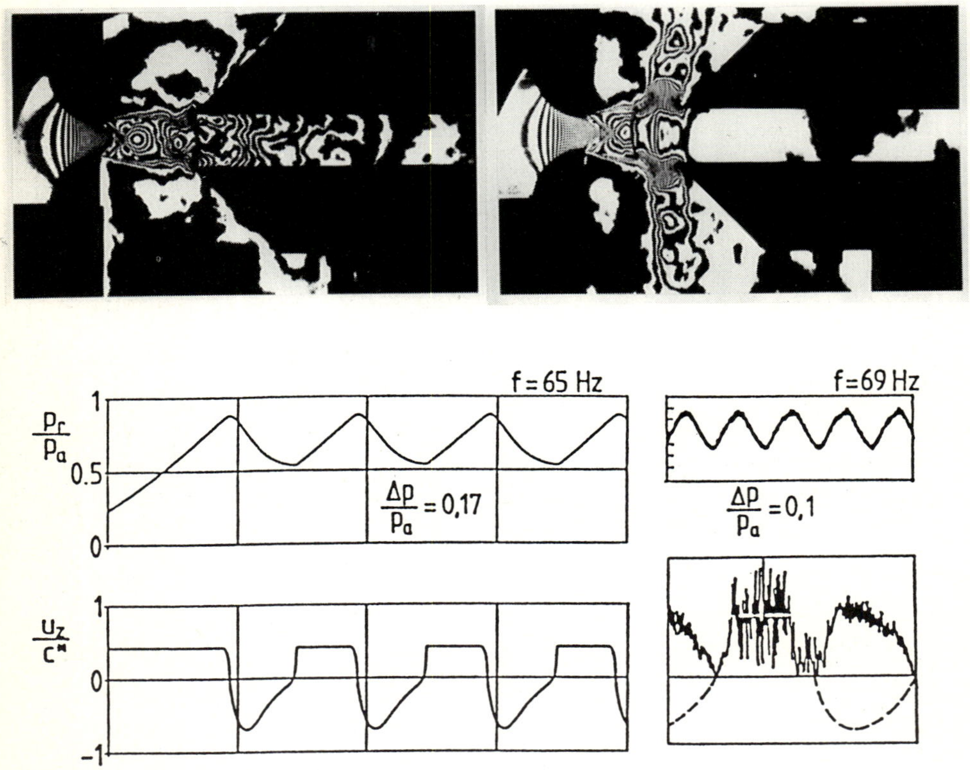

Fig. 10: Oscillations of the pressure and velocity in a Hartmann-Oscillator neck. Comparison between calculated (left) and measured (right) results.

Self-excited oscillations in transonic flows also occur when the heat addition by condensation of the gas is sufficiently high. These oscillations were first observed for the water condensation in supersonic airflows [40]. It is most probable that oscillations of this kind also occur in other transonic two-phase flows with phase transition.

Another kind of self-induced unsteady behaviour is observed when co-rotating vortices are interacting in the flow field. Especially at transonic speeds the interactions become strong and the radiated sound can be of high amplitude. This is one of the mechanisms for sound generation in transonic shear layers. For illustration Fig. 11 shows the interaction of two strong ring vortices generated by two shock waves emanating from the mouth of a pipe.

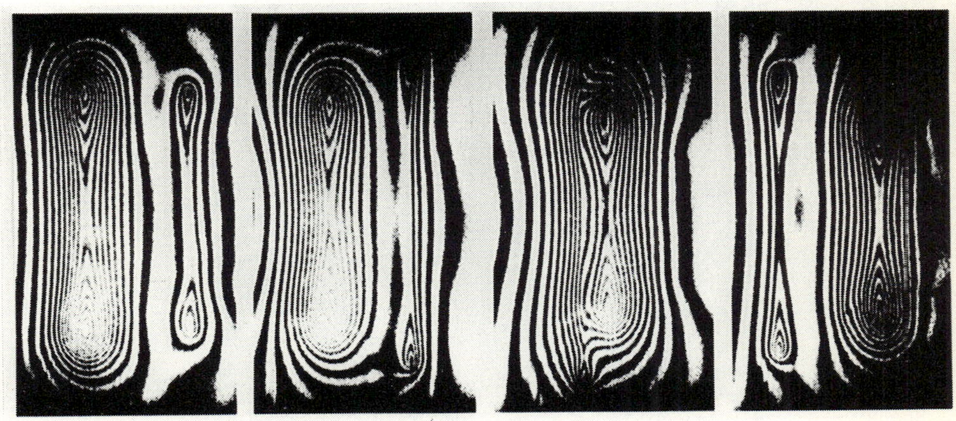

Fig. 11: Interaction of two ring vortices generated by a double shock wave emanating from a pipe: the second (stronger) vortex is creeping through the preceeding one. This process is continously repeated several times.

Also the shock turbulence interaction is a case where the unsteady behaviour is locally self-excited and can be treated in an inviscid model. Main effects are scattered sound waves and a strong deflection of the shock wave if the vortices are strong enough (see Fig.2b).

IV. Self-excited oscillations with dominant influence of viscosity

In some cases of transonic flow oscillations and unsteady behaviour the influence of viscosity in boundary layers and shear layers plays a dominant role in the mechanism. In these cases the viscosity cannot be neglected for an exact treatment. Occasionaly, it is possible to model the viscosity influence by physical models to avoid the difficult treatment of the full compressible Navier-Stokes equations [41-78]. One case where this is not really possible, is the turbulence in transonic boundary layers. Some related papers on this topic can be read in this book.

The transonic vortex shedding from cylinders is not understood completely. There are, of course, some model calculations which fix the point of flow separation to the shock waves terminating the oscillating supersonic flow fields. Fig. 12 shows two typical interferograms of an oscillating transonic flow field and the vortex shedding [41].

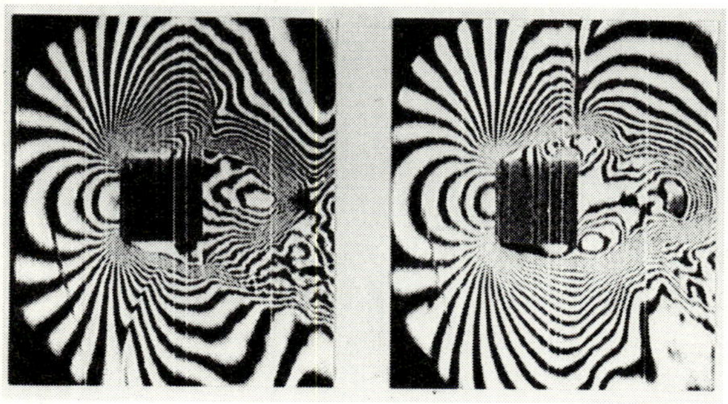

Fig. 12: Transonic vortex shedding and oscillations on a square cylinder (M = 0.6; cylinder: d = 20 mm; interframe time: t = 0.15 ms).

Another oscillation which is especially violent if the flow field is transonic is the so-called reattachment oscillation. This oscillation is observed when a separated flow reattaches under special circumstances, so that the separation bubble possibly develops a very different pressure in the closed, reattached state than in the completely separated flow. A typical example is the flow in a duct with a sudden enlargement of cross-section. Here, the increase of cross-section in steps normally induces separation with a following free expansion of the flow which eventually then reattaches to the wider section of the duct. If reattached, the pressure in the base area is very low and is corresponding to the pressure inside the supersonic jet. If detached from the

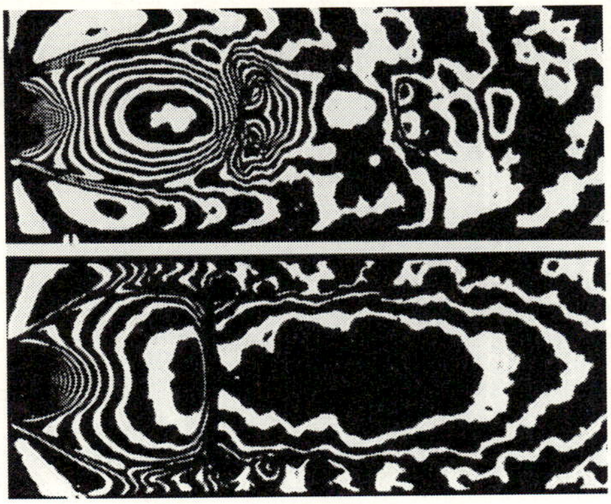

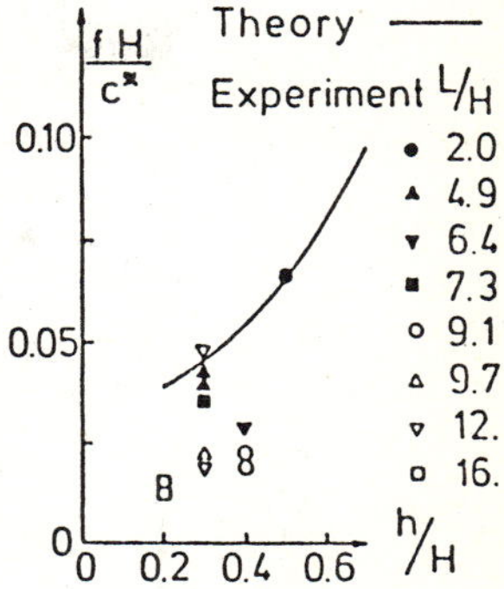

Oscillation frequencies for different duct geometries

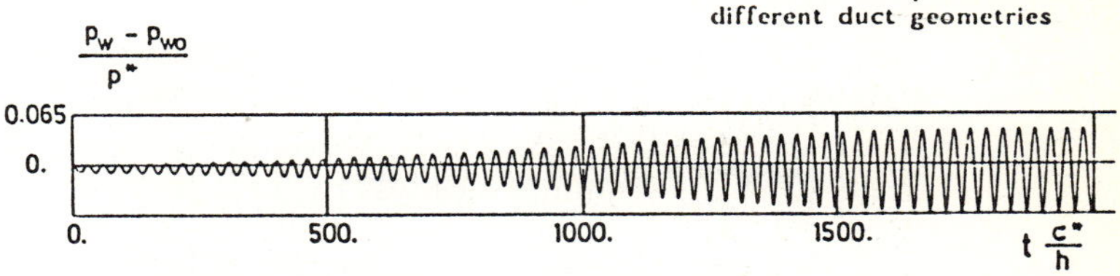

Fig. 13: The interferograms show the base pressure oscillation in a rectangular duct. The mathematical model for the transonic flow field shows periodic oscillations with a limiting cycle (h nozzle, H duct width; L duct length; f = frequency p = pressure) [44].

walls, the pressure in the base area is of the order of the pressure downstream of the terminating shock waves. So-called base pressure oscillations alternate between these states which are shown in Fig. 13. A mathematical treatment models just what was described above, with a coupling equation for the subsonic gap flow between the base area and the downstream flow field. The frequency and the amplitude of the oscillations are in agreement with the experimental observations [42-44].

A special case where the interaction between shock waves and shear layers and boundary layers produces very regular oscillations is the so-called "intake buzz". A sketch of the related intake configuration and some pressure traces are given in Fig. 14 [51]. The shear layer from the crossing point of the inclined shock wave from the intake cone and the normal shock wave is interacting with the lip to generate very regular oscillations in a certain regime of the flow rate. With reduced flow rate this oscillation is exchanged by the more severe pressure fluctuations of the shock boundary layer interactions taking place at the nose cone.

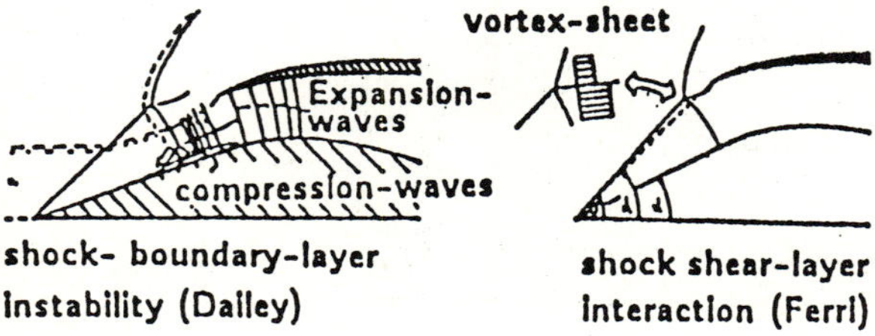

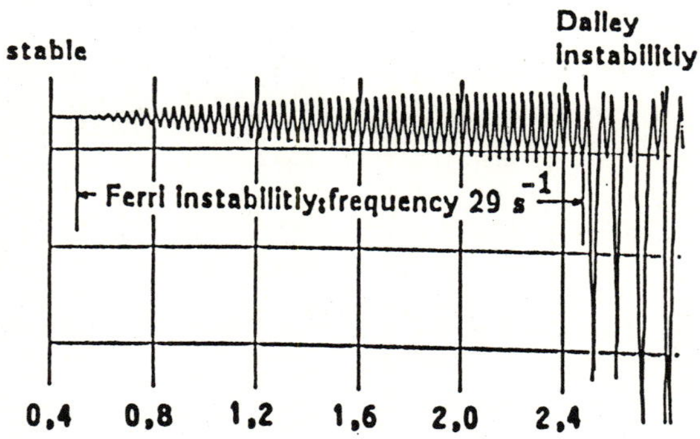

Fig. 14: "Inlet Buzz", trace of the time-dependent pressure in the combustion chamber. Time, choking and distance shock-lip increase with the numbers on the abscissa. (M = 2.225; circular intake with D = 111.5 mm diameter; total length L = 2.0 m; DFVLR-Köln Nr. 87-4-27; results of Krohn, Jungclaus and Weinreich) [51].

In case of transonic airfoil flow this instability has become very famous as the so-called "buffeting" [47-57]. This phenomenon is also discussed in a paper by Basler and Stanewski in another article in this book. Therefore, the discussion of the unsteady shock boundary layer interaction can here be restricted to the example given above and a discussion of the buffet instability mechanism in a one-dimensional model of a Laval nozzle flow.

Some gasdynamic considerations on the instability limits of the flow in a Laval nozzle with a shock-induced separation have already been published by the author in the foregoing "Symposium Transonicum II". Recently performed numerical calculations of a model flow and also further stability considerations have confirmed these stability limits as the range of absolute instability for a normal shock separating the boundary layer [76]:

$$\text{Shock Mach Number} \quad M_s \leq (k+3)/2 = 1.48 \ (\text{for } k = 1.4)$$

By the stability considerations and the numerical calculations a range of dynamic instability could be found which exceeds this limit depending on the downstream nozzle configuration. In Fig. 15 the range of dynamic instability for a reflexion coefficient $0 = r = -1$ is shown to reach up to a Mach number $M_s = 1.74$. Above that Mach number also dynamic instability can occur depending on the impedance of the nozzle exit [28].

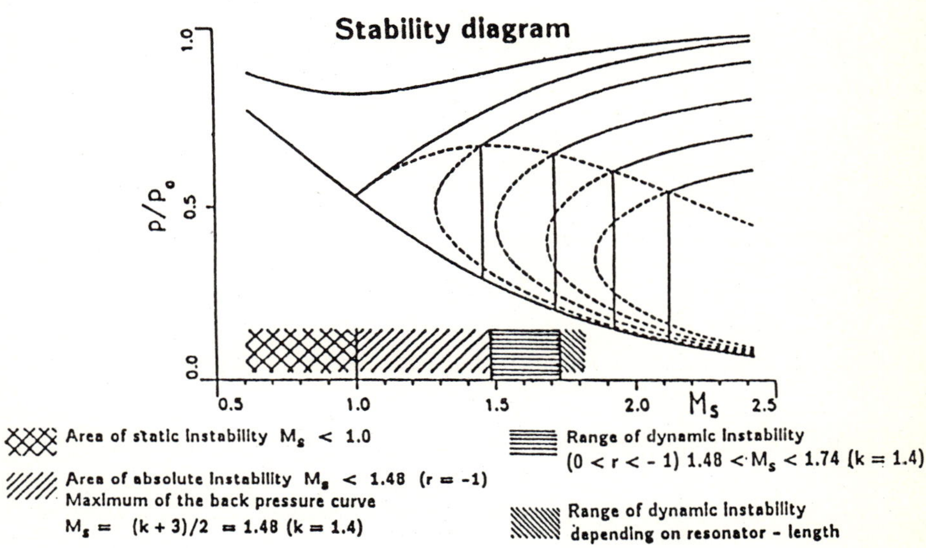

Fig.15: Pressure-Mach number diagram for a transonic flow with shocks. Stability regimes depending on exit reflection coefficient r for the case of shock boundary layer separation are shaded.

Although these considerations are made for a Laval nozzle flow, it is believed that the absolute instability limits also apply to flows in other geometrical configurations if the flow is separated by a normal shock wave. The range of dynamic instability is once more dependent on the geometrical conditions in the flow field downstream of the shock wave. Therefore, it has no general upper limit. But, nevertheless, it is very probable that an additional range of dynamic instability above that of the absolute instability exists for all flow geometries.

References

1. Ballhaus, W.F. and P.M. Goorjian: Implicit Finite-Difference Computations of Unsteady Transonic Potential Flows About Airfoils, Including the Treatment of Irregular Shock-Wave Motions, AIAA Paper, 77-205 (1977).

2. Borland, C.J., D.P. Rizetta and H. Yoshihara: Numerical Solution of Three-Dimensional Unsteady Transonic Flow over Swept Wings. AIAA Journal Vol. 20 No. 3, 340-347 (1982).

3. Caradonna, F.X. and M.P. Isom: Numerical Calculation of Unsteady Transonic Potential Flow over Helicopter Rotor Blades, AIAA Journal, Vol. 14 No. 4, 482-488 (1976).

4. Chyu, W.J., S.S. Davis and K.S. Chang: Calculation of Unsteady Transonic Flow over an Airfoil. AIAA Journal Vol. 19, No. 6, 684-690 (1980).

5. Dowell, E.H.: A Simplified Theory of Oscillating Airfoils in Transonic Flow, Proc. Symp. Unsteady Aerodyn. Tucson, Vol II, 655-679 (1975).

6. Goldstein, M.E., W. Braun and J.J. Adamczyk: Unsteady Flow in a Supersonic Cascade with Strong In-Passage Shocks, J. Fluid Mech., Vol. 83 No. 3, 569-604 (1977).

7. Landahl, M.: Unsteady Transonic Flow, Pergamon Press, (1961).

8. Levy Jr., L.L: Experimental and Computational Steady and Unsteady Transonic Flows about a Thick Airfoil. AIAA Journal Vol. 16, No. 6, 564-572 (1978).

9. Magnus, R. and H. Yoshihara: The Transonic Oscillating Flap, AIAA Paper, 79-327 (1976).

10. McCroskey, W.J.: Unsteady Airfoils. Ann. Rev. Fluid Mech. 14, 285-310 (1982).

11. Meier, G.E.A., H.-M. Lent and K.F. Löhr: Sound generation and Flow interaction of Vortices with an Airfoil and a flat plate in Transonic Flow. (1987).

12. Rein, M., G. Grabitz and G.E.A. Meier: Instationäre Strömung in einer Lavaldüse mit zeitabhängigem Querschnitt. Z. angew. Math. Mech. 68, pp 337 (1988).

13. Seegmiller, H.L., J.G. Marvin and L.L. Levy Jr.: Steady and Unsteady Transonic Flow. AIAA Journal Vol. 16, No. 12, 1262-1270 (1978).

14. Tijdeman, H. and R. Seebass: Transonic Flow past oscillating Airfoils. Ann.Rev. Fluid Mech. 12, 181-222 (1980).

15. Traci, R.M., E.D. Albano and J.L. Farr: Small-Disturbance Transonic Flows about Oscillating Airfoils and Planar Wings, AFFDL-TR-75-10 (1975).

16. Williams, M.H., S.R. Bland and J.W. Edwards: Flow instabilities in transonic small-disturbance theory. AIAA Journal 23, 1491-1496 (1985).

17. Sears, W.R.: Aerodynamics, Noise, and the Sonic Boom. AIAA Journal Vol. 7, No. 4, 577-586 (1969).

18. Lent, H.-M.: Schallentstehungsmechanismen in transsonischen Strömungen beim Auftreffen von Wirbeln auf eine Profilvorderkante. MPI Strömungsf. Bericht 5 (1986).

19. Meier, G.E.A. and R. Timm: Unsteady Vortex Airfoil Interaction. Conf. Proc. 386 Unsteady Aerodynamics-Fundamentals and Applications to Aircraft Dynamics (1985).

20. McCroskey, W.J. and G.R. Srinivasan: Transonic interaction of unsteady vortical flows. NASA tech. Memorandum 86658, (1984).

21. Möhring, W.F., E.A. Müller and F.F. Obermeier: Schallerzeugung durch instationäre Strömung als singuläres Problem. ACUSTICA Vol.21 (1969).

22. Nakagawa, T., G.E.A. Meier, R. Timm and H.-M. Lent: Vortex Shedding of a Square Cylinder in front of a Slender Airfoil at High Reynolds Numbers, Part 2: Compressibility Effect. MPI Strömungsf. Bericht 24, (1985).

23. Nakagawa, T., G.E.A. Meier, R. Timm and H.-M. Lent: Compressible flows in the wakes of a square cylinder and thick airfoil arranged in tandem. Proc. R. Soc. Lond. 411, 379-394 (1987).

24. Adamson, T.C.: Unsteady transonic flows in two-dimensional channels. J. Fluid Mech 52, 437-449 (1972).

25. Messiter, A.F and T.C. Adamson Jr.: Forced oscillations of transonic channel and inlet flows with shock waves. AIAA Journal 22, 1590-1599 (1984).

26. Rein, M.: Berechnung einer instationären transsonischen Strömung mit Verdichtungsstößen und Ablösung in Lavaldüsen. MPI Strömungsf. Bericht 14, (1984).

27. Rein, M., G. Grabitz and G.E.A. Meier: Nonlinear Wave Propagation in Transonic Nozzle Flows. To be published in Journ. of Sound and Vibration (1988).

28. Ehrenfried, K.: Dynamik der Lavaldüsenströmung bei unterschiedlichen Randbedingungen. Diplomarbeit (1987).

29. Mabey. D.G. and J.R. Chambers: Report on Unsteady Aerodynamics-Fundamentals and Application to Aircraft Dynamics. AGARD Report No. 222 (1986).

30. Bisplinghoff, R.L. and H. Ashley: Principles of Aeroelasticity, J. Wiley and Sons, Inc., (1962).

31. Bodapati, S. and C.S. Lee: Unsteady Wake Measurements of an Oscillating Flap at Transonic Speeds. Conf. proc. AIAA-84-1563 (1984).

32. Horiuti, K., W.J. Chyu and D.A. Buell: Unsteady, Transonic Flow Computations for an Airfoil with an Oscillating Flap. Conf. proc. AIAA-84-1562 (1984).

33. Brocher, E., A. Maresca, and M.-H.Bournay: Fluid dynamics of the resonance tube. J. Fluid Mech. 43, 369-384 (1970).

34. Eastep, F.E. and J.J. Olsen: Transonic Flutter Analysis of a Rectangular Wing with Conventional Airfoil Sections. AIAA Journal Vol. 18, No. 10, 1159-1164 (1980).

35. Hartmann, J., and B. Trolle: Modus operandi of the air jet-pulsator. Dan. Mat. Fys. Medd. 10 (1930).

36. Jungowski, W.M., and G.E.A Meier: Planar jets impinging on various obstacles and some modes of flow oscillations. Mitt. Max-Planck-Institut für Strömungsforschung, Göttingen, 78 (1984).

37. Jungowski, W.M. and G. Grabitz: Self-sustained oscillation of a jet impinging upon a Helmholtz resonator. J. Fluid Mech. 179, 77-103 (1987).

38. Rockwell, D. and E. Naudascher: Review: Self-sustaining oscillations of flow past cavities. Trans. ASME, Vol. 100, 152-165 (1978).

39. Thompson, P.A.: Jet-driven tube. AIAA J. 2, 1230-1233 (1964).

40. Zierep, J. and S. Lin: Ein Ähnlichkeitsgesetz für instationäre Kondensationsvorgänge in der Laval-Düse. Forschung im Ingenieurwes. 34, 97-99 (1968).

41. Nakagawa, T., G.E.A. Meier, R. Timm and H.-M. Lent: Influence of Shape on Vortex Shedding from Two-Dimensional Cylinders at High Reynolds Number and High Subsonic Mach Number. MPI Strömungsf. Bericht 22, (1985).

42. Anderson, J.S., W.M. Jungowski, W.J. Hiller and G.E.A. Meier: Flow Oscillations in a Duct with a Rectangular Cross-Section, J. Fluid Mech. Vol 79, Pt 4 796-784 (1977).

43. Jungowski, W.M. : On the pressure oscillating in a sudden enlargement of a duct section. Fluid Dyn. Trans. 3, 735-741 (1967).

44. Meier, G.E.A., G. Grabitz, W.M. Jungowski, K.J. Witzcak and J.S. Anderson: Oscillations of the Supersonic Flow Downstream of an Abrupt Increase in the Duct Cross-Section. AIAA Journal Vol. 18, No. 4, pp 394 (1979).

45. Roshko, A. and G.J. Thomke: Observations of turbulent reattachment behind an axisymmetric downstream-facing step in supersonic flow. Douglas Report SM-4 3069 (1965).

46. Szumowski, A.P., W.C. Selerowicz and G.E.A. Meier: Oscillations of a subsonic flow in an abruptly expanding circular duct. Lecture Notes in Physics (Flow of Real Fluids), Springer Verlag 235, 28-40 (1985).

47. Bogar, T.J.: Structure of self-excited oscillations in transonic diffuser flows. AIAA Journal 24, 55-61 (1986).

48. Bohning, R. and J. Zierep: Stoß-Grenzschichtinterferenz bei turbulenter Strömung an gekrümmten Wänden mit Ablösung. Z. Flugwiss. Weltraumforsch. 6 Heft 2, 68-74 (1982).

49. Dailey, C.L.: Supersonic diffuser instability. J. Aero. Space Sci. 22, 733-749 (1955).

50. Davis, S.S. and G.N. Malcolm: Transonic Shockwave/Boundary-layer interactions on an oscillating airfoil. AIAA Journal 18, 1306-1312 (1980).

51. Krohn, E.-O., G. Jungclaus and H.-L. Weinreich: Überschalleinlauf-Instabilitäten ("Inlet Buzz"). To be published in Z.f.W (1989).

52. Liou, M.S: Analysis of viscous-inviscid interactions in transonic internal flows with a shock wave and mass transfer. AIAA paper 81-0004 (1981).

53. Rein, M., G. Grabitz and G.E.A. Meier: Strömungsschwingungen in Lavaldüsen. ZFW 10, 1-4 (1986).

54. Sajben, M., T.J. Bogar and J.C. Kroutil: Forced oscillation experiments in supercritical diffuser flows. AIAA Journal 22, 465-474 (1984).

55. Salmon, J.T., T.J. Bogar and M. Sajben. Laser doppler velocimeter measurements in unsteady, separated, transonic diffuser flows. AIAA Journal 21, 1690-1697 (1983).

56. Schadow, K.C., J.E. Crump and F.S. Blomshield: Cumbustion instability in a research dump Cumbustor: inlet shock oscillations. CPIA Pub 347 Vol. III, 357-370 (1981).

57. Szumowski, A.P. and W.C. Selerowicz: Oscillations of a subsonic flow in an abruptly expanding circular duct. Lect. Notes Phys., Vol. 235, 28-40 (1985).

58. Anderson, J.S., W.M. Jungowski, G.E.A. Meier and K.J. Witczak: Self-Excited Oscillations in a Duct Flow with oblique shock waves. Proc. Inst. Acoust. (1978).

59. Anderson, J.S., W.M. Jungowski, W.J. Hiller and G.E.A. Meier: Flow oscillations in a duct with a rectangular cross-section. J. Fluid Mech. Vol. 79/4, 769-784 (1977).

60. Anderson, J.S., G. Grabitz, G.E.A. Meier, W.M. Jungowski and K.J. Witczak: Base pressure oscillations in a rectangular duct with an abrupt enlargement. Arch. Mech. 30/4-5, 337-351 (1978).

61. Basler, D.: Experimentelle Untersuchungen der Ausbreitung stoßinduzierter Störungen an transsonischen Profilen (1987).

62. Ho, C.-M. and N.S. Nosseir: Dynamics of an impinging jet. Part 1, J. Fluid Mech. Vol. 105, 119-142 (1981).

63. Meier, G.E.A. and A.P. Szumowski: Oscillatory Instabilities in Internal Transonic Flow. To be published, (1988).

64. Neuwerth, G.: Akustische Rückkopplungserscheinungen am Unter- und Überschallfreistrahl, der auf einen Störkörper trifft. DLR-FB 72-72, (1972).

65. Powell, A.: On the Mechanism of choked Jet Noise, Proc. Phys. Soc., Vol 66B, 1039-1056 (1977).

66. Rockwell, D. and E. Naudascher: Self-sustained oscillations of impinging free shear-layers. Ann. Rev. Fluid Mech. Vol. 11, 67-94 (1979).

67. Szumowski, A.P. and G.E.A. Meier: Schwingungen der Überschallströmung in einem Kanal mit Querschnittssprung. MPI Strömungsf. Bericht 23 (1978).

68. Wagner, F.-R.: Zum Schall- und Strömungsfeld eines axialsymmetrischen Freistrahls beim Auftreffen auf eine Wand. Z. Flugwiss. Vol. 19 No. 1, 30-44 (1971).

69. Anderson, J.S., W.M. Jungowski, W.J. Hiller and G.E.A. Meier: Flow oscillations in a duct with a rectangular cross-section. J. Fluid Mech. 79, 739-748 (1977).

70. Bogdanov, S.L., J.E. Vas: Preliminary investigations of spiked bodies at hypersonic speeds. J. Aero. Space Sci. 26, 65-74 (1982).

71. Délery, J.M.: Shock-wave boundary-layer interaction. AGARD-AG 280 (1986).

72. Dvořák, R.: On the unsteady boundary layer-shock wave interaction in the lower transonic region. Arch. Mech. 16, 211-222 (1964).

73. Ferri, A. and A.M. Nucci: The origin of aerodynamic instabilities of supersonic inlets at subcritical conditions. NACA RM L 50 K 30 (1951).

74. Hasan, M.A.Z. and A.K.M Hussain: The "whistle-nozzle" phenomenon. J. Fluid Mech. 115 134, 433-458 (1983).

75. Meier, G.E.A. and W.J. Hiller: An experimental investigation of unsteady transonic flow by high-speed interferometric photography, AGARD Conf. Proc. No. 35/9 (1968).

76. Meier, G.E.A.: Shock-induced oscillations on a Laval nozzle. IUTAM Symp. Transonicum II, Springer Verlag, 252-261 (1976).

77. Rein, M., G. Grabitz and G.E.A. Meier: Strömungsschwingungen in Lavaldüsen. ZFW 10, 1-4 (1986).

78. Selerowicz, W.C., A.P. Szumowski and G.E.A. Meier: Experimentelle Untersuchungen schwingender Strömungen in einer Rohrdüse mit Querschnittssprung. Max-Planck-Institut für Strömungsf. Göttingen, Bericht 2 (1987).

Unsteady Aeroelastic Computations for Flexible Configurations at Transonic and Supersonic Speeds

V. Shankar and H. Ide

Rockwell International Science Center
Thousand Oaks, California, USA

Abstract

An aeroelastic package consisting of 1) an aerodynamic solver, 2) a structural response model, and 3) a grid generation/update program has been developed to study both static and dynamic response of flexible aerospace configurations. The aerodynamics at present is a full potential solver but can be an Euler/Navier–Stokes solver also. The structural response is based on a generalized modal representation. The grid generation package not only sets up the initial grid, but also updates the grid position in time to follow the instantaneous position of a flexible configuration using a "grid speed" procedure.

The aeroelastic package can handle 1) static rigid, 2) dynamic rigid, 3) static flexible, and 4) dynamic flexible cases. Also, the aerodynamic solver is based on an unsteady formulation and can handle subsonic, transonic, and supersonic flow conditions. Results are presented for rigid and flexible configurations at different Mach numbers ranging from subsonic to supersonic conditions. The dynamic response of a flexible wing below and above its flutter point is demonstrated.

Introduction

For analysis/design of flexible aerospace configurations, it is important to include the coupling between aerodynamics and structures, especially in the transonic/low supersonic regimes where shock loadings are critical, to account for static and dynamic aeroelastic effects. The status of CFD methods, full potential to Navier–Stokes, is fairly mature for rigid configurations, but computations of flexible shapes using nonlinear aerodynamic codes have been rather limited due to large computer execution and memory requirements. Use of the transonic small disturbance theory[1,2] and full potential methods[3–7] for aeroelastic computations offer economy of computation, though the range of their applications may be limited. However, the mechanics of coupling aerodynamics and structures, once perfected for full potential methods, can easily be extended to Euler and Navier–Stokes models.

The aeroelastic methodology of this paper has three modules: 1) a full potential based flow solver, 2) a structural response model based on generalized modal functions,

J. Zierep, H. Oertel (Eds.)
Symposium Transsonicum III
IUTAM Symposium Göttingen, 24.-27. 5. 1988

and 3) a grid package for update of geometry and grid at each time level. The first module, the full potential solver, employs the following concepts: 1) internal Newton iterations for time accuracy and computational efficiency, enabling large Courant number steady/unsteady computations; 2) implicit boundary conditions consistent with the approximate factorization scheme; 3) flux biasing techniques based on sonic conditions for treatment of hyperbolic regions with shocks and sonic surfaces; 4) local time linearization for density; 5) unsteady wake treatment; and 6) nonreflecting outer boundary conditions. The second module (aeroelastic model) includes both static and dynamic options by solving the generalized structural equations of motion.

The grid package module is based on a three–dimensional elliptic grid solver for steady grids and a grid speed procedure for advancing the grid in time.

The purpose of this paper is to present a full potential aeroelastic package that can handle both static and dynamic aeroelastic effects. There are two steps in achieving this objective: 1) development of a robust time–accurate aerodynamic algorithm, and 2) development of an aeroelastic model that solves for the structural response of a flexible configuration due to nonlinear aerodynamic forces.

Results are presented for both static and dynamic flexible cases.

Formulation

The conservation law form of the time–dependent full potential equation in the Cartesian coordinate system x,y,z, and t is given by

$$\rho_t + E_x + F_y + G_z = 0 \tag{1}$$

where $E = \rho u$, $F = \rho v$, $G = \rho w$, and ρ is the density expressible in terms of the velocity potential ϕ through assumptions of isentropicity and irrotationality; u, v, and w are Cartesian velocity components along x, y, and z, respectively.

Under a body–fitted coordinate transformation represented by

$$\tau = \tau, \zeta = \zeta(x,y,z,t), \eta = \eta(x,y,z,t), \xi = \xi(x,y,z,t),$$

Eq. (1) can be written as

$$\bar{\rho}_\tau + \bar{E}_\zeta + \bar{F}_\eta + \bar{G}_\xi = 0 \tag{2}$$

where

$$\bar{\rho} = \frac{\rho}{J} \qquad \bar{E} = (\rho\zeta_t + E\zeta_x + F\zeta_y + G\zeta_z)/J$$
$$\bar{F} = (\rho\eta_t + E\eta_x + F\eta_y + G\eta_z)/J \qquad \bar{G} = (\rho\xi_t + E\xi_x + F\xi_y + G\xi_z)/J$$

and J, Jacobian for the transformation, is given by $J = \partial(\tau,\zeta,\eta,\xi)/\partial(t,x,y,z)$. The density ρ is given by Bernoulli's law

$$\rho^{\gamma-1} = 1 - \frac{\gamma-1}{2}M_\infty^2\left[2\phi_\tau + (U+\zeta_t)\phi_\zeta + (V+\eta_t)\phi_\eta + (W+\xi_t)\phi_\xi - 1\right]. \quad (3)$$

The contravariant velocities U,V,W are given by $U = \zeta_t + u\zeta_x + v\zeta_y + w\zeta_z$, $V = \eta_t + u\eta_x + v\eta_y + w\eta_z$, and $W = \xi_t + u\xi_x + v\xi_y + w\xi_z$.

Associating the subscripts i,j,k with the ξ,η,ζ directions, a numerical approximation to Eq. (2) can be written in the semidiscrete conservation law form given by

$$\left(\frac{\rho}{J}\right)_\tau + \left(\hat{E}_{i,j,k+1/2} - \hat{E}_{i,j,k-1/2}\right) + \left(\hat{F}_{i,j+1/2,k} - \hat{F}_{i,j-1/2,k}\right) + \left(\hat{G}_{i+1/2,j,k} - \hat{G}_{i-1/2,j,k}\right) = 0 \quad (4)$$

where $\hat{E}$,$\hat{F}$,$\hat{G}$ are representative fluxes approximating the real fluxes $\bar{E}$,$\bar{F}$,$\bar{G}$ of Eq. (2). Depending on the type of flow (subsonic, transonic, or supersonic) at the half node points (i,j,$k \pm 1/2$; i,$j \pm 1/2$,k; $i \pm 1/2$,j,k), the fluxes $\hat{E}$, $\hat{F}$, and $\hat{G}$ are properly defined.

The objective is to solve Eq. (4) in terms of the velocity potential ϕ.

Let Eq. (4) be represented as

$$R(\phi) = 0 \quad (5)$$

where ϕ is the unknown to be solved at every grid point (i,j,k) in the current $(n+1)$ time plane. The Newton iteration for solution to Eq. (5) is

$$R(\phi_*) + \left(\frac{\partial R}{\partial \phi}\right)_{\phi=\phi_*} \Delta\phi = 0 \quad (6)$$

where ϕ_* is the currently available ϕ at the $(n+1)$ level and $\Delta\phi = \phi - \phi_*$. At convergence, $\Delta\phi$ will approach zero driving ϕ to ϕ^{n+1}. Execution of Eq. (6) requires an initial guess for ϕ at $(n+1)$ to initiate the Newton iteration process. Now, the treatment of the discretized equation, Eq. (4), to the form of Eq. (6) in terms of $\Delta\phi$ is described.

Treatment of $\frac{\partial}{\partial\tau}\left(\frac{\rho}{J}\right)$ in Eq. (1)

$$\frac{\partial}{\partial\tau}\left(\frac{\rho}{J}\right) = \frac{(a_1 - b_1)\left\{\left(\frac{\rho}{J}\right)^{n+1} - \left(\frac{\rho}{J}\right)^{n}\right\} - b_1\left\{\left(\frac{\rho}{J}\right)^{n} - \left(\frac{\rho}{J}\right)^{n-1}\right\}}{a_1\Delta\tau_1 - b_1(\Delta\tau_1 + \Delta\tau_2)} \quad (7)$$

where $a_1 = (\Delta\tau_1 + \Delta\tau_2)^2$, $b_1 = \Delta\tau_1^2$, $\Delta\tau_1 = \tau^{n+1} - \tau^n$, $\Delta\tau_2 = \tau^n - \tau^{n-1}$.

The unknown quantity in Eq. (7) is ρ^{n+1}. Following Eq. (6), this is written as

$$\rho = \rho(\phi_*) + \left(\frac{\partial\rho}{\partial\phi}\right)_{\phi=\phi_*} \Delta\phi \tag{8}$$

where $\Delta\phi = \phi - \phi_*$, and

$$\left(\frac{\partial\rho}{\partial\phi}\right)_{\phi=\phi_*} = \left[-\frac{\rho}{a^2}\left\{\frac{1}{\Delta\tau_1} + U\frac{\partial}{\partial\zeta} + V\frac{\partial}{\partial\eta} + W\frac{\partial}{\partial\xi}\right\}\right]_{\phi=\phi_*} \tag{9}$$

is a differential operator. The density quantities ρ^n and ρ^{n-1}, appearing in Eq. (7), are not linearized as they are known. Since ρ^{n+1} is iterated to convergence using Eq. (8), conservation is maintained when $\Delta\phi \to 0$.

Treatment of Spatial Flux

Equation (4) requires evaluation of $\hat{E}$, $\hat{F}$, and $\hat{G}$ at various spatial half node points. As mentioned earlier, $\hat{E}$ represents $\bar{E}$ appearing in Eq. (2). For transonic flows, the fluxes are defined to be

$$\hat{E} = \tilde{\rho}\frac{U}{J} \ ; \ \hat{F} = \tilde{\rho}\frac{V}{J} \ ; \ \hat{G} = \tilde{\rho}\frac{W}{J} \tag{10}$$

where $\tilde{\rho}$ is an upwind biased value of density, designed to produce the necessary artificial viscosity for treatment of hyperbolic regions. Expanding the flux representation in Eq. (10) in terms of $\Delta\phi$, one gets

$$\hat{E} = \frac{\tilde{\rho}}{J}\left\{\zeta_t + a_{11}(\phi_* + \Delta\phi)_\zeta + a_{12}(\phi_* + \Delta\phi)_\eta + a_{13}(\phi_* + \Delta\phi)_\xi\right\} \tag{11}$$

where $\tilde{\rho} = \tilde{\rho}(\phi_*)$, and a_{11}, a_{12}, a_{13} are transformation metrics.

The biased value of density is given by

$$\tilde{\rho} = \frac{1}{q}\left[(\rho q) \pm \left\{\frac{U}{Q}\frac{\partial}{\partial\zeta} + \frac{V}{Q}\frac{\partial}{\partial\eta} + \frac{W}{Q}\frac{\partial}{\partial\xi}\right\}(\rho q)^-\right] \tag{12}$$

where $Q = \sqrt{U^2 + V^2 + W^2}$.

The quantity $(\rho q)^-$ appearing in Eq. (12) is defined to be

$$\begin{aligned}(\rho q)^- &= \rho q - \rho^* q^* && \text{if } q > q^* \\ &= 0 && \text{if } q \le q^*\end{aligned} \tag{13}$$

The quantities $\rho^* q^*$, ρ^*, and q^* represent sonic values of the flux, density, and total velocity, respectively. These sonic conditions are given by (using the density and speed of sound relationships)

$$(q^*)^2 = \frac{1 + \frac{(\gamma-1)}{2} M_\infty^2 (1 - 2\phi_\tau - 2\zeta_t\phi_\zeta - 2\eta_t\phi_\eta - 2\xi_t\phi_\xi)}{\frac{\gamma+1}{2} M_\infty^2} \tag{14}$$

$$\rho^* = (q^* M_\infty)^{2/(\gamma-1)}.$$

Note that for steady flows, the sonic conditions ρ^* and q^* are only a function of the freestream Mach number, and for a given flow they are constants. For unsteady flows, ρ^* and q^* need to be computed everywhere due to the presence of ϕ_τ and other unsteady terms in Eq. (14).

The fluxes $\hat{F}$ and $\hat{G}$ are modeled similar to Eq. (11). Thus, every term in Eq. (4) has been written in terms of ϕ_* and $\Delta\phi$. Grouping all these terms, one can write Eq. (4) in the form of Eq. (6).

An implicit model for solving Eq. (6) is developed using an approximate factorization form for $\left(\frac{\partial R}{\partial \phi}\right)$. At the body surface, the boundary condition $v = \eta_t + f(\phi_\zeta, \phi_\eta, \phi_\xi) = 0$ is imposed within the implicit framework. Details on the construction of the implicit treatment are given in Ref. [6].

Grid Update

For unsteady computations involving body motion, the grid is allowed to move in time to follow the surface topology. Unsteadiness can be introduced by forced oscillation (pitch, plunge, or combination of both) of a rigid configuration or by dynamic structural response of a flexible configuration. The latter type of unsteady computations is performed for prediction of flutter boundaries. An unsteady calculation is usually started from an initial condition that corresponds to the steady state static rigid solution performed using a static rigid grid. The first step in unsteady computations is to compute surface grid speeds x_τ, y_τ, and z_τ at every surface grid point. The grid speeds corresponding to outer boundary grid points are taken to be zero maintaining a fixed location for the outer boundary. The grid speeds at any interior field grid are then obtained by interpolating the body value and the zero outer boundary value along a constant coordinate line. Knowing grid speeds, the location of a grid point at the next time level is obtained from $x^{n+1} = x^n + x_\tau \Delta\tau$; $y^{n+1} = y^n + y_\tau \Delta\tau$; $z^{n+1} = z^n + z_\tau \Delta\tau$. For dynamic rigid unsteady calculations, the surface grid speeds are known from any prescribed body motion. For flexible computations, surface grid speeds are computed from solving an appropriate aeroelastic structural equation of motion.

Aeroelastic Model

The aeroelastic model is based on the generalized modal approach. In the physical space, the structural equation is written as

$$[m]\ddot{z} + [c]\dot{z} + [k]z = \{f\} \tag{15}$$

where m, c, k, and f represent the mass, damping, stiffness, and force, respectively. Corresponding to each natural frequency ω_i (eigenvalues of Eq. (15)), there exists an eigenvector ϕ_i which represents the amplitude of deflection of the various points of the system. For the first n modes of natural frequency, one can compute the n left and right eigenvectors forming the generalized mode shapes. Using the generalized mode shape, one can rewrite Eq. (15) as

$$M\{\ddot{q}\} + [C]\{\dot{q}\} + [K]\{q\} = \{F\} \tag{16}$$

where M, C, and K are generalized mass, damping, and stiffness matrices and $\{F\}$ is the generalized force. These are defined by $M = \phi^T m\phi$; $K = \phi^T k\phi$, $C = \phi^T c\phi$, $F = \phi^T f$ and $z = \phi q$, where q is the generalized deflection.

Static Flexible Case

The flexibility of the structure comes into the calculation through stiffness (K) and generalized mode shapes (ϕ). The steps involved in computing the static flexible equilibrium aerodynamics are given below:

1. Solve rigid aerodynamics for lift force.
2. Solve $Kq = F \Rightarrow \{z\} = [\phi]\{q\}$.
3. Increment $\{z\}$ in smaller steps and compute
4. Repeat Step 2.
5. Continue until $\{z\}$ and lift converge.

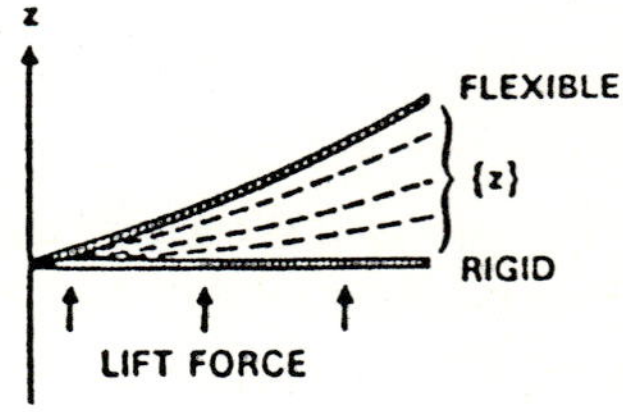

Dynamic Flexible Case

For dynamic response, Eq. (16) is solved. The time integration procedure for Eq. (16) will be included in the full paper. The aerodynamics and the structural response are computed in a time accurate fashion using internal Newton iterations. The various steps involved in this computation are:

1. Compute rigid aerodynamics.
2. Compute static flexible for a given dynamic pressure Q.

3. Set up an initial perturbance by perturbing either q or $\dot{q}$ of any mode.
4. Solve $M\ddot{q} + C\dot{q} + Kq = F \Rightarrow \{z\}$.
5. Compute x_τ, y_τ, z_τ on the surface.
6. Define x_τ, y_τ, z_τ for each field grid point allowing their values to go to zero at outer boundary.

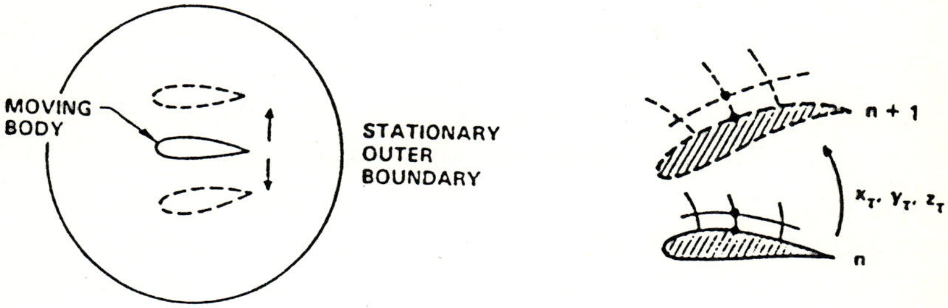

7. Compute new grid location $x^{n+1} = x^n + x_\tau \Delta\tau, \cdots$ and new metrics.
8. Converge aerodynamics using internal Newton iteration.
9. Repeat Step 4. (For prediction of flutter Q gradually vary Q and monitor generalized modal deflections.)

Results

Using the unsteady, aeroelastic method of this paper, results are generated for static rigid (steady), dynamic rigid (forced oscillation), static flexible and dynamic flexible (unsteady) cases.

For computation of steady flows, the grid and metrics are computed only once and the internal Newton iteration for time accuracy is not performed because the transients in reaching the time asymptotic steady state are not of any importance. For unsteady computations, the grid and metrics are computed for each time level and internal Newton iterations (two to four) are performed to allow larger ΔT values.

Results for static rigid and dynamic rigid cases are not included in this paper and can be found in Refs. [4–7].

Static Flexible

Figure 1 shows the planform shape of a flexible fighter wing with control surfaces. Figure 2 shows results for a static flexible computation. The deflected shape of the flexible wing for a dynamic pressure of 288 psf is shown. For the wing to be aeroelastically stable, the tip airfoil undergoes a nose down rotation, causing the tip load to decrease.

A comparison of lift coefficients between rigid and flexible calculations is shown in Table 1.

M	α	Q(psf)	$C_{L_{test}}$	$C_{L_{flex}}$	$C_{L_{rigid}}$
1.05	1.6	335	0.0785	0.0800	0.1450
1.15	0.5	230	0.0433	0.0400	0.0685
0.90	0.0	220	0.0446	0.0441	0.0637

Table 1 Comparison of Lift Coefficients between Rigid and Flexible Calculations

Figure 3 shows a flexible computation with both leading and trailing edge controls deflected. The correlation of lift variation with angle of attack compares well with experimental data[8].

Typical surface pressure distributions for $M_\infty = 0.9$, $\alpha = 0.048$, and $Q = 1.74$ psi with clean wing (no control surfaces are deflected) at two different span stations are shown in Fig. 4a. Also, pressure distributions for the similar flow condition, except Q is slightly lower, with the trailing edge outboard (TEO) control surface deflection ($\delta_{\text{TEO}} = 5.0$ downward direction) are shown in Fig. 4b. Pressure distributions for supersonic flow conditions are shown in the following figures. Pressure distributions for the flow condition of $M_\infty = 1.15$, $\alpha = 1.66$, and $Q = 2.35$ psi with clean wing are shown in Fig. 5a. Pressure distributions with $\delta_{\text{TEO}} = 2.0$ for the same flow condition are shown in Fig. 5b.

The change in aerodynamic forces (lift and moment) with single control surface deflection for $M_\infty = 0.9$ and $M_\infty = 1.15$ is plotted. The C_L and C_M versus δ for both TEI and TEO in the subsonic regime are shown in Fig. 6a. Similar plots for the supersonic flow condition are shown in Fig. 6b.

The control surface effectiveness for roll moment was also investigated. The variation of roll moment coefficient for a unit control surface deflection with Q for $M_\infty = 0.9$ and $M_\infty = 1.15$ are shown in Fig. 7. In all these, the correlation of calculated results with data is good.

<u>Dynamic Flexible</u>

Figure 8 shows results for dynamic aeroelastic computations at three different dynamic pressures (Q). For $Q < Q_{\text{Flutter}}$, the structure is aerodynamically stable as shown by the decaying amplitudes of various quantities. The value for the flutter Q was obtained by the nonlinear aerodynamics/aeroelastic code through numerical search. Exactly at the flutter Q, the dynamic response of the structure does not decay. For $Q > Q_{\text{Flutter}}$, the structure is aerodynamically unstable as shown by the growing amplitudes of various quantities. A typical dynamic simulation such as the one shown in Fig. 8, for a given Q, takes approximately 4 minutes of CPU time on a CRAY/XMP for 35,000 grid points.

Conclusions

A three-dimensional unsteady full potential algorithm is presented. The method employs a triple approximate factorization scheme coupled with internal Newton iterations for time accuracy and computational efficiency. The hyperbolic regions with shocks and sonic surfaces are treated using a flux biasing procedure based on sonic reference conditions. The implicit treatment of the unsteady equation allows for large times to be taken, even for unsteady computations, while maintaining time accuracy. Unsteady results are presented for forced oscillation of rigid configurations and for dynamic aeroelastic response of flexible geometries. Prediction of flutter boundaries at transonic Mach numbers is demonstrated. Future extensions of the capability will include an interacting unsteady boundary layer to account for viscous effects.

References

1. G.P. Guruswamy. Transonic aeroelasticity of wings with active control surfaces. Paper No. 87-0709, ASME 28th Structures, Structural Dynamics and Materials Conference, Monterey, CA, April, 1987, pp. 16–30.
2. C.J. Borland, R.G. Thorne, and L.R. Yeagley. Transonic unsteady aerodynamics for aeroelastic applications. Vol. 1, AFWAL-TR-80-3107, May, 1981.
3. J.B. Malone, N.L. Sankar, and W.A. Sotomayer. Unsteady aerodynamic modeling of a fighter wing in transonic flow. AIAA Paper No. 84-1566, AIAA 17th Fluid Dynamics, Plasmadynamics, and Lasers Conference, June 25–27, 1984.
4. V. Shankar, H. Ide, J. Gorski, and S. Osher. A fast, time-accurate unsteady full potential scheme. AIAA Paper No. 85-1512-CP, AIAA 6th Computational Fluid Dynamics Conference, Cincinnati, July 15–17, 1985.
5. V. Shankar and H. Ide. Aeroelastic computations of flexible configurations. Symposium on Advances and Trends in Computational Structural Mechanics and Fluid Dynamics, Washington, D.C., October 17–19, 1988.
6. V. Shankar, H. Ide, and T. Goebel. Unsteady full potential computations for complex configurations. AIAA Paper No. 87-0110, AIAA 25th Aerospace Sciences Meeting, Reno, Nevada, January 12–15, 1987.

7. H. Ide and V. Shankar. Unsteady full potential aeroelastic computations for flexible configurations. AIAA Paper No. 87-1238, AIAA 19th Fluid Dynamics, Plasmadynamics, and Lasers Conference, Hawaii, June 8–10, 1987.

8. G.D. Miller and T.R. Logan. AFW technology wind tunnel test results from first tunnel entry. Rockwell Report NA/86/1399L, AFWAL Contract F33615-85-C-3209, CDRL 007, February, 1987.

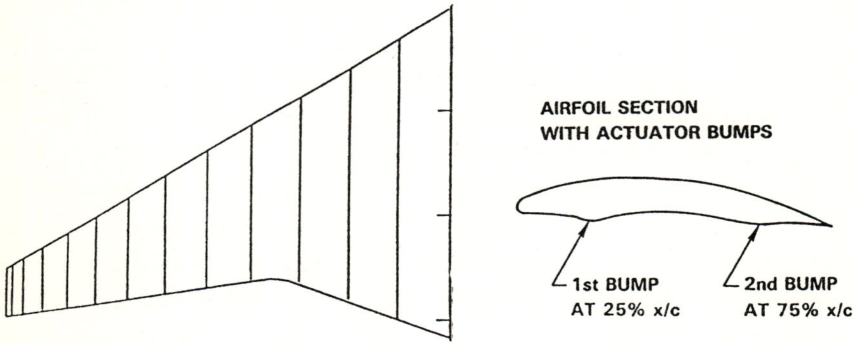

Fig. 1 Planform shape of a flexible fighter wing.

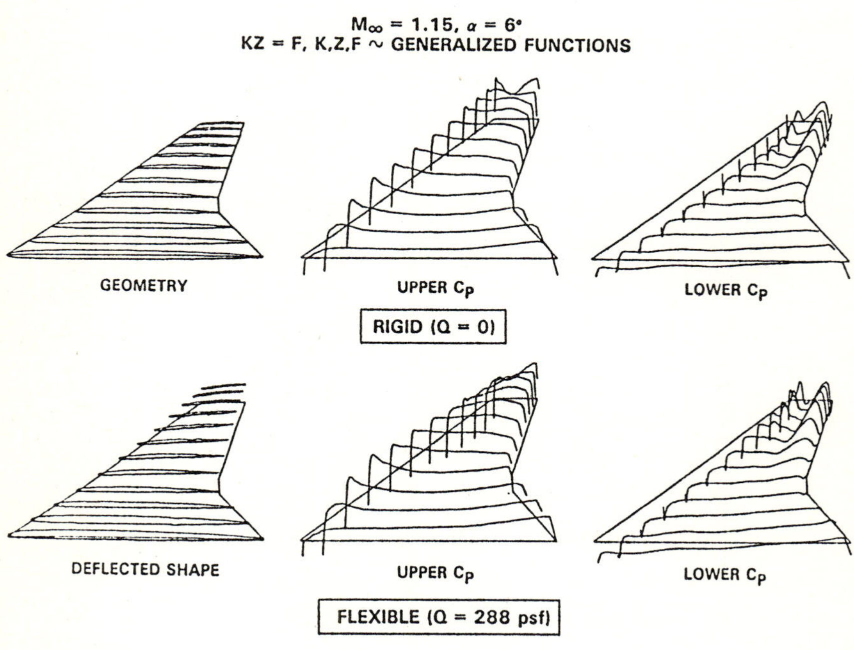

Fig. 2 Static flexible computation.

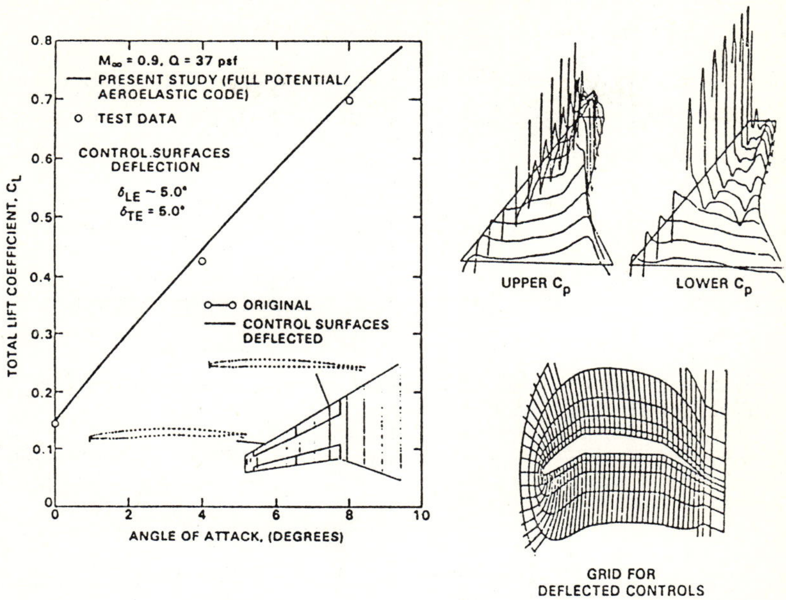

Fig. 3 C_L versus α correlation for a flexible wing with deflected controls.

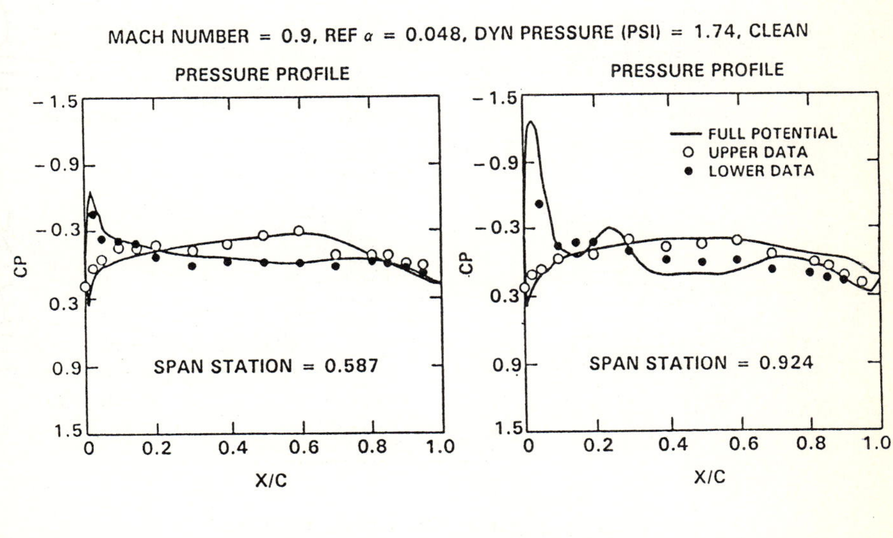

Fig. 4a Pressure distributions for a flexible wing (AFW) at $M_\infty = 0.9$, $\alpha = 0.048$, $Q = 1.74$ psi with undeflected controls

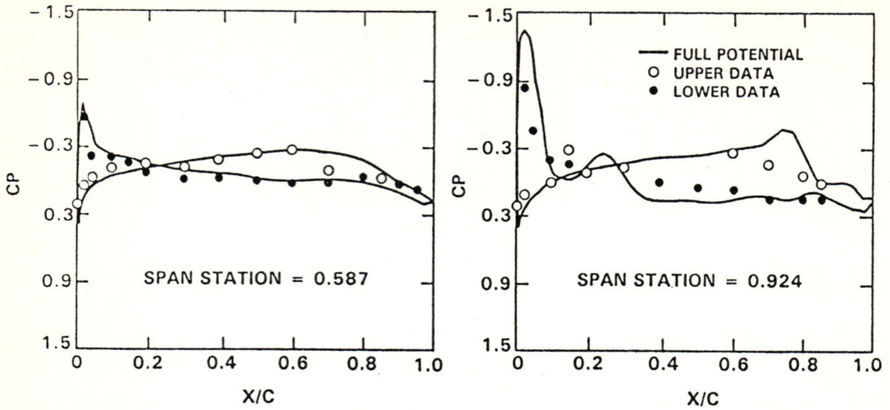

Fig. 4b Pressure distributions of AFW model for $M_\infty = 0.9$, $\alpha = 0.048$, $Q =$ 1.054 psi with $\delta_{\mathrm{TEO}} = 5.0°$ (downward).

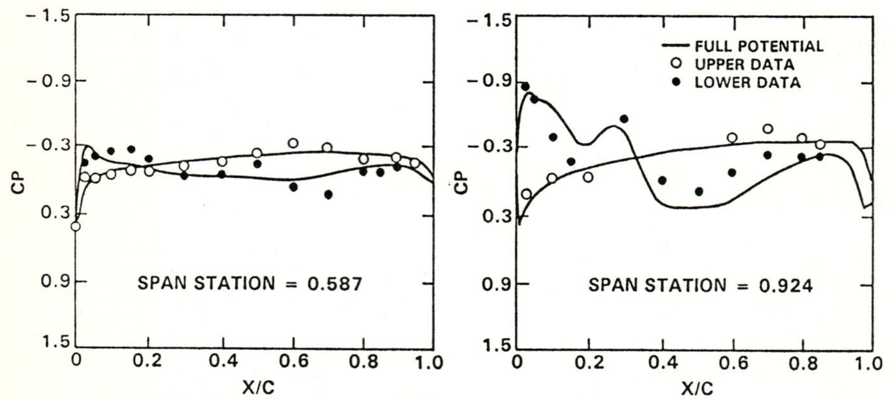

Fig. 5a Pressure distributions of AFW model for $M_\infty = 1.15$, $\alpha = 1.66$, $Q =$ 2.35 psi with undeflected controls.

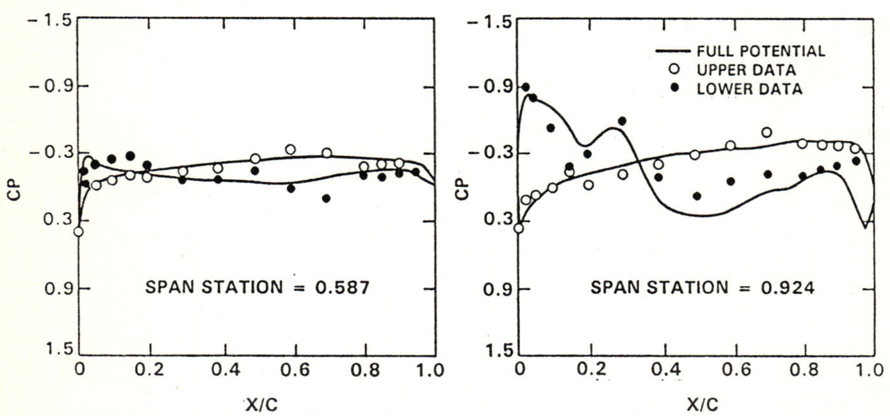

Fig. 5b Pressure distributions of AFW model for $M_\infty = 1.15$, $\alpha = 1.66$, $Q =$ 2.35 psi with $\delta_{\mathrm{TEO}} = 2.0$.

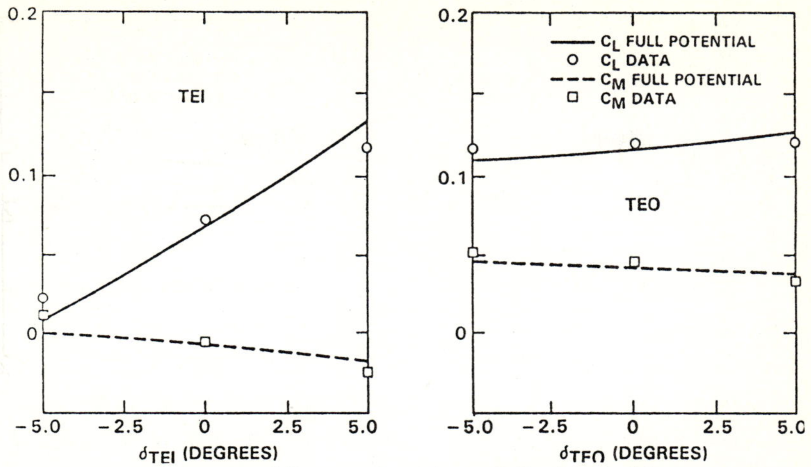

Fig. 6a C_L and C_M change with control surfaces deflections. $M_\infty = 0.9$, $\alpha = 0°$, $Q = 1.05$ psi.

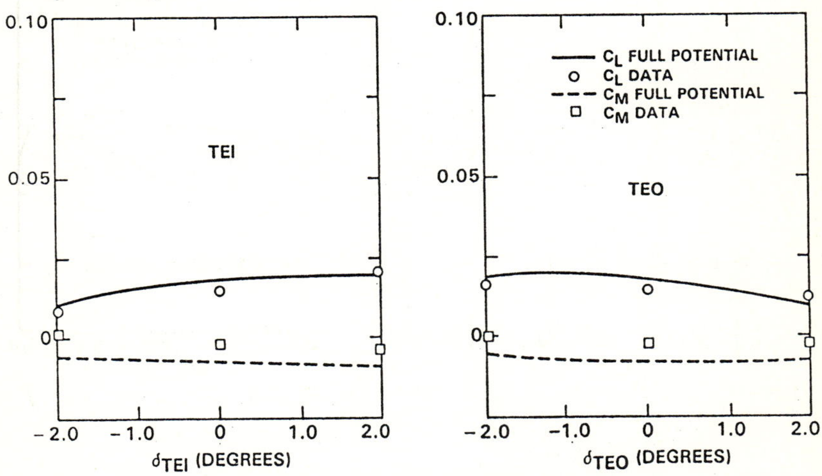

Fig. 6b C_L and C_M change with control surfaces deflections. $M_\infty = 1.15$, $\alpha = 0°$, $Q = 1.74$ psi.

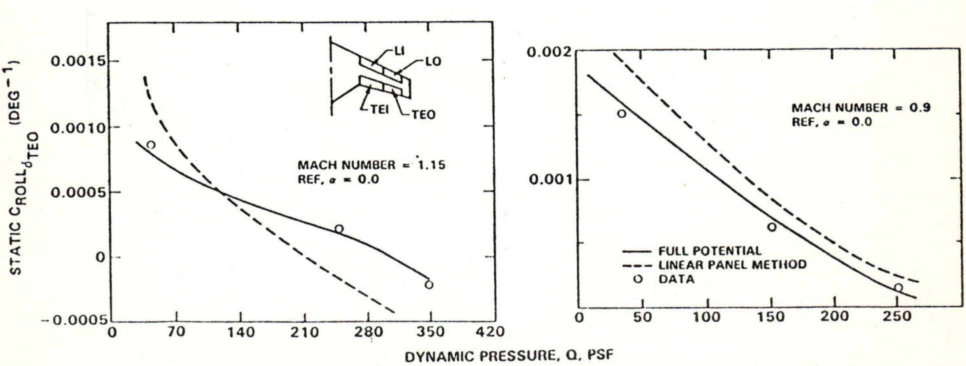

Fig. 7 Roll moment change with the dynamic pressure for two Mach numbers.

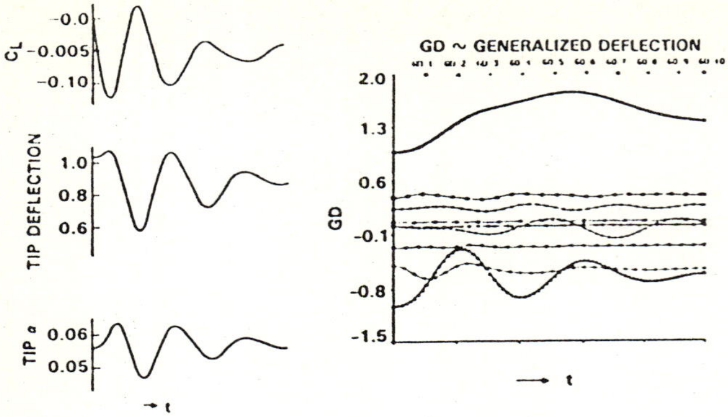

M∞ = 1.15, α = 0°, Q = 360 psf — BELOW FLUTTER

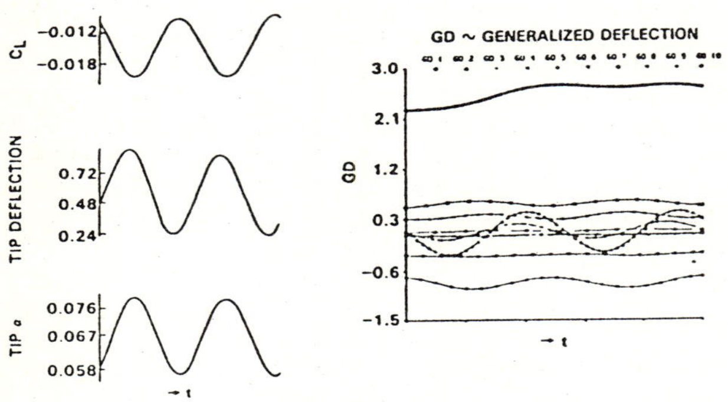

M∞ = 1.15, α = 0°, Q = 490 psf — FLUTTER POINT

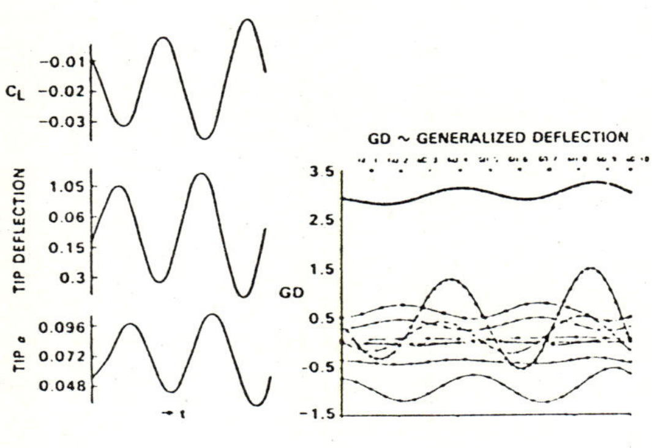

M∞ = 1.15, α = 0°, Q = 550 psf — ABOVE FLUTTER

Fig. 8 Dynamic flexible computations for different dynamic pressures: a) below flutter, b) at flutter point, and c) above flutter point.

Mechanism and Reynolds Number Dependence of Shock-Induced Buffet on Transonic Airfoils

E. Stanewsky and D. Basler

DFVLR
Institut für Experimentelle Strömungsmechanik
D-3400 Göttingen, FRG

Summary:
Tests were carried out at transonic speeds to investigate the effect of the Reynolds number on the buffet process. In these tests the airfoil model investigated was equipped with regular pressure orifices, dynamic pressure transducers and surface hot-film sensors. The unsteady flow field was observed by high-speed real-time holographic interferometry. The analysis of the results revealed that within the domain of heavy buffet the shock oscillation frequency decreased with increasing Reynolds number - in accordance with the effect of viscosity on the extent of the shock-induced separation bubble at steady conditions - while prior to buffet onset the shock oscillation frequency increased with Reynolds number. The investigation will be continued with the Reynolds number variation eventually being carried up to full-scale conditions.

1. INTRODUCTION

The flow about a transonic airfoil or wing is characterized at the design condition by an extended region of supersonic flow on the upper surface generally terminated by an isentropic recompression or a weak normal shock. At Mach numbers or angles of attack above the design point, shocks of increasing strength will appear and shock-induced separation will eventually develop leading to total separation, i.e., separation extending from the foot of the shock to the trailing edge, as Mach number or incidence is further raised. Once separation reaches the trailing edge, shock oscillations may ensue as a result of the interaction between trailing edge flow and shock location, a condition commonly referred to as airfoil or wing buffet, Figure 1.

The flow development about transonic airfoils or wings may be very sensitive to changes in the Reynolds number - or more general, to changes in the state and condition of the boundary layer upstream of the upper surface shock - especially in the presence of strong shock waves and separation [1] [2]. This is indicated in Figure 2 where the lift coefficient at buffet onset is shown as function of the Reynolds number in a Reynolds number range representative of wind tunnel and flight conditions. The knowledge of such a dependence is, of course, essential when predicting the flight performance of an aircraft from low Reynolds number wind tunnel

J. Zierep, H. Oertel (Eds.)
Symposium Transsonicum III
IUTAM Symposium Göttingen, 24.-27. 5. 1988

tests. Since only sparse information is available concerning the Reynolds number dependence of the actual buffet process, i.e., the effect of viscosity on shock oscillation frequency and amplitude, exploratory tests were performed in a limited Reynolds number range to determine this dependency; the investigation will eventually be carried up to full-scale flight Reynolds numbers.

2. THE EXPERIMENT

The supercritical airfoil CAST 7 / DOA1 was utilized for the present experiment. The flow about this airfoil is known to be sensitive to viscous changes and a large body of data is available for this airfoil at steady flow conditions [3] [4]. The airfoil model with a chord of c = 100 mm was instrumented as follows, Figure 3: surface pressure orifices to determine the average pressure distribution, surface flush-mounted dynamic pressure transducers to register the time- and space-dependent pressure fluctuations and surface hot-film sensors mainly utilized to detect transition and separation locations. Density distributions in the (unsteady) flow field and "flow visualization" were provided by a holographic high-speed real-time interferometer whose set-up relative to test section and model is illustrated in Figure 4; its main components are: a 4 W Argon Laser, an acousto-optic output coupler utilized to generate high intensity, short duration light pulses, a beam splitter for dividing the laser beam into an object and a reference beam, a mirror system to adjust the length of the reference beam, the holographic plate and the high speed movie camera. A detailed description of the system is given in Reference [5].

The tests were conducted in a transonic test section with slotted top and bottom walls, especially designed to fit an existing (hypersonic) Ludwieg-tube facility [5], which provided a run time of 0.3 s. Characteristic data of the set-up were: tunnel height to chord ratio $h/c = 3$, tunnel width to chord ratio $B/c = 1.5$, test section length $l = 1$ m, wall open area ratio $\tau = 3.2$ percent. The tunnel width to chord ratio selected was found to provide "two-dimensional" flow conditions over 50 percent of the model span. The variables of the investigation were the freestream Mach number, the angle of attack and the state and condition of the boundary layer varied by changing Reynolds number and, in some instances, by artificially fixing boundary layer transition.

3. ANALYSIS OF VISCOUS EFFECTS ON BUFFET

3.1 THE BUFFET MECHANISM

Figure 5 shows typical (time-averaged) pressure distributions for an increasing angle of attack going from a pre-buffet state to a condition beyond buffet onset. There are, in the present context, two characteristic features of these distributions to be considered: the trailing edge pressure decreases rapidly with increasing angle of attack while, at the same time, the shock moves swiftly upstream. Decreasing trailing edge pressure indicates that separation is developing at or has reached the trailing edge. As a consequence, conditions under which the flow leaves the trailing edge region are being altered - in essence a decambering of the airfoil takes place - which causes a reduction in circulation and a corresponding upstream shift of the shock. The close coupling between conditions at the trailing edge and shock location is believed to be part of the mechanism driving the shock oscillations responsible for buffet.

The latter is demonstrated for the actual buffet process in Figure 6 where the shock location and the boundary layer thickness at the trailing edge, normalized by the boundary layer thickness for attached flow conditions, are presented as function of time; both, shock location and boundary layer thickness were obtained from interferograms. One observes - without going into the details of disturbance propagation - that the most forward shock location corresponds closely to a state (in time) where the boundary layer at the trailing edge reaches its maximum thickness, while the most aft position of the shock is associated with an attached or nearly attached boundary layer at the trailing edge. The shock is close to its most rearward location very weak thus reducing the load on the boundary layer and correspondingly its overall thickness (also see Fig. 11 for the variation in the extent of the supersonic region during one cycle in the shock oscillation).

3.2 VISCOUS EFFECTS ON OSCILLATION FREQUENCY

Shock oscillation frequency shall be considered to show the effect of viscous conditions on the buffet process. This frequency is plotted in Figure 7 as function of the Reynolds number for various freestream Mach numbers and angles of attack well within the buffet domain. One observes that the (reduced) frequency decreases with increasing Reynolds number. This holds for both angles of incidence and the three Mach numbers considered. It is postulated - and some evidence will be presented below - that the reduction in frequency is mainly due to the following development: The boundary layer thickness (or momentum thickness) upstream of the shock reduces with increasing Reynolds number. This slows down the development of shock-induced separation, i.e., the propagation speed of shock generated disturbances decreases, resulting in the observed reduction in shock oscillation frequency.

The data points for free and fixed transition at a Mach number of M_∞ = 0.760 and a Reynolds number of Re = 6×10^6 do not seem to support this postulation since the boundary layer for fixed transition is likely to be much thicker than for free transition, still the buffet frequency is noticeably lower. Here, pressure distributions corresponding to these data points may provide an explanation, Figure 8: In the case of free transition the shock is located further downstream thus reducing the distance between shock and trailing edge - hence the higher frequency. Taking the shock location into account by basing the reduced frequency on the distance between the (average) shock location and the trailing edge, it can be seen, Figure 9, that the thicker boundary layer upstream of the shock is indeed associated with the higher frequency which is further evidence in support of the mechanism described above, by which viscosity manifests its influence on buffet.

To demonstrate the dependence of the development of the shock-induced separation bubble on viscous parameters upstream of the shock, results obtained in boundary layer measurements on two transonic airfoils are depicted in Figure 10 in the form of the extent of the separation bubble, normalized by the momentum thickness immediately upstream of the shock, as function of the shock-upstream Mach number M_1, the latter representative of the shock strength [1]. One observes, Figure 10a, that separation develops noticeably slower with increasing shock-upstream Mach number for the thinner boundary layer, here obtained by fixing transition relatively far aft. The boundary layer parameter identified as actually dominating this development is the initial momentum thickness, θ_1, Figure 10b. (Also note the independence of the onset of separation on viscous conditions.)

The parameter of the outer inviscid flow dominating the development of shock-induced separation is the Mach number immediately upstream of the shock or the shock strength. The relation between shock strength, represented by the height of the shock above the airfoil surface, shock location and boundary layer thickness at the trailing edge is depicted in Figure 11 for one shock oscillation cycle. During the upstream movement, shock strength increases initially and a thickening of the boundary layer at the trailing edge can be observed. At one stage in this forward movement, shock strength starts to decline while the trailing edge boundary layer thickness continues to grow rapidly to a maximum nearly coinciding with the most forward shock location; shock strength continues to decrease during the downstream movement of the shock until the boundary layer is fully reattached. The present plot, besides demonstrating the role of the shock-upstream Mach number or shock strength within the buffet process, provides a good indication of the time-delay between the generation of a disturbance at the shock and its arrival at the trailing edge.

3.3 FREQUENCY BEHAVIOR IN THE VICINITY OF THE BUFFET BOUNDARY

The preceding discussion concerned freestream conditions well within the buffet domain ($\alpha = 4°$ and $5°$). At an angle of attack of $\alpha = 3°$,

Figure 12, one is close to the buffet boundary and one may, for instance dependent on Mach or Reynolds number, face a pre- or post-buffet-onset state. One observes in Fig. 12 that at the Mach number of $M_\infty = 0.78$, the reduced frequency decreases with increasing Reynolds number, similar to the behavior at the higher incidences, while at freestream Mach numbers of $M_\infty = 0.760$ and 0.740, the trend reverses: the reduced frequency increases with Reynolds number. Consulting the right-hand diagram of Fig. 12 and using the criterion $c_{pTE} \leq 0.05$ to indicate the buffet boundary (c_{pTE} = trailing edge pressure coefficient), it seems that at least at $M_\infty = 0.740$ a pre-buffet state prevails while at $M_\infty = 0.780$ one is well within the buffet domain; at the freestream Mach number of $M_\infty = 0.760$, the situation is marginal. Here, a plot of the angle of attack dependence of the shock oscillation frequency for this Mach number is more instructive: at $\alpha = 3°$ and a Reynolds number of $Re = 6x10^6$ one is still within the buffet domain, while at the higher Reynolds number the buffet boundary has not yet been passed, indicative of the Reynolds number dependence of buffet onset.

It is, at present, not quite certain, why the shock oscillation frequency increases with Reynolds number at pre-buffet conditions since here regions of (shock-induced) separation and attached flow may be present on the airfoil simultaneously. One cause is, nevertheless, likely to be related to the dependence of the convection velocity of disturbances on the average velocity in the viscous layer downstream of the shock which increases with Reynolds number due to fuller boundary layer profiles and a reduction in the extent of separated regions.

The relation between the shock oscillation frequency, the amplitude of the shock movement and the shock location relative to the buffet boundary is shown in Figure 13. Two characteristics should be noted:

1. Full buffet conditions are reached only after the shock has moved upstream from its most rearward position by almost 10 % and the trailing edge pressure coefficient has dropped to $c_{pTE} = -0.05$.
2. The amplitude of the shock movement decreases with increasing frequency; the amplitude seems to be determined mainly by the shock location on the airfoil and the change in shock location for a given disturbance (corresponding to a given $\Delta\alpha$) at the trailing edge.

4. CONCLUSION AND FUTURE TESTS

Exploratory tests were carried out to determine the influence of the Reynolds number on buffet. It was found that at fully separated conditions, buffet frequency decreases with increasing Reynolds number or decreasing initial boundary layer thickness mainly due to the effect of the latter on the development of shock-induced separation. At pre-buffet conditions, shock oscillation frequency increases with Reynolds number which is believed to be a result of the fuller boundary layer profiles downstream of the shock, hence greater disturbance convection velocities, associated with the higher Reynolds number. Future tests, to be carried out in a high Reynolds number transonic cryogenic Ludwieg-tube facility, shall establish details of the mechanism driving the shock oscillation, identify boundary layer parameters of dominant influence on buffet and establish the Reynolds number dependence of buffet up to flight conditions.

REFERENCES

1. Laster, M.L. (Editor): Boundary Layer Simulation and Control in Wind Tunnels. AGARD-AR-224, April 1988.
2. Elsenaar, A.; Binion, T.W., Jr. and Stanewsky, E.: Reynolds Number Effects in Transonic Flow. AGARDograph to be published 1988.
3. Barche, J. (Editor): Experimental Data Base for Computer Program Assessment. AGARD-AR-138, 1979.
4. Elsenaar, A. (Chairman): Two-dimensional Transonic Testing Methods. Final Report of the GARTEUR Action Group AD (AG-02), GARTEUR/TP-011, July 1981.
5. Basler, D.: Experimental Investigation of the Convection of Shock-induced Disturbances on Transonic Airfoils. DFVLR-FB 87-28, July 1987.

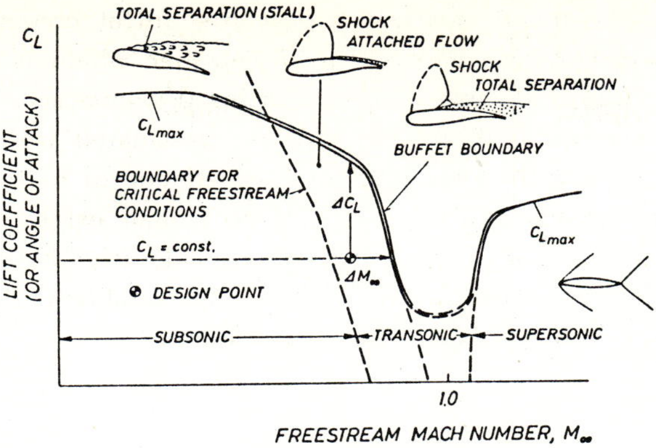

Fig. 1: The buffet boundary

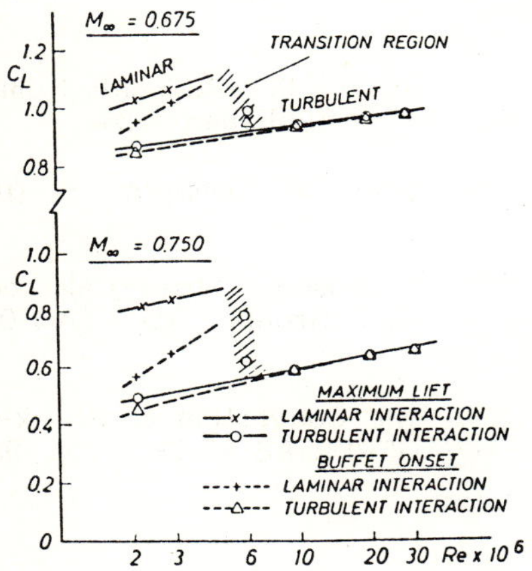

Fig. 2: Reynolds number dependence of maximum lift and buffet onset

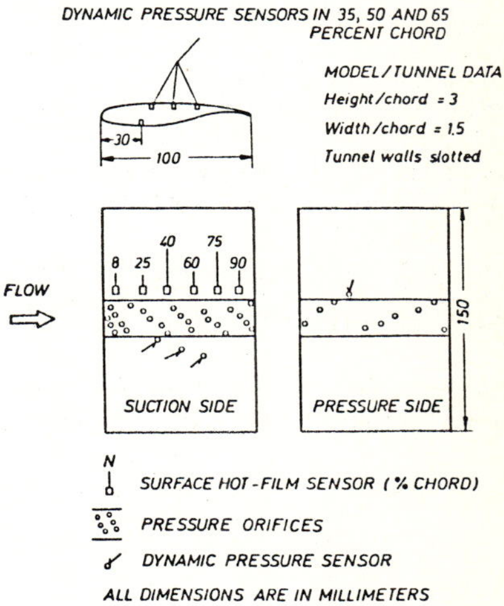

Fig. 3: Airfoil CAST 7 / DOA1: Model instrumentation

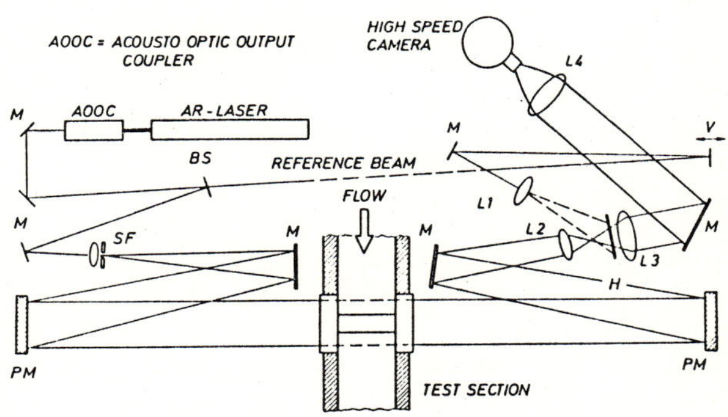

Fig. 4: The holographic high-speed real-time interferometer

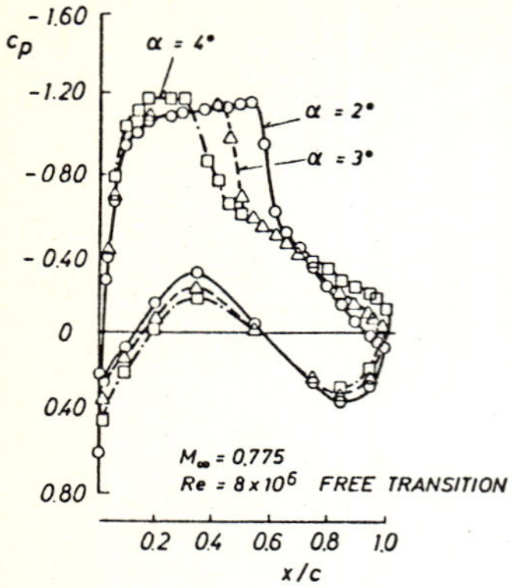

Fig. 5: Pressure distributions for pre- and post-buffet onset conditions

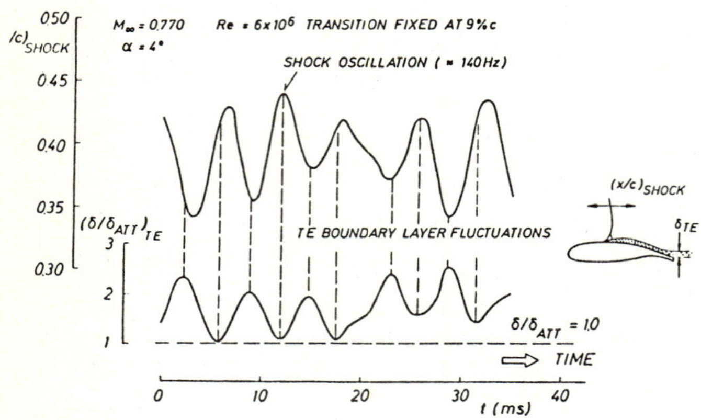

Fig. 6: Shock oscillation and trailing edge boundary layer fluctuations

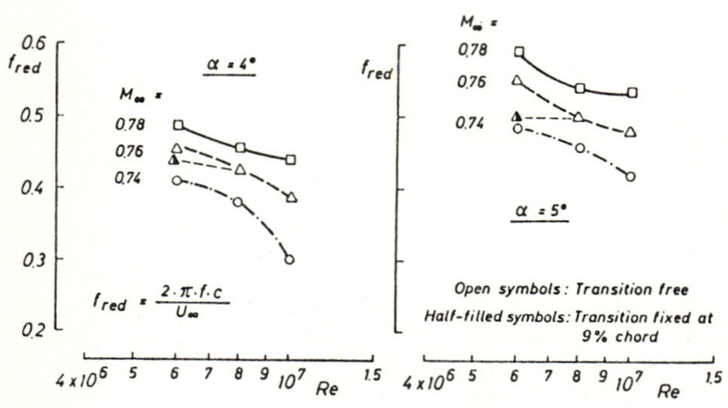

Fig. 7: Dependence of the reduced shock oscillation frequency, f_{red}, on Reynolds number

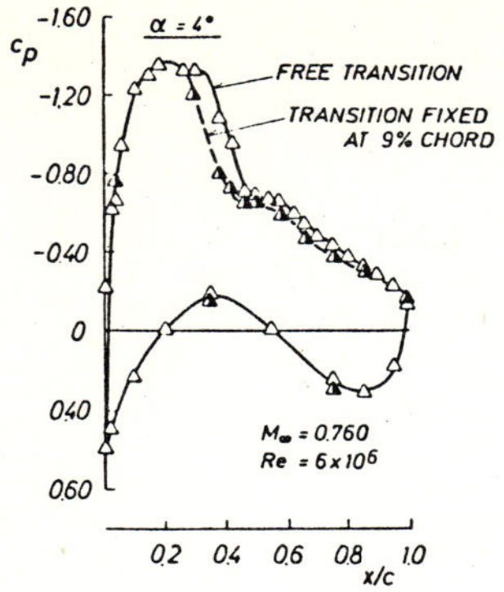

Fig. 8: Effect of early transition on the time-averaged pressure distribution

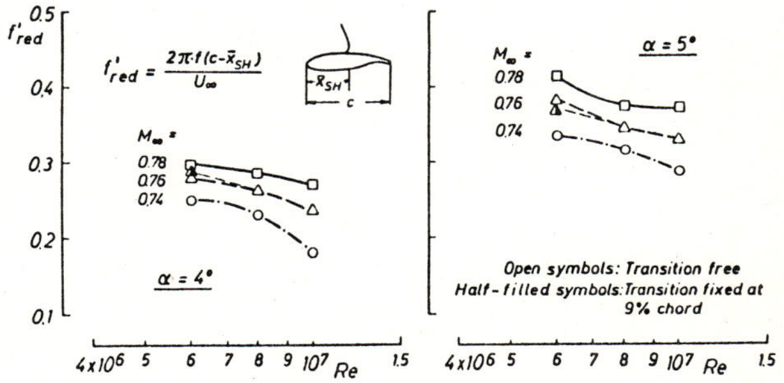

Fig. 9: Dependence of the reduced shock oscillation frequency, f'_{red}, on Reynolds number

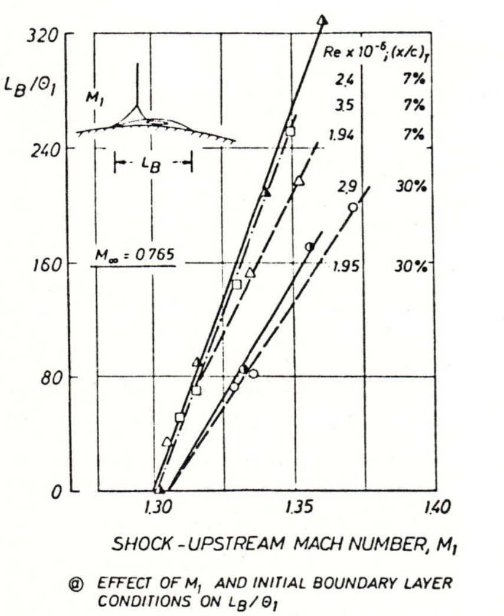

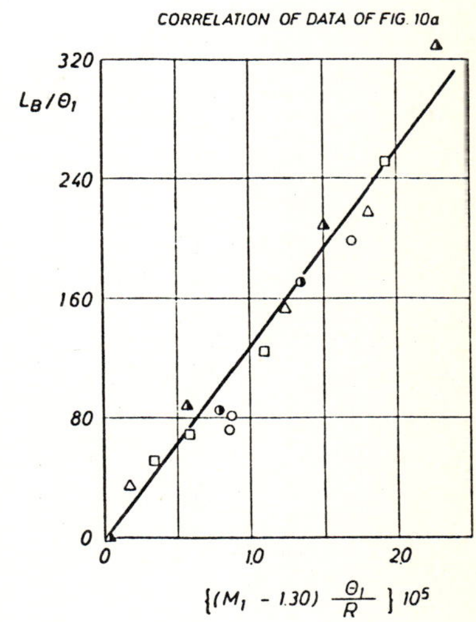

Fig. 10a: Extent of the shock-induced separation bubble

Fig. 10b: Concluded
R = Average radius of upper surface contour between shock and trailing edge

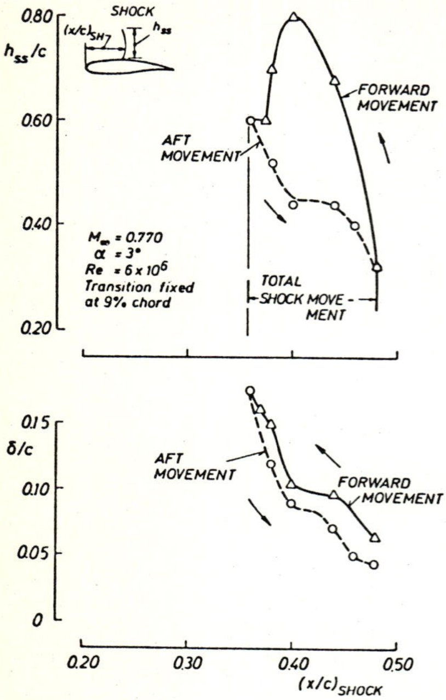

Fig. 11: Shock movement and shock strength

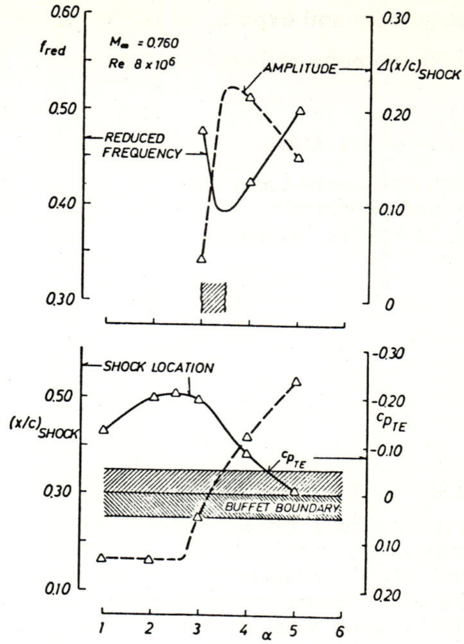

Fig. 13: Relation between frequency, amplitude and shock location relative to the buffet boundary

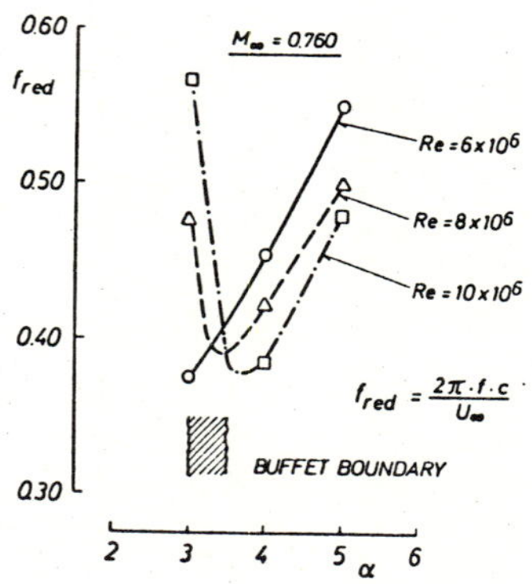

Fig. 12: Effect of angle of attack on shock oscillation frequency, f_{red}

Computational and Experimental Research on Buffet Phenomena of Transonic Airfoils

N. Hirose and H. Miwa

National Aerospace Laboratory
7-44-1 Jindaiji-Higashi
Chofu, Tokyo 128, JAPAN

Abstract

Buffet phenomena on supercritical airfoils have been investigated both experimentally and numerically. The experiments were performed for NACA0012 and KORN 75-06-12 airfoils using high-speed Schlieren VTR and pressure transducers. The analysis shows significant effects of the geometries and the Reynolds number. A 2-D N-S code was used to predict these phenomena. The result shows the macro-scale effects of airfoil geometry, Reynolds number, and transition. A fine strucure of the motion such as the periodic trailing edge pressure history was captured. The basic frequency of the buffet motion agrees well with the experiment.

1. Introduction

Buffet phenomena in transonic flow are characterized by a very complicated flow regime with shock wave oscillation, shock-wave induced boundary layer separation and separated vortex shedding behind the shock wave. This unsteady aerodynamic oscillation produces large unsteady aerodynamic forces which are not desirable. Therefore, the prediction and investigation in detail of these phenomena are quite important.

Experimental investigation is very difficult because heavy unsteady loading on the model and support is dangerous, and high quality measurements of unsteady pressure and flow field are required. A few research works have been reported in the past. McDevitt, Levy and Deiwert [1-3] made a biconvex circular-arc airfoil experiment and compared it with Navier-Stokes computations. By the way, Deiwert's computation was the first Navier-Stokes analysis for practical applications at high Reynolds number and computed on ILLIAC-IV. Levy and Bailey [4] then computed the buffet boundary for the NACA65-213 and GARABEDIAN-KORN airfoils. Roos [5] made a buffet experiment for Whitcomb and NACA0012 airfoils and investigated the unsteady pressure field.

In Japan, supercritical airfoils have been tested in the High-Reynolds Number 2-D Wind Tunnel (HR2DWT) since its construction in 1978 [6]. Also, a Navier-Stokes code was developed and code verification was made for the NACA0012 and several supercritical airfoils [7]. For one practical application airfoil, the buffet boundary as well as static force data showed good agreement [8]. Computation gave a buffet frequency of 45*Hz* while the energy power spectra measured in the wind tunnel showed a distinct peak at about 60*Hz* nearly equal to the computed result.

In the present paper, a further experimental and computational investigation was made on the buffet phenomena of NACA0012 and the GARABEDIAN-KORN 75-06-12 airfoils. In the experiment, effects of Reynolds number on conventional and supercritical airfoils were examined. In the computations, the effects of Reynolds number, angle of attack, and transition models on buffet mode were examined. Also, comparison with experiment was made for the KORN airfoil.

2. Experimental Research

2.1 Wind Tunnel and Procedures

Experiments were conducted in the NAL HR2DWT, a blow-down flow facility for airfoil testing. The test section is 30*cm* wide and 100*cm* high. A 25*cm*-chord airfoil model is installed between the optical windows and supported by the rigid steel blocks outside the flow passage (Fig.1). The wind tunnel provides a maximum Reynolds number of $40 \cdot 10^6$ (at $M = 0.8$), a Mach number range from 0.2 to 1.15 and a running time of 9 to 100 sec.

Airfoil sections used were the NACA0012 and the GARABEDIAN-KORN 75-06-12. Mach number range was 0.5 to 0.85, Reynolds numbers were 6, 15 and $30 \cdot 10^6$. Pressure fluctuations at every 10% chord station were measured by small pressure transducers (KULITE XCQ-080) connected by an adapter to the pneumatic connector located aside of the model. Static pressure holes of diameter 0.6*mm* were utilized for this purpose. A total of 10 points was measured (Fig.2). The transducer signals were recorded by a 14 channel analog data recorder. The pressure fluctuation at one wake point and one side-wall point above the trailing edge and the strain at the model's support section were also measured. The frequency response between the pressure hole and the transducer was checked beforehand. A maximum frequency of 200*Hz* can be measured with an accuracy of 90%. The R.M.S. level of the pressure fluctuation normalized to the dynamic pressure of the free stream, power spectra, and the cross-correlations with the trailing edge at the reference point were computed to see the time-space correlation of the separated vortices.

A high-speed camera and a high-speed Video Tape Recorder (200 frames/sec.) were also used for the optical measurements of shock wave movement and flow separation patterns.

2.2 Results and Discussion

(Shock Wave Location) Shock wave location obtained from VTR images is plotted against angle of attack with Mach number M as parameter for the NACA0012 and KORN airfoils, respectively, in Figs.3 (a) and (b). The solid line represents the Reynolds number , $Re = 30 \cdot 10^6$ and the dashed line $Re = 6 \cdot 10^6$. The shaded area shows unsteady shock wave oscillation, i.e. buffet area. At full buffet point, the amplitude of the shock wave movement increases with increasing Reynolds number for both airfoils. The increment is greater for the NACA0012 than for the KORN airfoil. The Mach number dependence is different for the two airfoils. In the case of the NACA0012, the amplitude of oscillation increases with Mach number while it decreases in the case of the KORN airfoil. The largest oscillation occurs on the supercritical airfoil at $M = 0.7$.

At high Reynolds numbers, the shock wave is located rearward for both airfoils when the angle of attack is below the point of buffet onset but the situation reverses when buffet starts in the case of the NACA0012 (Fig.4 a & b).

(Strain and Trailing Edge Pressure Fluctuation) The strain at the model's support section and the trailing edge pressure fluctuation C_{PTE}' are closely related. The start of model vibration, i.e. buffet onset, corresponds with the start of C_{PTE}'. Significant buffet exists for the whole range of Mach numbers in the NACA0012 case but it almost disappears when the Mach number is increased to 0.85 in the KORN case.

(Presurre Fluctuation on Airfoil Surface) Fig.5 shows the pressure fluctuation distribution at $Re = 30 \cdot 10^6$ with angle of attack as a parameter for both airfoils. The shock wave location is also shown. Pressure fluctuation peak point and shock location should coincide but they differ at the angle of attack required for buffet onset. This was caused by the shock wave location optically measured along a line of 10*mm* above the surface. A local minimum point of fluctuation can be observed just behind the shock wave which indicates the existence of a separation bubble.

For low Reynolds number, the pressure fluctuation is larger in the KORN case than in the NACA0012 case. As the Reynolds number increases, the pressure fluctuation decreases more in the KORN case. This can be clarified by the trailing edge pressure fluctuation C_{PTE}' shown in Fig.6. Distinctive difference between the two airfoils is observed.

(Power Spectra and Cross-Correlations) Fig.7 shows power spectra and cross-correlations for the NACA0012 airfoil at $M = 0.70$, $\alpha = 6°$, and $Re = 15 \cdot 10^6$. At low Reynolds numbers, the peak of the power spectra lies at very low frequency (2.5*Hz*). As the Reynolds number increases, the peak position moves rearward and the peak frequency increases to 62.5*Hz*. The spectrum becomes broad banded with various frequency components which means that various sizes of vortices exist. The time lag of the peak position of the cross-correlation is negative. This indicates that separated vortices move streamwise from the foot of the vortices. Correlation is high for high Reynolds numbers.

Power spectra and cross-correlations for the KORN airfoil are shown in Fig.8 at $M = 0.75$, $\alpha = 4°$, and $Re = 30 \cdot 10^6$. At low Reynolds numbers, the peak of the power spectra lies at very low frequency (5*Hz*) as in the NACA0012 case but with broad-band both in frequency and space. As the Reynolds number increases, a sharp peak of mono-frequency (60*Hz*) is observed and periodic cross-correlation is obtained as shown in the corresponding figure. The time history of the pressure fluctuation in Fig.9 shows that periodically organized separated vortices are formed behind the shock wave when the Reynolds number is high. Fig.10 shows this vortex shedding; the Strouhal number n is about $0.06 \sim 0.08$, depending on the angle of attack, being nearly independent of Reynolds number.

3. Computational Research

3.1 Navier-Stokes Code

A two-dimensional code based on an Implicit Approximate Factorization scheme for time-averaged thin-layer Navier-Stokes equations was used in the computational research [7]. The turbulence model applied is the conventional Baldwin-Lomax algebraic model [9]. The details of the numerical scheme and the turbulence model can be found in the references. For the computation of flow with laminar to turbulent transition (termed as "transitional flow"), Baldwin's criterion of transition: *max.* $\mu_t < C_{MUTM} = 14$, was employed. A "fully turbulent" computation was done assuming that the flow was turbulent from the leading edge to both sides. A relaxation mode of eddy viscosity is also employed in the wake region.

Computing and Flow Conditions

Computations were carried out for the NACA0012 airfoil to see the effects of Reynolds number, angle of attack, and transition models on buffet. A C-type mesh of 133*x*49 points covering the computing region's size of 10 chord lengths of the airfoil in each direction was used. 93 points lie over the airfoil. The minimum normal size of the mesh is $1 \cdot 10^{-5}$, the streamwise size at the leading edge is 0.007.

The flow conditions were $M = 0.75$, $\alpha = 0 - 6^{\circ}$, $Re = 1, 10, 50 \cdot 10^{6}$. Fully turbulent flow and transitional flow were compared.

KORN airfoil computations were made using a similar mesh with a leading edge streamwise width of 0.0025 to cover the subtle geometry of the supercritical airfoil. Flows at $Re = 6 \cdot 10^{6}$ and transitional models were computed for $M = 0.60 - 0.95$ and angles of attack to cover the buffet region.

3.3 Results and Discussion

NACA0012 Analysis

(Lift Characteristics) At low Reynolds numbers the transition point is located in front of the shock wave at midchord and this strongly affects pressure and friction force distributions as well as lift characteristics, as compared with the fully turbulent case. Fig.11 shows C_l vs. α at $Re = 1 \cdot 10^{6}$. Transitional flow (△) gives higher lift at the nonlinear range, $\alpha \geq 2^{\circ}$, than fully turbulent flow (○). The buffet onset point is the same at $\alpha = 5^{\circ}$, but the range of C_l oscillations differs. Fig.12 compares Mach number contours at fully turbulent (a) and transitional (b) cases at $\alpha = 4^{\circ}$. The difference in boundary layer development and shock position can be seen easily. At $Re = 10 \cdot 10^{6}$, the transition point is located before 0.04*C*. Therefore, the boundary layer development differs little from the fully turbulent case. The C_l vs α curves almost coincide with each other. The buffet onset point is 6°.

(Unsteady Force Characteristics) Unsteady aerodynamic forces were plotted against time in Fig.13. The symbols ○ . □ and △ represent C_l, C_d, and C_m, respectively. (a) is the case: $\alpha = 5°$, fully turbulent, and $Re = 1 \cdot 10^6$. Only one parameter was changed in the other cases. (b): transitional, (c): $Re = 50 \cdot 10^6$, (d): $\alpha = 6°$, transitional. As opposed to the conclusion from the previous result [8], it was found that the periodic oscillation curves are not simply sinusoidal. At low Reynolds numbers where transition effects are significant, the curve is complicated.

Although a more detailed study is required, the obtained result roughly shows that:

as Re increases	$\Rightarrow T_P$ increases	(T_P :basic period)
fully turb. → trans.	$\Rightarrow T_P$ increases	
as α increases	$\Rightarrow T_P$ decreases	
as Re increases	$\Rightarrow \alpha_{\text{buffet onset}}$ increases and ΔC_l , ΔC_d decreases	
fully turb. → trans.	$\Rightarrow \Delta C_l$, ΔC_d decreases	

Typical $T_P = 22$ ($\alpha = 5°$, fully turb.) to 31 ($\alpha = 5°$, trans.)
4-5 times longer than in supercritical airfoil case ($T_P = 5.54$) [8]

(Pressure Fluctuation caused by Transition) When the transition point is located at mid-chord in low Reynolds number flows, pressure fluctuation starts to appear at $\alpha = 4°$ before buffet onset can be observed at small C_l oscillation. In high Reynolds number flows, it disappears. To confirm that this is a physical and not a numerical oscillation, the transition point X_{tr} as well as C_{PTE}, C_d, and C_p at $x_{lower} = 0.19$, and x_{upper} , $x_{lower} = 0.75$ were plotted against time in Fig.14 ($\alpha = 4°$, transitional). X_{tr} on the lower surface moves from 0.40 to 0.56C with period $p_{short} = 0.47$. C_{PTE} and C_d synchronously oscillate with the same period. C_p at $x_{lower} = 0.19$ where the flow is laminar and C_p at $x_{lower} = 0.75$ where it is turbulent, are also synchronized, and the amplitudes show large differences. C_p also has a long period mode oscillation of $p_{long} = 2.89$. Such unsteadiness appears on the supercritical lower surface at $\alpha \geq 1°$ and cannot be found in the supercritical upper surface and high Reynolds number flow in which transition occurs near the leading edge.

When buffet starts, such small fluctuation is covered by large scale vortex shedding. Trailing edge pressure, however, seems to retain some influence. Fig.15 shows C_{PTE} for fully turbulent and transitional cases. High frequency modes appear in the latter case. A supercritical airfoil case is also shown for comparison.

(Vortex Shedding Pattern) Time-dependent vortex shedding was examined by plotting velocity vectors and variable contours at discretized phase angles. The scale of vortex and pressure fluctuation was greater in fully turbulent flow than in transitional flow.

The vortex at the foot of the shock wave grows larger as the shock wave moves forward until the separated region is large enough to stop the shock wave movement. Just after the shock wave stops and begins to move backward, the vortex leaves the airfoil floating down the stream. C_{PTE} reaches a minimum and oscillates. After the vortex is shed, the flow behind the shock wave becomes attached and the shock wave is pulled back. C_{PTE} regains positive pressure.

3.3.2 KORN Airfoil Analysis

(Lift Characteristics) Fig.16 (a) shows lift characteristics vs α with Mach number as a parameter. At the buffet region, maximum and minimum C_l are plotted to represent the amplitude of buffet. This amplitude nearly corresponds to the width of the shock wave because lift is mostly created by the supercritical region in front of the shock wave. The amplitude is larger at low Mach number and almost dies out when $M \geq 0.85$. Buffet disappears at $M = 0.90$. Buffet range in terms of angle of attack is plotted against Mach number in Fig.16 (b). This result is in excellent agreement with the experimental result.

(Buffet Patterns) Unsteady forces were plotted against time in Fig.17 with typical Mach contours at several Mach numbers. The buffet pattern and the shock wave shape become quite different as the Mach number increases. At low Mach numbers, buffet is similar to the low speed stall phenomena. Large scale vortices appear over the entire upper surface. The turbulence model for the separated region greatly influences the solution and the present results only give the qualitative pattern of separation with too large eddy viscosity. At $M = 0.75$, the shock wave is located on the relatively flat mid-chord surface. Shock-induced separation does not extend that much. The oscillation is simply sinusoidal as in the previous supercritical airfoil case. For higher Mach numbers, a fish-tail shock wave is located near the trailing edge and buffet disappears.

When compared in detail, the shock wave is wider in computation than in experiment. Present computations have poor resolution for shock waves. The turbulence model has a large effect. In the experiment it is difficult to find the farthest edges of the shock waves. The transition point is not clear. These and other factors should be examined in the future.

4. Concluding Remarks

Differences in various aspects were observed between NACA0012 and the KORN airfoils in experimental research. The airfoil geometry and the resultant pressure distribution are the reasons for differences. In addition, supercritical airfoils give more favorable buffet characteristics than conventional ones. Computations revealed other aspects of buffet phenomena. Macro-scale characteristics such as buffet boundary and shock wave location agreed well with experiment. Unsteady flow characteristics, however, exhibit both agreement and disagreement. For example, the effect of Reynolds number on periodicity on the NACA0012 is opposite. Two factors influencing the computed result are: evaluation of the transition point and of the model for the separated region. A refined model and a fine mesh analysis should be done in the future. An advanced experiment is scheduled in order to attain good two-dimensionality and direct pressure measurements on the surface.

Acknowledgement

The authors express their thanks to Division Chief Mr. K. Takashima, Dr. I. Kawamoto, Messrs. Y. Oguni, M. Sato, H. Kanda, and N. Kawai, and J. Miyakawa

(MHI) and Prof. R. Kawamura and Mr. Y. Matsumoto (Nihon Univ.) for their contributions in experiments, computations, and helpful discussions.

References

[1] McDevitt, J.B., Levy jr., L.L. and Deiwert, G.S., AIAA J.14, 606-613, (1976).
[2] Levy jr., L.L., AIAA J.16, 564-572, (1978).
[3] Seegmiller, H.L., Marvin, J.G. and Levy jr., L.L., AIAA J.16, 1262-1269, (1978).
[4] Levy jr., L.L. and Bailey, H.E., AIAA J.19, 1488-1490, (1981).
[5] Roos, F.W., J. Aircraft 17, 781-788, (1980).
[6] Takashima, K., ICAS-Paper #82-5-4-4, (1982).
[7] Kawai, N. and Hirose, N., NAL-TR-816, (1984).
[8] Miyakawa, J., Hirose, N. and Kawai, N., AIAA Paper 85-5025, (1985).
[9] Baldwin, B.S. and Lomax, H., AIAA Paper 78-257, (1978).

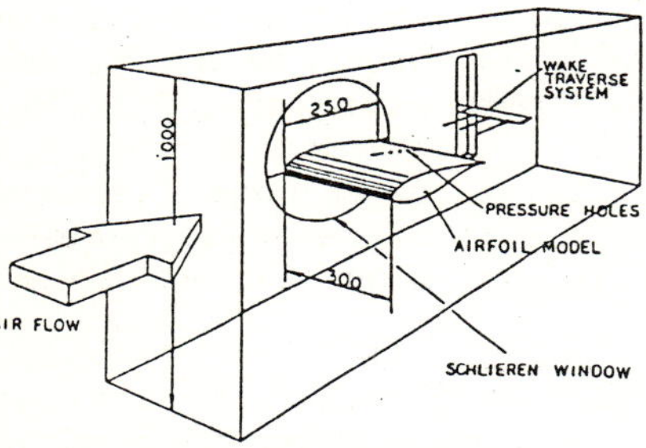

Fig. 1 HR2DWT(schematic)

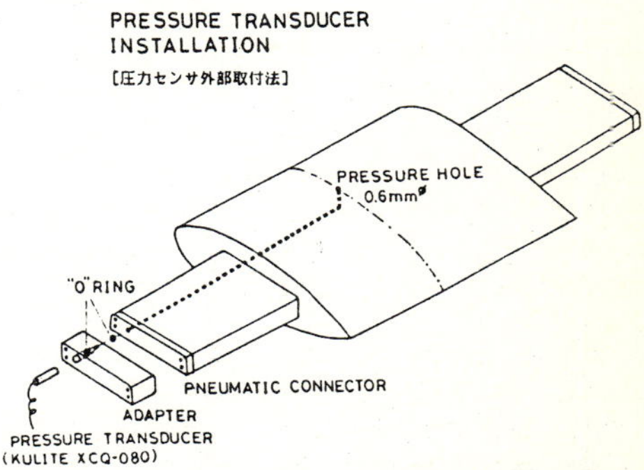

Fig. 2 Model and pressure transducer installation

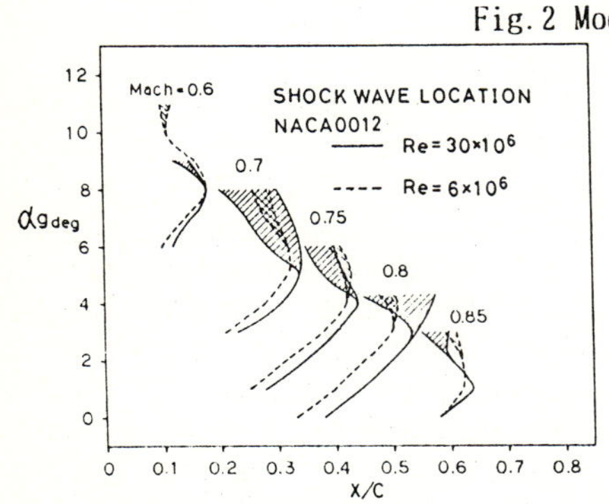

(a) NACA0012 airfoil

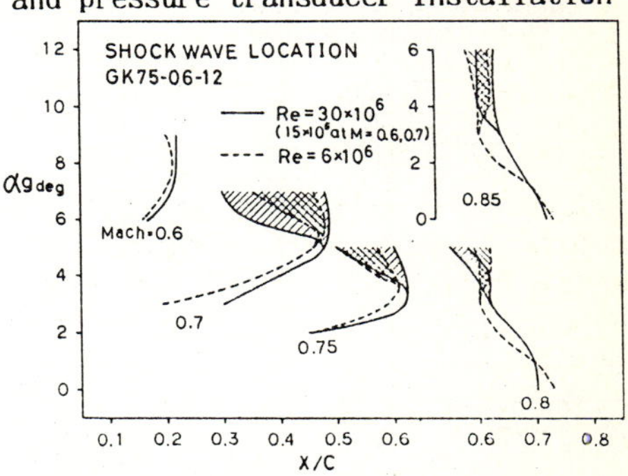

(b) KORN airfoil

Fig. 3 Shock wave locations

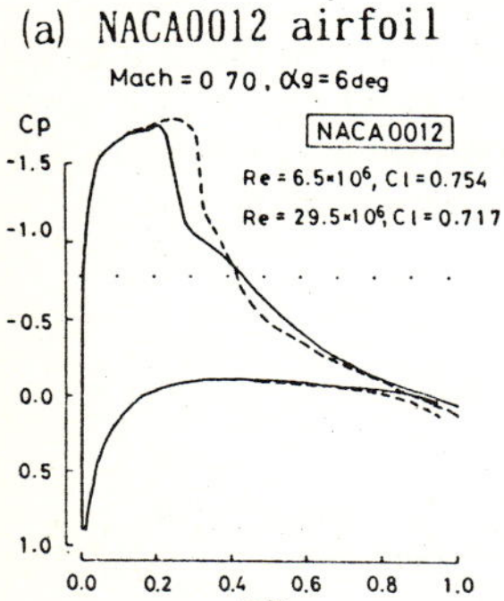

(a) NACA0012 airfoil, M=0.70, α=6°

(b) KORN airfoil, M=0.75, α=4°

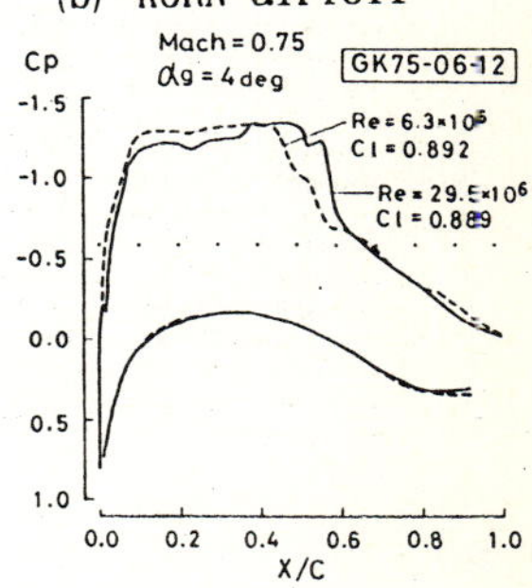

Fig. 4 Surface pressure distributions at buffet range

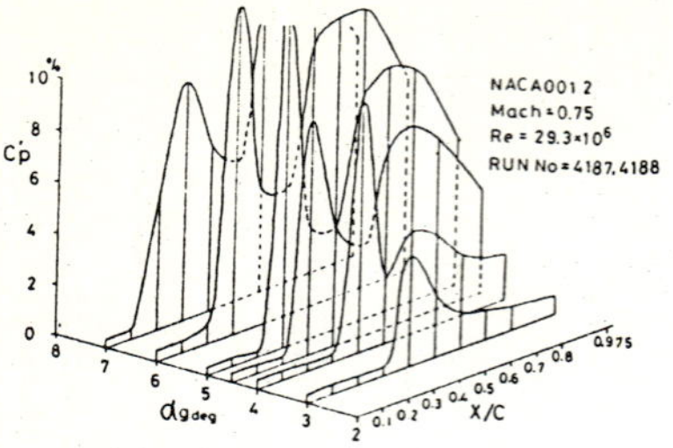

(a) NACA0012 airfoil

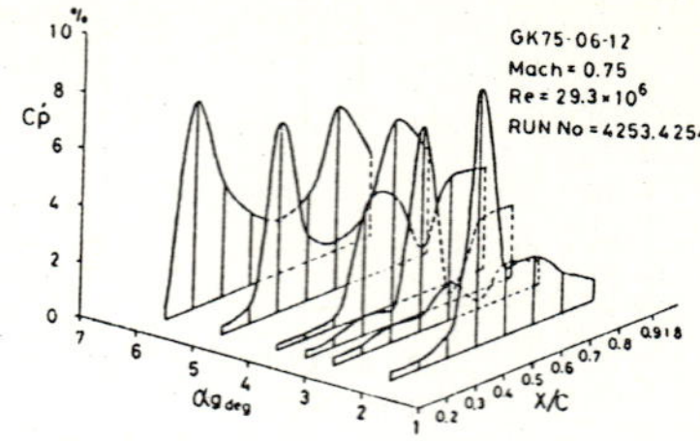

(b) KORN airfoil

Fig. 5 Pressure fluctuation distributions

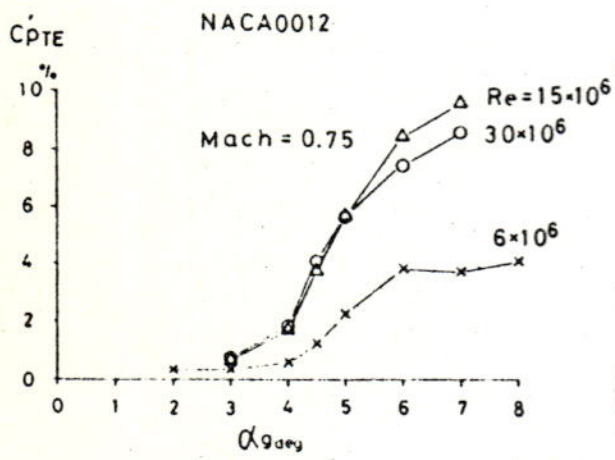

(a) NACA0012 airfoil , M=0.75

(b) KORN airfoil , M=0.75

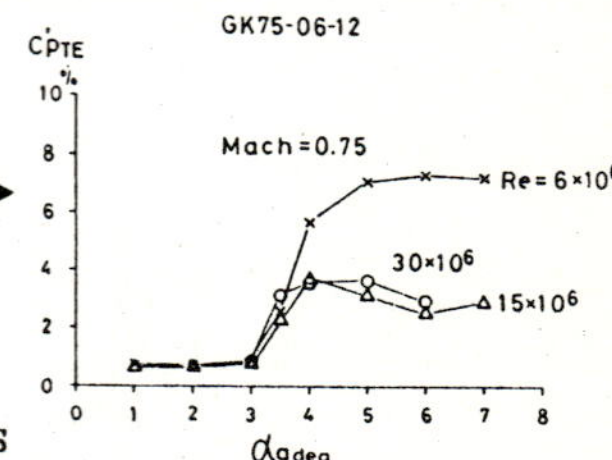

Fig. 6 Trailing edge pressure fluctuations

NACA0012 airfoil

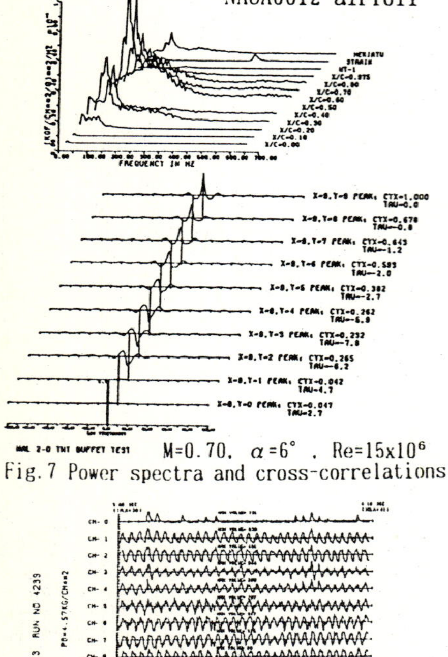

M=0.70, α=6° , Re=15x10^6

Fig. 7 Power spectra and cross-correlations

KORN airfoil

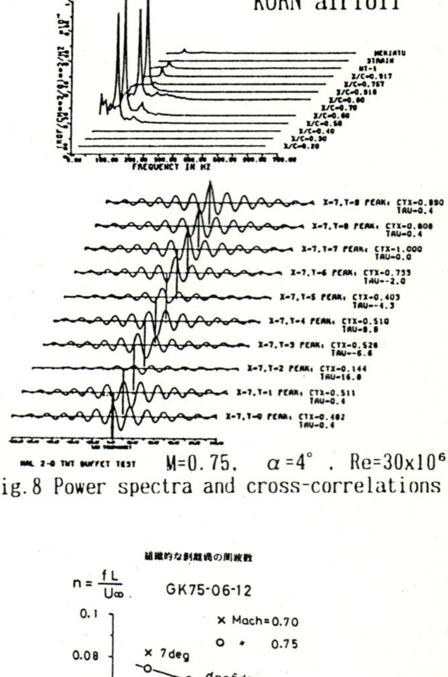

M=0.75, α=4° , Re=30x10^6

Fig. 8 Power spectra and cross-correlations

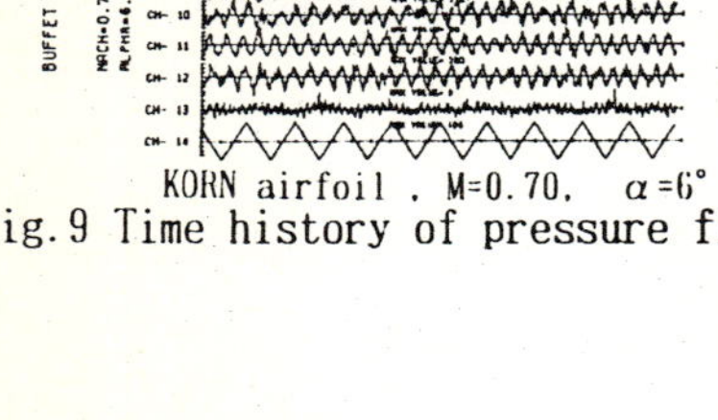

KORN airfoil , M=0.70, α=6°

Fig. 9 Time history of pressure fluctuation

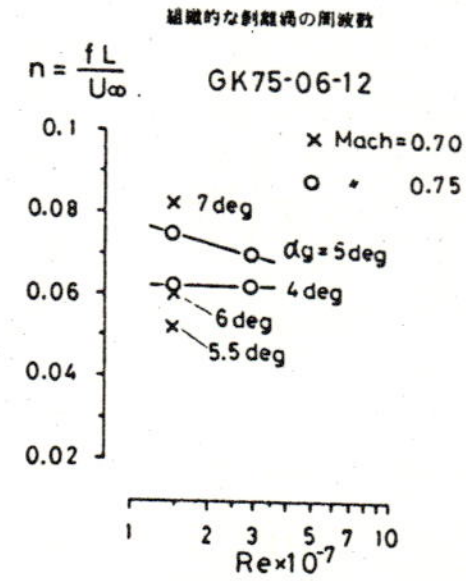

Fig. 10 Strouhal number of separated vortices KORN airfoil

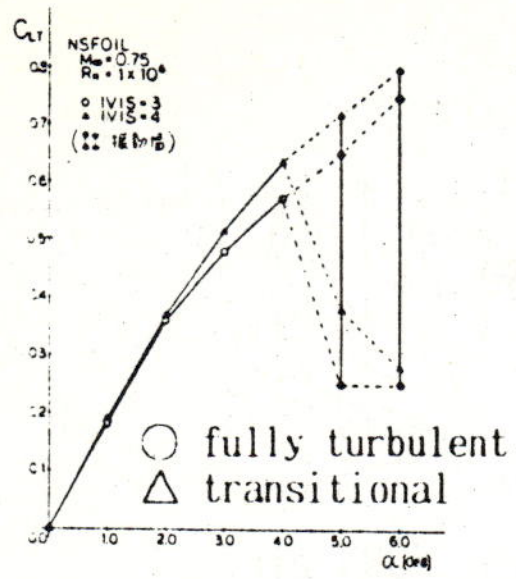

Fig. 11 C_l vs. α, NACA0012 Re=$1x10^6$

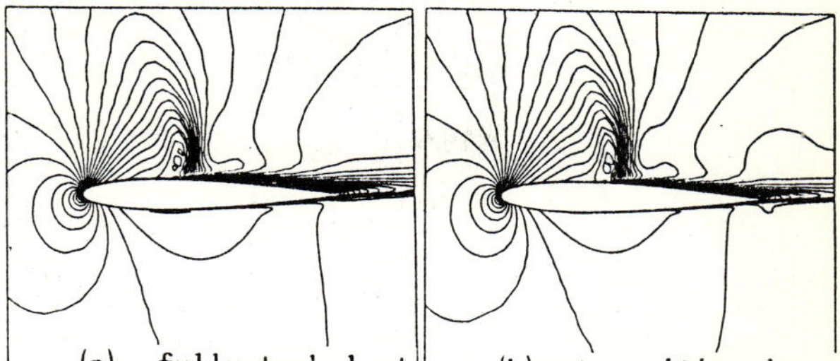

(a) fully turbulent (b) transitional

Fig. 12 Mach number contours, NACA0012, α = 4°, Re=$1x10^6$

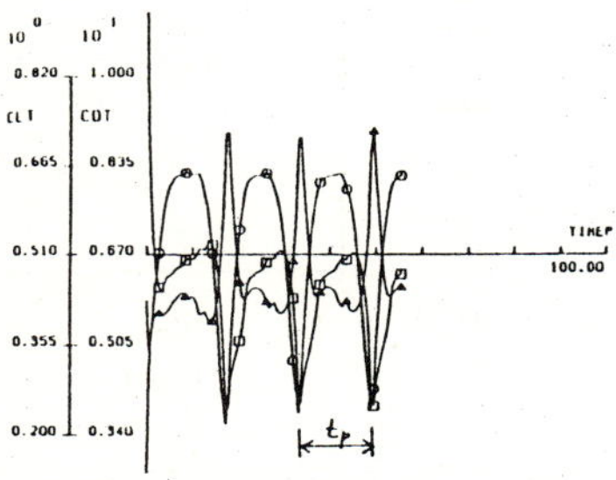

(a) α = 5°, fully turbulent, Re=$1x10^6$

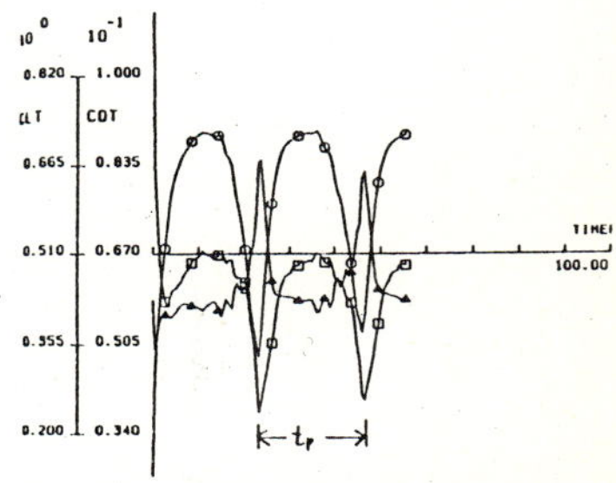

(b) transitional, Re=$1x10^6$

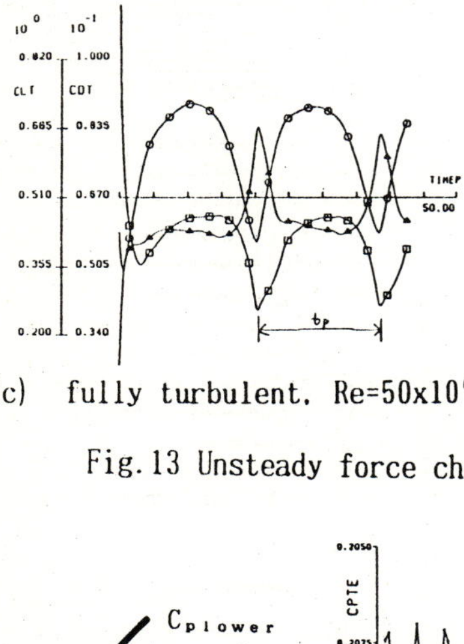

(c) fully turbulent, Re=$50x10^6$

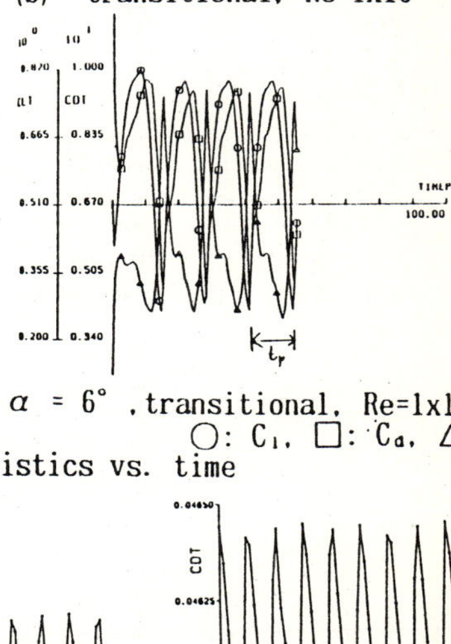

(d) α = 6°, transitional, Re=$1x10^6$

○: C_l, □: C_d, △: C_m

Fig. 13 Unsteady force characteristics vs. time

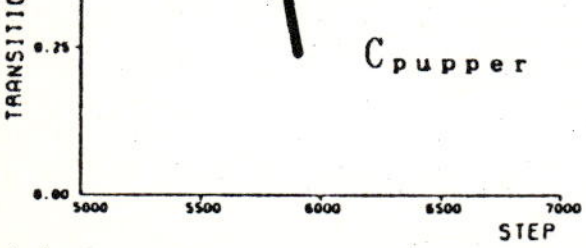

(a) transition point X_{tr}

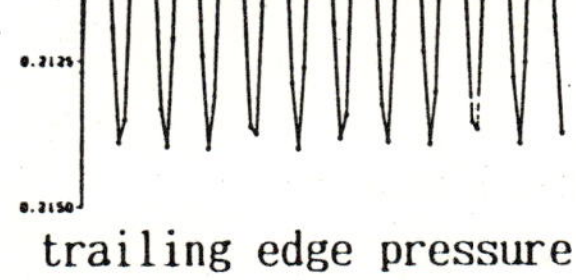

(b) trailing edge pressure C_{PTE}

(c) drag coefficient C_d

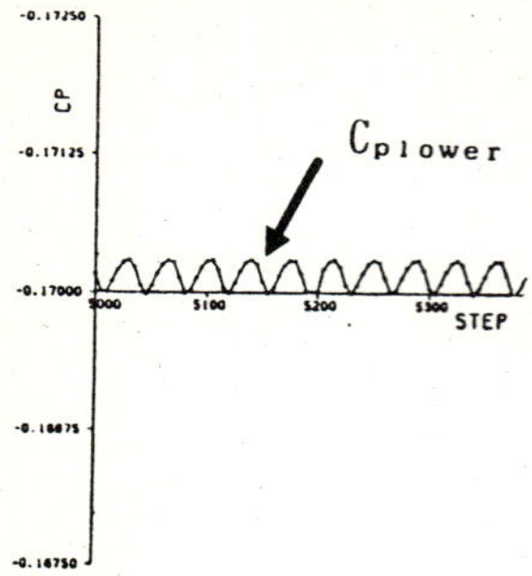

(d) C_p at x_{lower}=0.19

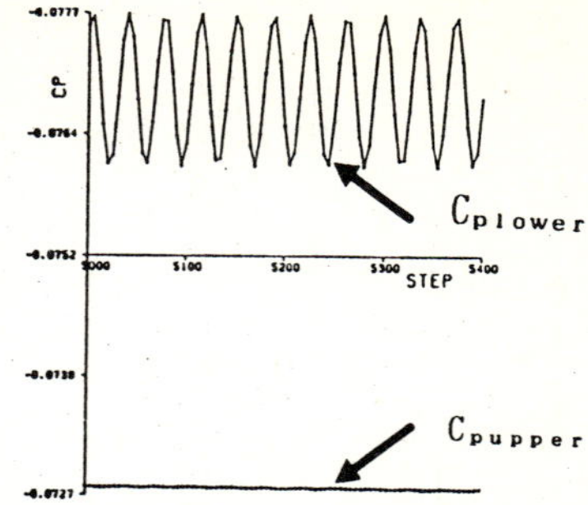

(e) C_p at x_{lower}=0.75

Fig.14 Unsteady fluctuations vs. time

NACA0012, M=0.75, Re=1x10^6, α=4° , transitional

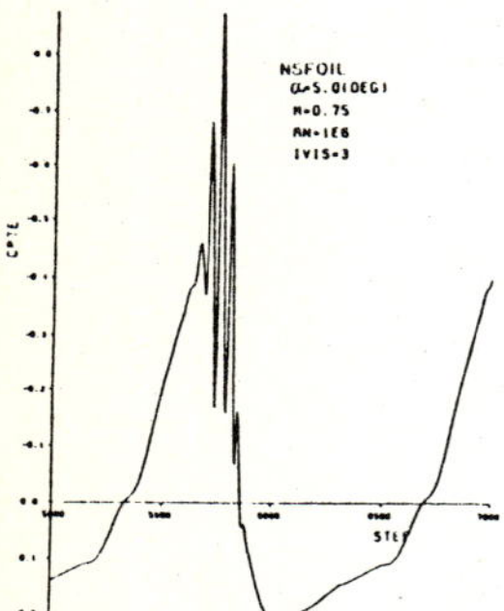

(a) fully turbulent

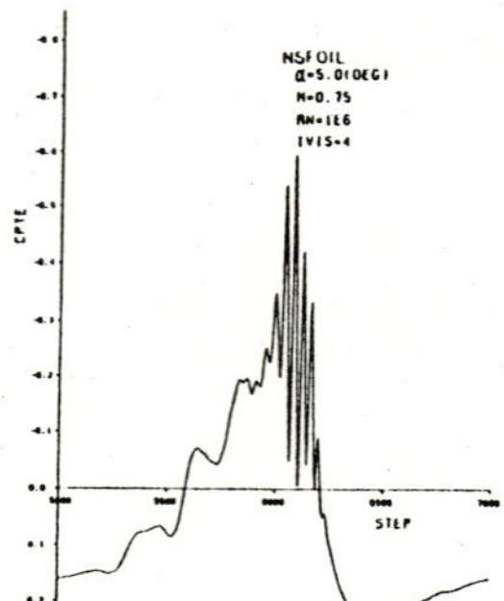

(b) transitional

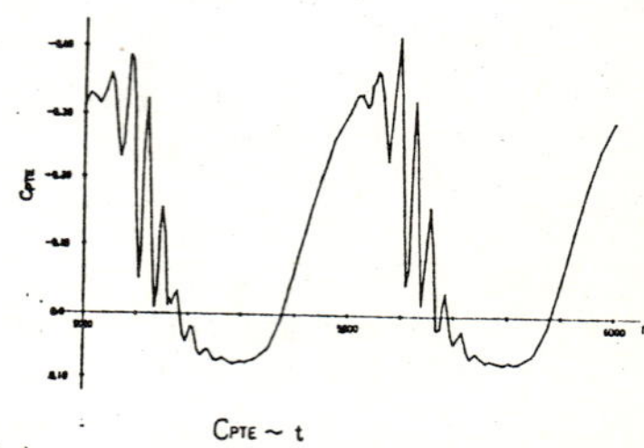

(c) a supercritical airfoil [8]

NACA0012, M=0.75, Re=1x10^6, α=5°

Fig.15 Trailing edge pressure C_{PTE} vs. time

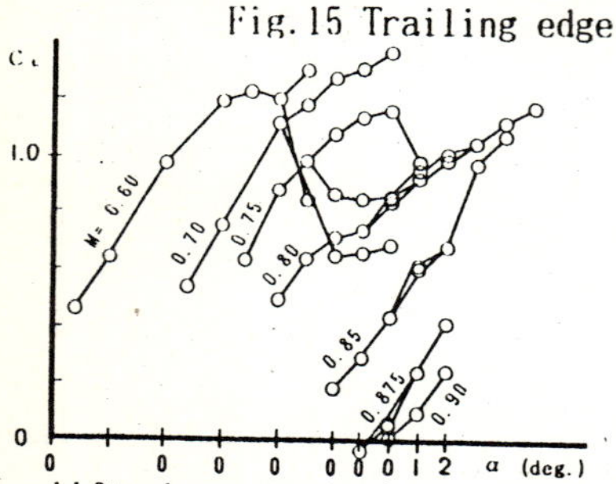

(a) lift characteristics vs. α

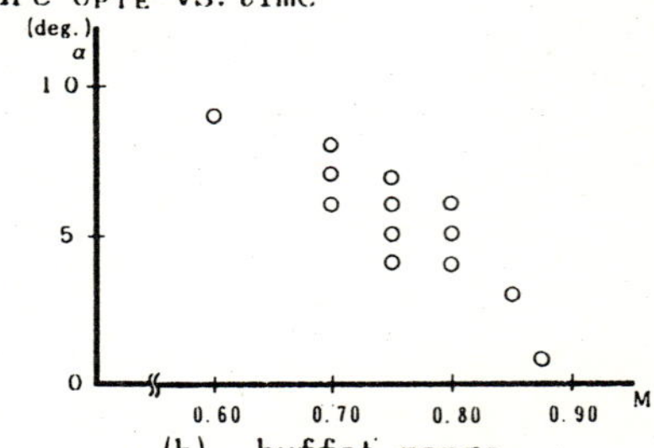

(b) buffet range

Fig.16 Buffet characteristics of KORN airfoil

Re=6x10^6, transitional

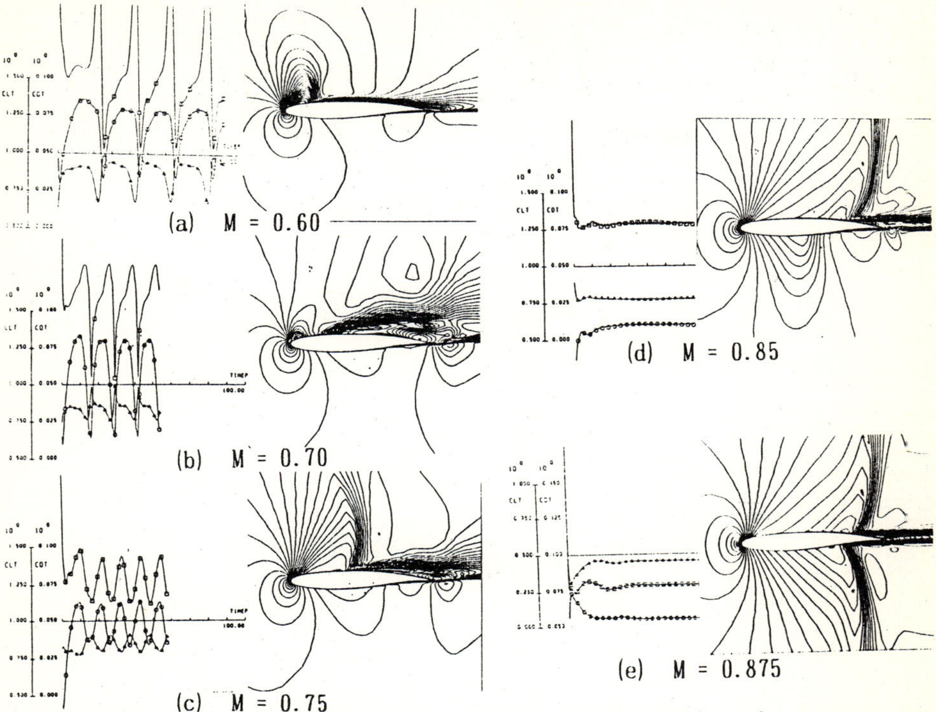

Fig. 17 Unsteady force characteristics vs. time and Mach number contours KORN airfoil

Concluding Remarks

H.G. Hornung

Graduate Aeronautical Laboratories
California Institute of Technology
Pasadena, California 91125
USA

1. Introduction

The following remarks represent a subjective view of this symposium. The boundary conditions imposed on them also make the resolution poor.

Research workers in transonics have their love-hate relationships with the subjects

- Separation
- Transition
- Turbulence
- Vortices
- Shocks

These friends or enemies, as the case may be, are tackled with the weapons

- CFD
- Asymptotics
- Experiment

for the purpose of

- Understanding
- Prediction
- Design

preferably in that order. They pose enormous and intricate difficulties and pitfalls, as this symposium has amply shown.

What is new relative to 13 years ago? 13 years ago there were

- no Euler or "Navier-Stokes" computations
- no 3-D compressible stability computations
- no three-component transonic LVD measurements

J. Zierep, H. Oertel (Eds.)
Symposium Transsonicum III
IUTAM Symposium Göttingen, 24.-27. 5. 1988

- no cryogenic or adaptive-wall wind tunnels

A consequence is that high-Reynolds-number investigations are now possible, that the third dimension has been made much more accessible and that both experimental, and numerical investigations have become much more extensive and much more expensive. This makes the need for coordination and planning of joint efforts almost imperative, and to this end there is no question that IUTAM provides an excellent forum.

In compiling the program of this IUTAM symposium, deliberate omissions were made in order to avoid shallowness in the contents. Thus, the problems of wind tunnel technology, a subject of considerable interest as a problem per se, has been omitted, though very important developments have taken place in the last 13 years.

2. The Sessions

2.1 CFD

There has been obvious and significant progress in this field. Attempts to define the conditions of computations more accurately, such as the F5 project, are to be commended. A lesson that was learned from this project may be summarized by the motto of the famous aircraft designer Kelly Johnson: "KISS" (keep it simple stupid), and that as much as possible of the existing experience should go into such programs at the planning stage.

A disappointing feature of this whole field is the enormous and complex dependence of results on the nature of the mesh. Also there is still only very sketchy understanding of the mechanisms of vorticity generation in Euler computations.

We have seen some impressive success of viscous/inviscid interaction techniques in the problem of buffet and some interesting effects due to heat addition.

Much needs to be done in the field of CFD to begin to come to grips with 3-D viscous flows with shocks and vortices. It is a view held by some at the symposium that this is a question of more powerful supercomputers.

After the strong rise of Euler computations in the early 80s, a slight indication of cold feet about this technique may be detected.

2.2 Asymptotics

I found all the contributions interesting. Results of this kind always represent an increase, albeit small, in understanding. However, a definite need exists to make the results more amenable to those not working directly in the field. To this end a few diagrams showing results might have been helpful.

On a lighter note, it was amusing that Julian Cole used green rather than black boxes and told us that "unspeakable practices and unnatural acts" were what asymptoticists hid behind their "matching activities late at night". I am not sure whether I understood H.K. Cheng's comment correctly, that asymptotic techniques should be restricted to slender bodies.

After the considerable mesh-dependence of the results of CFD, we experienced in this session the audience-dependence of the subject of Guiraud's talk.

2.3 Stability and Transition

While in stability theory and calculations very significant progress is to be observed, transition remains a bag of worms. For example, at NASA Langley the effect of 30 parameters is being studied by an army of researchers, and an enormous body of experience has been collected. However, predictions are still only conclusive in very limited situations.

Some efforts in nonlinear stability theory and in numerical simulation are interesting.

In this field, good detailed experiments are needed, and, here too, coordination of the planning would make joint efforts profitable.

2.4 Vortical Flows

It is clear from the presentations that the flow over delta wings is much more complex than was thought as recently as in 1983. Many shock features occur, even at subsonic free stream, that were not previously expected and are now partially understood.

Much is still confused, however. An example is the computed total pressure loss in the vortex core:

Murman: "The boundary layer effect is weak." (Euler)

Rizzi : "The boundary layer effect is strong." (Navier-Stokes)

Here too, computers are regarded to be still not sufficiently powerful. Also, still more detailed experiments are considered necessary. I thought the presentation by Elsenaar to be one of the highlights of the symposium.

To summarize this session in a lighter vein, I paraphrase Mohammed Hafez: "Much is still highly Dubois."

2.5 Shock-Boundary-Layer Interaction

The gross features of the effects of porous walls, which appeared five times, seem now to be understood. It has almost become a band-wagon. Perhaps, as Zierep suggests, the specialists in this subfield will meet in a workshop mode in the near future.

We have seen here, as in other separated flows, three-dimensional effects are very dramatic. Also, the turbulence model used in numerical computations clearly has a very important effect on the results. We have seen very nice results on unsteady SBLI from a cryogenic adaptive-wall tunnel. I regard the presentation of Délery as one of the highlights of the symposium.

2.6 Unsteady Flows

This session should perhaps have included two earlier unsteady-flow papers that were presented in other sessions for context. Two papers of the session dealt with the Reynolds number dependence of buffet which seems to be a subtle effect not independent of the airfoil used.

We also saw impressive success of relatively simple computations of aeroelastic effects.

The extensive review and classification of the different kinds of unsteady phenomena in transonics and the pertinent examples presented by Meier were another highlight.

3. The Format of the Symposium

Apart from the following major criticism, I liked the format of the symposium very much. The criticism is that we need much more discussion time! Only in those sessions where we had cancellations did any significant discussion take place. The value of such a symposium is drastically reduced without the discussion, which after all is the main reason for us to get together from all parts of the world. Probably, none of us will forget the admonishing words of session chairman Sune Berndt who was forced by the temporal constraints to say sternly:
"No questions!"

4. Message

It seems to me that this symposium told us that we should make more often such coordinated efforts as the F5 wing or the Vortex Flow Experiment which combine experiment and CFD, and that as much as possible of previous experience in the special field be used in the planning stage.

Also, the delightful comment by Earl Murman that "What we are doing is fun, it might even be important" may probably be given less conditional character, for, as has been made amply clear by a lucid, though rapid, presentation by Paul Rubbert, it IS important.

Finally, let me wish you all a safe trip home.